Sustainable Agriculture and Farming

Sustainable Agriculture and Farming

Edited by **Thelma Bosso**

Callisto Reference
New York

Published by Callisto Reference,
106 Park Avenue, Suite 200,
New York, NY 10016, USA
www.callistoreference.com

Sustainable Agriculture and Farming
Edited by Thelma Bosso

International Standard Book Number: 978-1-63239-729-4 (Hardback)

Printed in China.

Contents

Preface

This book outlines the processes and applications of sustainable agriculture and farming in detail. It will unfold the innovative aspects of this area. As a practice of farming, sustainable agriculture, refers to the use of natural resources to produce crops in such a way that it has minimal negative effects on the environment and yet, is able of producing good quality crops and livestock. It incorporates techniques like crop rotation, soil enrichment, cover crops, soil management, etc. to maximize produce in a harmless way. It works towards minimizing the use of pesticides, insecticides, fertilizers, to reduce the wastage of water, limit the use of non-renewable energy, etc. This text will talk in detail about the various techniques and concepts related to this area. From theories to research to practical applications, case studies related to all contemporary topics of relevance to sustainable agriculture have been included in this book. It will provide in-depth knowledge about the new advances being made in this field. Researchers and students in this field will be greatly assisted by this book.

Significant researches are present in this book. Intensive efforts have been employed by authors to make this book an outstanding discourse. This book contains the enlightening chapters which have been written on the basis of significant researches done by the experts.

Finally, I would also like to thank all the members involved in this book for being a team and meeting all the deadlines for the submission of their respective works. I would also like to thank my friends and family for being supportive in my efforts.

Editor

1

Clipped Cowpea (*Vigna unguiculata* (L.) Walp) Fodder Management: A Potential for Soil Organic Matter Enrichment in Degraded Savannah Soils of Nigeria

Banake E. Sambo[1] & Ehiabhi C. Odion[2]

[1] Department of Crop Science, Federal University Dutse, P.M.B.7156, Jigawa State, Nigeria

[2] Department of Agronomy, Institute for Agricultural Research, Ahmadu Bello University, P.M.B.1044, Samaru, Zaria, Nigeria

Correspondence: Banake E. Sambo, Department of Crop Science, Federal University Dutse, P.M.B.7156, Jigawa State, Nigeria. E-mail: banelisam@yahoo.com

Abstract

Severe nutrient depletion of the soils of the savannah tropics of Africa over the years has made it difficult to improve the productivity of varieties of crops using cultural practices alone. But interestingly, it has been observed that when a cowpea (*Vigna unguiculata* (L.) Walp) crop is cut (clipped) before senescence; it can regenerate after defoliation (provided there is enough soil moisture). And when the clipped organic fodder is added and/or ploughed back into the soil, it enriches the soil organic matter (SOM) content that in turn enhances crop productivity. Thus, this study was carried out with the objective of determining the influence of intra-row spacing, clipping height and time on the productivity of cowpea and SOM; at the Institute for Agricultural Research, Ahmadu Bello University, Samaru, Zaria, Nigeria; during the 2002-2005 wet seasons. The experimental lay out was a Randomized Complete Block Design (RCBD), replicated three times. The collected data was analyzed statistically using the analysis of variance test (ANOVA); and the means separation was done using the Duncan Multiple Range Test (DMRT). Results showed that the textural class of the experimental site soil was loam silt; with a conducive pH of (6.6 in 2002 and 5.5 in 2005), for crop growth. The soil organic carbon content (SOC), nitrogen (N) and cation exchange capacity (CEC) were very low (0.30 g kg^{-1}, 0.88 g kg^{-1} and 4.90 respectively) in 2002. These increased to 10.37 g kg^{-1}, 2.2 g kg^{-1} and 11.10 respectively in 2005. Total rainfall in 2002 was 1007.9mm and 871.5mm in 2005. Mean air temperature, relative humidity and sunshine hours in 2002 ranged between 21.0-31.5 °C, 16.0-90.6 and 4.5-8.5 respectively; while mean air temperature and relative humidity ranged between 24.7-32.0 °C and 40.5-85.3% respectively in 2005. Total harvested clipped fodder yield was 15t ha^{-1}; and this was added to the soil; and it effectively increased SOM content by about 42%. Consequently, it was concluded that the adoption of this innovative clipping management technology, holds great potential of improving soil pH, increasing soil CEC, SOM and crop productivity generally, for the low technology, resource poor, and subsistence farmers in the region. Without doubt, overcoming SOM decline is a major component in the development of more sustainable agro-systems.

Keywords: Clipping, fodder, production, potential, nutrient, depletion, soil organic matter, sustainability

1. Introduction

The cowpea (*Vigna unguiculata* (L.) Walp) crop is prominent in the farming systems of the semi-arid tropics, where they are grown mainly for both fodder (haulm) and for their mature seeds (Onwueme & Sinha, 1991; Odion & Singh, 2005b).

Generally, cowpea in the northern guinea savannah zone is sown in mid-July (Onwueme & Sinha, 1991), where its productivity has been variedly reported to be dependent on the time of planting, amount and distribution of rainfall. Cowpea requires a moderate amount of rainfall for good production. If rainfall is too heavy, the plant tends to be leafy and produce few pods. Cowpea is reported to be grown successfully on a wide variety of soils provided they are well drained; though light sandy loam soils are more suitable than heavy soils (Onwueme, 1979, Onwueme & Sinha, 1991). The crop is also, sometimes grown on poor acid soils for soil improvement (Onwueme & Sinha, 1991).

In the savanna area, dual purpose cowpea types are sown later in the season and at wider (90 × 30 cm, 75 × 20 cm) spacing's. As a consequence of the short growing season often experienced in the region, the late planting of the dual purpose cowpea often results in its poor vegetative growth with little or no grain production. As a result, it has been difficult to improve the productivity of these varieties using cultural practices alone. However, it has been discovered and observed that if the cowpea plant is cut (clipped) at 7-10 Weeks after Planting (WAP), before senescence, it can regenerate after defoliation (and provided there is enough soil moisture), and produce seeds (Singh, 1993; Odion & Singh, 2005b). The cowpea plant is able to maintain growth and to survive under dry semi-desert soils conditions due to the fact that it is drought resistant and grows deep roots (Singh, Mohan Raj, Dashiell, & Jackai, 1997). Thus, Odion and Singh (2005b), Sambo, Odion, Aliyu, and Labe (2013), concluded that the reproductive growth of the dual-purpose cowpea varieties was enhanced in their trials through clipping management, in spite of the close intra-row spacing under which the crop was grown. Hence, they postulated that it is possible that after clipping, the re-growth from clipped plots behaved like the semi-erect grain type of cowpea that are grown at this closer intra-row spacing's.

Since the cowpea crop when cut at the vegetative growth phase can regenerate provided there is enough soil moisture, thus it is possible to grow cowpea for "green manure" by clipping its foliage and allowing the clipped crop plants to grow to maturity for seeds (Odion & Singh, 2005b). The advantage of this practice, is in the ability of low resources, low input and low technology based farmers to improve on the soil organic matter (SOM) and/or nutrient status of their soils - bearing in mind that most African soils are reported to be typically impoverished and seriously deficient in nutrients - a situation compounded by the more or less permanent agricultural system practiced in the semi-arid tropics which results in severe nutrient depletion in sub-Saharan Africa (Zake, 1993).

But given that a proportionately large amount of available nutrients are retained in the vegetative cover of crops and SOM accumulates on the surface of the soil; the continuous practice of removing and exporting the top vegetative cover of crops from the farmlands, which is a common feature of the farming systems of the semi-arid regions, implies that the fertility of the soil declines rapidly as SOM is not maintained at an appropriate level. However, green manuring, which places, adds and/ or incorporates fresh green non-woody plant, rich in water, sugars, starches, protein and nitrogen into the soil (Kahnt, 1983); as organic fertilizer, could enhance the growth and productivity of crops (Karunairanjan, 1980). The green-manure crop could supply organic matter (OM) as well as additional nitrogen (N) particularly if it is a legume crop, which has the ability to fix atmospheric nitrogen (Patnaik, 2004). It also could protect the soil against erosion and leaching. Usually, green manures are either cut or brought from outside or grown in the field with the intention that they will be ploughed in as manure (Yegna Narayan Aiyer, 1975; Reijntjes, Haverkort, & Waters-Bayer, 1992).

The maintenance of OM is of fundamental importance to the productivity of tropical soils. If soil fertility in the tropics is to be maintained, fertilization must first and foremost be organic. What this means in practical terms is that, a proper management of tropical soils which have only recently come under cultivation must aim to maintain both the structure as well as the content of organic matter. In soils that have been cultivated for a long time, the aim should be to increase the organic mass and to improve their structure (Muller-Samann & Kotschi, 1994).

Thus far, in many tropical cropping systems, little or no agricultural residues are returned to the soil. This is reported to lead to a decline in SOM (Lal, 1986; Bouwman, 1990; Post & Mann, 1990; Woomer & Ingran, 1990); which frequently results in lower crop yields (Lal, 1986); or plant biomass productivity (Woomer & Ingran, 1990).

Recent studies have focused on the effect of green clipped crop materials and placing or incorporating such plant residues into the soil for fertility maintenance and or improvement. Since cowpea can regenerate after defoliation, it might be possible to grow the crop of dual purpose cowpea for both grain and fodder and through clipping management, placement and or incorporation, for fertility maintenance and/or improvement of SOM content (Odion & Singh, 2005a; Odion, Asiribo, Ogunlela, Singh, & Tarawali, 2007).

Cowpea fodder can potentially serve as a source for both soil nutrients and organic matter as well as for other beneficial agricultural purposes which hitherto have been constraints to the improvement and sustainability of crops production in the savanna; and the remedies had been both expensive to the resource poor farmers and have to be gotten from outside the farm. However, the process of ameliorating the negative effects of SOM loss for sustain fertility and crop productivity are now within the farmers' reach and may not be as expensive as the imported technologies (Smalling & Nandwa, 1996; Henao & Baanante, 1999). In the light of the foregone, this study was carried out with the objective of determining the influence of intra-row spacing, clipping height and time on SOM of the dual purpose cowpea (*Vigna unguiculata* (L.) Walp).

2. Materials and Methods

Field experiments were carried out on the research farm of the Institute for Agriculture Research, Samaru, Nigeria, during the 2002, 2003, 2004 and 2005 cropping seasons. Samaru (11°.11'N, 07°.38'E and 686m above sea level) is located in the northern Guinea savanna agro-ecology of Nigeria (Keay, 1959). Usually, rainfall becomes established in the region between mid-May and early June and peaks in July/August. Total annual rainfall ranges between 883-1062mm, with an average of 945.20 mm. The dry season starts at about mid-October and extends to the end of April. The mean minimum and maximum temperatures during the rainy seasons range between 14-22 °C and 29-34 °C, respectively. The textural class of the soil of the experimental site was classed loam silt; with a pH of 6.6 in 2002 and o 5.5 in 2005. The soil organic carbon content (SOC), nitrogen (N) and cat-ion exchange capacity (CEC) were very low (0.299 g kg^{-1}, 0.087 g kg^{-1} and 4.90) in 2002. Phosphorus (6.51mg kg^{-1}) and potassium (0.10cmol kg^{-1}) were equally low. The SOC and CEC increased to 10.37, 2.2 and 11.10 in 2005, respectively. Total rainfall in 2002, 2003, 2004 and 2005 was 1007.9mm, 1135.4mm, 1019.0mm and 871.5mm respectively. In 2002, 2003, 2004 and 2005, mean air temperature and relative humidity (RH) ranged, 21.0-31.5 °C and 16.0-90.6%, 23.4-31.2 °C and 16.8-78.4%, 25.9-27.9°C and 60.4-78.4% respectively. While sunshine hours in 2002, 2003, 2004 ranged 4.5-8.5, 5.5-8.1, 5.5-7.5 respectively.

The treatments comprised of three intra-row spacing, 15.0, 30.0 and 45.0cm on ridges 75cm apart; three clipping heights (no clipping control, 12.5cm and 25.0cm); and three clipping periods (64,74 and 84 days after planting – DAP) respectively. Factorial combinations of these treatments were laid out in a Randomized Complete Block Design (RCBD), and replicated three times.

The land was ploughed and harrowed twice at the time of land preparation, before planting, using mechanical power to give a good soil tilt. It was then ridged at 75cm spacing and demarcated into a gross and net plot size of 9.0 m^2 and 4.5 m^2 respectively.

The variety used (Kano 1696/ Sampea 6) is a highly photosensitive dual purpose cowpea, which is usually planted late in the season for fodder as well as grains when possible. It is late maturing (100-120 days), and has a spreading habit; growing vegetative with little grain yield (250-1000 kg ha^{-1}) production. It produces long pods, extra large white seed with black eye, rough seed coat texture and has good palatability.

Fertilizer was applied at the recommended rate of 10 kg N ha^{-1}, 36 kg P_2O_5 ha^{-1} and 20 kg K_2O ha^{-1} (Enwezor, Udo, Usoroh, Ayotade, Adepetu, & Chude, 1989). Weeds were controlled at planting using Gramazone (Paraquat) as a pre and post emergence herbicide; supplemented with manual weeding operations. Post-emergence protection against insect pest and fungal attacks were provided using the insecticides and fungicides as and when necessary.

The fields were planted on 13th July, 17th July, 7th June and 17th June, in 2002, 2003, 2004 and 2005 respectively. The cowpea crop was clipped (harvested) at 64, 74 and 84 days after planting (DAP). The first clipping was done on 15th September, 2002; 19th September, 2003; 10th August, 2004, and 17th August, 2005, respectively. The clipped fodder was placed on the plots to decay and act as a source of organic manure; while the clipped plants were left to grow to maturity alongside the control (unclipped) plots for further yield analysis.

Four plants per plot were sampled for the determination of the following: Total clipped fodder yield: The clipped fodder yield was obtained by weighing and recording the total clipped crop residue (fodder) from each plot, and this was recorded and converted to total fodder yield in tons per hectare (t ha^{-1}) basis.

Soil Organic Matter - SOM: Soil nutrient (status) analysis was undertaken using a 5 g air dry soil. SOM content was calculated according to the procedures as described by (Bray & Kurtz, 1945; Black, 1965; Day, 1965; Bremmer, 1965; International Institute for Tropical Agriculture, 1975; Kundsen, Peterson, & Paratt, 1982).

The data collected was compiled and analyzed statistically using the analysis of variance test (F-test) as described by (Snedecor and Cochran 1967); and the means were compared using the Duncan Multiple Range Test-DMRT (Duncan, 1955).

3. Results

3.1 Harvested Fresh Fodder Yield of Cowpea

In 2002, fresh fodder yield significantly decreased with increasing intra-row spacing from 15 to 45 cm. While in 2005 and the combined analysis for 2002 - 2005, fresh fodder yield was significantly higher at 15cm intra-row spacing than at 30 and 45 cm intra-row spacing; which were statistically the same with each other (Table 1). Clipping height had no significant influence on harvested fresh fodder yield throughout the study period. In all the years, except in 2003, harvested fresh fodder produced from plots clipped at 74 and 84 DAP were statistically the same; but significantly higher than harvested fresh fodder obtained from plots clipped at 64 DAP. In 2003, harvested fresh fodder yield obtained at 64 and 74 DAP were statistically the same; but significantly higher than that produced at 84 DAP.

Table 1. Harvested fresh fodder yield (t ha^{-1}) as influenced by intra-row spacing, clipping management and time treatments of dual purpose cowpea, at Samaru, Zaria, Nigeria

Harvested Fresh Fodder Yield (t ha^{-1})					
Treatment	**2002**	**2003**	**2004**	**2005**	**Combined 2002-2005**
Intra-row spacing(cm)					
15	39.53a	1.19	12.75	19.03a	18.13a
30	26.55b	1.17	12.95	11.81b	13.12b
45	15.72c	1.03	12.67	12.98b	10.60b
SE ±	2.90	0.17	0.86	1.97	0.93
Clipping height (cm)					
0 (Control-unclipped)	-	-	-	-	-
12.5	26.93	1.29	12.83	16.48	14.38
25.0	27.60	0.98	12.74	12.74	13.52
SE ±	2.90	0.17	0.86	1.97	0.93
Clipping time (DAP)					
64	17.64b	0.78b	11.45	9.33b	9.80b
74	33.66a	0.98b	13.10	19.50a	16.81a
84	30.49a	1.64a	13.80	15.00a	15.23a
SE ±	2.90	0.17	0.86	1.97	0.93
Interaction					
I*H	NS	**	NS	NS	NS

Means followed by different letter (s) are significantly different at ≤ = 0.05; using Duncan's Multiple Range Test (DMRT). ** = Highly significant. NS = Not Significant.

The interaction of treatments on fresh fodder yield was significant in all the years investigated, except the combined analysis for 2002-2005. Clipping height and clipping time interaction on fresh fodder yield was significant in 2004 (Table 2). At 64 DAP, plants clipped at 25.0 cm height produced significantly higher fresh fodder yield than plants clipped at 12.5cm height and the control plots. Similarly, plants clipped at 12.5 cm, recorded significantly higher fresh fodder yield than those of the control plots. But at 74 DAP, fresh fodder yield of plants clipped at 12.5 cm and the control plants were statistically similar; but significantly higher than those clipped at 25.0 cm height. While at 84 DAP, clipping height did not significantly affect the amount of fresh fodder produced. The highest mean harvested fresh fodder yield was recorded when crop plants were clipped at 74 DAP and at 12.5 cm clipping height.

Table 2. Interaction of clipping height and time on fresh fodder yield (t ha^{-1}) of dual purpose cowpea, in 2004, at Samaru, Zaria, Nigeria

	Clipping height (cm)		
	Control (unclipped)	**12.5**	**25.0**
Clipping time (DAP)			
64	11.45bc	8.43d	14.47a
74	13.10abc	16.10a	10.11d
84	13.82ab	13.95ab	13.64abc
SE± 1.49			

Means followed by different letter (s) are significantly different at ≤ = 0.05; using Duncan's Multiple Range Test (DMRT).

3.3 Soil Organic Matter (SOM) Content

In 2004, 2005 and the combined analysis for 2002-2005, SOM content consistently, significantly increased with increase in intra-row spacing from 15 to 45 cm. While, in 2003, SOM obtained at 15cm and 30cm intra-row spacing were statistically similar, but significantly higher than the SOM contents obtained at 45cm intra-row spacing. The influence of clipping height on the SOM content was significant in 2003, 2004, 2005 and the combined analysis for 2002-2005; where the SOM content of plots clipped at 12.5 and 25 cm heights were statistically the same, but significantly higher than the SOM content of the control plots. Clipping time had a significant effect on SOM content only in 2002; where the SOM content of plots clipped at 74 and 84 DAP were statistically at par; but significantly higher than for plots clipped at 64 DAP. There was a significant interaction of the treatments on SOM content in 2003, 2004, 2005 and the combined analysis for 2002-2005 (Table 3).

Table 3. Percent soil organic matter (SOM) content as influenced by intra-row spacing, clipping management and time treatment of dual purpose cowpea at Samaru, Zaria, Nigeria

	Soil Organic Matter (SOM) content				
(%)					
Treatment	**2002**	**2003**	**2004**	**2005**	**Combined 2002-2005**
Intra-row spacing (cm)					
15	0.527	1.607a	1.488a	1.672a	1.323a
30	0.514	1.602a	1.459b	1.593b	1.292b
45	0.511	1.200b	1.046c	1.509c	1.066c
SE ±	0.009	0.004	0.005	0.011	0.004
Clipping height (cm)					
0 (Control-unclipped)	0.503	0.866b	0.830b	1.155b	0.838b
12.5	0.520	1.773a	1.582a	1.807a	1.420a
25	0.528	1.769a	1.581a	1.812a	1.422a
SE ±	0.009	0.004	0.005	0.011	0.004
Clipping time (DAP)					
64	0.503	1.461b	1.329	1.578	1.217
74	0.516	1.472a	1.331	1.598	1.229
84	0.532	1.475a	1.333	1.698	1.259
SE ±	0.009	0.004	0.005	0.011	0.004
Interaction					
I*H	NS	**	**	**	**

Means followed by different letter (s) are significantly different at ≤ = 0.05; using Duncan's Multiple Range Test (DMRT). ** = Highly significant. NS = Not Significant.

Clipping height and time interaction on SOM was significant in the combined analysis for 2002 - 2005 (Table 4). At 15 cm intra-row spacing, plants clipped at 12.5 cm and 25.0 cm height produced statistically similar, but significantly higher SOM than the control plants. Similarly, plants clipped at 12.5 cm, recorded significantly higher fresh fodder yield than those of the control plots. On the other hand, crops clipped at 15 and 30 cm intra-row spacing produced statistically similar SOM content. However, a further increase in intra-row spacing to 45 cm resulted in a significant reduction in the SOM contents of the plots. Generally, the highest mean SOM content were recorded when crops were clipped at either 12.0 or 25.0 cm and at 15 cm intra-row spacing (Table 4).

Table 4. Interaction of clipping height and intra-row spacing on percent soil organic matter (SOM) content of dual purpose cowpea, combined for 2002 - 2004, at Samaru, Zaria, Nigeria

	Clipping height (cm)		
	Control (unclipped)	**12.5**	**25.0**
Intra-row Spacing (cm)			
15	0.919c	1.52a	1.52a
30	0.875c	1.49a	1.50a
45	0.722d	1.23b	1.23b
SE± 0.12			

Means followed by different letter (s) are significantly different at ≤ = 0.05; using Duncan's Multiple Range Test (DMRT).

4. Discussions

4.1 Harvested Fresh Fodder Yield of Cowpea

The favorable growth of cowpea could be attributable to the soil texture, favorable pH, high temperatures and the good amount of rainfall recorded during the period viz. the soil moisture conditions required to support good crop growth; except in 2003, when the harvested fresh fodder yield was remarkably low; largely due to observed excessive flooding of the plots. Because cowpea is reported to be a warm-weather annual crop, capable of withstanding heat better than most other legumes and at the same time, drought-resistant. It is known to be grown successfully on a wide variety of soils provided they are well drained. Moisture deficiency has been noted to have an adverse effect mainly on vegetative growth. (Onwueme, 1979; Onwueme & Sinha, 1991).

The favorable response of cowpea to these climatic and environmental conditions facilitated the production of large amounts of plant biomass. When this green crop material was clipped and the harvested material added to the soil (green manuring), which is primarily a form or method of biological soil fertility maintenance, it proved to be beneficial in improving the soil physical and chemical properties of the soils and eventually becoming a source of SOM enrichment and nutrients supply (Lee & Wani, 1988; Lal, Wilson, & Okigbo, 1979; Mulongoy & Akobundu, 1990; IITA, 1992). Indeed, this was clearly shown in the improvements brought about in the soil CEC during the period. Without a doubt, the gain in SOM has been found to have considerable impact on the buffering capacity and nutrient content of the soil viz. improving the soil CEC by about a third. Certainly the soil CEC is reportedly a chemical property important to soil fertility; accordingly, soils with high CEC can sustain intensive, continuous cultivation; while those with low CEC would undergo rapid physical and chemical degradation (Chude, Amapu, Malgwi, & Ano, 2002).

Harvested fresh fodder yield of the dual purpose cowpea was highest at 15cm intra-row spacing (Table 1). Definitely, if rapid production of the green crop biomass is desired (i.e. when yield is the product of growth of vegetative material), then denser sowing is preferable, as higher yield per area is achieved, as rapidly as possible, through greater population pressures; for maximum radiation interception. (Gardner, Pearce, & Mitchell, 1985; Muller-Samann & Kotschi, 1994).

Clipping management practices have been reported to be more common with grasses and forage legumes (Chapparo & Sollenberger, 1991). In the present study, the adoption and application of this clipping management practice showed that it facilitated the production of large amounts (15t ha^{-1}) of plant material (green fresh fodder) on-farm; which can be put to various uses. The observed high increase in fresh fodder yield (62%) at 74 DAP, has been reported to coincide with the period of the crops maximum vegetative growth (Musa, 1990).

4.2 Soil Organic Matter (SOM) Content

Soil organic matter (SOM) content decreased with increasing intra-row spacing and adoption of clipping treatment over the control (Table 3 and 4). Indeed, this could be attributed to the high volume of added plant organic material obtained at high plant densities; and is in conformity with the findings of Murthy and Hirekerur (2004), which highlighted that the vegetation determines the quantities and quality of organic material added each year. In addition, Adams, Bamford, and Early (1998) reported that most SOM is concentrated on the top soil because most of the roots occur in the zone and the plant residues tend to be added to the surface, forming the leaf litter layer. In this study, SOM content added on clipped plots increased by about 41%. Indeed, this result is supported by the findings of Jones (1971), who in a study in the dry savanna climate of Samaru, Nigeria, reported that the C-content of a sandy loam (Alfisol) was improved from 0.45% to 0.67% in 9 years; signifying a 66% increase.

4.3 Soil Organic Matter, Soil Fertility, Organic Agriculture and Sustainability

Many agricultural farmlands in the savannah are degraded due to various factors ranging from intensive weathering, high rainfall, temperature and decomposition rates, soil erosion, leaching, deterioration of the soil structure – low organic matter content (SOM), low CEC, and low moisture retention capacity of soils, resulting in low soil fertility.

Organic manure (OM) often includes wastes from both plants and animals whether composted or fresh but intended for use in farming. It is often brought into the field to improve the soils physical, chemical and biological properties, so as to enhance crop performance. Organic matter plays a key role in soils and the significance of organic mass in tropical soils is greater than any other soil characteristics apart from moisture (Asadu, Ezeaku, & Nnaji, 2004). Thus, the maintenance of organic matter is of fundamental importance to the productivity of tropical soils. If soil fertility in the tropics is to be maintained, fertilization must first and foremost be organic. What this means in practical terms is that, a proper management of tropical soils must aim to maintain both the structure as well as the content of organic matter; and for soils that have been cultivated for a long time, the aim should be to increase the organic mass and to improve structure (Muller-Samann & Kotschi, 1994). Without a doubt, organic farming avoids or largely excludes the use of synthetic fertilizers and pesticides and as far as possible promotes the purposeful maintenance and replenishment of soil fertility; as in crop rotation, crop residues and animal manures to supply plant nutrients, and to control weeds, insects and other pests (Wikipedia, 2005).

Yet in many tropical cropping systems, little or no agricultural residues are returned to the soil leading to a decline in soil organic matter (Lal, 1986; Bouwman, 1990; Post & Mann, 1990; Woomer & Ingran, 1990); and lower crop yields (Lal, 1986), or plant biomass productivity (Woomer & Ingran, 1990). It is therefore clear that if soil fertility in the tropics is to be sustained, then, the amount of soil organic matter will need to be maintained at the level of economic yields and not degraded through cultural practices.

5. Conclusion

It can be concluded that, cowpea possesses a great potential for improving soil organic matter (SOM) in degraded tropical soils; provided appropriate cultural and management practices which could lead to improvements in the soil pH and CEC (physical and chemical conditions) are adopted. In this regard, for the low technology, resource poor, and subsistence farmers of the region, the adoption of this innovative clipping management technology enhanced the production of large amounts of green plant biomass (fodder) on-farm/in-situ; which could be put to various uses: it could be incorporated into the soil as organic (green) manure for improving the soil organic matter (SOM) status. Certainly, in the savannah tropics where most soils are inherently low in soil nutrient content/fertility amongst other diverse challenges militating against boosting crop productivity - overcoming soil organic matter (SOM) decline is a major component in the development of more sustainable agro-systems; which encompasses the successful management of resources for agriculture to satisfy human needs, while maintaining or enhancing the quality of the environment and conserving natural resources in a sustainable manner.

References

Adams, C. F., Bamford, K. M., & Early, M. P. (1998). *Principles of Horticulture* (3rd ed.) Butterworth-Heinmann, London, U.K.

Asadu, C. L. A., Ezeaku, P. I., & Nnaji, G. U. (2004). The Soils of Sub-Saharan African and Management Needs for Sustainable Farming. In M. A. Badejo, & A. O. Togun (Eds.), *Strategies and Indices of Sustainable Agriculture in the Tropics* (Vol.2, pp. 1-27), Ibadan, Nigeria: College Press and Publishers Ltd.

Black, C. A. (1965). Methods of Soil Analysis. 11 Chemical and Microbial Properties (p. 1578). Madison, Wisconsin, American Society of Agronomy.

Bouwman, A. F. (1990). Exchange of Green House Gases Between Terrestrial Ecosystems and the Atmosphere. In P. L. Woomer & Swift, M. J. (Eds.), The Biological Management Tropical Soil Fertility. John Wiley and Sons, USA. A Co-Publication with the Tropical Soil Biology and Fertility Programme (TSBF) and Sayce Publishing (U.K).

Bray, R. H., & Kurtz, L. T. (1945). Determination of Total Organic and Available Forms of P in soil. *Soil Science,*

59, 39-45. http://dx.doi.org/10.1097/00010694-194501000-00006

Bremmer, J. M. (1965). Total Nitrogen. In C. A. Black (Ed.), Methods of Soil Analysis (Part 2, pp. 511-576). American Society of Agronomy.

Chaparro, C. J., & Sollenberger, L. E. (1991). Productivity and Persistence of Moth Elephangrass under Clipping Management (p. 141). Agronomy Abstracts. 1991, Annual Meeting.

Day, P. L. (1965). Particle Fractionation and Particle Size Analysis. In C. A. Black (Ed.), Method of Soil Analysis part 1. Physical and Mineralogical Properties (pp. 545-567). American Society of Agronomy.

Chude, V. O., Amapu, I. Y., Malgwi, W. B., & Ano, D. A. (2002). Train-the-Trainers Workshop on Interpretation of Soil Test Data and Fertilizer Recommendations for State ADP Facilitators (Crops) Engaged in SPFS Projects. PCU/IAR/NRCRI: 21-22 and 28-29th August, 2002.

Duncan, D. B. (1955). Multiple Range and Multiple F-tests. *Biometrics, 11*, 1-42. http://dx.doi.org/10.2307/3001478

Enwezor, W. O., Udo, E. J., Usoroh, N. J., Ayotade, K. A., Adepetu, J. A., Chude, V. O., & Udegbe, C. I. (1989). Fertilizer Use and Management Practices for Crops in Nigeria (pp. 1-168). Fertilizer Procurement and Distribution Division of the Federal Ministry of Agriculture, Water Resource, Rural Development. Lagos, Nigeria.

Gardner, F. P., Pearce, R. B., & Mitchell, R. L. (1985). Physiology of Crop Plants (pp. 76-97). AMES, IOWA, U.S.A: Iowa State University Press. http://dx.doi.org/10.1104/pp.77.3.540

Henao, J., & Baanante, C. (1999). Nutrient Depletion in the Agricultural Soils of Africa 2020 Vision Brief 62. Washington D.C. International Food Policy Research Institute.

International Institute for Tropical Agriculture. (1975). Selected Method for Soil and Plant Analysis (p. 44). International Institute for Tropical Agriculture (IITA), Ibadan, Nigeria. Annual Report.

International Institute for Tropical Agriculture. (1992). Sustainable Food Production in Sub-Saharan Africa 1. IITA contributions. International Institute for Tropical Agriculture - IITA, Ibadan Nigeria.

Jones, M. J. (1971). The Maintenance of Soil Organic Matter under Continuous Cultivation at Samaru, Nigeria. *Journal of Agricultural Science, 77*, 473-482. http://dx.doi.org/10.1017/S0021859600064558

Kahnt, G. (1983). Grundungung, Z.verb. Auflage. DLG-varlag, Frankfurt/M. P. In K. M. Muller-Samann & J. Kotschi (Eds.), Sustaining Growth: Soil Fertility Management in Tropical Small Holdings. CTA/GTZ-Weikersheims Margraf

Karunairanjan, R. D. (1980). Grundungung In den Tropen. *IFOAM, 35*, 18-21.

Keay, R. W. J. (1959). *An Outline of Nigerian Vegetation.* (3rd ed.) Government Printer, Lagos. In B. T. Kang, I.O. Akobundu, V. M. Manyong, R. J. Carsky, N. Sanginga & E. A. Kueneman (Eds.), Moist Savannas of Africa; Potentials and Constraints for Crop Production (pp. 31-48). Proceedings of an IITA/FAO Workshop held from 19-23 September, Cotonou, Republic of Benin.

Kundsen, D., Peterson, G. A., & Paratt, P. F. (1982). Lithium, Sodium and Potassium. In Pagered, A. L. (Ed.), Method of Soil Analysis (pp. 225-246). American Society of Agronomy.

Lal, R. (1986). Soil Surface Management in the Tropics for Intensive Land Use and High Sustained Productivity. In P. L. Woomer & M. J. Swift (Eds.), The Biological Management of Tropical Soil Fertility. John Wiley & Sons, U.S.A. A Co. Publication with the Tropical Soil Biology and Fertility Programme (TSBF) and Sayce Publishing (U.K.).

Lal, R., Wilson, G. F., & Okigbo, B. N. (1979). Changes in Properties of an Alfisol Produced by Various Grasses and Legume Crop Covers. *Soil Science, 127*(6), 377-382. http://dx.doi.org/10.1097/00010694-197906000-00009

Lee, K. K., & Wani, S. P. (1988). Soil Fertility and Fertilizer Management in Semi arid Tropical India. Colloguium on Soil Fertility and Fertilizer Management in Semi arid Tropical India. Proceedings held 10/17-18/88, in Hyderabad, India. IFDC-Nlabama, U.S.A.

Mulongoy, K., & Akobundu, I. O. (1990). Agronomic and Economic Benefit of Nitrogen Contributed by Legume in Live-Mulch and Alley Cropping Systems. In International Institute for Tropical Agriculture: Sustainable Food Production in Sub-Saharan Africa 1. IITA's Contributions. IITA, Ibadan, Nigeria. http://dx.doi.org/10.1007/978-1-4684-6432-0_54

Muller-Samann, K. M., & Kotschi, J. (1994). Sustaining Growth. Soil Fertility Management in Tropical Small Holding.CTA; GTZ-Weikersheim: Margra; Germany.

Murthy, R. S., & Hirekerur, L. R. (2004). Soils. In C. S. Viswanath (Ed.), Handbook of Agriculture. (pp. 20-72). New Delhi: Indian Council of Agricultural Research.

Musa, U. T. (1990). Effect of Seed Size and Phosphorus Level on the Growth and Yield of Cowpea (*Vigna unguiculata* (L.) Walp) varieties. M.Sc. Agronomy Seminar. Department of Agronomy, Ahmadu Bello University, Zaria, Nigeria.

Odion, E. C., & Singh, B. B. (2005a). Production Dual Purpose Cowpea Varieties at Two Clipping Heights in the Sudan Savanna of Nigeria: 1. Fodder Production in Pure Stand. *Journal of Arid Agriculture, 15*, 9-14.

Odion, E. C., & Singh, B. B. (2005b). Production of Dual Purpose Cowpea Varieties at two Clipping Heights in the Sudan Savanna of Nigeria 2. Pod and grain Production in Pure Stand. *Journal of Arid Agriculture, 15*, 15-19.

Odion, E. C., Asiribo, O. E., Ogunlela, V. B., Singh, B. B., & Tarawali, S. A. (2007). Strategies to Improve and Sustain Food Production Capacity in the Savannah: The Role of Leguminous Fodder Crops in Maintaining Soil Fertility and Health. *Journal of Food, Agriculture and Environment, 5*(2), 338-344.

Onwueme, I. C. (1979). Cassell's Tropical Agriculture Series, Book 2 Science (pp. 56-61). Macmillan Publishing Co. Inc. N.Y..

Onwueme, I. C., & Sinha, T. D. (1991). *Field Crop Production in Tropical Africa CTA*, Ede, The Netherlands.

Patnaik, N. (2004). Soil Fertility and Fertilizer Use. In C. S. Viswanath (Ed.), Handbook of Agriculture (pp. 203-247). New Delhi: Indian Council of Agricultural Research.

Post, W. M., & Mann, L. K. (1990). Changes in Soil Organic Carbon and Nitrogen as a Result of Cultivation. In P. L. Woomer & M. J. Swift (Eds.), The Biological Management of Tropical Soil Fertility. John Wiley and Sons, U.S.A. A Co. Publication with the Tropical Soil Biology and Fertility Programme (TSBF) and Sayce Publishing (U.K).

Reijntjes, C., B. Haverkort, B., & Waters-Bayer, A. (1992). Farming for the Future: An Introduction of Low-External-Input. ETC/ILEIA. The Macmillan Press Ltd., London and Basingstoke.

Sambo, B. E., Odion, E. C., Aliyu, L., & Labe, D. A. (2013). Cowpea (*Vina unguiculata* (L.) Walp) clipping management technology 2: A potential for sustained yield and food security in the savannah regions of Nigeria. *Journal of Agricultural and Crop Research, 1*(4), 55-60.

Singh, B. B. (1993). Cowpea Breading. Archival Report 1988-92. Grain legume improvement programme, crop improvement division, IITA, Ibadan, Nigeria.

Singh, B. B., Mohan Raj, D. R., Dashiell, K. E., & Jackai, L. E. N. (1997). Advances in Cowpea Research. Co-publication of International Institute of Tropical Agricultural Sciences (JIRCAS). IITA Ibadan, Nigeria.

Smalling, E. M. A., & Nandwa, S. M. (1996). Soil Fertility in Africa is at Stake (p. 36). Agronomy Abstracts.

Snedecor, G. W., & Cochran, W. G. (1967). *Statistical methods 6th edition*. (p. 456). U.S.A: Iowa State University Press.

Wikipedia. (2005). Organic farming. From Wikipedia, the free encyclopedia. Retrieved from on December 18, 2005, from http://wikipedia.org/organicfarming

Woomer, P. L., & Ingran, J. S. I. (1990). The Biology and Fertility of Soils. TSBF Report 1990, Nairobi Kenya TSBF. John Wiley and Sons, U.S.A. Co Publication with the Tropical Soil Biology and Fertility Programme (TSBF) and Sayce Publishing (U.K).

Yegna Narayan Aiyer, A. K. (1975). *Principles of Crop Husbandry in India*, (5th ed.) The Bangalore Press India.

Zake, J. Y. K. (1993). Overcoming Soil Constraints of Crop Production. In Y. Ahenkorah, E. Ownsan – Bennoah, & G. N. N. Dowuona (Eds.), Sustaining Soil Productivity Intensive African Agriculture: Overcoming Soil Constraints (pp 57-63). Seminar Proceedings, Accra, Ghana.

2

A Stochastic Frontier Analysis of Technical Efficiency of Maize Production Under Minimum Tillage in Zambia

John Ng'ombe[1] & Thomson Kalinda[1]

[1] Department of Agricultural Economics & Extension, University of Zambia, Lusaka, Zambia

Correspondence: John Ng'ombe, Department of Agricultural Economics and Extension, University of Zambia, Lusaka, Zambia. E-mail: njohnnedson@gmail.com

Abstract

Minimum tillage and other conservation agriculture practices are not only associated with income gains but are also claimed to be the panacea to the declining agricultural productivity and soil degradation problems in Africa and across the world. The few studies on technical efficiency related to the agricultural sector performance in Zambia have not attempted to determine how technically efficient smallholder farmers that produce maize under minimum tillage are. This study used stochastic frontier analysis based on both the half-normal and exponential model distributions on 2008 cross-sectional nationally representative data of 160 smallholder maize farm households that adopted minimum tillage in Zambia. Results indicate that maize farmers face increasing returns to scale (1.074) implying that there were opportunities for them to improve their technical efficiency as they were operating in stage I of their production functions. The half-normal and exponential model distributions indicate average technical efficiency scores of 60 and 71.7 percent, respectively. Their respective lowest efficiency scores were 9.3 and 8.5 percent. The highest efficiency scores for the half-normal and exponential model distributions were 89.3 and 90.9 percent. Maximum likelihood estimation results show that marital status, level of education of household head, square of household size, off farm income, agro-ecological region III, distance to vehicular road and access to loans are statistically significant factors that affect technical efficiency of smallholder maize farmers that practice minimum tillage in Zambia. The study calls for increased infrastructural development through construction of improved road network, schools and colleges in remote areas as a means to increasing accesss to knowledge and other agricultural services in order to enhance their technical efficiency levels. It also recommends promotion of minimum tillage practices in recommended agro-ecological regions to improve their technical efficiency. The study further acclaims for increased access to loans by smallholder maize farmers that practice minimum tillage as this would in one way induce them to invest in improved varieties and equipment that would help enhance their technical efficiency in Zambia.

Keywords: minimum tillage, smallholder maize farmers, technical efficiency, stochastic frontier analysis, Zambia

1. Introduction and Background

Climate change is a global issue that has taken centre stage in discussions by many scientists and policy makers. Its perceived negative impacts on food production coupled with current decline in agricultural productivity have contributed to the promotion of many climate adaptation strategies across Sub-Saharan Africa (SSA) and the world over. One of the most notable of these climate change adaptation strategies in Zambia is minimum tillage (MT), a component of conservation agriculture (CA) (Kuntashula et al., 2014). MT and other CA practices are not only associated with income gains but are also claimed to be the panacea to the declining agricultural productivity and soil degradation problems in Africa (Pieri et al., 2002; Haggblade & Tembo, 2003; Kabwe & Donovan, 2005; Giller et al., 2009). As a result, many governments and international development institutions in Africa and Zambia in particular have invested substantially in the promotion of these improved technologies that are thought to result in efficient use of resources and improved productivity (Haggblade & Tembo, 2003; Kassie et al., 2008).

MT is a dry-season land preparation practice that mainly involves minimum soil disturbance, either through planting in basins or ripping. Pieri et al. (2002) contended that MT involves planting of the crop into the soil's vegetative cover or crop residue without any disturbance or with less breaking to the soil surface. The key principle of MT is that it requires restriction of soil disturbance to a precise area where the crop is sown resulting into a minimum soil turnover of around 10% of the area for farming (FAO, 2011). This improves soil structure, increases water infiltration thereby facilitating plant growth and root development and thus enhancing productivity (Haggblade & Tembo, 2003).

As earlier mentioned, farmers have an option to practice ripping as a form of MT. Ripping is restricted to farmers that have access to animal traction or mechanical power. A study by Siziba (2008) showed that a ripper is used to rip into the soil to make a single furrow where seeds and fertilizers can be placed manually. Haggblade et al. (2010) observed that the foundation of ripping technology in Zambia is based on earlier efforts in agricultural engineering that concentrated on the development of ox-drawn ripping equipment for animal draft low-tillage systems in Magoye, Zambia which led to the development of the Magoye ripper in 1986. The Magoye ripper was invented in a way to help facilitate animal draught low tillage systems and water conservation, an alternative to the commonly known ox-plow. Haggblade and Tembo (2003) pointed out that the Magoye ripper requires less animal draught power than the ox-plough as it contains smaller dishes and has no mold board that an ordinary ox-plough has.

According to a study by Haggblade and Tembo (2003), the origin of MT in Zambia dates back to the late 1980s and early 1990s. For example, one of the donor-financed non-governmental organizations (NGOs) called the Co-operative League of the United States of America (CLUSA) required its farmers to plant in CF basins as a pre-condition to receiving input credit in the early 1990s (Haggblade & Tembo, 2003). Since then, smallholder farmers in Zambia have continued to adopt MT though its adoption has remained low. For example, a countrywide study by Ngoma et al. (2013) observed that only about 3.9% among Zambia's population of smallholder farmers practiced MT in 2012. Furthermore, a study by Kuntashula et al. (2014) indicated an adoption rate of about 12% from six selected districts in Zambia's agro-ecological region (AER) II in 2012. Zambia has three AERs. AER I mainly covers the Southern part of the country and is characterized by rainfall patterns of between 400 mm to 800 mm per year. AER II covers the Central and Eastern parts of the country with rainfall patterns between 800 and 1000 mm per year. AER III covers the Northern parts of the country with rainfall above 1000 mm per year (Haggblade & Tembo, 2003).

The low adoption rates of MT have been in the midst of numerous compelling benefits that are claimed to arise from its adoption in relation to efficient use of resources. For example, Haggblade and Tembo (2003) pointed out that planting in basins and rip lines harvest water in years of sporadic rainfall and ensure the precise application of fertilizer and other inputs next to the plants. This facilitates plant growth and development contributing to improved productivity. Haggblade and Tembo (2003) further acknowledged that by allocating land preparation to the dry season, in advance of the rains, MT redistributes heavy labor as well as animal and mechanized draft requirements out of the peak planting period. A study by Haggblade et al. (2010) confirmed that increased attention to use of MT also offers high productivity gains. Furthermore, Kassie et al. (2010) observed that MT increases agricultural productivity particularly from its ability to conserve soil moisture in dry environments.

Since MT practices are associated with these productivity gains, there is compelling interest to measure and understand its technical efficiency in the face of declining agricultural productivity among smallholder maize farmers in Zambia. For example, the few studies on technical efficiency related to the agricultural sector performance in Zambia have not attempted to determine how technically efficient smallholder maize farmers that practice MT have been (see for example; Mwape, 1988; Chiona, 2012; Steven Kabwe, 2012; Michael Kabwe, 2012; Chiona, 2014; Musaba & Bwacha, 2014; Namonje, 2014). Mwape (1988) focused on the relative economic and allocative efficiency of emergent and commercial maize farms in Zambia while Kabwe (2012) assessed technical, allocative and economic efficiency of smallholder maize producers in the Chongwe district. Studies by Chiona (2012) and Kabwe (2012) focused on the technical and allocative efficiency of smallholder maize farmers and factors that affect allocative, technical and economic efficiecy of cotton smallholder farmers in Zambia, respectively. A study by Namonje (2014) on effect of late delivery of subsidized fertilizer on smallholder maize productivity and production also assessed the technical efficiency of smallholder farmers in Zambia.

Furthermore, a study by Chiona et al. (2014) focused on the technical efficiency of smallholder maize farmers of Central Zambia while Musaba and Bwacha (2014) studied the technical efficiency of smallholder maize farmers in Zambia's Masaiti district. In other countries, similar studies were done by Henderson and Kingwell (2001) in Australia, Gouse et al. (2006) in South Africa and Langemeier (2010) in the United States of America (USA). As mentioned before, none of the recent work in Zambia on technical efficiency has focussed on smallholder maize farmers that practice MT. This knowledge gap was the main motivation for this study.

Therefore, this study contributes to the analysis of technical efficiency of the agricultural sector in Zambia by determining the technical efficiency of smallholder maize farmers who practice MT using the stochastic frontier analysis (SFA). The study further identifies the factors that affect their technical efficiency. There is no doubt that evaluating technical efficiency of smallholder maize farmers that adopt MT in Zambia is vitally important because it could guide MT resource utilization and may therefore lead to considerable resource savings, which could have important implications for both policy formulation and farm management (Bravo-Ureta & Riegler, 1991).

1.1 The Conceptual Model

Technical efficiency is defined as the ability to operate on the production frontier or isoquant frontier (Effiong & Onyenweaku, 2006; Greene, 2008). To build on this definition, this study is underpined by the model of Farrel (1957) that intuitively explains the measure of technical efficiency in simplicity. According to Farrel (1957), a firm is considered to have successfully achieved technical efficiency if it has produced a large amount of output given a correctly measured set of inputs. In his illustration, Farrel (1957) assumed that there is a given firm with a known efficient production function producing a single product while using two resources *X* and *Y* under constant returns to scale. These assumptions are necessary as they avail us with all the relevant information to be presented in a simple isoquant diagram shown in Figure 1. The point *P* represents the inputs of two resources per unit of output that the firm is observed to use. The isoquant *SS'* represents the various combinations of the two resources that a perfectly efficient firm might use to produce the unit output. Now the point *Q* represents an efficient firm using two resources in the same ratio as *P*.

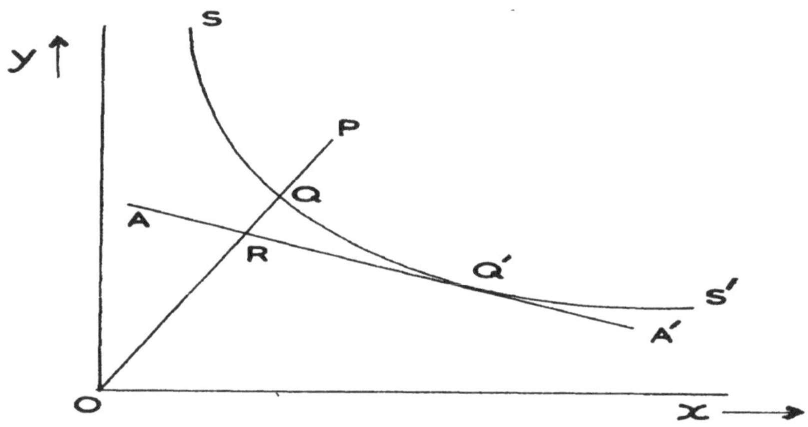

Figure 1. Farrel's measure of technical efficiency

Source: Farrel (1957).

As shown in Figure 1, the firm could produce the same output as *P* using only a fraction of *OQ/OP* as much of each resource. It could also be thought of producing *OQ/OP* times as much output from the same inputs. From this analysis, Farrel (1957) defined the firm's technical efficiency as *OQ/OP*, which naturally is true because this ratio has the properties that a measure of technical efficiency obviously needs. It takes the value unit for a perfectly efficient firm and would become indefinitely large. Moreover, so long as the isoquant *SS'* has a negative slope, an increase in the resource per unit output of one factor would imply lower efficiency, when other factors would be held fixed. However, Farrel (1957) postulated that one also needs to know the extent to which the firm uses the various resources in best proportions in view of their prices.

Therefore, following the same model of Farrel (1957), if the isocost line *AA'* has a slope equal to the ratio of prices of two resources, *Q'* and not *Q* is the optimal method of production. In as much as both points represent 100 percent technical efficiency, the costs of production at *Q'* would only be a fraction *OR/OQ* of those at point *Q*. It is therefore natural to define this ratio as the price efficiency of *Q*. Furthermore, Farrel (1957) explained that if the observed firms were to change the proportions of their inputs until they were the same as those represented by *Q'* while keeping technical efficiency constant, their costs would be reduced by an amount *OR/OQ* as long as resource prices did not change. It would be therefore reasonable to let this ratio measure the price efficiency of the observed firm too.

The argument drawn from this model seems not to be entirely conclusive as it would be impossible to say what would happen to the firms' technical efficiency if they changed the proportions of their inputs. However, with this qualification, the model seems to be the best measure of technical efficiency available (Farrel, 1957). Furthermore, the model also has the desirable property of giving the same price efficiency to the firms using resources in the same proportions. If the observed firms were perfectly efficient, both technically and in respect of prices, their costs would be a fraction *OR/OP* of what they are in fact are. According to Farrel (1957), it is convenient to call this ratio the overall efficiency of the firm and one may note that it is equal to the product of technical and price efficiencies.

1.2 The Stochastic Frontier Analysis

The previously discussed model by Farrel (1957) motivated several authors to improve the theory of measurement of technical efficiency and two methods have since been widely used to measure efficiency. The two methods are data envelopment analysis (DEA) and stochastic frontier method or analysis. The former is a nonparametric approach that developed out of mathematical programming techniques while the latter is a parametric approach that estimates technical efficiency within a stochastic production function model (Chakraborty et al., 2001). In this study, the stochastic frontier analysis was used in preference to the DEA framework because it posseses the following attractive features that make it worthy to be used as an alternative to the DEA framework (Greene, 2008).

Firstly, the stochastic frontier analysis posseses the 'stochastic' aspect unlike the DEA framework that allows it to handle appropriately measurement problems and other stochastic influences that would otherwise show up as causes of inefficiency for smallholder maize farmers. Secondly, the stochastic frontier analysis is suitable for such kind of data as were used in this study that possess unmeasured but surely substantial cross country heterogeneity. It provides a means to accommodating such heterogeneity which the DEA framework does not (Greene, 2008). Thirdly, the programming procedures under DEA framework are not based explicitly on an assumed statistical model and thus the properties of the estimators are therefore ambiguous (Greene, 2008). In short, the DEA framework's estimators from mathematical programming have the notable disadvantage that they do not naturally produce standard errors for the coefficients, which the stochastic frontier analysis is able to, so inference is excluded in the former, something that is not preferred in this study.

The main pioneers of the stochastic frontier analysis were Aigner et al. (1977) and Meeusen and van den Broeck (1977) with the motivation that the unconventionality from the frontier of the production activity could not be under total control by the economic agent in this case the smallholder maize farmer under study. Their work was an extension of the study by Farrel (1957) from which they formulated the stochastic production frontier so as to bridge the gap between theory and empirical work. Aigner et al. (1977), Meeseen and Van den Broeck (1977) and Battese and Corra (1977) postulated that the error term of the stochastic frontier production function is composed of two independent components of which the stochastic noise term would allow statistical tests of hypotheses related to the production structure and level of efficiency. The two error component stochastic frontier production function suggested to represent technical inefficiency was employed as shown below:

$$Q_i = f(X_i;\beta)\exp(v_i - u_i) \tag{1}$$

where Q_i is the scalar output of the ith smallholder farm, $\mathbf{X}_i$ is the vector of resources and β is a vector of parameters. v_i is the disturbance term assumed to be independently and asymmetrically distributed $(-\infty \leq v_i \leq \infty)$ and it captures the effects of random shocks outside the smallholder farmers' control. Such random shocks include unfavourable external effects, unpredictable events such as bad weather, equipment failure and injuries among others. u_i captures the factors that are under the control of the smallholder maize farmers. In this paper, u_i is assumed to be independently and asymmetrically distributed as normal-half-normal and normal-exponential (Aigner et al., 1977; Meeusen & van den Broeck, 1977). Other possible specifications of the distribution of u_i were suggested by Greene (1980a) and Lee (1983) and are still being used in most empirical work. This study assumed the two distributions because the normal-half-normal distribution is unduly narrow and therefore need for more robust results (Greene, 2008).

Empirically, technical efficiency of a smallholder maize farmer of the ith household is defined as the ratio of the observed amount of output (Q_i) to the corresponding potential or frontier amount of output (Q_i*) conditioned on the amount of resources used by that farmer. This definition is represented by Equation (2) below.

$$Technical\ Efficiency_i = \frac{Q_i}{Q_i^*} \tag{2}$$

Substituting Equation (1) into Equation (2) yields the following;

$$Technical\ Efficiency_i = \frac{f(X_i,\beta)\exp(v_i - u_i)}{f(X_i,\beta)\exp(v_i)} = \exp(-u_i) \quad (3)$$

The main thrust of this analysis is from Jondrow et al. (1982). From their work, the density function of u_i is as follows;

$$f(u_i) = \frac{1}{\sqrt{2\Pi}}(\frac{1}{\sigma_n})\exp(\frac{-u^2}{2\sigma_n^2});\ \mathrm{u_i} \geq 0 \quad (4)$$

For v_i, the density function is as follows;

$$f(v_i) = \frac{1}{\sqrt{2\Pi}}(\frac{1}{\sigma_v})\exp(\frac{-v^2}{2\sigma_v^2});\ \mathrm{u_i}\ \ (-\infty \leq \mathrm{v} \leq \infty) \quad (5)$$

And the density function of Q_i which is the joint density of (u_i and v_i) is given by;

$$f(Q_i) = 1 \Big/ \{\sigma\sqrt{\frac{\Pi}{2}}\} \exp(\frac{\varepsilon^2}{2\sigma^2})[1 - F(\frac{\varepsilon}{\sigma})(\frac{\gamma}{1-\gamma})\quad -\infty \leq \mathrm{u_i} \leq \infty \quad (6)$$

where *F(.)* is the cummulative distribution of the standard normal random variable while ε_i, σ^2, and γ are defined as follows;

$$\varepsilon_i = v_i - u_i$$
$$\sigma^2 = \sigma_v^2 + \sigma_u^2$$
$$\gamma = \sigma_u^2 \Big/ \sigma^2 \quad (7)$$

where γ lies between 0 and 1. Its values that are near 1 mean that the random component of the inefficiency effects significantly contributes to analysis of production systems. The loglikelihood of the sample is given below;

$$L(Q_i : \theta) = \Pi[\frac{1}{\sigma\sqrt{(\frac{\Pi}{2})}}\exp(\frac{\varepsilon^2}{2\sigma^2})(1-F)\{(\frac{\varepsilon}{\sigma})(\frac{\gamma}{1-\gamma})] \quad (8)$$

where θ is the parameter to be estimated and it is equivalent to the production parameters σ^2 and γ (Jondrow et al., 1982). Jondrow et al. (1982) observed that measurement of u_i for individual observations is derived from the conditional distribution of u_i and ε_i. If the normal and half-normal distributions of v_i and u_i, are respectively given, the conditional mean of u_i on ε_i and can be derived as follows following Jondrow et al. (1982).

$$E(u_i \mid \varepsilon_i) = \int u_i f(u_i \mid \varepsilon).du_i \quad (9)$$

where $f(u_i| \varepsilon_i) = f(u_i; \varepsilon_i)/\mathrm{f}(\varepsilon_i)$

The density function of u_i given ε_i and using Equations (4) and (5) is equivalent to the following after re-parameterization.

$$f(u_i;\varepsilon_i) = \frac{1}{\sqrt{2\Pi}}\frac{\sigma}{\sigma_v\sigma_u}\exp[\frac{-\sigma_u^2}{2\sigma_u^2\sigma_v^2}(u_i + \sigma_u^2/\sigma^2)^2]\ 1\Big/1 - F(.) \quad (10)$$

where *F(.)* is the cummulative normal density function. Therefore, Jondrow et al. (1982) concluded this with the expected value of u_i as follows:

$$E(u_i;\varepsilon_i) = (-\sigma_u\sigma_v/\sigma)[f(.)/(1-F(.)) - (\varepsilon)/\sigma\sqrt{\gamma/(1-\sigma} \quad (11)$$

Where *f(.)* represents the values of the standard normal density function. The estimates of $E(u_i \mid \varepsilon_i)$ can be obtained by the method of Maximum Likelihood Estimation (MLE) which is used to obtain estimates of γ, σ_v, and σ_v by evaluating Equation (10). As a result, the technical efficiency of an individual smallholder farmer can be calculated as follows:

$$Technical\ Efficiency_i = \exp(u_i ; \varepsilon_i) \tag{12}$$

For the exponential case, u_i is assumed to follow one-parameter exponential with the following density function: $f(u_i) = \exp(\frac{-u_i}{\sigma_v}) / \sigma_u$. The following results were obtained by Jondrow et al. (1982) as the resulting expected value of u_i,

$$E(u_i \mid \varepsilon_i) = \sigma_v [\frac{f(\frac{\varepsilon_i}{\sigma_v} + \frac{\sigma_v}{\sigma_u})}{1 - F(\frac{\varepsilon_i}{\sigma_v} + \frac{\sigma_v}{\sigma_u})} - \frac{\varepsilon_i}{\sigma_v} + \frac{\sigma_v}{\sigma_u}] \tag{13}$$

Where the *F(.), f(.)* and the parameters σ_u σ_v and ε_i are as previously defined. The results of the expected value of u_i are obtained as in Equation (12).

2. Data and Methods

2.1 Data Types and Sources

This study used data from a survey of households in Zambia conducted in 2008 by the Central Statistical Office (CSO) with financial and technical support from the Food Security Research Project (FSRP) and the Ministry of Agriculture and Co-operatives (MACO). The data were nationally representative and collected as part of the Rural Incomes and Livelihood Survey (RILS) and were the third Supplemental Survey to the 1999/2000 Post Harvest Survey (PHS) that was conducted in August/September 2000. The objectives of the survey were several. But this study mainly used the data as they contained necessary information about farmer adoption of CF practices and general perceptions. A total sample of 7,825 smallholder farm households was selected using a stratified three-stage sampling procedure. However, only 160 households were relevant for this study as these were the only smallholder maize farm households that adopted MT practices in Zambia. An adopter of MT in this study was considered as a farm household that planted a maize crop on any of their pieces of farm land and practiced MT in the 2006/2007 farming season.

2.2 The Empirical Model

To estimate Equations (12) and (13), this study follows the approach similar to the work by Battese and Coelli (1995) on panel data in which they estimated technical inefficiency from the stochastic frontier. In their study, technical inefficiency was simultaneously explained by a set of variables that were representative of the firm specific characteristics and time. The method dodges away from the inconsistency problems that come with the two-stage approach used in previous empirical works (for example; Kalirajan, 1981; Pitt & Lee, 1981; Kumbhakar & Hjalmarsson, 1995; Kabede, 2001). This method is a one step approach in which the inefficiency term is explicitly expressed as a function of the farmer or firm characteristics and the random error term (Kumbhakar et al., 1991; Reifschneider & Stevenson, 1991). Since the major interest of the study was technical efficiency measurement, the Cobb-Douglas production function was employed because of its capacity to provide an adequate representation of the production technology (Binam et al., 2004). It was specified as follows:

$$\ln Q_i = \beta_0 + \beta_1 \ln X_{1i} + \beta_2 \ln X_{2i} + \beta_3 \ln X_{3i} + \beta_4 \ln X_{4i} + v_i + u_i \tag{14}$$

where Q_i = total maize output in kilograms per hectare, β_i = the k×1 vector of parameters to be estimated and *X*'s are resources used to produce total maize output which include capital, land, labour, and seeds. They are also defined in Table 1. *ln* is the natural logarithm. v_i is the statistical noise term while u_i represents farm specific characteristics that are related to the inefficiency term and is estimated as shown in Equation (15).

$$u_i = \delta_0 + \delta_1 Z_1 + \delta_2 Z_2 ... + \delta_k Z_k \tag{15}$$

δ_i 's are the parameters to be estimated while Z_i's are the characteristics that are hypothesized to affect technical inefficiency. These are also defined in Table 1. Estimation of both Equations (14) and (15) was simultaneously done using MLE in Stata 13.

3. Results and Discussions

3.1 Descriptive Statistics

The variables used in this study are presented in Table 1. Column (1) shows the symbols representing the variables as indicated in Equations (14) and (15). Column 2 shows the definitions of the same variables. Columns (3), (4), (5) and (6) respectively indicate the variable means, standard deviation, minimum and maximum values. For the stochastic production function variables, the average total maize output was about 2 metric tons per hectare (ha) while the lowest was 11 kilograms per ha. The highest total maize output recorded was 48.432 metric tons per ha. The average amount of farm land cultivated was 1.61 ha with the lowest land cultivated being 0.2 ha while the largest area was 8 ha. This confirms that the farmers involved were smallholder farm households. The average capital owned by the smallholder farm household was ZMW625 while the highest amount owned was ZMW16, 700. Average man-days recorded in the season were about 31 days while average amount of seeds used were about 108 kilograms (kg) per ha.

Table 1. Summary statistics for variables used in the study

(1)	(2)	(3)	(4)	(5)	(6)
Variable	**Variable description**	**Mean**	**Std. Dev**	**Minimum**	**Maximum**
Output and input variables					
Q	Total maize productivity (kg/ha)	2032.55	4020.31	11.0.	48432.15
X_{1i}	Land in hactares	1.61	1.55	0.20	8
X_{2i}	Capital in ZMK	625.51	1821.39	0	16700
X_{3i}	Labour (mandays)	31.64	14.77	5.18	73.57
X_{4i}	Seeds in kg	107.99	36.89	3	214
Farm specific Variables					
Z_{1i}	Age in years	50.68	15.58	23	90
Z_{2i}	Gender of head (=1 if male, 0 otherwise)	0.72	0.45	0	1
Z_{3i}	Marital status (=1 if single, 0 otherwise)	0.17	0.38	0	1
Z_{4i}	Primary education (=1 if yes, 0 otherwise)	0.77	0.42	0	1
Z_{5i}	Basic education (=1 if yes, 0 otherwise)	0.23	0.42	0	1
Z_{6i}	Secondary education (=1 if yes, 0 otherwise)	0.07	0.25	0	1
Z_{7i}	Tertiary education (=1 if yes, 0 otherwise)	0.03	0.16	0	1
Z_{8i}	Household size	5.41	2.54	1	13.00
Z_{9i}	Off farm income per adult equivalent (ZMW)	1971.44	11200	6.207.06	138000
Z_{10i}	Agro-ecological region 1 (=1 if yes, 0 otherwise)	0.25	0.45	0	1
Z_{11i}	Agro-ecological region 2 (=1 if yes, 0 otherwise)	0.57	0.50	0	1
Z_{12i}	Agro-ecological region 3 (=1 if yes, 0 otherwise)	0.18	0.39	0	1
Z_{13i}	Extension service (=1 if received, 0 otherwise)	0.20	0.40	0	1
Z_{14i}	Livestock holding in total livestock units (TLU)	2.79	5.93	0.02	53.74
Z_{15i}	Distance to vehicular road in kilometres	6.33	11.78	0	80
Z_{16i}	Distance to input market in kilometres	33.85	36.10	0	150
Z_{17i}	Distance to product market in kilometres	16.77	23.54	0	126
Z_{18i}	Fertilizer (=1 if had access, 0 otherwise)	0.30	0.46	0	1
Z_{19i}	Belong to farmer association (=1 if yes, 0 otherwise)	0.07	0.23	0	1
Z_{20i}	Access to loans (=1 if yes, 0 otherwise)	0.26	0.44	0	1
Sample size					**160**

For farm specific characteristics, Table 1 indicates that the smallholder farm household heads had an average age of about 50 years old. The youngest farm household head was 23 years old while the oldest was 90 years old. About 72 percent of household heads were males while only about 17 percent of the household heads were single by marital status. The rest were considered to be married or have been married before. Table 1 further indicates that about 77 percent of the household heads had attended primary education while about 23 percent and 7 percent had attended basic and secondary education, respectively. Furthermore, the summary statistics indicate that about 3 percent of the household heads had attended tertiary level of education. Average household size was about 5 while the maximum number of household members recorded was 13, the lowest being 1. Table 1 further indicates that the average off farm income per adult equivalent was ZMW1, 971, the highest amount recorded as ZMW138, 000.

In terms of location by AER, 25 percent of smallholder maize farmers that practiced MT were located in AER I while 57 percent and 18 percent were located in AER II and AER III, respectively. AER I and II are low rainfall and drought prone areas while AER III are high rainfall areas in Zambia (Haggblade & Tembo, 2003). The average distances to vehicular road, input and output markets were about 6, 34 and 17 kilometers, respectively. Furthermore, the longest distances smallholder maize farmers would travel to access vehicular road network, input and output markets were 80, 150 and 126 kilometres, respectively. For access to fertilizer, about 30 percent of the smallholder maize farmers had access to it. Some of the farm households in the data belonged to farmer associations and Table 1 indicates that about 7 percent of farmers in the sample had such a previledge. Furthermore, Table 1 shows that about 26 percent of smallholder maize farm households that practiced MT had access to loans, while the rest did not. The last row shows that the sample size used in this study was 160.

3.2 Empirical Results

The MLE estimates of the stochastic frontier analysis are reported in Table 2. Column (1) presents the variables for both the stochastic production frontier and the technical inefficiency functions while Columns (2) and (4) present their estimates of the coefficients under half-normal and exponential model distributions, respectively. Standard errors under half-normal and exponential model distributions are respectively reported in Columns (3) and (5). Variance parameters are presented after variables in the technical inefficiency function while the log-likelihood values are presented in the last row for both model distributions.

The log-likelihood functional values and parameter estimates for both the stochastic production and technical inefficiency functions are relatively the same for both distributions. However, the variance parameters are not similar in their magnitudes though they are all statistically significant at 1 percent. As Kabede (2001) rightly suggested, the reason for the differences in their magnitudes could be due to the different specification of distributions of the inefficiency term which should be suitable for the data setting under study. The generalized likelihood ratio test defined by the chi-square χ^2 for presence of technical inefficiency indicates that the null hypothesis of absence of technical inefficiency ($\gamma = 0$) is strongly rejected at 5 percent significance.

$$[\chi^2, (5\%, 4) = 10.4].$$

As shown in Table 2, estimates of λ are large for both half-normal and exponential model distributions (1.746 and 0.817, respectively) and statistically significant at 1 percent indicating that both models were of good fit. σ's (0.873 and 0.623) were statistically significant at 1 percent implying that the conventional production function would not be a satisfactory representation of the data used. γ represents the ratio of variance of farm specific technical efficiency to the variance of total maize output. The estimates of γ for half-normal and exponential model distributions were 0.753 and 0.400 respectively. These results mean that under the half-normal model distribution, more than 75.3 percent of the variation in total maize output was due to differences in their technical efficiency.

Table 2. Maximum likelihood estimates of stochastic production frontier and technical inefficiency models

(1)	(2)	(3)	(4)	(5)
Model	Half-normal		Exponential	
	Coefficient	Std. Error	Coefficient	Std. Error
Dependent variable	Ln(total maize output/hectare)		Ln(total maize output/hectare)	
Stochastic production frontier function				
Constant	1.949	1.182	1.838*	1.104
Ln(land in hactares)	-1.85***	0.16	-1.85***	0.140
Ln(capital in ZMW)	0.008	0.010	0.010	0.010
Ln(labour in mandays)	2.612***	0.830	2.621***	0.800
Ln(seed in kg)	0.306***	0.089	0.291***	0.085
Technical inefficiency function				
Age in years	-0.004	0.004	-0.003	0.004
Gender (=1 if male, 0 otherwise)	0.007	0.131	0.019	0.126
Marital status (=1 if single, 0 otherwise)	0.292*	0.154	0.288*	0.152
Primary education (=1 if yes, 0 otherwise)	-0.202	0.186	-0.160	0.171
Basic education (=1 if yes,0 otherwise)	-0.230*	0.136	-0.199	0.132
Secondary education (=1 if yes, 0 otherwise)	-0.152	0.265	-0.168	0.259
Tertiary education (=1 if yes, 0 otherwise)	-0.767*	0.441	-0.727*	0.386
Household size	0.0334	0.075	0.033	0.074
Square of household size	-0.925***	0.359	-0.920***	0.347
Off farm income (ZMW)	35.3***	0.654	35.0***	0.581
AER I (=1 if yes, 0 otherwise)	0.0572	0.240	-0.0329	0.238
AER II (=1 if yes, 0 otherwise)	0.167	0.230	0.102	0.219
AER III (=1 if yes, 0 otherwise)	0.546**	0.264	0.499**	0.241
Extension service (=1 if yes, 0 otherwise)	-0.185	0.143	-0.158	0.139
Livestock holding in tlu	-0.026	0.034	-0.020	0.034
Distance to vehicular road (km)	0.010**	0.005	0.010**	0.005
Distance to input market (km)	-0.003	0.002	-0.003	0.002
Distance to product market (km)	-0.002	0.002	-0.001	0.002
Fertilizer access (=1 if yes, 0 otherwise)	0.171	0.140	0.157	0.137
Belong to farmer association (=1,0 o/w)	0.103	0.243	0.084	0.229
Access to loans (=1 if yes, 0 otherwise)	-0.126***	0.036	-0.128***	0.038
Constant	-0.554	0.623	-1.847	0.494
Variance parameters				
Lambda (λ)	1.746***	0.359	0.817***	0.148
Sigma square (σ^2)	0.763***	0.258	0.394***	0.054
Sigma_u (σ_u)	0.758***	0.236	0.397***	0.098
Sigma_v (σ_v)	0.434***	0.129	0.486***	0.060
Gamma (γ)	0.753		0.400	
Log likelihood	-151.10		-149.71	

***Significant at 1 percent; **Significant at the 5 percent level *Significant at the 10 percent level.

However, the exponential model results suggest that at least 40 percent of variation in their farm output is due to differences in their technical efficiency. This result should not raise eyebrows as results from the two models still indicate closely related levels of the average technical efficiency (60 and 71.7 percent) of the same sample (see Table 3).

The coefficients of the stochastic production frontier functions represent output elasticity values. They are similar in their magnitudes and signs for both model distributions. For instance, the output elasticity values for land, labour and seeds are statistically significant at 1 percent significance level in both cases. The output elasticity value of land means that if the smallholder maize farmers that practiced MT increased land allocation by 10 percent, this would result in 18.5 percent fall in the total maize output holding other factors fixed. The positive output elasticity value for labour means that if the smallholder maize farmers were to increase their man-days by 10 percent, this would be associated with about 26 percent rise in total maize output holding other factors constant. The farming season in Zambia is long enough to accommodate this observation.

In Table 1, it was indicated that the lowest number of days smallholder maize farmers that practice MT would work on their farm lands were about 5 man-days. This was perhaps due to illness or other factors the farm household members were exposed to. The positive output elasticity value for seeds means that if smallholder maize farmers increased their amount of seeds by 10 percent, this would be associated with about 3 percent increase in total maize output, *ceteris paribus*. Notice that the types of seeds considered in this study were improved varieties and the result is not surprising as improved varieties are meant to improve output (Kabede, 2001).

The results on labour and seeds are consistent with the results by Gouse et al. (2006) and Langemeier (2010). Results on seeds and and labour corroborate with those obtained by Chiona et al. (2014). The output elasticity value for capital was so low in magnitude that it was statistically insignificant. This was surprising as capital is expected to be a statistically significant input in production. The reason might be because only a few smallholder maize farmers had enough capital for farm investments. As seen from Table 1, the farm household with lowest capital recorded nothing and the average amount was ZMW635 only.

Among the results reported, the result for labour is the highest (2.610). This is not startling as the labour input has the most influence on any farm output. This result is consistent with Kabede (2001) who observed that farming activity is traditonal and labour intensive and thus availability of labour determines what can be produced at a farm. Studies on CA practices in Zambia have pointed out that CA practices include MT and that they duly require labour input for positive results (Haggblade & Tembo, 2003; Nyanga, 2012; Ng'ombe et al., 2014). Thus production functions of smallholder maize farm households that practice MT in Zambia are dorminated by the labour input.

Moreover, the sum of the output elasticity values in Table 2 represents returns to scale (RTS) of 1.074. This entails that maize production by smallholder farmers in Zambia was in stage I of the neoclassical production function. This implies that there is need for the farmers to continue increasing their production as they were still short of their production frontiers. This was confirmed by the generalized loglikelihood ratio test discussed above. The result further implies that if the smallholder maize farmers who practiced MT were to increase their resources by one percent, their total maize output would rise more than proportionally.

3.3 Factors That Affect Technical Efficiency

Sources of technical efficiency are the variables that appear in the inefficiency function of Table 2. Marital status is statistically significant at 10 percent under both model distributions. Its positive coefficient means that smallholder maize farm household heads that are single would more likely reduce technical efficiency at the farm, holding other factors fixed. This implies that married household heads would on the other hand increase the farm's technical efficiency. The reason could be because the household heads and their spouses would be in a better position to advise each other on what agricultural technologies to adopt and share information about how best they can apply them at their farm. Besides, smallholder farm households whose heads are married might have more labour available than those headed by single heads, an input that was observed to have most contribution in the stochastic frontier production function.

The dummy variable basic education is statistically significant at 10 percent under the half-normal model distribution only. But the sign is negative in both cases. The result means that smallholder maize farmers that have received education at basic level would be more technically efficient than those that have not. This result is consistent with Belbase and Gravowski (1985), Kalirajan and Shand (1985), Bravo-Ureta and Pinheiro (1997) and Kabede (2001). However, they contrast those results by Musaba (2014) who observed an opposite relation between education and technical efficiency. Furthermore, similar results are obtained as indicated by the dummy variable tertiary education. The negative coefficient also means that smallholder maize farmers that practice MT would be more technically efficient if they received tertiary education than otherwise, holding other factors constant. The reason might be that such farmers would more likely follow and apply the recommended principles underlying MT at their farms and result in efficient resource use. These results are analogous to those under the exponential model distribution case.

Household size is not statistically significant at affecting technical inefficiency. However, the variable square of household size is statistically significant at 1 percent under both model distributions. This result means that as the size of the smallholder maize farm household increases, it does not affect technical efficiency until a certain number when it positively affects it, *ceteris paribus*. This result shares the opposite version of the law of diminishing returns in production but it corroborates with the findings by Ajewole and Folayan (2008).

Off farm income is statistically significant at 1 percent under both model distributions. Its positive coefficient means that holding other factors constant, increase in off farm income of a smallholder maize farm household that practices MT would reduce its technical efficiency. A more plausible explanation is that smallholder maize farmers that are involved in off farm income generating activities would allocate less time on MT practices such as dry season land preparation but concentrate more on the latter. In other words, such households would trade off their managerial input from MT to off farm income generating activities. In such cases, attention to MT practices is reduced and hence negatively affecting their technical efficiency at a farm. This finding is consistent with the results by Abdulai and Eberlin (2001), Ajewole and Folayan (2008) and Chiona (2014).

Biophysical conditions were also considered to affect technical efficiency of smallholder maize farm households that practice MT in Zambia. For example, the coefficient of AER III was positive and statistically significant under both model distributions. This finding means that when other factors are held constant, smallholder maize farm households that practice MT and are located in AER III would be technically less efficient than those located in Zambia's other AERs. The coefficients for AER I and AER II were both positive but statistically insignificant. Zambia's AER III is the highest rainfall area and previous studies contend that such an area is unsuitable for MT practices (Pieri et al., 2002; Haggblade & Tembo, 2003; Nyanga et al., 2011; Ng'ombe et al., 2014).

Distance to a vehicular road is statistically significant at 5 percent under both model distributions. Distance to a vehicular road in this study refers to the amount of distance smallholder maize farmers that practice MT have to travel to reach a road that is passable by vehicles. The longer the distance is, the more remote the area would be considered. Thus the variable distance to vehicular road was used in this study as a proxy for remoteness. The positive coefficient implies that keeping other factors constant, if smallholder maize farm households that practice MT are located in remote areas, they would be more technically inefficient than those in non-remote areas. This could be because access to agricultural services such as field days, field visits by MT advocates, smooth communication and extension services among others would be limited. This would in one way cut them off from access to valuable information on MT and thereby negatively affecting their technical efficiency.

Furthermore, a smallholder farm household's access to loans was statistically significant for both model distributions. This means that when other factors are held constant, smallholder maize farmers that have access to loans or credit would be more technically efficient than those who do not. This result is consistent with the findings by Nwagbo (1989), Desai and Mellor (1993), Obwona (2006), Ajewole and Folayan (2008). A more suitable explanation for this could be as suggested by Desai and Mellor (1993) that if farm level credit is properly extended and managed, it enhances diversified systems of agriculture which stabilize and perhaps increase agricultural productivity.

3.4 Distribution of Technical Efficiency of MT Smallholder Farm Households in Zambia

Results of the distribution of technical efficiency scores of smallholder maize farm households that practiced MT are reported in Table 3. Column (1) shows the interval of technical efficiency scores in which households lie. Column (2) shows the number of households under each interval while column (3) shows their respective percentages. Column (4) shows their cummulative percentages. Column (1) through to column (4) fall under the half-normal model distribution case while the other columns in Table 3 fall under the exponential model distribution case. Columns (5), (6) and (7) respectively show the number of households, their percentages and cummulative percentages under each interval.

Table 3. Distribution of technical efficiency of smallholder maize farmers that practice MT in Zambia

Model	Half-normal case			Exponential case		
(1)	(2)	(3)	(4)	(5)	(6)	(7)
Model	Half-normal			Exponential		
Technical efficiency	Number of household s	Percent	Cummulative percent	Number of households	Percent	Cummulative percent
< 20%	1	0.63	0.63	1	0.63	0.63
20.01- 30%	6	3.75	4.38	4	2.5	3.13
30.01- 40%	11	6.88	11.25	1	0.63	3.75
40.01- 50%	20	12.5	23.75	3	1.88	5.63
50.01- 60%	32	20	43.75	16	10	15.63
60.01- 70%	38	23.75	67.5	28	17.5	33.13
70.01 - 80%	43	26.88	94.38	57	35.63	68.75
80.01 - 90%	9	5.63	100	49	30.63	99.38
90 -100%	0	0		1	0.63	100
Total	160	100		160	100	
Mean	0.600			0.717		
Std. Dev	0.158			0.134		
Minimum	0.093			0.085		
Maximum	0.893			0.909		

Results indicate that under the half-normal case, 43 smallholder maize farm households that practiced MT in Zambia were between 70 and 80 percent technically efficient. For the exponential model distribution, 57 households were in that interval. For the exponential case, only 1 smallholder maize farm household was between 90 and 100 percent technically efficient while under the half-normal, there was none with such a level. Both cases indicate that there was only 1 smallholder maize farm household with technical efficiency less than 20 percent. The distributions of the technical efficiency scores are fairly normal for both cases. The half-normal model results shows that on average, at least 60 percent of the smallholder maize farm households were technically efficient in Zambia while the exponential model shows an average score of 71.7 percent.

The estimated range of technical efficiency scores for the half-normal model distribution is 80 percent (0.093 to 0.893 percent) while for the exponential case, it is 82.4 percent (0.085 to 0.909 percent). Values from both model distributions are seemingly comparable though their distributions are different. However, both results are fairly and normally distributed as observed before. They are similar to results obtained by Chiona et al. (2014).

4. Conclusions and Policy Implications

This study used the stochastic frontier analysis to estimate the levels of technical efficiency of smallholder maize farmers that practice MT in Zambia. The Cobb-Douglas production function was the main type of production function used with the inefficiency term assumed to be half-normally distributed in one analysis and exponentially distributed in another. Data used were 2008 nationally representative with 160 smallholder maize farm households that adopted MT in the 2006/2007 farming season in Zambia.

Results showed that among the four main factors of production (land, labour, capital and seeds) used, labour was the most used input and that capital was not statistically significant though its effect on total maize output was positive. Returns to scale were increasing implying that if the smallholder maize farmers increased their inputs by one percent, total maize output would increase more than proportionally. Efficiency results indicated that individual differences in their technical efficiency levels at their farms partly contributed to variation in their total maize output.

The half-normal model distribution results of the stochastic frontier production function showed that more than 60 percent of the smallholder maize farmers that practiced MT were technically efficient while the exponential model distribution results indicated closely related results of 71.7 percent. The most technically inefficient farmer under the half-normal model recorded an efficiency score of 9.3 percent while the most technically efficient farmer was about 89.3 percent technically efficient. For the exponential model, the most technically inefficient farmer had an efficiency score of 8.5 percent while the most technically efficient farmer was 90.9 percent technically efficient. These results show that smallholder maize farm households that practice MT in Zambia still do not operate from their frontier levels. Despite the average levels of their technical efficiency scores being higher 60 percent, they have about 30 to 40 percent levels of efficiency to exhaust.

From the potential factors that were sought to affect technical efficiency of smallholder maize farmers that adopted MT in Zambia, marital status, level of education, square of household size, off farm income, AER III, distance to vehicular road and access to loans were statistically significant. Results showed that married household heads were more technically efficient than single ones. This indicated that marriage is an important institution in Zambia that enhances technical efficiency of smallholder maize farmers that practice MT. Levels of education at basic and tertiary level positively contributed to technical efficiency of smallholder maize farmers. This implies that smallholder maize farmers that practice MT should be encouraged to receive at least basic level of education and as much as college education where they would perhaps be exposed to relevant knowledge or acquire skills that would improve their efficiency levels when executing MT practices. But that would be necessary if basic schools and colleges filled with skilled staff are present in remote areas where it is generally expected that illiteracy levels are high.

The statistical significance of the square of household size means that as household size increases, there is a certain number that it reaches that is pivotal to improving levels of technical efficiency. This result encourages the need for smallholder maize farmers that practice MT to work together in numbers; something that would increase amount of labour available during their practice of of MT farming activities. A large household with members in their productive ages might increase supply of labour for agricultural activities, an input which was considered the most contributing factor to total maize output in the production function in this study. Off farm income positively affected technical inefficiency among MT smallholder farmers. This implies that MT competes for labour used for off farm income generating activities and therefore this study calls for decreased divided attention by smallholder maize farmers who adopt MT in order for them to be more technically efficient.

Furthermore, smallholder maize farmers that adopted MT and were located in AER III were more technically inefficient than those located in other AERs in Zambia. This calls for need for promotion of MT in AER I and II, areas with erratic rainfall patterns in Zambia as previously mentioned. Large distance to vehicular road positively affected technical inefficiency implying that there is need for more infrastructural development in the transport sector so that smallholder maize farmers that practice MT in remote areas are not disadvantaged from access to agricultural information and other agricultural services by MT promoting agencies such as Conservation Farming Unit and others in Zambia. The significance of access to loans implies that access to such forms of credit was necessary for smallholder maize farmers as this would probably encourage them to invest in MT modern equipment such as the Magoye ripper or be able to purchase improved seeds and fertilizer to improve their efficiency. This revelation opens up the ground to call for financial institutions and government to increase access to loans to smallholder maize farmers that practice MT in Zambia.

Acknowledgements

The authors appreciate the contributions made by Indaba Agricultural Policy Research Institute (IAPRI) and Michigan State University (MSU) through release of their databases. Without their valuable support, this paper would not have been completed.

References

Abdulai, A., & Eberlin, R. (2001). Technical efficiency during economic reform in Nicaragua: evidence from farm household survey data. *Economic Systems, 25*, 113-125. http://dx.doi.org/10.1016/s0939-3625(01)00010-3

Aigner, D. J., Lovell, C. A. K., & Schmidt, P. (1977). Formulation and estimation of stochastic frontier production models. *J. Econ., 6*, 21-37. http://dx.doi.org/10.1016/0304-4076(77)90052-5

Ajewole, O. C., & Folayan, J. A. (2008). Stochastic Frontier Analysis of Technical Efficiency in Dry Season Leaf Vegetable Production among Smallholders in Ekiti State, Nigeria. *Nigeria. Agricultural Journal, 3*(4), 252-257. http://medwelljournals.com/abstract/?doi=aj.2008.252.257

Battese, G. E., & Coelli, T. J. (1995). A model for technical inefficiency effect in stochastic frontier production for panel data. *Empirical Economics, 20*, 325-332. http://dx.doi.org/10.1007/bf01205442

Battese, G. E., & Corra, G. S. (1977). Estimation of a production frontier model: with application to the pastoral zone of Eastern Australia. *Australian Journal of Agricultural and Resource Economics, 21*(3), 169-179. http://dx.doi.org/10.1111/j.1467-8489.1977.tb00204.x

Belbase, K., & Grabowski, R. (1985). Technical efficiency in Nepalese agriculture. *The Journal of Developing Areas*, 515-526.

Binam, J. N., Tonyè, J., Nyambi, G., & Akoa, M. (2004). Factors affecting the technical efficiency among smallholder farmers in the slash and burn agriculture zone of Cameroon. *Food Policy, 29*(5), 531-545. http://dx.doi.org/10.1016/j.foodpol.2004.07.013

Bravo - Ureta, B. E., & Pinheiro, A. E. (1997). Technical, economic, and allocative efficiency in peasant farming: evidence from the Dominican Republic. *The Developing Economies, 35*(1), 48-67. http://dx.doi.org/10.1111/j.1746-1049.1997.tb01186.x

Bravo-Ureta, B. E & Riegler, L (1991). Dairy Farm Efficiency Measurement Using Stochastic Frontier and Neoclassical Duality. *American Journal of Agricultural Economics, 73*, 421-428. http://dx.doi.org/10.2307/1242726

Chakraborty, K., Biswas, B., & Lewis, W. C. (2001). Measurement of technical efficiency in public education: A stochastic and nonstochastic production function approach. *Southern Economic Journal*, 889-905. http://dx.doi.org/10.2307/1061576

Chiona, S. (2011). Technical and Allocative Efficiency of Smallholder Maize Farmers in Zambia. Master of Science Thesis. University of Zambia, Lusaka. Retrieved from http://dspace.unza.zm:8080/xmlui/handle/123456789/1813?show=full

Chiona, S., Kalinda, T., & Tembo, G. (2014). Stochastic Frontier Analysis of the Technical Efficiency of Smallholder Maize Farmers in Central Province, Zambia. *Journal of Agricultural Science, 6*(10), 108-118. http://dx.doi.org/10.5539/jas.v6n10p108

Desai, B. M., & Mellor, J. W. (1993). *Institutional Finance for Agricultural Development: An Analytical Survey of Critical Issues* (Food Policy Review 1). Washington: International Food Policy Research Institute.

Effiong, E. O., & Onyenweaku, C. E. (2006). Profit efficiency in broiler production in Akwa Ibom State, Nigeria. *International Journal of Agriculture and Rural Development, 7*(1), 72-79. http://dx.doi.org/10.4314/ijard.v6i1.2588

Farrell, M. J. (1957). The measurement of productive efficiency. *Journal of the Royal Statistical Society. Series A (General)*, 253-290. http://dx.doi.org/10.2307/234310

Food and Agriculture Organization (FAO). (2011). *FAO Stat.* Retrieved on January 12, 2014, from http://faostat.fao.org/site/291/default.aspx

Giller, K. E., Witter, E., Corbeels, M., & Tittonell, P. (2009). Conservation agriculture and smallholder farming in Africa: the heretics' view. *Field crops research, 114*(1), 23-34. http://dx.doi.org/10.1016/j.fcr.2009.06.017

Gouse, M., Piesse, J., & Thirtle, C. (2006). Output and labour effects of GM maize and minimum tillage in a communal area of KwaZulu Natal. *Journal of Development Perspective, 2*, 35-48.

Greene, W. H. (2008). The Econometric Approach to Efficiency Analysis. *The Measurement of Productive Efficiency and Productivity Chang*e, 92-250. http://dx.doi.org/10.1093/acprof:oso/9780195183528.003.0002

Greene, W. (1980a). Maximum Likelihood Estimation of Econometric Frontier Functions. *Journal of Econometrics, 13*, 27-56. http://dx.doi:10.1016/0304-4076 (80)90041-x

Henderson, B. B., & Kingwell, R. S. (2001). An Investigation of the Technical and Allocative Efficiency of Broadacre Farmers. In *2002 Conference (46th), February 13-15, 2002, Canberra* (No. 125109). Australian Agricultural and Resource Economics Society.

Haggblade, S., Kabwe, S., & Plerhoples, C. (2011). Productivity impact of conservation farming on smallholder cotton farmers in Zambia.

Haggblade, S., Tembo, G., Kaboré, D., Reij, C., Ajayi, O. C., Franzel, S., … Place, F. (2010). Sustainable Soil Fertility Management Systems. In S. Haggblade & P. Hazell (Eds.), *Successes in African Agriculture: Lessons for the Future*. Baltimore: Johns Hopkins University Press.

Haggblade, S., & Tembo, G. (2003). *Development, Diffusion, and Impact of Conservation Farming in Zambia.* Food Security Research Project.

Jondrow, J., Knox Lovell, C. A., Materov, I. S., & Schmidt, P. (1982). On the estimation of technical inefficiency in the stochastic frontier production function model. *Journal of econometrics, 19*(2), 233-238. http://dx.doi.org/10.1016/0304-4076 (82)90004-5

Kabwe, M. (2012). Assessing Technical, Allocative and Economic Efficiency of Smallholder Maize Producers Using the Stochastic Frontier Approach in Chongwe District, Zambia. Master of Science Dissertation, University of Pretoria, Pretoria, South Africa.

Kabwe, S., & Donovan, C. (2005). *Sustained use of Conservation Farming practices among small scale farmers in Zambia.* Michigan State University, USA.

Kabwe, S. (2012). *Factors Affecting Efficiency of smallholder cotton producers in Zambia* (M.Sc dissertation, University of Zambia).

Kalirajan, K. (1981). An econometric analysis of yield variability in paddy production. *Canadian Journal of Agricultural Economics/Revue canadienne d'agroeconomie, 29*(3), 283-294. http://dx.doi.org/10.1111/j.1744-7976.1981.tb02083.x

Kalirajan, K. P., & Shand, R. T. (1985). Types of education and agricultural productivity: a quantitative analysis of Tamil Nadu rice farming. *The Journal of Development Studies, 21*(2), 232-243. http://dx.doi.org/10.1111/j.1744-7976.1981.tb02083.x

Kalirajan & Shand (1985). Types of education and agricultural productivity: A quantitative analysis of Tamil Nadu Rice farming. *Journal of Development Studies, 21*(2), 232-243. http://dx.doi.org/10.1080/00220388508421940

Kebede, T. A. (2001). Farm household technical efficiency: A stochastic frontier analysis. *A Study of Rice Producers in Mardi Watershed In the Western Development Region of Nepal. Departemen of Economics and Social Sciences, Agricultural University of Norway.*

Kassie, M., Zikhali, P., Pender, J., & Köhlin, G. (2010). The economics of sustainable land management practices in the Ethiopian Highlands. *Journal of agricultural economics, 61*(3), 605-627. http://dx.doi.org/10.1111/j.1477-9552.2010.00263.x

Kassie, M., Pender, J., Yesuf, M., Kohlin, G., Bluffstone, R., & Mulugeta, E. (2008). Estimating returns to soil conservation adoption in the northern Ethiopian highlands. *Agricultural Economics, 38*(2), 213-232. http://dx.doi.org/10.1111/j.1574-0862.2007.00295.x

Kumbhakar, S. C., & Hjalmarsson, L. (1995). Labour - use efficiency in Swedish social insurance offices. *Journal of Applied Econometrics, 10*(1), 33-47. http://dx.doi.org/10.1002/jae.3950100104

Kumbhakar, S. C. (1991). Estimation of technical inefficiency in panel data models with firm-and time-specific effects. *Economics Letters, 36*(1), 43-48. http://dx.doi.org/10.1016/0165-1765(91)90053-N

Kumbhakar, S., Ghosh, S., & McGuckin, J. (1991). A Generalized Production Frontier Approach for Estimating Determinants of Inefficiency in U.S. Dairy Farms. *Journal of Business and Economic Statistics, 9*, 279-286. http://dx.doi.org/10.1080/07350015.1991.10509853

Kuntashula, E., Chabala, L. M., & Mulenga, B. P. (2014). Impact of Minimum Tillage and Crop Rotation as Climate Change Adaptation Strategies on Farmer Welfare in Smallholder Farming Systems of Zambia. *Journal of Sustainable Development, 7*(4), 95-110. http://dx.doi.org/10.5539/jsd.v7n4p95

Langemeier, M. (2010). Relative technical and cost efficiency of no-till farms. *Journal of International Farm Management, 5*(2), 1-11. www.ifmaonline.org

Lee, L. (1983). A Test for Distributional Assumptions for the Stochastic Frontier Function. *Journal of Econometrics, 22*, 245-267. http://dx.doi:10.1016/0304-4076 (83)90102-1

Meeusen, W., & Van den Broeck, J. (1977). Efficiency Estimation from Cobb – Douglas Production Functions with Composed Error. *International Economic Review, 18*, 435-444. http://dx.doi.org/10.2307/2525757

Musaba, E., & Bwacha, I. (2014). Technical Efficiency of Small Scale Maize Production in Masaiti District, Zambia: A Stochastic Frontier Approach. *Journal of Economics and Sustainable Development, 5*(4), 104-111.

Mwape, F. (1988). *Relative Economic and Allocative Efficiency of Emergent and Commercial Maize Farms in Zambia.* Ph.D Thesis, University of Manitoba, Winnipeg, Canada. Retrieved from http://mspace.lib.unmanitoba.ca/jspui/handle/1993/16767

Namonje, T. S. (2014). Does Late Delivery of Subsidized Fertilizer Affect Smallholder Maize Productivity and Production? *"Employment Strategies in the Developing World - How to create sufficient, productive and decent jobs"* Lusaka 18th September 2014.

Ngoma, H., Mulenga, B. P., & Jayne, T. S. (2013). Determinants and Extent of Use of Minimum Tillage Practices among Zambian Smallholder Crop Farmers from 2008 to 2012.

Ngombe, J., Kalinda, T., Tembo, G., & Kuntashula, E. (2014). Econometric analysis of the factors that affect adoption of conservation farming practices by smallholder farmers in Zambia. *Journal of Sustainable Development, 7*(4), 124-138. http://dx.doi.org/10.5539/jsd.v7n4p124

Nyanga, P. H., Johnsen, F. H., Aune, F. H., & Kalinda, T. H. (2011). Smallholder Farmers' Perceptions of Climate Change and Conservation Agriculture: Evidence from Zambia. *Journal of Sustainable Development, 4*(4), 73-85. http://dx.doi.org/10.5539/jsd.v4n4p73

Nyanga, P. H. (2012). Factors influencing adoption and area under conservation agriculture: a mixed metthods approach. *Sustainable Agriculture Research, 1*(2), 27. http://dx.doi.org/10.5539/sar.v1n2p27

Nwagbo, E. C. (1989). Impact of Institutional Credit on Agriculture in Funtua Local Government area of Katsina State, Nigeria. *Samaru J. Agric. Res., 6*, 75-86.

Obwona, M. (2006). *Determinants of technical efficiency differentials amongst small and medium scale farmers in Uganda: A case of tobacco growers* (Vol. 152). African Economic Research Consortium.

Onyenweaku, C. E., & Effiong, E. O. (2006). Technical efficiency in pig production in Akwa Ibom State, Nigeria. *International Journal of Agriculture and Rural Development, 6*(1), 51-57. http://dx.doi.org/10.4314/ijard.v6i1.2588

Pieri. C., Evers, Landers, G., Connell, P., & Terry, E. (2002). *No-Till Farming for Sustainable Rural Development.* Agriculture & Rural Development Working Paper. Washington, DC 20433. Retrieved from http://www.betuco.be/CA/Notillage%20Farming%20for%20 Sustainable%20 Development.pdf

Pitt, M. M., & Lee, L. F. (1981). The measurement and sources of technical inefficiency in the Indonesian weaving industry. *Journal of development economics, 9*(1), 43-64. http://dx.doi:10.1016/0304-3878 (81)90004-3

Reifschneider, D., & Stevenson, R. (1991). Systematic departures from the frontier: a framework for the analysis of firm inefficiency. *International Economic Review*, 715-723. http://dx.doi.org/10.2307/2527115

Schmidt, P., & Knox Lovell, C. A. (1979). Estimating technical and allocative inefficiency relative to stochastic production and cost frontiers. *Journal of econometrics*, *9*(3), 343-366. http://dx.doi:10.1016/0304-4076 (79)90078-2

Siziba, S. (2008). Assessing the adoption of conservation agriculture in Zimbabwe's smallholder sector. PhD Thesis. University of Hohenheim, Germany.

3

Consumers' Willingness to Pay for Cassava Flour Inclusion in Bread, a Case Study of Lagos State, Nigeria

Erih E. A.[1], Oni O. A.[1] & Fashogbon A. E.[2]

[1] Department of Agricultural Economics and Extension, Delta State University, Delta State, Nigeria

[2] Department of Agricultural Economics, University of Ibadan, Oyo State, Nigeria

Correspondence: Erih E. A., Department of Agricultural Economics and Extension, Delta State University, Delta State, Nigeria. E-mail: eguonoerih@gmail.com

Abstract

The study was conducted to address the issue of consumers' willingness to pay for inclusion of cassava flour in bread in Lagos State. The data were collected randomly from a sample of 300 respondents in the state through the use of a well structured questionnaire. The contingent valuation method was adopted to estimate both the mean willingness to pay of consumers and the factors that affect their willingness to pay and these were analyzed using the bivariate probit model. The most significant model was the 10percent cassava flour inclusion as it has the highest number of significant variables (eight) followed by 20percent cassava flour inclusion with seven significant variables and 30 and 40percent cassava flour inclusion with six significant variables each. The factors that influenced consumers' willingness to pay for CCW bread were the respondent's age, gender, respondents' awareness, married respondents, respondents with head position,and bread share which is the proportion of respondents income spent on bread to total income. The mean willingness to pay for bread with an inclusion of cassava flour at 10, 20, 30, and 40percent cassava flour inclusion were ₦172.70, ₦165.00, ₦154.07, and ₦142.78 respectively for initial bid while the mean willingness-to-pay at 10, 20, 30, and 40percent cassava flour inclusion were ₦180.20, ₦150.41, ₦143.35, and ₦127.36 respectively for subsequent bid. Arising from the result of this study, birth control policies were recommended among respondents in the study area.

Keywords: bivarate probit model, composite cassava wheat bread, contingent valuation, willingness to pay

1. Introduction

In most countries of the world, in recent years, partly as a result of structural adjustment policies and pressures from multilateral financial institutions, various governments have increasingly focused on identifying other food components that can be added as a component in the bread making process. The rising cost of wheat importation for the production of bread, a well-established and accepted food product relished by the general public, has been a matter of great concern to the Nigerian government in recent times. This has led to the increased interest in cassava flour which has been identified as a close, cost-effective substitute to wheat flour in bread production.

Willingness to pay measures the resources individuals are willing and able to give up (Golan & Kuchler, 1999). Also, WTP is the maximum amount of money an individual is willing to pay for a commodity; therefore, WTP is an indicator of the value of the commodity to that individual. Willingness to pay measures is considered useful for several reasons. First, they can directly inform policy makers by providing information about how much people value some goods or services and can thus inform the pricing of these goods or services (Hanley *et al.*, 2003). Second, WTP measures can be important inputs in economic evaluations such as cost benefit analyses (Loomes, 2001; Oliver et al., 2002; Negrín et al., 2008). Third, WTP measures can be a convenient tool to make relative comparisons and rankings of the desirability of goods and services. Alternatively, WTP measures can be derived from discrete choice models estimated using either revealed preference data or data from discrete choice experiments (DCEs). In these cases, the WTP for an alternative attribute can be calculated as the ratio of the attribute coefficient to the price coefficient (Train, 2003).

Responses from people who report a high level of certainty about their willingness to pay exhibit significant anomalies that increases as uncertainty increases (Watson & Ryan, 2006).

Nigeria is rarely mentioned when it comes to cassava export even though it is the world's largest producer of cassava. This is because most of the staples produced in the country do not reach the global market due to the very poor infrastructure in Nigeria. Post harvest losses of staples are extremely high in Nigeria (IITA, 2010), worsened by poor post harvest, handling and marketing strategies. The report from International Institute of Tropical Agriculture IITA, however indicate that Nigeria can produce very good bread from a combination of 30% cassava flour and 70% wheat flour. This however depends on sustained willingness to pay for composite cassava wheat (CCW) Bread by consumers. Examining the consumers' perception and willingness to pay for CCW bread in Lagos State is therefore inevitable. The study therefore seeks to provide answers to the following research questions generated.

i. What are the factors that influence consumers' willingness to pay for bread with an inclusion of cassava flour?

ii. What is the mean willingness to pay for bread with an inclusion of cassava flour?

The consumer's WTP for a good is a fundamentally behavioural concept. The behaviour in question is that of buying (or selling) a good. How much one is willing to pay (or accept) for a good at a particular point in time will depend on a variety of factors, including of course, the expected intrinsic value.

In general, the willingness to pay a price premium decreases as the price premium increases, consistent with the law of demand. In consumer behavior theory, consumers make their own decisions to balance the marginal health utility and marginal price of one unit of quality-food product, a simple framework was used to analyze consumer behavior towards food products, which includes the willingness to pay a price premium. Consumers' willingness to pay (WTP) is a key concept in the management of sustainability because it is a prerequisite for producers in organizations. There are different types of WTP that materialize differently in different circumstances and influence the consumers' decision-making process.

WTP is an important concept for the literature that approaches sustainability from an instrumental stakeholder theory perspective. It has its roots in economics, where willingness to pay is defined as the maximum amount of money that an individual is willing to sacrifice to obtain a good or service (Freeman, 2003).

2. Review of Empirical Studies on Consumers' Willingness to Pay for Food Products

Grassi (2010), worked on "Public and Private Provision under Asymmetric Information: Ability to Pay and Willingness to Pay". He modeled the interaction between the public sector and the market and analyzed whether a mixed system of provision helps the public sector with the problem of affordability. He found out that willingness to pay was independent on the budget constraint, and then on ability to pay, Consumers with high willingness to pay may not afford the good at a given price, Consumers cannot borrow money if needed, The market allocation is inefficient and that the public sector has a budget, but it is insufficient to supply all consumers for free. It observes consumers' wealth and implements a policy to maximize the sum of consumers' utilities subject to the wealth constraints. He considered two optimal policies: rationing and subsidization. First it studied the public supplier as the sole provider of the good. Any rationing policy that exhausts the budget is optimal. The optimal subsidy scheme requires cross subsidization: rich consumers pay a price greater than marginal cost, and some poor consumers pay less than marginal cost. The budget and the revenue collected from rich consumers funds the subsidies for poor consumers. He then characterized the equilibrium of a simultaneous moves game where the public sector interacts with a firm in the provision of the good.

Motivated by the far-reaching benefits of the new cassava policy on the substitution of wheat flour with cassava flour in bread production to the economy, a logistic regression model was employed to examine households' perception and willingness to pay for bread with cassava flour inclusion in Osogbo Metropolis, Osun State, Nigeria. Findings suggest that bakery owners adopting the use of cassava flour in bread production pay careful attention to the taste, packaging, size, colour and price of the bread since these variables affect the buying decisions of the consumers. While consumers' willingness to pay a premium varied with degrees of cassava flour inclusion, households' willingness to pay showed a negative relationship with the premium price and a positive, significant relationship with household income and share of bread in total household food expenditure. Thus, proper attention should be given to price stabilization of bread with cassava flour inclusion as well as the design of empowerment programmes targeted at increasing household income.

3. Materials and Methods

3.1 Study Area

The study was carried out in Lagos State Nigeria. Lagos state is in the southwest geopolitical zone. It falls on latitude 6.523° North and longitude 3.54° West. The highest maximum temperature ever recorded in Lagos was 37.3 °C (99.1° F) and the minimum 13.9 °C (57.0° F) (Lagos Meteorological Organization, 2012). The main source of livelihood of people living in Lagos is income from private establishments in the state. Most commercial and financial business is carried out in the central business district situated on the island. Lagos has one of the highest standards of living as compared to other cities in Nigeria as well as Africa and it is one of the fastest growing cities in the world, experiencing a population increase of about 275,000 persons per annum (Rasaki, 2012). The state was chosen because it is Nigeria's economic and commercial capital. Primary data was utilized for this study. The primary data was obtained through the use of well structured questionnaire. The questionnaire was pretested to remove any ambiguity. Primary data was collected from the consumer households in the state. The study employed a two stage sampling procedure. At the first stage, random sampling of three (3) LGAs was done, whereby one local government area in Lagos Island and two local governments in Lagos main land were selected. In the second stage eighty respondents were randomly selected in Lagos Island local government area, one hundred respondents were randomly selected from Ikeja LGA while one hundred and twenty were selected from Alimosho LGA on the basis of sampling proportionate to size. This leaves the total number of respondents selected for this study at three hundred.

3.2 Methods of Data Analysis

The analytical techniques adopted in this study include descriptive statistics such as tables, percentages and frequencies to analyze the socio-economic characteristics of consumers with respect to their preference and willingness to pay pattern for cassava inclusion in bread in the study area. The consumers' preference pattern, proxied by their willingness to pay will be broken down by Bivariate Probit Model.

Contingent valuation method was used to generate choice pattern and the responses were later analyzed using Bivariate Probit Model, which was used to determine the factors that influence consumers' willingness to pay for cassava inclusion in bread. In estimating the mean willingness to pay of consumers and potential consumers of Composite Cassava Wheat Bread, a double-bounded contingent valuation model was used in which the respondents were asked a series of questions that progressively narrowed down to their willingness to pay. This method has been shown to generate more efficient estimates than those based on a single question or those that ask open-ended question about willingness to pay (Watson & Ryan, 2006).

The probit model r category is built from a latent regression in the same manner as the binomial probit model. We begin with $y^*_i = \boldsymbol{\beta}^T\mathbf{x}_i + \boldsymbol{\varepsilon}_i$ where $\mathbf{x}$ is a vector of predictor variable for the i th observation and $\boldsymbol{\beta}^*$ is the unknown parameter. As usual, y^* is unobserved variable, that follow as: (Greene, 2005).

$$y = 0 \text{ if } \gamma_0 < y^* \leq \gamma_1$$
$$y = 1 \text{ if } \gamma_1 < y^* \leq \gamma_2$$

.

.

.

$$y = r - 1 \text{ if } \gamma_{r-1} < y^* \leq \gamma_r$$

The probability for each observed response has r category, i.e.:

$$P(y = 0) = P(\gamma_0 < y^* \leq \gamma_1) = \Phi(\gamma_1 - \boldsymbol{\beta}^T\mathbf{x}) - \Phi(\gamma_0 - \boldsymbol{\beta}^T\mathbf{x})$$
$$P(y = 1) = P(\gamma_1 < y^* \leq \gamma_2) = \Phi(\gamma_2 - \boldsymbol{\beta}^T\mathbf{x}) - \Phi(\gamma_1 - \boldsymbol{\beta}^T\mathbf{x})$$

.

.

.

$$P(y = r\text{-}1) = P(\gamma_{r-1} < y^* \leq \gamma_r) = \Phi(\gamma_1 - \boldsymbol{\beta}^T\mathbf{x}) - \Phi(\gamma_{r-1} - \boldsymbol{\beta}^T\mathbf{x})$$

Bivariate probit model (r x c) is a probit model which involves two response variables, i.e.

$y_1^* = \boldsymbol{\beta}^T{}_1\mathbf{x} + \boldsymbol{\varepsilon}_1$ and $y_2^* = \boldsymbol{\beta}^T{}_2\mathbf{x} + \boldsymbol{\varepsilon}_2$. The first variable has r category that is

$$y_1 = 0 = \text{if } \gamma_0 < y_1^* \leq \gamma_1$$
$$y_1 = 1 = \text{if } \gamma_1 < y_1^* \leq \gamma_2$$

.

.

.

$$y = r\text{-}1 = \text{if } \gamma_{r-1} < y_1^* \leq \gamma_r$$

Whereas, the second has c category, that is

$$y_2 = 0 \text{ if } \delta_0 < y_2^* \leq \delta_1$$

$$y_2 = 1 \text{ if } \delta_1 < y_2^* \leq \delta_2$$

.

.

.

$$y_3 = c\text{-}1 \text{ if } \delta_{c-1} < y_2^* \leq \delta_c$$

Variables (y_1^*, y_2^*) that satisfy normal bivariate distribution can be written as $(y_2^*\ y_2^*) \sim N(\boldsymbol{\mu},\Sigma)$.

Bivariate normal density function (y_1^*, y_2^*) is:

$$f(y_1^*,y_2^*) = \frac{1}{2\pi|\Sigma|^{1/2}} exp\left[-\frac{1}{2}\begin{pmatrix} \mathbf{y}^*_1 & -\boldsymbol{\beta}^T{}_1\,\mathbf{x} \\ \mathbf{y}^*_2 & -\boldsymbol{\beta}^T{}_2\,\mathbf{x} \end{pmatrix} \Sigma^{-1} \begin{pmatrix} \mathbf{y}^*_1 & -\boldsymbol{\beta}^T{}_1\,\mathbf{x} \\ \mathbf{y}^*_2 & -\boldsymbol{\beta}^T{}_2\,\mathbf{x} \end{pmatrix}\right]$$

The probability of bivariate normal density function (y_1^*, y_2^*) with thresholds γ and δ is as follows:

$$P(y_1^*,<\gamma\, y_2^*\leq\delta)=\int_{-\infty}^{\gamma} 1 \int_{-\infty}^{\delta} 1\ f(y_1^*, y_2^*)\ dy_1^*,\ dy_2^*$$

The average, true willingness-to-pay value ($\mu TWTP$) is calculated in the following manner (Haab & McConnell, 2002): where β_0 is the regression constant value, and $\beta 1$ the regression coefficient value for the proposed willingness-to-pay value in the bivariate probit regression model. The explanatory variables are the initial (*BID1*), and the follow-up willingness-to-pay values (*BID2*) that were proposed to respondents in the survey.

In the bivariate probit regression model, dependent variables represent the respondent's answers to the initial (*RESP1*) and the follow-up willingness-to-pay value (*RESP2*). These are binary variables that take the value 0 if the respondent accepts the proposed value and 1 otherwise.

i.e. resp 1 (r) = Consumers' willingness to pay for bread with cassava flour inclusion (willing to pay =1, 0 = otherwise)

resp 2 (c) = Consumers' willingness to pay for bread with cassava flour inclusion (willing to pay =1, 0 = otherwise) r x c)

The following can be classified as the determinants of willingness-to-pay (explanatory variables):

$x_{1=}$ age (*years*);

$x_{2=}$ sex (female=1, 0 = otherwise);

$x_{3=}$ household size (number);

$x_{4=}$ Tertiary Education (yes=1, 0=otherwise);

$x_{5=}$ Respondent's monthly income (*Naira*);

$x_{6=}$ Awareness of cassava bread (aware=1, 0= otherwise); and

$x_{7=}$ Marital Status (married=1, 0= otherwise)

$x_{8=}$ Household head position (head position=1, 0= otherwise)

$x_{9=}$ dependency ratio (dependant=1, 0= otherwise)

$x_{10=}$perception of respondents to cassava bread

$x_{11=}$ Share of Bread in total household food expenditure

$\rho_{i\,=}$ Error

For objective 2, following the result estimates from factors driving WTP using the bivariate Probit model, Krinsky and Robb Procedure (1986) used a bootrap estimate to generate the mean willingness to pay. This study will also follow this methodology.

4. Discussion

Table 1. Socio-economic characteristics of respondents

Variables	Frequency	Percentage
Gender		
Male	154	51.33
Female	146	48.67
Total	300	100.00
Marital Status		
Single	53	17.67
Married	239	79.67
Divorced	1	0.33
Widowed	7	2.33
Total	300	100.00
Household size		
1-2	28	9.33
3-4	74	24.67
5-6	140	46.67
7-8	54	18.00
9-max	4	1.33
Total	300	100
Mean 5.0133		
SD 1.8103		
Monthly Income(₦)		
Min-20000	29	9.67
20001-40000	53	17.67
40001-60000	92	30.67
60001-80000	55	18.33
80001-100000	22	7.33
100001-Max	49	16.33
Total	300	100.00
Mean ₦79,086.17		
SD ₦80,270.43		
Min ₦8,000		
Max ₦500,000		

Source: Field Survey, 2013.

4.1 Bread Consumption Pattern

Table 2 presents the food expenditure, non food expenditure and bread expenditure in relation to the different socioeconomic characteristics. The findings reveal a higher expenditure by the males than the females. From the table, it can be seen that the mean food, non food and bread expenditure for male and female respectively are ₦1711.78, ₦504.06, ₦142.55 and ₦1602.28, ₦480.79 and ₦119.85. The expenditure of food, non-food and bread was highest in the age category of 21-30 years which was ₦2649.23, ₦714.55 and ₦250.34 respectively. The married people had more expenditure than the singled, widowed or divorced. This was reflected in their mean food, non-food and bread expenditure as ₦3039.67, ₦832.57, ₦287.34 respectively. This means that married people made more expenses than others as is expected. The household heads made the highest expenditure than any other members of the households as their mean expenditure were ₦1908.41, ₦621.31 and ₦139.02 for food, non-food and bread respectively as compared to ₦1279.52, ₦396.14 and ₦94.82 respectively for the mean expenditure of spouses.

Table 2. Distribution of consumer's expenditure based on their socioeconomic characteristics

		Food Expenditure ₦		Non-food Expenditure ₦		Bread Expenditure ₦	
Socioeconomic Characteristics		Mean	Standard Deviation	Mean	Standard Deviation	Mean	Standard Deviation
Gender	Male	1711.78	1736.06	504.06	550.22	142.55	235.91
	Female	1602.28	1731.13	480.79	621.47	119.85	147.75
Age Category	21-30	2649.23	3020.07	714.55	981.93	250.34	362.94
	31-40	1481.27	1225.12	442.67	478.84	118.69	131.36
	41-50	1150.38	727.99	378.51	292.25	71.86	57.09
	51-60	1328.39	745.63	387.46	305.50	88.61	71.71
	61 and above	1951.67	1359.02	664.12	492.50	131.72	114.98
Marital Status	Married	3039.67	3305.58	832.57	1131.34	287.34	393.35
	Single	1342.96	904.43	418.40	344.58	96.32	90.27
	Divorced	342.85	321.56	357.14	325.43	28.57	24.53
	Widowed	2161.91	938.17	477.14	118.43	167.62	148.26
Household Position	Child	1892.04	1954.45	541.67	567.19	157.81	253.84
	Spouse	1279.52	837.55	396.14	313.88	94.82	91.46
	Head	1908.41	2688.58	621.31	1180.36	139.02	160.84
	Relative	1104.98	1241.28	545.49	438.64	263.72	342.65
HhSize Category	Minimum-2	818.43	533.18	302.65	223.59	37.82	26.11
	3-4	842.99	431.69	251.09	180.27	73.84	60.45
	5-6	1149.71	535.76	352.45	225.41	86.69	74.56
	7-8	1945.16	1052.59	579.17	446.94	129.46	107.89
	9-max	5137.5	3575.14	1458.93	1306.68	485.54	467.29
Educational Level	Primary	1200.24	825.71	475.52	258.94	120.19	104.54
	Secondary	1670.24	1423.59	507.32	526.87	129.96	200.51
	Tertiary	1590.42	1859.95	464.55	649.47	125.02	183.70
	No Formal Education	3273.81	3616.38	780.65	574.28	290.03	394.89
Religion	Christian	1695.79	1837.96	485.17	611.92	129.89	174.42
	Islam	1563.64	1496.38	477.93	462.53	142.71	262.79
	Tradition	1796.98	1645.62	711.73	814.15	112.42	86.75
	Eckankar	983.33	225.46	302.22	175.16	36.67	3.33

4.2 Factors That Influences Consumers' Willingness to Pay for Bread With an Inclusion of Cassava Flour

Table 3 shows the direct effect of the explanatory variables on consumers' willingness to pay for cassava flour inclusion in bread. The rho which is the correlation coefficient is 0.947. It shows a strong relationship between the first response and the second response. The probability Chi-square shows that the equation is significant at 1percent.

The result reveals that six out of the eleven variables are statistically significant in response 1 and these are the bid or premium price, age, sex, marital status, head position and bread share. However, in the second response, only four variables are significant and they are bid or premium price, sex, awareness and head position. This result is in agreement with the opinion Adepoju and Oyewole (2013) that factors such as bid or premium price, age, sex, marital status, head position and bread share affect consumers' willingness to pay for cassava flour inclusion in bread significantly. The findings reveal that premium price (bid), marital status and household income exerted significant negative influence on households' willingness to pay for bread with cassava flour inclusion respectively. However, the coefficient of age is positive which is against studies reviewed; (Yusuf et al., 2007). It is significant at 10percent in the first bid, implying that age has a positive impact on consumers willingness to pay for cassava flour inclusion in bread at 10percent inclusion level in the first bid and encourages consumers to make better decisions as regards to willingness to pay for cassava flour in bread. This means that as the age of consumers increases, the more willing they are to pay for the first bid of cassava bread at 10percent cassava flour inclusion level.

Feminine gender is positive and significant at five percent level. This implies that females are more willing to pay for 10 percent cassava flour inclusion than males.

Awareness of cassava bread is positive and significant at 10percent in the second bid. This implies that people who were aware of cassava bread are more willing to pay for 10percent cassava flour inclusion than those who are not.

Respondents who are married have a negative relationship in the first bid only and it is significant at 10percent. This means that married respondents are not WTP for 10 percent cassava flour inclusion.

Being the head of the house is positive and significant at 10percent and 5percent in the first and second bid respectively. This implies that household heads are more willing to pay for 10 percent cassava flour inclusion than non household heads.

Table 3 also shows the direct effect of the explanatory variables on consumers' willingness to pay for 20percent cassava flour inclusion in bread. The rho which is the correlation coefficient is 0.907. It shows a strong relationship between the first response and the second response. The probability Chi-square shows that the equation is significant at 1percent and so the null hypothesis is rejected that the socio-economic variables would not influence consumers' WTP for 10 and 20 percent cassava inclusion.

The result reveals that six out of the eleven variables are statistically significant in response 1 and these are the bid, sex, household size, head position, dependency ratio and bread share. However, in the second response, only three variables are significant and they are the bid, sex, and head position.

Feminine gender is positive and significant at 1percent and five percent level of significance in the first and second bid respectively. This implies that females are also more willing to pay for 20 percent cassava flour inclusion as they are for 10percent cassava flour inclusion than males.

Awareness of cassava bread not significant at 20 percent unlike at 10 percent. This implies that people who were aware of cassava bread are not WTP for 20percent cassava flour inclusion unlike at 10 percent cassava flour inclusion. The married factor has a no significant relationship on WTP to for 20 percent cassava flour inclusion. Being the head of the house is positive and significant at 5percent and 10percent in the first and second bid respectively. This implies that household heads are more willing to pay for 20percent cassava flour inclusion than non household heads as well as for 10percent cassava flour inclusion.

Results also shows that higher household size significantly reduce willingness to pay for 20percent inclusion at 5percent level of significance.

Table 3. Seemingly unrelated bivariate probit model estimates of factors driving WTP for 10% and 20% inclusion of cassava flour in bread

	10%						**20%**					
	Response 1			**Response 2**			**Response 1**			**Response 2**		
Predictor	**Coefficient**	**Std. Err.**	**P>|z|**	**Coefficient**	**Std. Err.**	**P>|z|**	**Coef.**	**Std. Err.**	**P>|z|**	**Coef.**	**Std. Err.**	**P>|z|**
Bid (₦)	-0.031***	0.004	0.000	-0.039***	0.006	0.000	-0.064***	0.011	0.000	-0.0251***	0.0109	0.021
Age (years)	0.004*	0.007	0.581	0.002	0.008	0.817	0.008	0.008	0.340	0.0017	0.0086	0.844
Sex	0.771**	0.317	0.015	0.647**	0.330	0.050	0.971***	0.338	0.004	0.7658**	0.3458	0.027
Hh size	-0.087	0.068	0.202	-0.028	.071	0.690	-0.154**	0.073	0.036	-0.0911	0.0792	0.250
Years of education	-0.010	.022	0.661	-0.036	.023	0.113	-0.013	0.022	0.552	0.0185	0.0210	0.459
Monthly income	-3.38e-07	1.01e-06	0.738	-1.30e-06	1.13e-06	0.250	-1.47e-07	1.04e-06	0.887	-1.90e-07	1.16e-06	0.869
Awareness	0.069	0.163	0.067	0.293*	0.177	0.097	0.100	0.164	0.541	-0.0600	0.1838	0.744
Married	-0.478*	0.250	0.056	-0.289	0.256	0.259	-0.268	0.250	0.285	-0.1963	0.2685	0.465
Headposition	0.532*	0.322	0.099	0.706**	0.340	0.038	0.690**	0.340	0.042	0.5710*	0.3488	0.096
DepRatio	1.001	0.611	0.101	0.982	0.670	0.143	1.508**	0.634	0.017	1.1369	0.7036	0.106
Perception	-0.217	0.215	0.312	-0.194	0.229	0.399	-0.156	0.215	0.468	-0.1444	0.2388	0.545
Breadshare	-2.606*	1.553	0.093	-2.757	1.852	0.137	-3.359**	1.650	0.04	-1.4027	1.8236	0.442
Constant	5.594***	1.194	0.000	7.119***	1.401	0.000	10.333***	1.884	0.000	3.2671	2.1377	0.126
Rho	0.947	0.030					0.907	0.044				
Likelihood-ratio test of rho=0: $chi^2(1)$ = 118.557 Prob> chi^2 = 0.0000							Likelihood-ratio test of rho=0: $chi^2(1)$ = 85.314 Prob> chi^2 = 0.0000					
Number of obs = 300							Number of obs = 300					
Log likelihood = -254.52971							Log likelihood = -256.084					
Wald chi^2 (24) = 77.37							Wald chi^2(24) = 61.61					
Prob > chi^2 = 0.0000							Prob> chi^2 = 0.0000					

Note: ***=significant at 1 percent, **=significant at 5 percent, *=significant at 10 percent,

Bid 1=₦160, Bid 2= ₦180 (10% & 20%).

Table 4 shows the direct effect of the explanatory variables on consumers' willingness to pay for cassava flour inclusion in bread. The rho which is the correlation coefficient is 0.902. It shows a strong relationship between the first response and the second response. The probability Chi-square shows that the equation is significant at 1percent. The result reveals that five out of the eleven variables are statistically significant in response 1 and these are the bid, sex, married, head position and dependency ratio. However, in the second response, only three variables are significant and they are bid, sex and head position. Feminine gender is positive and significant at five percent level of significance in the first and second bids. This implies that females are also more willing to pay for 30 percent cassava flour inclusion than males as well as they are for 20 and 10 percent cassava flour inclusion.

Awareness of cassava bread not significant at 30 and 20 percent unlike at 10 percent cassava flour inclusion. This implies that people who were aware of cassava bread are not willing to pay for both 30 and 20percent cassava flour inclusion unlike at 10 percent cassava flour inclusion. The married factor has a negative relationship on willing to pay and it is significant at five percent level of significance in the first bid. It implies that married respondents are less willing to pay for 30 percent cassava flour inclusion in the first bid. Being the head of the house is positive and significant at 5percent in the first and second bids. This implies that household heads are more willing to pay for 30, 20 and 10percent cassava flour inclusion than non household heads.

Table 4. Seemingly unrelated bivariate probit model estimates of factors driving WTP for 30% & 40% Inclusion of cassava flour in bread

Prediction	**Coef.**	**Std.Err**	**P > |z|**	**Coef.**	**Std.Err**	**P > |z|**	**Coef.**	**Std.Err**	**P > |z|**	**Coef.**	**Std.Err**	**P > |z|**
	30%						**40%**					
	Response 1			**Response 2**			**Response 1**			**Response 2**		
Bid 1	-0.139***	0.016	0.000	-0.042***	0.010	0.000	-0.122***	0.015	0.000	-0.045***	-0.093	0.000
Age	0.012	0.009	0.158	0.009	0.009	0.294	0.0001	0.009	0.989	-0.009	0.0172	0.428
Sex	0.971**	0.376	0.010	1.066**	0.407	0.009	0.849**	0.350	0.015	0.937*	1.536	0.047
Hhs	-0.093	0.073	0.205	-0.059	0.079	0.456	-0.120	0.076	0.118	-0.075	0.0302	0.421
years of edu	-.006	0.025	0.822	0.014	0.027	0.605	0.000	0.025	0.993	0.040	0.048	0.258
monthlyinc~e	-1.86e-07	1.21e-06	0.877	-6.05e-07	1.27e-06	0.635	-7.09e-07	1.16e-06	0.541	7.62e-07	1.56e-06	0.559
awarenes_c~d	0.260	0.185	0.158	0.075	0.201	0.709	0.264	0.182	0.146	0.380	0.621	0.114
Married	-0.721**	0.288	0.012	-0.202	0.294	0.491	-0.640**	0.278	0.021	0.109	-0.094	0.768
headposit	0.838**	0.384	0.029	1.028**	0.415	0.013	0.656*	0.360	0.068	0.745	1.361	0.123
depRatio	1.299*	0.719	0.071	1.251	0.785	0.111	1.533**	0.713	0.032	1.370	2.931	0.138
perception	-0.175	0.242	0.471	0.039	0.264	0.883	-0.281	0.243	0.248	-0.552*	0.195	0.079
breadshare	-2.566	1.755	0.144	-0.692	1.970	0.725	-3.499	0.114	0.258	-3.035	0.116	0.296
Constant	2.931***	2.590	0.000	4.212**	1.904	0.027	17.834***	2.393	0.000	5.456**	22.523	0.010
Rho	0.9021	0.0704					0.892	0.096				

30%	40%
Likelihood-ratio test of rho=0: $chi^2(1)$ = 50.7848 Prob> chi^2 = 0.0000	Likelihood-ratio test of rho=0: chi2(1) = 35.915 Prob> chi^2 = 0.000
Number of obs = 300	Number of obs = 300
Log likelihood = -201.90137	Log likelihood = -186.174
Wald $chi^2(24)$ = 99.83	Wald $chi^2(24)$ = 91.99
Prob > chi^2 = 0.0000	Prob> chi^2 = 0.0000

Note: ***=significant at 1 percent, **=significant at 5 percent, *=significant at 10 percent,

Bid 1=₦150, Bid 2= ₦160 (30%)

Bid 1=₦140, Bid 2= ₦150 (40%).

Table 4 also shows the direct effect of the explanatory variables on consumers' willingness to pay for 40% cassava flour inclusion in bread. The rho which is the correlation coefficient is 0.892. It shows a strong relationship between the first response and the second response. It also shows that as the level of cassava flour is increasing in bread, the rho value is reducing but it is still within the range of 0.7 to 1.0 which shows a strong relationship. The probability Chi-square shows that the equation is significant at 1percent. The result reveals that five out of the eleven variables are statistically significant in response 1 and these are bid, age, sex, married, dependency ratio and head position. However, in the second response, only three variables are significant and they are bid, sex and perception.

Feminine gender is positive and significant at five percent level of significance in the first bid and at 10 percent level of significance in the second bids. This implies that females are also more willing to pay for 40 percent cassava flour inclusion than males as well as they are for 30, 20 and 10 percent cassava flour inclusion.

Awareness of cassava bread not significant at 40, 30 and 20 percent unlike at 10percent cassava flour inclusion. This implies that people who were aware of cassava bread are not willing to pay for 40, 30 and 20percent cassava flour inclusion unlike at 10 percent cassava flour inclusion.

The married factor has a negative relationship on willingness to pay and it is significant at five percent level of significance in the first bid. It implies that married respondents are less WTP for 40 percent cassava flour inclusion in the first bid as well as for 30 and 20percent cassava flour inclusion. However, at 10percent cassava flour inclusion, married people are more willing to pay for cassava flour inclusion than the unmarried.

Being the head of the house is positive and significant at 10percent in the first bid. This implies that household heads are more WTP for 40percent cassava flour inclusion as well as 30, 20 and 10percent cassava flour inclusion than non household heads.

Dependants also had a positive relationship at 5percent level of significance at 40percent level of cassava flour inclusion. This shows that dependants also have a positive impact on willingness to pay for cassava bread at 40percent cassava flour inclusions.

Respondent's perception is negative and significant as 10percent level of significance. This shows that their perceptions have a negative impact on willingness to pay for cassava bread at 40percent cassava flour inclusions.

The most significant model is the 10percent cassava flour inclusion. This is because it has the highest number of significant variables (ten) followed by 20percent cassava flour inclusion with nine significant variables and 30 and 40 percent cassava flour inclusion with eight significant variables each.

4.3 Estimating Consumers' Mean Willingness to Pay for Cassava Flour Inclusion in Bread

In estimating the consumers' mean willingness to pay, the bivariate probit model was used. The tables below reveal the consumers' willingness to pay and the mean willingness to pay for cassava-wheat bread in the study area. From literatures reviewed, estimates of implicit prices are made on 'ceteris paribus' basis, that is, they are estimates of the Willingness to Pay (WTP) of respondents and mean WTP for an increase in the attribute of concern, given that everything is held constant i.e. all other factors are constant (Bennett & Blamey, 2001).

According to Table 5, the average initial values of willingness-to-pay at 10percent cassava flour inclusion is ₦172.70 with the average highest value of willingness-to-pay being ₦177.61 while the average lowest value of willingness-to-pay is ₦166.37. Also, the average subsequent value (at second bid) of willingness-to-pay at 10percent cassava flour inclusion is ₦180.20, with the average highest value of willingness-to-pay being ₦185.34 and the average lowest value of willingness-to-pay being ₦171.78.

Table 5. Krinsky and Robb (95 %) confidence interval for WTP measures (Nb of reps: 2000 and Equation)

	10% inclusion					20% inclusion				
MEASURE Resp 1	WTP	LB	UB	ASL*	CI/MEAN	WTP	LB	UB	ASL*	CI/MEAN
MEAN/MEDIAN	172.70	166.37	177.61	0.0000	0.07	165.00	161.40	167.50	0.0000	0.04
MEASURE Resp 2	WTP	LB	UB	ASL*	CI/MEAN	WTP	LB	UB	ASL*	CI/MEAN
MEAN/MEDIAN	180.20	171.78	185.34	0.0000	0.08	150.41	130.17	170.65	0.0185	1.03

*: Achieved Significance Level for testing H0: WTP<=0 vs. H1: WTP>0

LB: Lower bound; UB: Upper bound.

Also, according to Table 5, the average initial values of willingness-to-pay at 20percent cassava flour inclusion is ₦165.00 with the average highest value of willingness-to-pay being ₦167.50 while the average lowest value of willingness-to-pay is ₦161.40. Also, the average subsequent value (at second bid) of willingness-to-pay at 20percent cassava flour inclusion is ₦150.41, respectively with the average highest value of willingness-to-pay being ₦170.65 respectively and the average lowest value of willingness-to-pay being ₦130.17.

Similarly, according to Table 6 the average initial values of willingness-to-pay at 30percent cassava flour inclusion is ₦154.07 with the average highest value of willingness-to-pay being ₦155.39 while the average lowest value of willingness-to-pay is ₦152.54. Also, the average subsequent value (at second bid) of willingness-to-pay at 30percent cassava flour inclusion is ₦143.35 respectively with the average highest value of willingness-to-pay being ₦153.05 and the average lowest value of willingness-to-pay being ₦117.90.

Table 6. Krinsky and Robb (95 %) Confidence Interval for WTP measures (Nb of reps: 2000 and Equation)

	30% inclusion					30% inclusion				
MEASURE Resp 1	WTP	LB	UB	ASL*	CI/MEAN	WTP	LB	UB	ASL*	CI/MEAN
MEAN/MEDIAN	154.07	152.54	155.39	0.0000	0.02	142.78	140.74	144.31	0.0000	0.03
MEASURE Resp 2	WTP	LB	UB	ASL*	CI/MEAN	WTP	LB	UB	ASL*	CI/MEAN
MEAN/MEDIAN	143.35	117.90	153.05	0.0005	0.25	127.36	88.38	139.40	0.0025	0.40

*: Achieved Significance Level for testing H0: WTP<=0 vs. H1: WTP>0.

LB: Lower bound; UB: Upper bound.

Finally, in Table 6, the average initial values of willingness-to-pay at 40percent cassava flour inclusion is ₦142.78 respectively with the average highest value of willingness-to-pay being ₦144.31 while the average lowest value of willingness-to-pay is ₦140.74. Also, the average subsequent value (at second bid) of willingness-to-pay at 40percent cassava flour inclusion is ₦127.36 respectively with the average highest value of willingness-to-pay being ₦139.40 and the average lowest value of willingness-to-pay being ₦88.38.

The respondents' average net monthly income is ₦79,086.17 and average household size is 5 members. A significant reduction in the average maximum willingness-to-pay value is noticed as the level of cassava flour inclusion increases from 10percent through 40percent which reduces by 18.7percent in the first bid i.e. from ₦177.61 to ₦144.31, and reduces by 24.8percent in the second bid i.e. from ₦185.34 to ₦139.40.

Therefore, the consumers are willing to pay more for a lower level of cassava flour inclusion in bread. i.e. ₦172.70 for 10% level of cassava flour inclusion, ₦165.00 for 20% level of cassava flour inclusion , ₦154.07 for 30% level of cassava flour inclusion and ₦142.78 for 40% level of cassava flour inclusion for the size of conventional bread whose mean price is ₦200.

5. Conclusion and Recommendation

According to the results of the survey, respondents' decisions on willingness to pay for cassava flour inclusion in bread were significantly affected by sex, marital status, household size, head position, dependency ratio and bread share, depending on the level of cassava flour inclusion in the bread. Finally, the adjusted, average individual value of willingness-to-pay was used to calculate the aggregate willingness-to-pay. A significant reduction in the average maximum willingness-to-pay value is noticed as the level of cassava flour inclusion increases. Therefore, based on the findings of this survey, it is recommended that family planning policy should be promoted among households in the study area since findings shows that higher family size reduces willingness to pay for cassava inclusion in bread.

References

Adepoju, A. O., &Oyewole, O. O. (2013). Households' Perception and Willingness to Pay For Bread With Cassava Flour Inclusion in Osogbo Metropolis, Osun State, Nigeria. *African Association of Agricultural Economists (AAAE) Fourth International Conference,* Hammamet, Tunisia.

Bennett, J., & Blamey, R. (2001). *The Choice Modeling Approach to Environmental Valuation* (pp. 324-351), UK: Edward Elgar publishing, Inc.

Bonti-Ankomah, S., & Yiridoe, E. K. (2006). Organic and conventional food: a literature review of the economics of consumer perceptions and preferences. Final Report. Organic Agriculture Centre of Canada. Nova Scotia Agricultural College, Truro, Nova Scotia, Canada.

Freeman, A. M. III. (2003). Economic valuation: What and Why. In P. A. Champ, K. J. Boyle & T. C. Brown (Eds.), *A primer on nonmarket valuation* (pp. 1-25). London: Kluwer Academic Publishers. http://dx.doi.org/10.1007/978-94-007-0826-6_1

Grassi, S. (2010). Public and Private Provision under Assymetric Information: Ability to Pay and Willingness to Pay. *Econometrica, 52*(2), 456-463.

Greene, W. H., Hensher, D. A., &Rose, J. M. (2005). Using Classical Simulation - Based Estimators to Estimate Individual WTP Values. A Mixed Logit Case Study of Commuters. *Applications of Simulation Methods in Environmental and Resource Economics, 32*(2), 17-33. http://dx.doi.org/10.1007/1-4020-3684-1_2

Haab, T. C., & McConnel, K. E. (2002). *Valuing Environmental and Natural Resources* (pp. 258-306). Business

& Economics. Edward Elgar Publishing.

Hanley, N., Ryan, M., & Wright, R. (2003). Estimating the monetary value of health care: Lessons from environmental economics. *Health Economics, 12*(1), 3-16. http://dx.doi.org/10.1002/hec.763

IITA. (2010). Annual Report of the International Institute of Tropical Agriculture .

Krinsky, I., & Robb, A. L. (1986). On Approximating the Statistical Properties of Elasticities. *Review of Economic and Statistics, 68*(1), 715-719. http://dx.doi.org/10.2307/1924536

Loomes, G. (2001). The use of cost-effectiveness thresholds outside the health sector. In *Cost effectiveness Thresholds: Economic and Ethical Issues, 16*(1), 37-43.

Meteorological Organization. (2012). Climatological Information for Lagos, Nigeria.

Negrín, M. A., Pinilla, J., & León, C. J. (2008). Willingness to pay for alternative policies for patients with Alzheimer's Disease. *Health Economics, Policy and Law, 3*(1), 257-275. http://dx.doi.org/10.1017/S1744133108004489

Oliver, A., Healey, A., & Donaldson, C. (2002). Choosing the method to match the perspective: Economic assessment and its implications for health-services efficiency. *Lancet, 359*(3), 1771-1774. http://dx.doi.org/10.1016/S0140-6736(02)08664-6

Rasaki, R. (2012). Managing Metropolitan Lagos. *New Scientist, 41*(1), 67-71.

Train, K. E. (2003). *Discrete Choice Methods with Simulation* (pp. 67-81). Cambridge University Press. http://dx.doi.org/10.1017/CBO9780511753930

Watson, V., & Ryan, M. (2006). Exploring preference anomalies in double bounded contingent valuation. *Health Economics, 67*(1), 13-18.

Yusuf, S. A., Salimonu, K. K., & Ojo, O. T. (2007). Determinants of Willingness to Pay for Improved House Solid Waste Management in Oyo State, Nigeria. *Research Journal of Applied Sciences, 2*(1), 233-239.

4

Effect of Some Preceded Peanut Cultivars on Wheat Yield and Agro – Economic Feasibility Under Two Cropping Systems in Sandy Soil

Abdel-Galil Mohamed Abdel-Galil[1], Sherif Ibrahim Abdel-Wahab[1] & Tamer Ibrahim Abdel-Wahab[1]

[1] Crop Intensification Research Department, Field Crops Research Institute, Agricultural Research Center, Giza, Egypt

Correspondence: Abdel-Galil Mohamed Abdel-Galil, Crop Intensification Research Department, Field Crops Research Institute, Agricultural Research Center, Giza, Egypt. E-mail: abdelgalil_abdelgalil@yahoo.com

Abstract

The cropping system is an approach to help farmer in decision making to remain sustainable in an ever-changing agricultural environment. A two-year study was carried out at Ismailia Agricultural Experiments and Research Station, ARC, Ismailia governorate, Egypt during 2011/2012 and 2012/2013 seasons to study the effect of preceded peanut cultivars on yield and profitability of wheat under conventional and inetsnive cropping systems in sandy soil. This experiment included six treatments which were the combinations of three peanut cultivars (Giza 4, Giza 5 and Ismailia 1) as preceding cultivars in the summer season and two cropping systems (conventional and intensive). A split plot design with three replications was used. The results indicated that peanut cv. Ismailia 1 increased available soil nitrogen (N) content which affected positively number of grains per spike and grain weight per spike and finally the economic yield. Growing fahl berseem in transition period between peanut and wheat enhanced available soil N content that increased grain yield per ha by about 1.00 percent as compared with those of conventional cropping system. Accordingly, intensive cropping system increased wheat grain yield and its attributes as compared with conventional cropping system. Peanut cultivars × cropping systems interaction had a significant effect on available soil N content and all the studied wheat traits except number of spikes/m^2. Intensive cropping system increased total and net returns as compared with conventional cropping system. The Egyptian farmers could achieve an increase in their income by US$ 755.9 per ha when using intensive cropping system which included peanut cv. Ismailia 1

Keywords: peanut cultivars, fahl berseem, wheat, cropping systems, economic return

1. Introduction

For the last several years in Egypt, special attention has been directed towards reclaiming desert soils, sandy or calcareous in nature. If Egyptian agricultural production must be intensified, ways of making maximum use of sandy soils must be pursued without soil degradation and pollution, which are often the result of intensive mineral N fertilizer application. The environmental problems associated with excessive use of N fertilizer is now a contentious issue and must be increasing number of crops per year outside the Nile Valley and Delta without soil degradation and pollution by using biological nitrogen fixation (BNF). However, BNF in legume nodules is severely affected by different environmental constraints under these soils conditions. Legumes contribute to a diversification of cropping systems and as N_2-fixing plant in sandy soil; it can improve the exchanges of nutrients and water in a sandy soil where plant roots thrive. N is the most important nutrient element required for crop production especially for cereals, which have been reported to dominate cultivated land in the world (Myers, 1988). Accordingly, BNF is a convenient source of N for sequential crop planting in Egyptian agriculture; its use is ultimately governed and regulated by economic and environmental considerations under sandy soils conditions.

Cropping index refers to the times of sequential crop planting in the same arable land in one year, usually defined as the ratio of the total seeding area to the arable land area (Liu, 2013). Wheat (*Triticum aestivum* L.) is the most important cereal crop because it is the staple food of the Egyptian people. Increasing wheat productivity in sandy soils is one of the main targets for crop intensification scientists where grain protein concentration and composition are major determinants of grain nutritional value (Feil, 1997) but N requires careful management when used for crop production on sandy soils because of the high potential for leaching losses of nitrate through these soils. For that should increase the number of crops per year with decreasing mineral N fertilizer use by BNF.

Several of the nutrients that are essential for growth of plants or bacteria play specific roles in symbiotic N_2-fixation through their effects on nodulation and the N_2- fixation process (O'Hara, 2001). It has been reported that the net benefits of legumes are often equivalent to the addition of 50-100 kg N per ha as fertilizer (Phoomthaisong et al., 2003), especially N deficiency is one of the major yield limiting factors for cereals (Shah et al., 2003). BNF occurs mainly through symbiotic association of legumes and some woody species with certain N_2-fixing microorganisms that convert elemental N into ammonia (Shiferaw et al., 2004).

On the other hand, peanut (*Arachis hypogaea* L.) is an important oil and protein crop; it contains about 40-50% oil, 25 - 30% protein, 20% carbohydrates and 5% fiber and ash, and makes a substantial contribution to human nutrition (Fageria et al., 1997). Peanut genotypes were differed in pod number per plant because it had good N fixing ability, and thus fixed N could contribute to agronomic performance in larger extent (Pimratch et al., 2004). The two local cultivars of peanut (Giza 4 and Giza 5) showed significant variation in most of the studied characters under sandy soil conditions (Abd El-Maksoud, 2008). Also, El-Saady et al. (2014) concluded that the differential benefits of peanut cultivars from available nutrients in soil, might be attributed to the variation in their root system volumes, as well as, their root exudations.

Accordingly, the cropping system adopted by the farmer in these soils must be physically viable, sustainable, less exhaustive acceptable to farming community and most important thing is that it should be economical. Peanut – wheat crop sequence is a conventional crop system in Ismailia governorate. Sandy soils requires proper management to offer optimum productivity of any cultivated crop, especially for the studied cropping systems here, which has been peanut followed by wheat where peanut and wheat cultivated areas reached about 44,653 and 251,759 thousands ha in new lands in 2012 with an average yields of 3.22 and 5.79 tons per ha, respectively, (Bulletin of Statistical Cost Production and Net Return, 2013 and 2014).

Although legume-cereal sequence increased biomass and grain yield of subsequent wheat (Hayat et al., 2008), however, this necessitates the inclusion of leguminous fodder crops as berseem (*Trifolium alexandrinum* L.), which capable of supplying green fodder after 70 days from sowing by one cut "fahl", after peanut harvest and before wheat sowing in our cropping system (peanut – fahl berseem – wheat) as these have the ability of enriching the N content of these soils by fixing N from the air, in addition to improving the productivity of soil. Some legumes are better at fixing N than others (Lindemann & Glover, 2013). So, the present study was, therefore, conducted to study the effect of preceded peanut cultivars on yield and profitability of wheat under conventional and intensive cropping systems in sandy soil.

2. Method

A two-year study was carried out at Ismailia Agricultural Experiments and Research Station, A.R.C., Ismailia governorate (Lat. 30° 35' 30" N, Long. 32° 14' 50" E, 10 m a.s.l.), Egypt during 2011/2012 and 2012/2013 seasons to study the effect of preceded peanut cultivars on yield and profitability of wheat under conventional and intensive cropping systems in sandy soil. Mechanical and chemical analyses of the soil (0-20 cm) were done by General Organization for Agricultural Equalization Fund, Agricultural Research Center, Giza, Egypt. The experimental soil had 12.16 percent clay, 2.20 percent silt and 85.64 percent sand, and loamy sand texture. Available soil N content was 5.0 percent in the first season and 10.0 percent in the second season before sowing peanut seeds (0-time). This experiment included six treatments which were the combinations of three peanut cultivars (Giza 4, Giza 5 and Ismailia 1) as preceding crops in the summer season and two cropping systems (conventional and intensive). The cropping systems were illustrated in Figure (1) as follows:

- Conventional cropping system (peanut / wheat crop sequence): planting peanut in the summer season then planting wheat in the winter season (pattern I).
- Intensive cropping system (peanut / fahl berseem / wheat crop sequence): planting peanut in the summer season then planting fahl berseem in the autumn season then planting wheat in the winter season (pattern II).

Wheat cultivar Sakha 94 was used. Peanut cultivars were sown on May 2nd and 8th at 2011 and 2012 summer seasons, respectively, while, fahl berseem cv. Gemmiza 1 was sown on September 11th and 13th at 2011 and 2012 seasons, while, wheat was sown on November 30th and 28th at 2011/2012 and 2012/2013 winter seasons, respectively. Peanut cultivars were uprooted on September 7th and 9th at 2011 and 2012 summer seasons, while, fahl berseem was cut on November 23rd and 21st at 2011 and 2012 winter season and wheat was harvested on May 6th and 9th at 2012 and 2013 summer season, respectively.

In the two seasons, peanut and berseem seeds were inoculated by *Bradyrhizobium* and *Rhizobium trifolii*, respectively, before seeding it. Peanut was seeded at 10 cm apart between hills with one plant per hill after thinning. In the two winter seasons, wheat grains and berseem seeds were drilled at the rate of 166.6 and 59.5 kg per ha, respectively. Normal practices for growing all crops were used as recommended in the area. Sprinkler irrigation was the irrigation system in the area. Calcium super phosphate (15.5% P_2O_5) at rate of 476 kg per ha and potassium sulfate (48.0 % K_2O) at rate of 238 kg per ha were applied during soil preparation for sowing peanut in the summer season. The previous rates of calcium super phosphate and potassium sulfate were also applied during soil preparation for sowing wheat in the winter season. Calcium sulfate at rate of 1190 kg per ha was applied for peanut after 35-40 days from peanut sowing. Mineral N fertilizer at rates of 107.1 and 238.0 kg per ha were applied during different growth stages of peanut and wheat, respectively. A split plot design replicated thrice was used. Peanut cultivars were randomly assigned to the main plots, while the cropping systems were allotted in subplots. Each plot contained five ridges, each ridge was 3.0 m in length, 0.6 m in width and the plot area was 9 m^2.

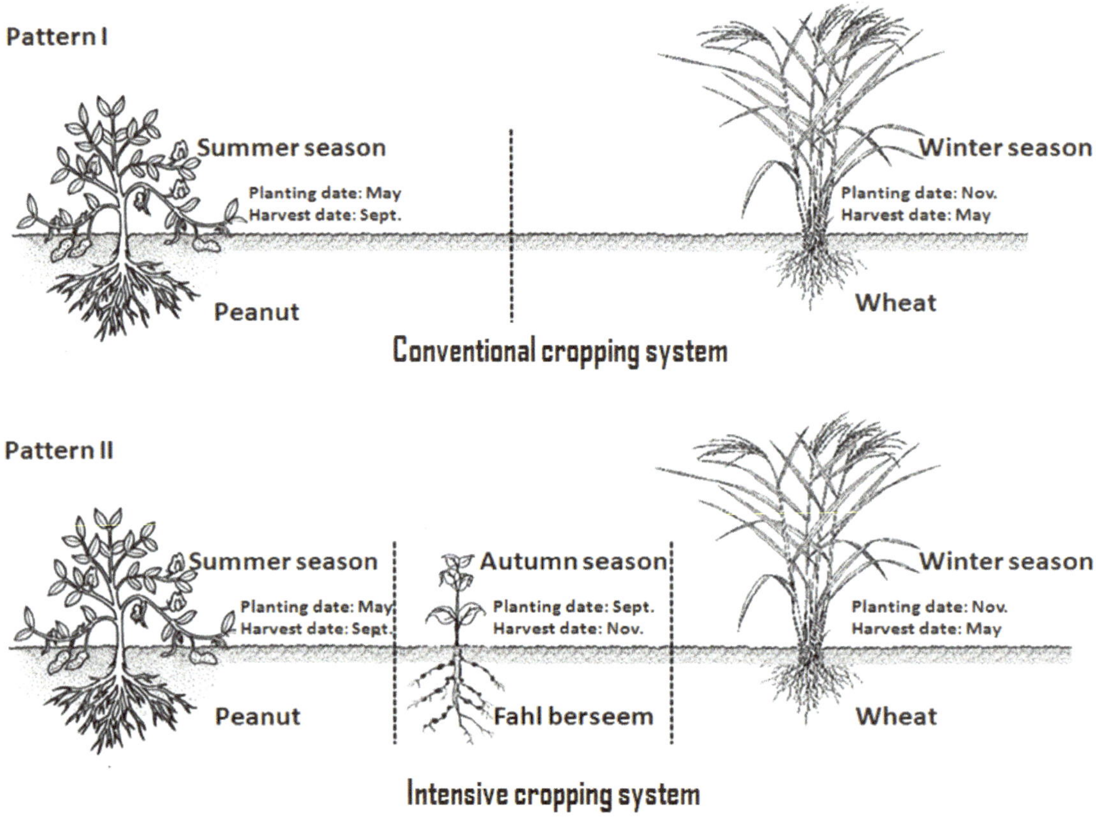

Figure 1. Conventional and intensive cropping systems under sandy soil conditions

2.1 The Studied Traits

2.1.1 Available Soil N Content

Soil samples were collected after harvest of peanut cultivars and after cutting fahlberseem by auger at the depth of 0-20 cm in the two growing seasons. These samples were air-dried, crushed, and sieved by 2-mm sieve and preserved for the chemical analysis. These analyses were performed in General Organization for Agricultural Equalization Fund, Agricultural Research Center, Giza, Egypt.

2.1.2 Wheat Yield and Its Attributes

These traits were measured at harvest on ten guarded plants of wheat from each plot: number of spikes per m^2, spike weight (g), number of grains per spike, 1000 – grain weight (g) and grain protein content (%). Protein percentage in wheat grains was determined by the improved kjeldahl method of (A.O.A.C., 2000) modified by distilling the ammonia into saturated boric solution and titration was carried out by using standard acid. Grain protein content was calculated by multiplying the total nitrogen by 5.7. These analyses were performed in General Organization for Agricultural Equalization Fund, Agricultural Research Center, Giza, Egypt. Grain yield of wheat per ha (ton) was recorded on the basis of experimental plot area by harvesting of all plants of each plot. In addition to estimate pod yield of peanut per ha (ton) and forage yield of berseem per ha (ton) on the basis of experimental plot area.

2.1.3 Economic Return

Peanut pods and wheat grains prices presented by (Bulletin of Statistical Cost Production and Net Return, 2013 and 2014) were used, while forage yield of berseem price presented by market price (2012) was used. Net return was calculated as follows: Net return = total costs – total return according to cropping system.

2.2 Statistical Manipulation

Analysis of variance of the obtained results of each season was performed. The measured variables were analyzed by ANOVA using MSTATC statistical package (Freed, 1991). Mean comparisons were done using least significant differences (L.S.D) method at 5% levels of probability to compare differences between the means (Gomez & Gomez, 1984).

3. Results

3.1 Available Soil N Content

3.1.1 Peanut Cultivars

Available soil N content was differed significantly after uprooting of peanut cultivars in combined data across 2011/2012 and 2012/2013 seasons (Table 1), where peanut cv. Ismailia 1 increased ($P \leq 0.05$) available soil N content by 60.37 and 46.77 percent as compared with available soil N content after uprooting of peanut cv. Giza 4 in the first and second season, respectively, while this increase was 16.43 and 13.75 percent as compared with available soil N content after uprooting of peanut cv. Giza 5 in the first and second seasons, respectively.

Table 1. Effect of peanut cultivars, cropping systems and their interaction on available soil N content in 2011/2012 and 2012/2013 seasons

Peanut cultivars	Growing season					
	2011/2012			2012/2013		
	Available soil N content (ppm)			Available soil N content (ppm)		
	C	I	Mean	C	I	Mean
Giza 4	18.0	35.0	26.5	23.0	39.0	31.0
Giza 5	28.0	45.0	36.5	32.0	48.0	40.0
Ismailia 1	35.0	50.0	42.5	41.0	50.0	45.5
Mean	27.0	43.3	35.1	32.0	45.6	38.8
L.S.D. at 0.05 Peanut cultivars			8.13			4.36
L.S.D. at 0.05 Cropping systems			5.78			2.91
L.S.D. at 0.05 Interaction			11.24			7.27

Note. C = Conventional cropping system, I = Intensive cropping system.

3.1.2 Cropping Systems

Available soil N content was affected significantly by the cropping systems in the two growing seasons (Table 1). Growing fahl berseem during the transition period between peanut and wheat increased ($P \leq 0.05$) available soil N content by 60.37 percent in the first season and 42.5 percent in the second season.

3.1.3 The Interaction Between Peanut Cultivars and Cropping Systems

Available soil N content was affected significantly by the interaction between peanut cultivars and the cropping systems in the two growing seasons (Table 1). The highest available soil N content was obtained by growing fahl berseem during the transition period between peanut cv. Ismailia 1 and wheat; meanwhile, the lower available soil N content was obtained by uprooting of peanut cv. Giza 4 as compared with the other treatments.

3.2 Wheat Yield and Its Attributes

3.2.1 Peanut Cultivars

Number of spikes per m^2, spike weight, number of grains per spike, 1000 – grain weight, grain yield per ha and grain protein content were affected significantly by peanut cultivars in the combined data across 2011/2012 and 2012/2013 seasons (Table 2). Growing wheat after peanut cv. Ismailia 1 caused significant increase in number of spikes per m^2, spike weight, number of grains per spike, 1000 – grain weight, grain yield per ha and grain protein content in comparison with growing wheat after the other peanut cultivars. Growing wheat after peanut cv. Ismailia 1 increased ($P \leq 0.05$) grain yield per ha by 1.16 and 5.67 percent as compared with those grown after peanut cv. Giza 5 and Giza 4, respectively.

Table 2. Effect of peanut cultivars, cropping systems and their interaction on yield and its attributes of wheat, combined data across 2011/2012 and 2012/2013 seasons

Peanut cultivars	Spikes/m^2 (no.)			Spike weight (g)			Grains/spike (no.)		
	C	I	Mean	C	I	Mean	C	I	Mean
Giza 4	367.66	375.66	371.66	1.86	1.89	1.87	37.51	37.84	37.68
Giza 5	384.00	396.66	390.33	1.98	2.06	2.02	39.94	40.16	40.05
Ismailia 1	387.33	398.00	392.66	2.06	2.09	2.07	40.23	40.52	40.37
Mean	379.66	390.11	384.88	1.96	2.01	1.98	39.23	39.51	39.37
L.S.D. at 0.05 Peanut cultivars			13.47			0.07			0.36
L.S.D. at 0.05 Cropping systems			6.02			0.04			0.19
L.S.D. at 0.05 Interaction			N.S.			0.11			0.52
Peanut cultivars	**1000 – grain weight (g)**			**Grain yield (ton/ha)**			**Grain protein content (%)**		
	C	I	Mean	C	I	Mean	C	I	Mean
Giza 4	38.92	39.08	39.00	4.90	4.95	4.93	8.26	8.34	8.30
Giza 5	40.24	40.50	40.37	5.13	5.17	5.15	8.60	8.69	8.64
Ismailia 1	40.46	40.62	40.54	5.18	5.24	5.21	8.70	8.76	8.73
Mean	39.87	40.07	39.97	5.07	5.12	5.09	8.52	8.59	8.55
L.S.D. at 0.05 Peanut cultivars			0.17			0.07			0.10
L.S.D. at 0.05 Cropping systems			0.19			0.03			0.05
L.S.D. at 0.05 Interaction			0.22			0.09			0.12

*Note.*C = Conventional cropping system, I = Intensive cropping system.

3.2.2 Cropping Systems

Number of spikes per m^2, spike weight, number of grains per spike, 1000 – grain weight, grain yield per ha and grain protein content were affected significantly by the cropping systems in combined data across 2011/2012 and 2012/2013 seasons (Table 2). Intensive cropping system caused significant increase ($P \leq 0.05$) in number of spikes per m^2, number of grains per spike, spike weight, 1000 – grain weight, grain yield per ha and grain protein content in comparison with conventional cropping system. Intensive cropping system increased grain yield per ha by about 1.00 percent as compared with those of conventional cropping system.

3.2.3 The Interaction Between Peanut Cultivars and Cropping Systems

Number of grains per spike, spike weight, 1000 – grain weight, grain yield per ha and grain protein content were affected significantly by the interaction between peanut cultivars and cropping systems except number of spikes per m^2 in the combined data across 2011/2012 and 2012/2013 seasons (Table 2). The highest values of spike weight, number of grains per spike, 1000 – grain weight, grain yield per ha and grain protein content were obtained by growing fahl berseem during the transition period between peanut cv. Ismailia 1 and wheat; meanwhile, the lowest values of spike weight, number of grains per spike, 1000 – grain weight, grain yield per ha and grain protein content were obtained by uprooting of peanut cv. Giza 4 as compared with the other treatments.

3.3 Economic Return

Intensive cropping system increased total and net returns by 35.23 and 42.09 percent, respectively, as compared with the other one (Table 3). Net return was varied between treatments from US$ 1530.9 (conventional cropping system by using peanut cv. Giza 4) to US$ 2739.8 (intensive cropping system by using peanut cv. Ismailia 1) per ha. Intensive cropping system by using peanut cv. Ismailia 1 increased net return by 38.10 percent as compared with conventional cropping system by using peanut cv. Ismailia 1. It is noticed that growing leguminous crop after peanut cv. Ismailia 1 is more profitable for Egyptian farmers.

Table 3. Economic return as affected by peanut cultivars, cropping systems and their interaction, combined data across 2011/2012 and 2012/2013 seasons

Treatments	Yield per ha (ton)		
	Peanut	Fahl berseem	Wheat
Conventional cropping system			
Peanut cv. Giza 4	2.45	---	4.90
Peanut cv. Giza 5	2.59	---	5.13
Peanut cv. Ismailia 1	2.78	---	5.18
Mean	2.60	---	5.07
Intensive cropping system			
Peanut cv. Giza 4	2.45	22.8	4.95
Peanut cv. Giza 5	2.59	23.3	5.17
Peanut cv. Ismailia 1	2.78	23.3	5.24
Mean	2.60	23.1	5.12

Treatments	Economic return (US$) per ha				
	Total return	Costs			Net return
		Peanut	Fahl berseem	Wheat	
Conventional cropping system					
Peanut cv. Giza 4	4402.2	1366.8	---	1504.5	1530.9
Peanut cv. Giza 5	4634.8	1366.8	---	1504.5	1763.5
Peanut cv. Ismailia 1	4855.2	1366.8	---	1504.5	1983.9
Mean	4630.7	1366.8	---	1504.5	1759.4
Intensive cropping system					
Peanut cv. Giza 4	6010.2	1366.8	890.8	1504.5	2248.1
Peanut cv. Giza 5	6274.5	1366.8	890.8	1504.5	2512.4
Peanut cv. Ismailia 1	6501.9	1366.8	890.8	1504.5	2739.8
Mean	6262.2	1366.8	890.8	1504.5	2500.1

Note. Prices of main products are that of 2012: Peanut pods: US$ 1060.95 per ton, Berseem green forage: US$ 69.71 per ton, Wheat grains: US$ 368.0 per ton.

4. Discussion

It seems that peanut cultivars had different ability of N fixation process under sandy soil conditions that led to different available soil N content after uprooting of these cultivars. These results may be a result of a mix of living rhizobia and dead peanut cultivars near the sandy soil surface where the populations of rhizobia is considered as a biological pool affect soil N dynamic. These results are in accordance with those obtained by Toomsan et al. (1995) who reported that there was a net contribution of N from N_2-fixing varieties of groundnut in all cases ranged from 13-100 kg N ha^{-1} when legume stover was returned to the soil. Also, Baldwin (2006) reported that the populations of microorganisms that make up the biological pool are the driving forces in soil nutrient dynamics. Together they also play a key role in building a soil structure that both retains and freely exchanges nutrients and water—a soil where plant roots thrive.

Naturally, available soil N content that is absorbed by peanut would be exhausted without replenishment from other N sources such as atmospheric N or more stable soil N. Accordingly, these results may be attributed to these cultivars have different ability of N fixation under sandy soil conditions. The main genetic potential for improving the amount of BNF in these cultivars depends upon factors that promote high photosynthetic productivity and efficient N-use with effective rhizobial associations under sandy soil conditions. N fixing ability in peanut varies widely, depending on peanut genotypes (Nambiar et al., 1983). It has been reported that various cultivars of grain legumes show significant differences regarding their ability to support BNF (Hardarson, 1993). At the time of pod fill, nodules on annual legumes generally lose their ability to fix N, because the plant feeds the developing seed rather than the nodule (Lindemann & Glover, 2013).

So, it may be possible that root system of peanut cv. Ismailia 1 developed and absorbed more moisture and nutrients from sandy soil as elongated roots cover more surface area in the experimental soil and as such fixed substantial quantity of N through nodulation on roots in comparison with peanut cv. Giza 4 and Giza 5. Consequently, peanut cv. Ismailia 1 had the highest assimilate partitioning that reflected on pod yield per ha in comparison with the other two cultivars (Figure 2), especially peanut cv. Giza 5 was superior than peanut cv. Giza 4 in pod yield kg per unit area (Asfour et al., 2009). Such effect was expected because of soil N content among these cultivars were differed in the experimental soil (Table 1), where there was significant correlation between fixed N and biomass production (Dinh et al., 2013).

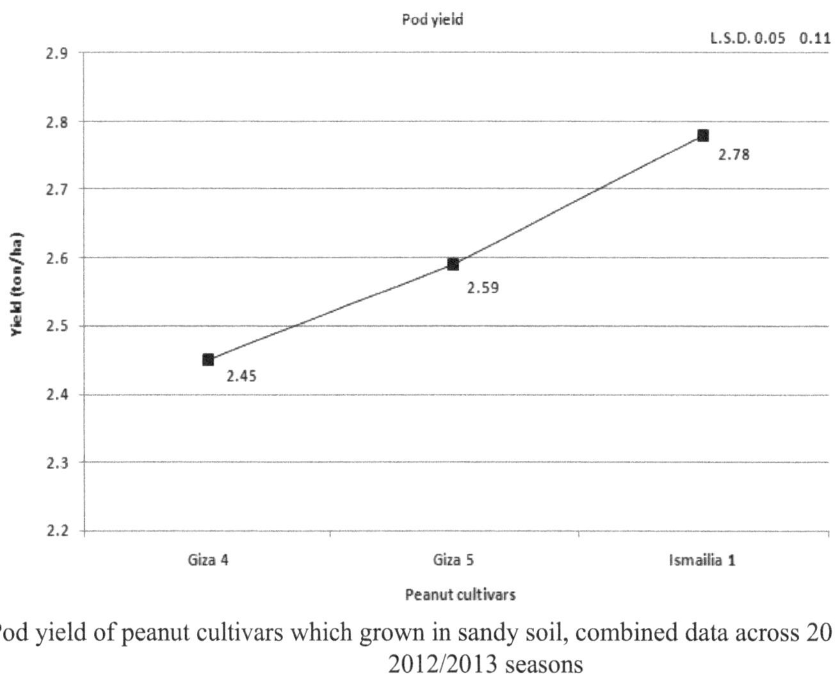

Figure 2. Pod yield of peanut cultivars which grown in sandy soil, combined data across 2011/2012 and 2012/2013 seasons

In view of the above, peanut cv. Ismailia 1 had the highest N fixing ability that led to positive residual effect on available soil N content and reflected on the yield attributes of wheat in comparison with the two peanut cultivars (Table 2), especially grains per spike had positive direct effect on grain yield (Khan et al., 2010). Also, grain weight per spike is very important yield components, which directly influence to yield and depends on grain number and grain chemical composition (Zecevic et al., 2010).

In regard to cropping systems, it is observed that fahl berseem was used as trap crop to capture left over N from the soil after peanut uprooting and improved total count of rhizobia that increased available soil N content and soil fertility. These results are in the same context with those by Baldwin (2006) who mentioned that from 40 to 75 percent of the total N contained in a legume cover crop is available in the soil for subsequent crops, depending on environmental conditions. He added that trap crops like small grains can be used to capture left over N from farm fields after a harvest of cash crops. Also, Pandey et al. (2008) concluded that massive loss of native soil N occurs in rice-wheat rotations when fields are left bare fallow during the dry-to-wet season transition period. Growing grain legumes and green manure crops during the transition season can immobilize soil N in the plant biomass in addition to adding N from the atmosphere. They added that reduced amounts of available soil N by growing crops during dry to wet season transition period result reduced N losses by both leaching and denitrification.

Significant increments in grain yield of wheat and its attributes under intensive cropping system can be attributed to the beneficial effect of nutrients released from the decomposition of the preceded legumes residues especially fahlberseem. Certainly, differences in genetic potential of these species played a major role in rooting pattern and N_2-fixing capacity that led to a variation in BNF under sandy soil conditions. Using leguminous crop as fahl berseem during transition period between peanut and wheat added more N to the sandy soil than conventional cropping system (Table 1). In this concern, Hayat et al. (2008) showed that the variation in grain yield production between two legumes can be attributed to differences in genetic potential.

This improvement had a great ecological importance. It is expected that if such improvement continues unabated year after year, the overall situation will improve considerably within a few years. The variations in biomass production among the legumes crops can be attributed to differences in genetic potential, plant architecture, rooting pattern and N_2-fixing capacity (Mandal et al., 1990). Exploitation of biological N is dependent on various factors like environment, nutritional status of soil and also the cropping pattern in use (Dakora & Keya, 1997). When the plant dies, the fixed N is released; making it available to other plants and this helps to fertilize the soil (Postgate, 1988). It is evident from the data that intensive cropping system could be played a major role in improvement of nutritional status of the experimental soil than the other one. Growing fahl berseem after peanut led to increase in soil N availability for wheat growth and development that affected positively the yield attributes of wheat (Table 2). These results are in harmony with those obtained by Banziger et al. (1994) who indicated that increasing N application induced increasing leaf area, tiller formation and leaf area index. This resulted in producing more dry matter and grain yield. Also, Nawab et al. (2006) showed that the increase in wheat yield after mung bean may be due to N fixation by the previous legume crop. Though groundnut is a legume and can fix N but the yields of wheat following groundnut is lower, which may be due to the greater quantity of Ca absorbed by groundnut or due to allelopathic effect of groundnut on wheat.

With respect to the interaction, the results imply that peanut cultivars responded differently ($P \leq 0.05$) to cropping systems for all studied traits except number of spikes/m^2.

5. Conclusion

According to results, it can be concluded that BNF by peanut cv. Ismailia 1 had significant effect on wheat grain yield per ha and this effect was increased by growing fahl berseem during transition period between peanut and wheat under sandy soil conditions. Also, intensive cropping system by using the local cultivar of peanut (Ismailia 1) was more profitable for Egyptian farmers.

References

A.O.A.C. (2000). *Association of Official Agricultural Chemists Official Methods of Analysis.*(15th A. O. A. C.), Washington, USA.

Abd El-Maksoud, M.F. (2008).Response of two peanut cultivars to row width and hill spaces in sandy soil.*Research Journal of Agriculture and Biological Sciences, 4*(5), 447-454.

Asfour, M.M., Abdalla, M.A., &Essa, A.H.A. (2009).Response of three peanut cultivars to gypsum.*Journal of Agricultural Sciences, Mansoura University, 34*(9), 9577-9588.

Baldwin, K.R. (2006). Crop Rotations on Organic Farms.North Carolina Cooperative Extension Service.College of Agriculture and Life Sciences.NC State University.

Banziger, M., Feil, B.,& Stamp, P. (1994).Competition between nitrogen accumulative and grain growth for carbohydrates during grain filling of wheat.*Crop Science, 34*, 440-446.http://dx.doi.org/10.2135/cropsci1994.0011183X003400020025x

Bulletin of Statistical Cost Production and Net Return.(2013).*Summer and Nili Field Crops and Vegetables and Fruit.* Agriculture Statistics and Economic Sector, Ministry of Egyptian Agriculture and Land Reclamation", Part (2), August 2013.

Bulletin of Statistical Cost Production and Net Return.(2014).*Winter Field Crops and Vegetables and Fruit.* Agriculture Statistics and Economic Sector, Ministry of Egyptian Agriculture and Land Reclamation", Part (1), February 2014.

Dakora, F.D., & Keya, S.O. (1997).Contribution of legume nitrogen fixation to sustainable agriculture in Sub-Saharan Africa.*Soil Biology and Biochemistry, 29*, 19-27. http://dx.doi.org/10.1016/S0038-0717(96)00225-8

Dinh, H.T., Kaewpradit, W., Jogloy, S., Vorasoot, N., &Patanothai, A. (2013).Biological nitrogen fixation of peanut genotypes with different levels of drought tolerance under mid-season drought.*SABRAO Journal of Breeding and Genetics, 45*(3), 491-503.

El-Saady, A.M., El-Sayed, A.A., Teilep, W.M., & El-Dahshouri, M.F. (2014).Response of some peanut (*Arachishypogaea* L.) cultivars grown in sandy soil to soil and foliar feeding with the different sources of phosphorus.*International Journal of Plant and Soil Science, 3*(6), 523-537. http://dx.doi.org/10.9734/IJPSS/2014/8089

Fageria, N.K., Baligar, V.C., & Jones, C.A. (1997). *Growth and mineral nutrition of field crop*(2nd Edition). New York, USA: Marcel Dekker Inc.

Feil, B. (1997). The inverse yield-protein relationship in cereals: Possibilities and limitations for genetically improving the grain protein yield.*Trends in Agronomy, 1*, 103-119.

Freed, R.D. (1991).*MSTATC Microcomputer Statistical Program.*Michigan State University, East Lansing, Michigan, USA.

Gomez, K.A., & Gomez, A.A. (1984).*Statistical Procedures for Agricultural Research.*(2nded.) Toronto, ON, Canada: John Willey and Sons.

Hardarson, G. (1993). Methods for enhancing symbiotic nitrogen fixation.*Plant Soil, 152*, 1-17.http://dx.doi.org/10.1007/BF00016329

Hayat, R., Ali, S., Siddique, M.T., &Tahir, H.C. (2008). Biological nitrogen fixation of summer legumes and their residual effects on subsequent rainfed wheat yield.*Pakistan Journal of Botany, 40*(2), 711-722.

Khan, A.J., Azam, F., & Ali, A. (2010).Relationship of morphological traits and grain yield in recombinant inbred wheat lines grown under drought conditions.*Pakistan Journal of Botany, 42*(1), 259-267.

Lindemann, W.C., & Glover, C.R. (2013).*Nitrogen Fixation by Legumes.*The College of Agricultural, Consumer and Environmental Sciences on the World Wide Web at aces.nmsu.edu.

Liu, X. (2013).*Cropping System in China* (pp. 12-15). Beijing: Agricultural Press.

Mandal, B.J., Dhara, M.C., Mandal, B.B., Das, S.K.,&Nandy, R. (1990).Rice, mung bean, soybean, peanut and black gram yields under different intercropping systems.*Agronomy Journal, 82*, 1063-1066.http://dx.doi.org/10.2134/agronj1990.00021962008200060006x

Myers, A. (1988). Cereal cropping.*Plant and Soil, 174*, 30-33.

Nambiar, P.C., & Dart, P.J. (1983).Factors influencing nitrogenase activity (acetylene reduction) by root nodules of groundnut, *Arachishypogaea* L. *Peanut Science, 10*, 26-29.http://dx.doi.org/10.3146/i0095-3679-10-1-8

Nawab, K. H., Amanullah, M.A.,& Ali, A. (2006).Response of wheat to farm yard manure, potassium and zinc under rainfed cropping patterns.*Middle East Journal of Scientific Research, 1*(1), 1-9.

O'Hara, G.W. (2001). Nutritional constraints on root nodule bacteria affecting symbiotic nitrogen fixation: a review.*Australian Journal of Experimental Agriculture, 41*, 417-433.http://dx.doi.org/10.1071/EA00087

Pandey, K.R., Shah, S.C., & Becker, M. (2008).Management of native soil nitrogen for reducing nitrous oxide emissions and higher rice production.*The Journal of Agriculture and Environment, 9*, 1-9.

Phoomthaisong, J., Toomsan, B., Limpinuntana, V., Cadisch, G., &Patanothai, A. (2003). Attributes affecting residual benefits of N2-fixing mung bean and groundnut cultivars.*Biology and Fertility of Soils, 39*(1),

16-24.http://dx.doi.org/10.1007/s00374-003-0676-4

Pimratch, S., Jogloy, S., Toomsan, B., Jaisil, P., Sikhinarum, J., Kesmala, T.,&Patanothai, A. (2004).Evaluation of seven peanut genotypes for nitrogen fixation and agronomic traits.*Songklanakarin Journal of Science and Technology, 26*(3), 295-304.

Postgate, J. (1998).*Nitrogen Fixation.*(3rded.) Cambridge University Press: Cambridge UK. Main website: WWW.CAMBRIDGEFORECAST.ORG

Shah, Z., Shah, S.H., Peoples, M.B., Schwenke, G.D., &Hrridge, D.F. (2003). Crop residue and fertilizer N effects on nitrogen fixation and yields of legume-cereal rotations and soil organic fertility.*Field Crops Research, 83*, 1-11.http://dx.doi.org/10.1016/S0378-4290(03)00005-4

Shiferaw, B., Bantilan, M.C.S., &Serraj, R. (2004).Harnessing the potential of BNF for Poor Farmers: Technological Policy and institutional constraints and research need.In R. Serraj(Ed.).Symbiotic Nitrogen Fixation; prospects for enhanced application in tropical agriculture (p. 3). New Delhi: Oxford & IBH publishing Co. Pvt. Ltd.

Toomsan, B., McDonagh, J.F., Limpinuntana, V., &Giller, K.E. (1995).Nitrogen fixation by groundnut and soyabean and residual nitrogen benefits to rice in farmers' fields in Northeast Thailand.*Plant and soil, 175*(1), 45-56.http://dx.doi.org/10.1007/BF02413009

Zecevic, V., Boskovic, J., Dimitrijevic, M.,&Petrovic, S. (2010). Genetic and Phenotypic variability of yield components in wheat (*Triticumaestivum*L.).*Bulgarian Journal of Agricultural Science, 16*(4), 422-428.

5

Evaluation of Tillage Practices for Maize (*Zea mays*) Grown on Different Land-Use Systems in Eastern Zambia

Paramu Mafongoya[1], Obert Jiri[1] & Mutondwa Phophi[1]

[1] University of KwaZulu-Natal, School of Agriculture, Earth and Environmental Science, Pietermaritzburg, 3029, South Africa

Correspondence: Obert Jiri, University of KwaZulu-Natal, School of Agriculture, Earth and Environmental Science, Pietermaritzburg, 3029, South Africa. E-mail: obertjiri@yahoo.co.uk

Abstract

Improved fallows of *Sesbania sesban* (Sesbania) have been known to improve soil physical and chemical properties and increase crop yield compared to traditional fallows. However, the effects of soil tillage practices after improved fallows on soil properties, weeds, labour and subsequent maize crop have not been assessed in Southern Africa. This study aimed to evaluate how tillage practices affect yield of maize and affect soil properties after two years of fallow and subsequent cropping phase. In this study, done at sites in eastern Zambia, maize yield from a two-year planted Sesbania, natural fallow, continuously fertilized and unfertilized maize were compared under conventional, flat till and zero tillage practices. A split plot experiment, with improved fallow systems in the main plot and the tillage practice in the subplot, was established at the sites. The results showed that the increases in grain yield under conventional tillage over zero tillage practice were 17.8% and 28.2% during 2000/2001 and 2001/2002 seasons, respectively, at Msekera. At Chadiza, the increases in grain yield under conventional tillage over zero tillage were 66.3% and 327.4% during 2000/2001 and 2001/2002 seasons, respectively. Greater maize yields were achieved under Sesbania planted fallows compared to the natural fallow and maize monoculture without fertilizer. Overall, zero tillage practice resulted in lower maize grain yield, higher bulk density, reduced water intake, higher weed infestation and high labour demand during weeding compared to conventional tillage.

Keywords: conventional tillage, flat till, grain yield, water intake, weeds, zero tillage

1. Introduction

The practice of intensive tillage can be effective for weed management. However, it increases the risk of soil erosion. Frequent tillage can increase land degradation, soil erosion and soil compaction, and these are major challenges that smallholder farmers are facing in crop production (Hamza & Anderson, 2005). The increase of land degradation has brought an interest of practicing conservation agriculture and there is need to increase the practice of conservation tillage practices, including conservation agricultural practices, amongst smallholder farmers in South Africa.

Conservation tillage is the use of agricultural practices that have minimal soil disturbances (Sarker et al., 2012). Conservation agriculture is defined as a farming system where there is a permanent soil cover, minimal soil disturbance and where crop rotation is practiced (Act et al., 2003). It also includes the use cover crops as a soil amelioration measure (Mariki, 2003). Small resource poor farmers who have adopted conservation tillage methods cite the reduction in labour inputs and drudgery as major drivers for adoption (Landers et al., 2002). The three of conservation agricultural practices pillars are sustainable and environmentally friendly, conserving natural resources such as soil nutrients and water (Chauhan et al., 2012). Conservation agriculture is also known to improve crop yields, reduce soil erosion, soil fertility improvement, and reduces labour requirements (Giller et al., 2009).

Weed management in conservation tillage practices is more problematic than in conventional tillage practices because weeds that are not buried deep in the soil have the ability to germinate and emerge vigorously (Singh et al., 2015), especially perennial weeds, and this results in loss of crop yield (Mader & Biere, 2011; Capinera, 2005). The use of herbicides is a common practice to control in conservation agriculture. However, the continuous application of herbicides with the same mode of action is problematic and could lead to herbicide resistance (Vencil et al., 2012). This is particularly a major problem on smallholder farmers practicing no-till who cannot afford herbicides. Although herbicide application can be effective in weed suppression, smallholder farmers should be aware that the effectiveness of herbicide can decrease with the amount of residues left on the soil surface after harvest (Karki & Shestha, 2014).

The use of cover crops can be used as an alternative to herbicide application by the smallholder farmers. Cover crops are important for sustainable crop production providing various benefits such as improvement of soil quality by improving biological, chemical and physical properties such as organic matter content (Dabney et al., 2001). Legume cover crops help in improving soil fertility (Matusso et al., 2014). Legume cover crops have the potential to utilize low available nutrients compared to cereal cover crops that utilize a lot of nitrogen (Fageria & Bailey, 2005). Smallholder framers can use legumes as cover crops because it results in high economic returns with minimal fertilisers application in maize production (Murungu, 2012). A number of legumes have been tested for weed suppression and can be applicable for smallholder farmers practicing conservation agriculture. Cowpea is known to be effective in weed suppression and can be suitable to be used in maize production. Its ability to suppress weeds is influenced by its growth habit. Cowpea develops over ground runners which can result in decreases light interception for weeds, thus reducing weed biomass and weed density (Bilalis et al., 2010). Cowpea has been found to be effective in suppressing amaranthus weed species (Zaviehmaradat et al., 2013). Grazing vetch is also known to be effective in weed suppression reducing weed species diversity. A study done in South Africa found that grazing vetch reduced weed density by 80% (Murungu et al., 2010).

In traditional shifting and semi-permanent hand-hoe tillage systems, zero or minimum tillage operations are common among small-scale farmers. This is due to labour constraints and lack of draught power. Farmers in eastern Zambia are not exceptional as they are faced with problems of shortage of labour during the growing season. For this reason, maize, a staple crop, is planted on flat land after the vegetation or crop residues are gathered and burned. Most resource poor farmers practice this system, traditionally known as "Galauza". In other cases, farmers leave fields fallow to natural vegetation for up to 5 years to restore soil fertility (Mafongoya et al., 2006). After this period, farmers gather the natural or crop residues and make ridges or mounds using hand hoes by covering the mulch on which crops are planted.

Labour shortage, especially at planting, make farmers to opt to plant on the flat. At Msekera, maize in improved fallow trials is planted on the flat after the soil has been tilled and later ridges are made during weeding when the maize is 50 cm high (Mafongoya et al., 1999). In Zambia, there has been an increased interest in conservation farming because of its benefits in soil erosion control, soil moisture conservation, soil structure improvement and increased net return to farmers. There is little quantitative data, however, that is known about the effect of this tillage system on yield of maize in the farming system of eastern Zambia.

There is need, therefore, to come up with a practice that is both economical and practical under resource poor farmers' depleted soils where such fallows have a potential. Improved fallows of Sesbania have also been known to improve the chemical and physical conditions of the soil (Kwesiga et al., 2005) as well as suppressing weeds during fallow phase. Therefore, the objectives of this study were to (i) evaluate how tillage practices affect yield of maize, (ii) determine the effects of tillage practices on soil properties after two years of fallow and subsequent cropping phase.

2. Materials and Methods

2.1 Site Description

The study was conducted at two sites in eastern Zambia, at Msekera Research Station (32°34' E, 13°38' S, altitude 1030 m above sea level) in Chipata and at Farmers' Training Center in Chadiza (32°30' E 14°03' S, altitude 1061 m above sea level) between 1997 and 2002. The top soil (0-20 cm) at Msekera has 1% carbon content, 25 % and 58 % clay and sand, respectively, and a pH ($CaCl_2$) of 4.8. They are classified as Typic Haplustalfs (USDA, 1975) or Ferric luvisols (FAO, 1988). Msekera receives an average rainfall of 1092 mm per annum (unimodal, Nov-April). At Chadiza the soils (0-20 cm layer) have 0.4% carbon content, and a pH ($CaCl_2$) of 4.2, 6 % and 71 % clay and sand, respectively. They are classified as Typic Haplustults (USDA, 1975) or Ferric acrisol (FAO, 1988). Chadiza receives an average rainfall of 900 mm per annum (unimodal, November - April).

2.2 Experimental Design

The experiment was carried out using a split plot design with three replications at each site. The main plots were: (1) maize (*Zea mays* L.) after a two year *Sesbania sesban* (prov. Chipata dam) (Sesbania) fallow, (2) maize after 2 years natural fallow, (3) maize monoculture with recommended fertilizer (112 kg N ha^{-1}, 18 kg P ha^{-1}, and 17 kg K ha^{-1}), and (4) maize monoculture without fertilizer. The sub-plots consisted of three practices of tillage: (1)

conventional tillage (farmers planting maize on ridges), (2) flat tilled (farmers planting maize on ploughed field with no ridges) and (3) zero tillage. The sub-plots measured 10 by 10 m.

2.3 Land preparation and Crop Management

Sesbania was planted in the field from nursery-raised, bare-rooted seedlings at the age of 5 weeks. The spacing between plants was 1.0 m by 1.0 m (10 000 plants ha^{-1}). Trees were felled to ground level after two years of growth in October 2000. Stumps and root systems were left in the soil. The above ground biomass of trees was measured at fallow clearing by separating the biomass components into foliage (leaves and twigs), branches and stems. These components were then weighed as green after which samples of each component were collected on plot basis and oven dried at 70°C to constant moisture. The plots with conventional tillage practice were prepared by covering the natural vegetation or crop residues with soil by making ridges as a common practice in eastern Zambia using hand hoes. The plots with flat tilled practice were ploughed by digging and burying the natural vegetation or crop residues on the surface to 20 cm depth with a hand hoe. On the zero tillage practice, a 3 cm diameter bamboo stick was used to open a fallow to a depth of 5 cm, where maize seed was placed at planting. Biomass production of natural fallow at the end of two years was assessed using four quadrants of 0.50 by 0.50 m (0.25 m^2) each (Klingman, 1971). Weeds, during the cropping phase, were noted before each weeding by the procedure mentioned above. Predominant weeds were *Acanthospermum hispidium* DC, A. *conyzoides, Bidens pilosa* L. and *Cassia obtusifolia* L. (Fabaceae). Weeds were controlled in the conventional tillage practice by re-ridging. In the flat tilled plots, the weeds were controlled by hand hoeing, and in the zero tillage plots by cutting the weeds at ground level with hand hoe. All plots were weeded twice during the crop season. Hybrid maize (*Zea mays* L. var. MM 604) was sown by hand in all tillage practices at 25 cm within the rows and 100 cm between the rows (44 444 plants ha^{-1}). Fertilizer to the 'fertilized maize control plots' was applied at the rate of 20, 18, and 17 kg N, P and K ha^{-1}, respectively, using Compound D at sowing and 92 kg N ha^{-1} using urea, four weeks after sowing. The maize was planted over 2 seasons, in the 2000/2001 and 2001/2002 seasons. Maize yield was measured at the end of each season.

2.4 Sample Collection and Analyses

Six replicate samples were taken from 0-20 cm soil depth in all plots for determination of total inorganic N. The first sampling was taken at fallow clearing (post-fallow preseason sampling, October 2000) and the second sampling was done in February 2001 (wet season sampling). Ammonium N was determined by colorimetric method (Anderson & Ingram, 1993). Nitrate concentrations were determined by cadmium reduction (Dorich & Nelson, 1984). The sum of NH_4^+-N and NO_3^--N constituted the total inorganic N.

Soil samples for determination of bulk density from all plots were collected using standard core rings (100 cm^3) from 0-20 cm soil layer at fallow clearing (October, 2000) and start of the second crop season (October, 2001) and oven-dried to constant weight at 105°C and weighed. Infiltration was monitored at fallow clearing towards the end of the dry season (October, 2000) and before start of the second cropping season (October, 2001) using the double ring infiltrometer (Bouwer, 1986). Measurements were recorded from 3 double rings inserted diagonally in a systematic design in the net plot for three hours at 0, 5, 10, 15, 20, 30, 45, 60, 90, 120, 150 and 180 minutes. The average readings were used to calculate infiltration rate per plot using Kostiakov (1932) model.

The data were subjected to analysis of variance (ANOVA) using the generalized linear model (Proc GLM) of the Statistical Analysis System (1996). The least significant difference (LSD) method was used at 5% to separate treatment means in case of a significant F-test (Gomez & Gomez, 1984).

3. Results and Discussion

3.1 Above Ground Tree Biomass

At both sites, no significant difference was recorded in above ground biomass in relation to tillage practice or fallow system (Table 1). Despite this, Sesbania had the highest standing total above ground biomass of 16.1 tha^{-1} under conventional tillage practice and 7.8 tha^{-1} under flat till practice at Chadiza and Msekera respectively (Table 1). Conventional tillage practice at Chadiza had biomass of 15.7 tha^{-1} and 0.4 tha^{-1} for wood and foliage (leaf + twigs), whereas wood biomass was 7.4 tha^{-1} and foliage (leaf + twigs) was 0.4 tha^{-1} for flat till practice at Msekera.

Table 1. Above ground biomass (tha^{-1}) at fallow clearance at Chadiza and Msekera as affected by fallow system and tillage practice in October 2001

Fallow System (FS)	Tillage practice		
	Conventional Tillage	Flat till	Zero tillage
Msekera			
Two years Sesbania fallow			
Leaf and twig	0.4	0.4	0.5
Stem	6.6	7.4	6.2
Total above ground biomass	7.0	7.8	6.7
Two years natural fallow			
Total above ground biomass	7.1	7.9	8.5
	$LSD_{(0.05\ level)}$: FS = NS, Tillage = NS, F x Tillage = NS		
Chadiza			
Two years Sesbania fallow			
Leaf and twig	0.4	1.0	0.3
Stem	15.7	12.0	8.3
Total above ground biomass	16.1	13.0	8.6
Two years natural fallow			
Total above ground biomass	6.1	6.4	6.2
	$LSD_{(0.05\ level)}$: FS = NS, Tillage = NS, FS x Tillage = NS		

LSD = least significant difference; FS = Fallow system; NS = not significant.

Table 2. Inorganic soil NO_3^--N and total inorganic-N (mg N kg^{-1}), at 0 – 20 cm depth, before sowing crop and during the wet season of first post-fallow crop (2000/2001) as affected by cropping system and tillage practice at Msekera

Cropping system (CS)	Tillage practice					
	Conventional tillage		Flat till		Zero tillage	
	Nitrate-N	Total inorganic-N	Nitrate-N	Total inorganic-N	Nitrate-N	Total inorganic-N
Inorganic nitrogen in October 2000						
Maize with fertilizer	5.12	6.99	1.90	3.22	2.60	4.24
Maize after Sesbania fallow	3.44	4.64	2.98	3.14	3.52	5.22
Maize after natural fallow	1.88	3.18	2.84	4.82	1.40	3.30
Maize without fertilizer	2.52	3.97	3.06	4.01	1.47	2.24
Inorganic nitrogen in February 2001						
Maize with fertilizer	15.16	19.64	1.81	4.99	1.94	6.23
Maize after Sesbania fallow	2.26	7.55	1.63	4.93	1.04	3.64
Maize after natural fallow	2.36	6.96	1.84	5.87	1.76	6.03
Maize without fertilizer	1.97	3.08	1.22	2.94	1.84	4.56

Nitrate-N: $LSD_{(0.05\ level)}$: CS = 1.82, Tillage = 0.72, CS x Tillage = 2.00
Total-N: $LSD_{(0.05\ level)}$: CS = 2.80, Tillage = 1.23, CS x Tillage = 3.18

LSD = least significant difference; CS = Cropping system; NS = not significant.

The above ground biomass reported in this study relates well to that reported by Kwesiga et al. (1995), Sileshi et al. (2006) and Mafongoya et al. (1999) under similar conditions. The high Sesbania biomass at Chadiza site was attributed to the good rainfall (1144.4 mm p.a.) of 1997/98 followed by another good season with a total of 1062 mm per annum in the 1998/1999 season. The other reason is that the type of soils at Chadiza has a top 40 cm sand layer followed by a clay subsoil which traps leached nutrients.

3.2 Top Soil Nitrogen Dynamics

At Msekera site, nitrate and total inorganic N in both October 2000 and February 2001 was highest under conventional tillage compared to zero or flat till practice (Table 2). The interaction between cropping system and tillage practice was significant ($P<0.05$). This could be attributed to the dry conditions experienced at the time of soil sampling. During the wet season sampling, significant differences were observed for cropping season and tillage practice ($P<0.05$).

At Chadiza site, there was no significant difference at both times of sampling in relation to soil nitrogen (Table 3). Despite this, zero tillage practice had generally lower concentrations of NO_3^- - N in the top 20 cm. This is in contrast to Khant (1971), who proposed greater N concentration in zero tillage plots due to less uptake and movement as a result of the absence of thorough land preparation. The low NO_3^- - N levels under Sesbania at both sites during wet season sampling could be a result of rapid N uptake by growing maize and rapid leaching of NO_3^- - N during high rainfall of 2000 (1342 mm p.a.). Okonkwo et al. (2008) reported similar results of NO_3^- - N being leached beyond rooting depth of maize.

Table 3. Inorganic soil NO_3^--N and total inorganic-N (mg N kg^{-1}), at 0 – 20 cm depth, before sowing crop and during the wet season of first post-fallow crop (2000/2001) as affected by cropping system and tillage practice at Chadiza

Cropping system (CS)	Tillage practice					
	Conventional tillage		Flat till		Zero tillage	
	Nitrate-N	Total inorganic-N	Nitrate-N	Total inorganic-N	Nitrate-N	Total inorganic-N
Inorganic nitrogen in October 2000						
Maize with fertilizer	4.91	7.63	2.20	5.68	2.04	4.12
Maize after Sesbania fallow	4.94	7.51	5.67	8.11	5.77	10.24
Maize after natural fallow	1.35	4.96	1.63	5.47	1.12	5.60
Maize without fertilizer	5.40	8.24	1.88	4.39	1.12	3.84
Inorganic nitrogen in February 2001						
Maize with fertilizer	1.49	7.30	3.1	6.22	2.9	6.58
Maize after Sesbania fallow	2.55	6.48	0.55	4.26	0.59	4.43
Maize after natural fallow	1.21	5.77	0.52	2.93	0.27	3.38
Maize without fertilizer	2.03	4.80	0.19	2.55	0.46	3.85
		Nitrate-N: $LSD_{(0.05\ level)}$: CS = NS, Tillage = NS, CS x Tillage = NS				
		Total-N: $LSD_{(0.05\ level)}$: CS = NS, Tillage = NS, CS x Tillage = NS				

LSD = least significant difference; CS = Cropping system; NS = not significant.

3.3 Cumulative Water Intake

Significant difference ($P<0.05$) was found in all cropping systems and tillage practices in both October 1999 and October 2000 seasons (Table 4). Cumulative water intake by the cropping system was 14.9% and 12.7% higher under conventional tillage than zero tillage practice in both years after three hours (Table 4). Natural fallow under conventional tillage practice had significantly higher cumulative water intake than under flat till or zero tillage practice in 1999 season (Table 4). This could be attributed to less runoff, high root mass, less compaction during the fallow period (Sjogren et al., 2010). Natural vegetation regrowth consists of many plant species with different types of root systems, which have the capacity to increase infiltration of water in the soil. Fallowing with various legumes and grass cover crops is known to improve soil infiltration (Chintu, 2004). Low cumulative water intake in maize with or without fertilizer on zero tillage practice could be attributed to deterioration of soil physical properties leading to high bulk density and reduced porosity. Similarly, Good and Beatty (2011) reported a decline in cumulative water intake in continuous maize with fertilizer, which they attributed to high soil bulk density, and penetrometer resistance under no till treatment. Generally, there was a decline in all cropping systems in cumulative water intake in the second post fallow season (October 2001) than the first post fallow season (October 2000). This decline could be attributed to break down of soil physical properties. The benefits accrued during fallowing are easily lost by cultivation (Wilkinson & Aina, 1976). This decline was more pronounced for natural vegetation fallow under conventional tillage practice and the least was for unfertilised, monocultivated maize under zero tillage practice. Continuous cultivation has been reported by several researchers (Liu et al., 2006) as being responsible for structural degradation, decrease in soil organic matter content.

Table 4. Cumulative water intake (mm) after 3 hours before sowing the first crop (October 2000) and before sowing second crop (October 2001) as affected by cropping system and tillage practice at Msekera

Cropping system (CS)	Tillage practice		
	Conventional tillage	Flat till	Zero tillage
Water intake in October 2000			
Maize with fertilizer	190.0	154.3	152.3
Maize after Sesbania fallow	255.0	251.7	234.7
Maize after natural fallow	306.0	256.7	262.7
Maize without fertilizer	140.3	130.3	126.0
	$LSD_{(0.05\ level)}$: CS = 7.1, Tillage = 4.2, CS x Tillage = 9.2		
Water intake in October 2001			
Maize with fertilizer	108.7	102.0	96.7
Maize after Sesbania fallow	170.7	160.0	111.0
Maize after natural fallow	158.3	179.0	174.3
Maize without fertilizer	95.0	94.7	93.0
	$LSD_{(0.05\ level)}$: CS = 17.5, Tillage = 12.6, CS x Tillage = 25.0		

LSD = least significant difference; CS = Cropping system.

3.4 Soil Bulk Density

The bulk density measured at fallow clearance was lowest under the maize planted after the natural fallow (1.14 gcm^{-3}) and Sesbania fallow (1.23 gcm^{-3}) flat till practice compared to maize monoculture with fertilizer (1.54 gcm^{-3}) and maize after natural fallow (1.53 gcm^{-3}) on zero tillage practice (Table 5, for Msekera site). The higher bulk density under zero tillage practice could be attributed to the non-incorporation of organic matter which was left on the soil surface. These soils are normally compacted if no tillage is used. Generally, soils with a bulk density higher than 1.6 g cm^{-3} are considered to restrict root growth (Cresswell and Hamilton (2002). Therefore, where minimum tillage or mixing of soil with organic matter is employed, bulk density is bound to be lowered. This is contrary to other researchers (Diana et al. 2008) who reported that the presence of residue on the soil surface is responsible for maintaining low soil bulk density. Bulk density measured after one year of cropping (October 2000) was lowest in maize planted after natural fallow (1.27 gcm^{-3}) and highest under maize monoculture with fertilizer (1.62 gcm^{-3}) under zero tillage practice (Table 5). In this study, bulk density measured after one year of cultivation led to progressive deterioration of the soil structure under all cropping treatments. The results from this experiment confirm the earlier findings by Liu et al (2006) who reported high bulk density after continuous monocropping. The increased bulk density could be linked to high soil compaction under the zero tillage practice, which could have impended root growth to exploit nutrients hence lower maize grain yields. No tillage, although advantageous through reduction of erosion and soil organic matter maintenance, could eventually lead to soil compaction with shallow rooting crops and insufficient residue return (Juo et al. 1996).

Table 5. Dry bulk density (g cm^{-3}) before sowing the first crop (October 2000) and before sowing second crop (October 2001) as affected by land-use system and tillage practice at Msekera

Cropping system (CS)	Tillage practice		
	Conventional tillage	Flat till	Zero tillage
Dry bulk density in October 2000			
Maize with fertilizer	1.30	1.36	1.54
Maize after Sesbania fallow	1.33	1.23	1.45
Maize after natural fallow	1.38	1.14	1.53
Maize without fertilizer	1.38	1.31	1.36
	$LSD_{(0.05\ level)}$: CS = NS, Tillage = 0.07, CS x Tillage = 0.16		
Dry bulk density in October 2001			
Maize with fertilizer	1.42	1.50	1.62
Maize after Sesbania fallow	1.45	1.37	1.57
Maize after natural fallow	1.48	1.27	1.59
Maize without fertilizer	1.52	1.45	1.47
	$LSD_{(0.05\ level)}$: CS = NS, Tillage = 0.06, CS x Tillage = 0.17		

LSD = least significant difference; CS = Cropping system; NS = not significant.

3.5 Maize Grain Yields

At Chadiza site, there was no interaction between cropping system and tillage practice with respect to maize grain yield in both crop seasons. However, the maize yields of maize monoculture with fertilizer during 2000 season under conventional tillage practice performed better than the rest of the cropping system and tillage practices (Table 6). During 2000 season the maize yields of fertilised monocultivated maize, maize after Sesbania fallow, maize after natural fallow and maize monoculture without fertilizer from conventional tillage practice were 49%, 46%, 42% and 47% above the zero tillage system respectively. On the other hand, maize yields from the flat till practice were not significantly different from the zero tillage practice. Similarly, maize yields during 2000/2001 season were highest in maize monoculture with fertilizer under conventional tillage compared to zero tillage or flat till practice (Table 6). The increases in grain yield under conventional tillage over zero tillage respectively were 66.3% and 327.4% during 2000/2001 and 2001/2002 seasons. In both cropping seasons at Chadiza site, zero tillage decreased maize grain yields compared to conventional tillage. This could be due to weed infestation, outbreak of *Cercospora* grey leaf spot disease during grain filling period and deterioration of soil properties under continuous zero tillage practice. Madal et al. (1994) reported higher yields under conventional tillage practice, which they associated with better root growth and higher water use. No till has also been reported to cause significant reductions in maize yield compared with conventional cultivation and deep tillage (Arora et al., 1991; Archarya & Sharma, 1994).

At Msekera, in both 2000/2001 and 2001/2002 seasons, maize yields were significantly different ($P<0.05$) among the cropping and tillage practices. No interaction between cropping and tillage practice with respect to maize grain yield in both seasons was recorded at Msekera site. In spite of this, the highest yields, irrespective of tillage practice were from maize monoculture with fertilizer in 2000 and 2001 season. The increases in grain yield under conventional tillage over zero tillage practice respectively were 17.8% and 28.2% during 2000 and 2001 season, respectively (Table 6).

Kwesiga et al. (2005) showed that improved fallow of Sesbania of one to three year duration has the capacity to increase yield of subsequent maize crops on N-deficient soils. Sesbania leaf biomass is higher in N and decomposes rapidly to supply N to maize crops in the first season. Mafongoya et al. (2006) reported similar benefits of Sesbania leaf biomass on subsequent maize grain yield. Whereas high maize yields in the control with fertilizer could be ascribed to N from fertilizer. Maize yields under conventional tillage practice surpassed yields from other tillage practices and this could be attributed to the improved soil fertility, concentration of organic matter along the ridge, and reduced weed infestation. On the other hand, low maize yields from zero tillage were a result of high weed infestation and pests/disease outbreaks (Sileshi, personal communication, Chitedze Research Station, Zambia).

Table 6. Maize grain yields (t ha^{-1}) at Chadiza and Msekera as affected by cropping system and tillage practice

Cropping system (CS)	Tillage practice		
	Conventional tillage	Flat till	Zero tillage
Chadiza 2001			
Maize with fertilizer	4.23	3.13	2.14
Maize after Sesbania fallow	3.76	3.77	2.01
Maize after natural fallow	1.07	0.86	0.60
Maize without fertilizer	1.42	1.03	0.75
$LSD_{(0.05\ level)}$: CS = 0.43, Tillage = 0.37, CS x Tillage = 0.74			
Chadiza 2002			
Maize with fertilizer	1.65	0.50	0.32
Maize after Sesbania fallow	0.99	0.99	0.32
Maize after natural fallow	0.74	0.37	0.26
Maize without fertilizer	0.55	0.21	0.03
$LSD_{(0.05\ level)}$: CS =0.41 Tillage = 0.36, CS x Tillage = NS			
Msekera 2001			
Maize with fertilizer	4.31	3.97	3.65
Maize after Sesbania fallow	3.35	3.28	3.39
Maize after natural fallow	2.12	2.06	1.92
Maize without fertilizer	1.15	0.96	0.86
$LSD_{(0.05\ level)}$: CS =0.50, Tillage = NS, CS x Tillage = NS			
Msekera 2002			
Maize with fertilizer	4.77	4.15	3.98
Maize after Sesbania fallow	1.69	1.70	1.67
Maize after natural fallow	1.10	1.02	0.93
Maize without fertilizer	0.67	0.53	0.38
$LSD_{(0.05\ level)}$: CS =0.28, Tillage = 0.32, CS x Tillage = NS			

LSD = least significant difference; CS = Cropping system; NS = not significant.

3.6 Dry Weed Biomass

Significant differences ($P<0.05$) were observed in weed infestation among the cropping systems and the tillage practices at fallow clearance and at the two weeding times (Table 7). Sesbania planted fallow had no weed biomass at fallow clearance compared to natural fallow. In general the highest and lowest weed cover was found in natural fallow and Sesbania fallow, respectively. Overall, the zero tillage practice and the natural fallow system had significantly high weed infestation at all times during the 2001/2002 season. The low weed infestation under Sesbania cropping system at fallow clearance could be attributed to its ability to suppress weeds in relation to other fallow systems. These results conform to Sileshi and Mafongoya (2003) findings under similar conditions. Significant difference in weed biomass was recorded for cropping system and tillage practice at 2 and 7 WAP (Table 7). Higher weed infestation occurred under natural fallow for zero tillage practice, compared to other cropping systems during the first and second weedings. Similarly, total weed biomass was also significantly affected by cropping system and tillage practice. The zero tillage practice under the natural fallow practice had the highest total weed infestation (Table 7). This could be as a result of weed seeds, which were still in the soil, which came up after the soil was slightly disturbed at weeding. Conventional tillage was able to suppress weeds more than other tillage practices throughout the subsequent weeding times during the crop growth. Whereas crop residues from maize monoculture with or without fertilizer, mostly consisted of stalks which were not able to suppress weeds (Rao, 1983). Böhringer (1991) reported that mulch morphology plays an important role in controlling weeds and facilitating hand hoe weeding.

Table 7. Total dry weed biomass production (kg ha^{-1}) as affected by cropping system and tillage practice at Msekera, Zambia

Cropping system (CS)	Tillage practice		
	Conventional tillage	Flat till	Zero tillage
Weed biomass at fallow clearance			
Maize with fertilizer	2440	2200	2667
Maize after Sesbania fallow	0	0	0
Maize after natural fallow	7063	7920	8450
Maize without fertilizer	2577	2713	2703
		$LSD_{(0.05\ level)}$: CS = 370, Tillage = 453,	CS x Tillage = 412
Weeds at 2WAP			
Maize with fertilizer	1920	2033	2200
Maize after Sesbania fallow	223	277	303
Maize after natural fallow	3133	3217	3333
Maize without fertilizer	1917	2030	2117
		$LSD_{(0.05\ level)}$: CS = 499,Tillage = 35,	CS x Tillage = 500
Weeds at 7WAP			
Maize with fertilizer	193	240	263
Maize after Sesbania fallow	157	240	327
Maize after natural fallow	1970	2060	2563
Maize without fertilizer	603	707	780
		$LSD_{(0.05\ level)}$: CS = 153, Tillage = 131,	CS x Tillage = NS
Total weed biomass			
Maize with fertilizer	26.7	47.3	43.3
Maize after Sesbania fallow	29.0	40.0	70.0
Maize after natural fallow	41.0	57.7	87.3
Maize without fertilizer	24.7	44.3	54.3
		$LSD_{(0.05\ level)}$: CS = 11.5, Tillage = 6.5,	CS x Tillage = 14.4

LSD = least significant difference; CS = Cropping system; NS = not significant.

Table 8. Labour requirements (man hours ha^{-1}) for different operations as affected by cropping system and tillage practice

Cropping system (CS)	Tillage practice		
	Conventional tillage	Flat till	Zero tillage
Labour at land preparation			
Maize with fertilizer	75.3	115.3	35.3
Maize after Sesbania fallow	103.3	103.7	5.3
Maize after natural fallow	77	112	60.7
Maize without fertilizer	73.0	103.7	46.0
	$LSD_{(0.05\ level)}$: CS = NS, Tillage = 22.6, CS x Tillage = NS		
Labour at planting			
Maize with fertilizer	26.0	48.0	54.3
Maize after Sesbania fallow	26.0	46.0	58.0
Maize after natural fallow	26.0	54.3	60.0
Maize without fertilizer	17.3	29.0	47.7
	$LSD_{(0.05\ level)}$: CS = NS, Tillage = 12.7, CS x Tillage = NS		
Labour at 1st weeding			
Maize with fertilizer	20.3	41.0	37.7
Maize after Sesbania fallow	23.3	35.7	64.7
Maize after natural fallow	36.7	54.7	89.9
Maize without fertilizer	20.3	38.7	50.3
	$LSD_{(0.05\ level)}$: CS = 12.3, Tillage = 6.3, CS x Tillage = 14.7		
Labour at 2nd weeding			
Maize with fertilizer	26.7	47.3	43.3
Maize after Sesbania fallow	29.0	40.0	70.0
Maize after natural fallow	41.0	57.7	87.3
Maize without fertilizer	24.7	44.3	54.3
	$LSD_{(0.05\ level)}$: CS = 11.5, Tillage = 6.5, CS x Tillage = 14.4		
Total Labour			
Maize with fertilizer	148.3	251.7	170.7
Maize after Sesbania fallow	181.7	224.7	198.0
Maize after natural fallow	181.0	278.7	265.7
Maize without fertilizer	135.3	215.7	198.3
	$LSD_{(0.05\ level)}$: CS = 34.4, Tillage = 33.4, CS x Tillage = NS		

LSD = least significant difference; CS = Cropping system; NS = not significant.

3.7 Total Labour Requirement

At land preparation, the zero tillage practice was the easiest to prepare and took less man hours compared to the flat till or conventional tillage practice (Table 8). The low labour for Sesbania under zero tillage practice could be attributed to low weed infestation as well as the improved soil structure, which made it easier to make fallows with a wooden peg. On the overall, the maize monoculture with fertilizer on the flat till tillage practice took

115.3 man hours ha^{-1} to prepare compared to 5.3 man hours ha^{-1} for Sesbania fallow zero tillage practice (Table 8).

Tillage practice had no significant difference ($P>0.05$) in the time it took to do the planting operation. Despite this, more time was spent in the natural fallow zero tillage practice compared to maize monoculture without fertilizer under conventional tillage practice during planting (Table 8). Under Sesbania zero tillage practice weeds germinated earlier than other cropping systems because of the improved fertility of the soil. Whereas under natural fallow there were a lot of weed seeds in the soil, which germinated after favourable conditions were met such as good rainfall and soil condition. The high weed infestation consequently led to increasing labour demand at weeding. Addati and Cassirer (2008) reported that farmers in Africa spent about 40% of their work hours weeding. This is because farmers using hand hoes for weeding would like to clean weed their small areas of land in order to get a good yield.

At 2 WAP the weed infestation was higher under the zero tillage practice and as such more time was spent for clean weeding in the maize after natural fallow and Sesbania fallow with zero tillage practices compared to maize mono culture with and without fertilizer under conventional tillage practice (Table 8). The reason has already been mentioned above.

After 7 WAP (second weeding) the maize after natural fallow and Sesbania fallow with zero tillage still had higher weed infestation as such more time was spent for clean weeding compared to the conventional tillage (Table 8). The major reason has also been mentioned before.

The total labour requirement for land preparation, planting and two weeding in one season under different cropping systems and tillage practice was highest under the maize after natural fallow with the flat till and zero tillage practices compared to the maize mono culture with or without fertilizer under conventional tillage (Table 8). Fertilized plots offered good crop stand, which eventually helped to suppress weeds and reduce labour demands. Low fertility under maize without fertilizer contributed to low weed infestation and less labour demand at weeding. Flat till and zero tillage practice required high labour input because traditionally most of the surface area is weeded, even in the case of scattered weed growth as reported by Vogel (1994).

Generally maize monoculture without fertilizer under conventional tillage practice resulted in low labour demand during land preparation, planting, first and second weeding. This could be ascribed to reduced biomass from maize stalks from previous season, which could have interfered with land preparation, planting and weeding operations.

4. Conclusions

This study illustrates the various tillage practices and their implication on labour in relation to maize production under different fallowing systems in smallholder farms. Zero tillage practice resulted in lower maize grain yield, higher bulk density, reduced water intake, higher weed infestation and high labour demand during weeding compared to conventional tillage. Despite zero tillage having less labour demand at land preparation, a farmer will need to invest in herbicides in order to control the weeds if this tillage practice is to be adopted. Conventional tillage improved the soil environment and resulted in increased maize yield in all cropping systems. Flat till practice has higher labour demand at land preparation in relation to other tillage practices and will cause a serious hindrance to households with shortage of labour.

References

Acharya, C. L., & Sharma, P. D. (1994). Tillage and mulch effects on soil physical environment, root growth, nutrient uptake and yield of maize and wheat on an Alfisol in north-west India. *Soil and Tillage Research, 32*(4), 291-302. http://dx.doi.org/10.1016/0167-1987(94)00425-E

ACT, C., & FAO, R. GTZ (2003) Conservation Agriculture, The future for Africa.

Addati, L., & Cassirer, N. (2008). *Equal sharing of responsibilities between women and men, including care giving in the context of HIV/AIDS*. Paper prepared for the Expert Group meeting on the equal sharing of responsibilities between women and men, including care giving in the context of HIV/AIDS, organized by the United Nations Division for the Advancement of Women, Geneva.

Anderson, J. M., & Ingram, J. S. I. (1993). *Tropical Soil Biology and Fertility: A Handbook of Methods* (2nd ed., p. 221). CAB International, Wallingford, UK.

Arora, V. K., Gajri, P. R., & Prihar, S. S. (1991). Tillage effects on corn in sandy soils in relation to water retentivity, nutrient and water management, and seasonal evaporativity. *Soil and Tillage Research, 21*(1), 1-21. http://dx.doi.org/10.1016/0167-1987(91)90002-F

Bilalis, D., Papastylianou, P., Konstantas, A., Patsiali, S., Karkanis, A., & Efthimiadou, A. (2010). Weed-suppressive effects of maize–legume intercropping in organic farming. *International Journal of Pest Management, 56*(2), 173-181. http://dx.doi.org/10.1080/09670870903304471

Böhringer, A. (1991). The potential of alleycropping as a labor efficient management option to control weeds: A hypothetical case. *Der Tropenlandwirt-Journal of Agriculture in the Tropics and Subtropics, 92*(1), 3-12.

Bouwer H. (1986). Intake rate: cylinder infiltrometer. In *Methods of soil analysis, part I. Physical and mineralogical methods. Agronomy Monograph no. 9* (2nd ed.). American Society of Agronomy, Soil Science Society of America.

Capinera, J. L. (2005). Symposium: Relationships between insect pests and weeds: an evolutionary perspective. *Weed Science, 53*, 892-901. http://dx.doi.org/10.1614/WS-04-049R.1

Chauhan, B. S., Singh, R. G., & Mahajan, G. (2012). Ecology and management of weeds under conservation agriculture: a review. *Crop Protection, 38*, 57-65. http://dx.doi.org/10.1016/j.cropro.2012.03.010

Chintu, R. (2004). Subsoil nitrogen dynamics as affected by planted coppicing tree legume fallows. *Experimental Agriculture, 40*, 327-340. http://dx.doi.org/10.1017/S0014479704001826

Cresswell, H. P., & Hamilton. (2002). Particle Size Analysis. In N. J. McKenzie, H. P. Cresswell, & K. J. Coughlan (Eds.), *Soil Physical Measurement and Interpretation for Land Evaluation* (pp. 224-239). CSIRO Publishing: Collingwood, Victoria.

Dabney, S. M., Delgado, J. A., & Reeves, D. W. (2001). Using winter cover crops to improve soil and water quality. *Communication in Soil Science and Analysis, 32*, 1221-1250. http://dx.doi.org/10.1081/CSS-100104110

Diana, G., Beni, C., & Marconi, S. (2008). Organic and mineral fertilization: Effects on physical characteristics and boron dynamic in an agricultural soil. *Communications in Soil Science and Plant Analysis, 39*, 1332-1351. http://dx.doi.org/10.1080/00103620802004037

Dorich, R. A., & Nelson, D. W. (1984). Evaluation of manual cadmium reduction methods for determination of nitrate in potassium chloride extracts of soil. *Soil Science Society of America Journal, 48*, 72-75. http://dx.doi.org/10.2136/sssaj1984.03615995004800010013x

Fageria, N. K., Baligar, V. C., & Bailey, A. (2005). Role of cover crops in improving soil and row crop productivity. *Communications in Soil Science and Plant Analysis, 36*, 2733-2757. http://dx.doi.org/10.1080/00103620500303939

FAO. (1988). Soil Map of the World-Revised Legend. *World Soil Resources Report* 60 FAO/UNESCO, Rome.

Good, A. G., & Beatty, P. H. (2001). Fertilizing Nature: A Tragedy of Excess in the Commons. *PLoS Biology, 9*, 8.

Gomez, K. A., & Gomez, A. A. (1984). *Statistical procedures for agricultural research* (2nd ed.). John Wiley and Sons Inc., New York.

Giller, K. E., Witter, E., Corbeels, M., & Tittonell, P. (2009). Conservation agriculture and smallholder farming in Africa: The heretics' view. *Field Crops Research, 114*, 23-34. http://dx.doi.org/10.1016/j.fcr.2009.06.017

Hamza, M. A., & Anderson, W. K. (2005). Soil compaction in cropping systems a review of the nature, causes and possible solutions. *Soil Tillage Research, 82*, 121-145. http://dx.doi.org/10.1016/j.still.2004.08.009

Juo, A. S. R., Franzluebberrs, K., Dabiri, A., & Ikhile, B. (1996). Soil properties and crop performance on a kaolinitic Alfisol after 15 years of fallow and continuous cultivation. *Plant and Soil, 180*, 209-217. http://dx.doi.org/10.1007/BF00015304

Karki, T. B., & Shrestha, T. (2014). Maize production under no-tillage system in Nepal. *World Journal of Agricultural Research, 2*, 13-17. http://dx.doi.org/10.12691/wjar-2-6A-3

Khant, G. (1971). Changes in N, P, K and C in three soil types after 5 year minimum cultivation. *Landwirtsch Forsch Sonderh, 26*, 273-280.

Klingman, D. L. (1971). Measuring weed density in crops. In L. Chiarappa (Ed.), *Crop Loss Assessment Methods*. 3.1.5/1-3.1.5/6. FAO, Rome.

Kostiakov, A. N. (1932). On the dynamics of the coefficient of water-percolation in soils and on the necessity for studying it from a dynamic point of view for purposes of amelioration. *Transactions of the sixth committee International Society of Soil Science*, Russian Part A: 17-21.

Kwesiga, F., Franzel, S., Mafongoya, P., Ajayi, O., Phiri, D., Katanga, R., ... Chirwa, T. (2005). Improved Fallows in Eastern Zambia: History, Farmer Practice and Impacts. A paper prepared for the IFPRI Workshop on "Successes in African Agriculture," Lusaka, Zambia, June 10-12, 2002

Kwesiga, F., Phiri, D., Mwanza, S., & Simwanza, P. C. (1995). Zambia/ICRAF Agroforestry Research Project. Annual Report. Chipata, Zambia. 80

Landers, J. N., Saturnino, H. M., & de Freitas, P. L. (2002). Organizational and Policy Considerations in Zero Tillage, in Saturnino H M and Landers J N (eds) *The Environment and Zero Tillage*, Brazil: Associacao de Plantio no Cerrado, Brasilia

Liu, X., Herbert, S. J., Hashemi, A. M., Zhang, X., & Ding, G. (2006). Effects of agricultural management on soil organic matter and carbon transformation – a review. *Plant Soil and Environment, 52*, 531-543.

Mandal, B. K., Saha, A., Dhara, M. C., & Bhunia, S. R. (1994). Effects of zero and conventional tillage on winter oilseed crop in West Bengal. *Soil Tillage Research, 29*, 49-57. http://dx.doi.org/10.1016/0167-1987(94)90101-5

Mader, P., & Berner, A. (2011). Development of reduced tillage systems in organic farming in Europe. *Renewable Agriculture and Food Systems, 27*, 7-11. http://dx.doi.org/10.1017/S1742170511000470

Mafongoya, P. L., & Bationo, A. (2006). Appropriate available technologies to replenish soil fertility in southern Africa. *Nutrient Cycling in Agroecosystems, 76*, 137-151. http://dx.doi.org/10.1007/s10705-006-9049-3

Mafongoya, P. L., Katanga, R., Mkonda, A., Chirwa, T. S., Chintu, R., & Matibini, J. (1999). Zambia/ICRAF Agroforestry Research Project, Annual Report. Chipata, Zambia, pp 55 -65

Mariki, W. (2003). The Impact of Conservation Tillage and Cover Crops on Soil Fertility and Crop Production in Karatu and Hanag Districts of Northern Tanzania, TFSC/GTZ Technical Report for 1999 – 2003, Arusha: TFSC

Matusso, J. M. M., Mugwe, J. N., & Mucheru-muna, M. (2014). Potential role of cereal-legume intercropping systems in integrated soil fertility management in smallholder farming systems of Sub-Saharan Africa. *Journal of Agriculture and Environmental Management, 3*, 162-174.

Murungu, F. S. (2012). Conservation Agriculture for smallholder farmers in the Eastern Cape Province of South Africa: recent developments and future prospects. *African Journal of Agricultural Research, 7*, 5278-5284.

Murungu, F. S., Chiduza, C., & Muchaonyerwa, P. (2010). Biomass accumulation, weed dynamics and nitrogen uptake by winter cover crops in a warm-temperate region of South Africa. *African Journal of Agricultural Research, 5*, 1632-1642.

Okonkwo, C. I., Mbagwu, J. S. C., & Egwu, S. O. (2008). Nitrogen Mineralization from Prunings of Three Multipurpose Legume and Maize Uptake in Alley Cropping System. *Agro-Science Journal of Tropical Agriculture, Food, Environment and Extension, 7*, 143-148. http://dx.doi.org/10.4314/as.v7i2.1596

Rao, V. S. (1983). *Principles of Weed Science* (p. 540). Oxford University Press and IBH, Oxford.

Rasaily, G., & Xiaodong, Q. (2012). Development Strategies of small scale conservation farming practices on two wheeled tractor in Bangladesh. *African Journal of Agricultural Research, 7*, 3747-3756.

SAS Institute. (1996). *SAS/STAT*, Release 6.12 SAS Institute Inc., Cary, NC.

Sileshi, G., & Mafongoya, P. L. (2003). Effect of rotational fallows on abundance of soil insects and weeds in maize crops in eastern Zambia. *Applied Soil Ecology, 23*, 211-222. http://dx.doi.org/10.1016/S0929-1393(03)00049-0

Sileshi, G., Kuntashula, E., & Mafongoya, P. L. (2006). Effects of improved fallows on weed infestation in maize in eastern Zambia. *Zambian Journal of Agricultural Science, 8*, 6-12.

Singh, H. P., Batish, D. R., & Kohli, R. K. (2003). Allelopathic Interactions and allelochemicals: new possibilities for sustainable weed management. *Critical Reviews in Plant Sciences, 22*, 239-311. http://dx.doi.org/10.1080/713610858

Sjogren, S., Keith, D., & Karlesson, A. (2010). Effect of Improved fallows with *Sesbania sesban* on maize productivity and *Striga hermonthica* in western Kenya. *Journal of Forestry Research, 21*, 379-386. http://dx.doi.org/10.1007/s11676-010-0085-0

USDA. (1975). *Soil Taxonomy, agricultural handbook number* 436, Soil Conservation Service, USA.

Vencill, W. K., Nichols, R. L., Webster, T. M., Soteres, J. K., Mallory-Smith, C., Burgos, N. R., ... McClelland, M. R. (2012). Herbicide resistance: toward an understanding of resistance development and the impact of herbicide resistant crops. *Weed Science (Special Issue), 60*, 2-30. http://dx.doi.org/10.1614/ws-d-11-00206.1

Vogel, H. (1994). Weeds in single-crop conservation farming in Zimbabwe. *Soil Tillage Research, 31*, 169-185. http://dx.doi.org/10.1016/0167-1987(94)90078-7

Wilkinson, G. E., & Aina, P. O. (1976). Infiltration of water into two Nigerian soils under secondary forest and subsequent arable cropping. *Geoderma, 50*, 51-59. http://dx.doi.org/10.1016/0016-7061(76)90070-7

Zaviehmavadat, L., Mazaheri, D., Majnon, H. N., & Rezaei, M. (2013). The effect of Maize and Cowpea Intercropping on weed control condition. *International Journal of Agronomy and Plant Production, 4*, 2885-2889.

6

Factors Determining the Profitability of Catfish Production in Ibadan, Oyo State, Nigeria

Oluwasola O.[1] & A. O. Ige[1]

[1] Department of Agricultural Economics, Obafemi Awolowo University, Ile-Ife, Nigeria

Correspondence: Oluwasola O, Department of Agricultural Economics, Obafemi Awolowo University, Ile-Ife, Nigeria. E-mail: oluwemimo_oluwasola@yahoo.com

Abstract

This study evaluated the socioeconomic factors influencing the profitability of catfish production in the city of Ibadan. Multistage sampling method was used to collect data from 120 fish farmers. Descriptive statistics, budgetary analysis and the multiple regression model were used to analyse the data obtained. The results showed that catfish production in Ibadan was male dominated as 80% of the fish farmers were men. The mean age of fish farmers was 44.3±12.0 years while as many as 78.3% of the farmers had post-secondary education. The mean family size was 5.2±1.9 while fish farmers were small operators with a mean farm size of 0.3±0.2 hectares. Fish farming is very recent as farmers had a mean farm experience of 6.9±6.5 years. Eighty per cent of the fish farmers got involved in fish farming for commercial reasons. The gross margin to catfish farming was ₦197,520.25 (US$ 987.60)/ha with a net income of ₦182, 573.04 (US$912.87)/ha. The budgetary analysis revealed that fish feed which constituted 79.18% of the total operating cost was the major cost item in catfish production. The regression analysis showed that fish farming experience, amount of labour used and quantity of feed used were significant determinants of net income in catfish production. The study concluded that there is the need to access fish farmers to substantially cheaper feed inputs to ensure the use of adequate quantity and quality of feed in catfish production. This will enhance output, productivity and net income in catfish enterprises.

Keywords: Catfish, feed, unemployment, nutrition, policies, disincentive, subsistence, livelihood

1. Introduction

The enormous opportunities presented by increasing urbanization in Nigeria can be exploited by policy makers and agricultural sector operators to meet the challenges of unemployment, poverty and inadequate nutrition that pervades Nigerian cities. This is because the cities offer large markets, cosmopolitan and affluent population that demands more and better nutrition, and growing industrial and service sectors that depend on the agricultural sector for raw materials. Recent figures on the level of unemployment in Nigeria which stood at 23.9% in 2011(National Bureau of Statistics {NBS}, 2012; CIA, 2014) as well as urban unemployment estimated at 29.5% in 2013 (This Day, 2015) has been unacceptably high. Coupled with this is the challenge of poverty which also stood at 69% (112.5 million people) of the estimated national population of 163 million (NBS, 2014a). The NBS (2014b) also puts youth unemployment at 54% (with 48.1% male and 51.9% female). Furthermore, the average protein intake in Nigeria is about 19.38 grams/caput/day which is far below the Food and Agricultural Organization's (FAO) requirement of 75 g/caput/day (FAO, 1995).

Although the intensity of urban growth can pose a major challenge to the development and growth of large scale farming, fish production is an enterprise that requires small areas of land and which can make use of the several stream channels for the sustainability of the urban environment. In Nigeria where most of the cities have large agricultural resources and potentials relative to the industrial and service sectors, fish farming can generate significant employment, provide income earning potentials for the urban population, enhance the socio-economic status of the farmer as well as generate foreign exchange (Oluwasola & Ajayi, 2013; Olagunju, et al., 2007; Adekoya & Miller, 2004; Eyo, 1992). As succinctly put by Pretty et al. (2003), with dwindling food production, degrading agricultural environment, widespread poverty and insecurity in Africa, fish farming provides the poor and hungry with a low cost and readily available strategy to increase food production using less land and less water without further damage to the environment. Fish farming also has enormous potentials of improving the nutritional standard of the masses of the people. Fish contains higher percentage of protein than meat and is important for its high nutritive value and significance in improving human health. Faced with increasing wealth, changing dietary patterns and urbanization, fish farming provides the key to meeting fish demand in the face of dwindling fish supply from captured sources. Fish farming is not just uniquely placed to reverse the declines in supplies experienced from capture fisheries but also has notable potentials for new livelihood opportunities, providing mechanism for lower priced fish, enhanced nutritional security and employment for poor communities (Jagger & Pender, 2001).

Prior to the 1990s, the development of fish farming in Nigeria was driven by socio-economic objectives including, nutrition improvement of rural communities, generation of additional family income, creation of employment and diversification of income generating activities. Fish farming was promoted by International Organizations and agencies as well as the governments of Nigeria at Federal, State and Local Government levels. In Nigeria, fish farming is predominantly an extensive land based system, practiced mainly at subsistence levels in fresh waters (Anyawu-Akeredolu, 2005). However, only a few species such as catfish and tilapia are cultured commercially (Komolafe and Arawomo, 2007) as commercial farming is yet to become widespread (Fagbenro, 2005). At present, most fish farmers operate small-scale farms ranging from homestead concrete ponds (25 - 40 meters) to small earthen ponds (0.02 - 0.2 hectares). The industry produced over 85,000 tonnes of fish in 2007 (FDF, 2008). FAO (2005b, 2006b) pointed out that Nigeria with extensive mangrove ecosystem and over 14 million hectares of inland water surface out of which 1.7 million are available and suitable for fish farming, should not have any major challenge in achieving sufficient and sustainable fish output to meet domestic demand. Yet, in spite of the nations enormous fish potentials and increasing interests in the subsector (Tobor, 1990; Shimang, 2005; Welcome, 1979; Kapetsky, 1981), the gap between the demand for fish in Nigeria (1.3 million metric tonnes annually) and its supply from domestic production (about 0.45 metric tonnes annually) has continued to widen (Oluwasola & Ajayi, 2013). This study thus seeks to evaluate the socioeconomic factors influencing the profitability of catfish production in the city of Ibadan with a view to determining its effects on employment generation, poverty alleviation and enhanced nutrition among the urban population. Specifically, the paper described the socioeconomic characteristics of fish farmers in the city of Ibadan, analyzed costs and returns to catfish farming and determined the factors affecting the profitability of catfish production.

2. Methodology

The study was carried out in the city of Ibadan, Oyo State in the in the southwestern geopolitical zone of Nigeria. The city is located on the coordinates of longitude $3^{0}5$ ' East of the Greenwich Meridian and latitude $7^{0}23$' North of the Equator. Administratively, the city is divided into eleven local government areas (LGAs) and has a population of 1,338,659 (NPC, 2006). Ibadan is the largest traditional city south of the Sahara (UN-Habitat, 2001) and is connected by road, rail and air transport to several cities in the southwestern geopolitical zone. The city enjoys two climatic seasons -the rainy season which lasts from April to October and the dry season which lasts from November to March. The city experiences a mean annual rainfall of 1,262mm and a temperature range of between 22.5 °C and 27.5 °C (Oguntoyinbo, 1994).

Primary data were used for the study. A multistage sampling procedure was used to select respondents for the study. In the first stage, the city of Ibadan was purposively selected because of the large and ready market for a burgeoning fish farming enterprise. At the second stage, six of the eleven LGAs including: Egbeda, Ido, Ibadan North-West, Ona-Ara, Ibadan North and Ibadan North East were randomly selected while from a list of catfish farmers obtained from the State Agricultural Development Programme (ADP), twenty catfish farmers were randomly selected from each LGA. In all, 120 catfish farmers were sampled for the study.

Data collected were analysed with descriptive statistics, budgetary techniques and the multiple regression model. Descriptive statistics including frequency counts, means and percentages were used to describe the socio-economic characteristics of the catfish farmers while the budgetary technique was used to determine the gross margin and profitability of catfish enterprises. The profit was estimated:

$$\pi_i = P_i Q_i - TCi \tag{1}$$

where,

π_i = net income (₦),

P_i = price per kg of fish produced (₦),

Q_i = Fish output (kg), and,

TCi = total costs of production (fixed cost {FC} plus variable cost {VC}) (₦).

Variable costs (VC) included in the analysis were expenditures on labour, feed, fingerlings, broodstock, fertilizer, lime, transportation, drugs and security. Items that could be used for more than a production cycle were classified as fixed costs (FC). These included nets, plastics and depreciation on pond construction, generator and pumping machines.

A multiple regression model was used to determine the factors affecting net income from catfish farming in the city of Ibadan. The model was implicitly specified as:

$$Y = f(X_1, X_2, X_3, X_4, X_5, X_6, X_7, \varepsilon_i) \quad (2)$$

where,

Y	=	Net income (₦ /ha)
X_1	=	Age of farmer (Years)
X_2	=	Family size
X_3	=	Farming experience (Years)
X_4	=	Size of fish pond (m^2)
X_5	=	Labour (man days)
X_6	=	Quantity of feed used (kg)
X_7	=	Educational level of respondent (Years)
ε_i	=	Error term

The implicit function was linearized and specified in a log-linear form (Joshi & Jha, 1992; Idumah, 2006) as:

$$LnY = \beta_0 + \beta_1 X_1 + \beta_2 X_2 + \dots\dots\dots\dots\dots + \beta_7 X_7 + \varepsilon_i \quad (3)$$

According to Greene (2005) and Gujarati (2003), the disturbance term is expected to fulfil all the assumptions of clssical regression model except that of homoscedasticity which breaks down when cross-sectional data are used (Ayanwale and Osotimehin, 2001). The technique of the ordinary least square (OLS) was then used to estimate the multiple regression equation. Rather than a straight line, geometric interpretation of the model involved a plane. The strength of the relationship between net income from catfish production and the specified explanatory variables was therefore measured by the R^2.

A priori expectations were for X_1, X_3, X_4, X_5, X_6 and X_7 to be positively associated with net income while X_2 could be positively or negatively associated with net income depending on whether the family is a production or consumption unit. The selection of the variables used in the regression model as well as the *a priori* expectations were based on the assumption that in traditional and near subsistent farming enterprises characterised by low resource inputs, age (X_1), family size (X_2), farming experience (X_3), labour (hired and family labour) (X_5) and educational level of respondents (X_7) are critical to output (Akinola & Adeyemo, 2008; Aihonsu, 2002; Oluwasola, 1996; Olayide et al., 1981) and the ability of farmers to take risks in adopting new innovations or technologies (Feder et al., 1985; Adesina & Zinnah, 1992;). Other studies on fish farming like (Olaoye et al., 2013; Ifejika et al., 2007; Olagunju et al., 2007) added other inputs such as feed (X_6) and pond size (X_4). The selection of these variables followed after these earlier studies. Three functional forms of the regression model - linear, semi-log and the double logarithm were fitted to the model but only the double logarithm model which provided the best fit and was in line with *a priori* expectation was selected and discussed.

3. Results and Discussions

3.1 Socio-economic Characteristics of Cat Fish Farmers

Table 1 shows that majority of the fish farmers (33.3%) aged between 41 and 50 years with an age range of between 23 and 65 years. The mean age of the farmers was 44.3±12.0 years. The fish farmers as indicated were in their prime age and hence, economically active (Aihonsu & Olatingiri, 2012; Fregene et al., 2011). Given necessary resources, these set of farmers could increase their productivity. They are also in the age where they could take risks that could increase output as well as income (Akinola & Adeyemo, 2008; Adesina & Zinnah, 1992; Feder et al., 1985). Eighty percent of the fish farmers were men while the remaining 20% were women. The dominance of men in fish production was also reported by Fregene et al. (2011). The dominance of the male gender could be due to a number of factors including the access to water points as access to land is through male lineage (Oluwasola, 1998) and is only by purchase that women can have access to such land resources. It could also be because of the drudgery of fish farming. However, the number of women involved is significant in a region where women are more involved in commerce than in production.

Table 1. Socio-economic characteristics of respondent farmers

Variable	Frequency	Percentage	Mean	Standard Deviation
Age (Years)				
≤	20	16.7		
31-40	29	24.2		
41-50	40	33.3	44.3	12
51-60	16	13.3		
≥61	15	12.5		
Gender				
Male	96	80.0		
Female	24	20.0		
Level of education				
Did not go to school	2	1.7		
Primary school education	1	0.8		
Secondary school education	23	19.2		
Tertiary education	94	78.3		
Marital Status				
Single	35	29.2		
Married	81	67.5		
Separated	4	3.3		
Family size				
≤5	59	49.2		
6 – 10	61	50.8	5.2	1.9
Farm Size (ha)				
≤ 0.5	110	91.7		
0.51 – 0.75	3	2.5		
0.76 - 1.00	6	5.0	0.3	0.2
≥ 1.01	1	0.8		
Farming experience of farmers (years)				
≤ 5.0	68	56.7		
5.1 – 10.0	35	29.2		
10.1 –15.0	5	4.2	6.9	6.5
15.1 - 20.0	7	5.8		
≥21	5	4.1		
Occupational status in fishing				
Primary occupation	53	44.2		
Secondary occupation	67	55.8		
Source of beginning capital				
Savings	69	57.5		
Family	30	25.0		
Friends	8	6.7		
Cooperatives	11	9.2		
Commercial bank	2	1.7		
Reasons for farming				
Subsistence	24	20.0		
Demand for fish	69	57.5		
Price	27	22.5		

Source: Field survey, 2014

As shown in Table 1, majority of the fish farmers were educated with as much as 78.3% having attended Universities, Polytechnics and Colleges of Education. Fregene et al. (2011), Fregene and Digun-Aweto (2008) also reported high levels of education among fish farmers in Oyo and Osun States of the country. Clearly, these set of farmers were educated farmers seeking to tap into the rich fish market to make financial gains. Only 1.7% of them did not go to school. Majority (67.5%) of them were married (Adewuyi et al., 2010) hence, could easily make use of family labour to perform critical farm tasks. It however also means that they will incur higher household expenditure that will reduce the quantum of income realizable from the enterprise. The mean family size was 5.2±1.9 while fish farmers were small operators with a mean farm size of 0.3±0.2 hectares. Fish farm sizes were small because most of the farmers had no access to credit funds to invest in the enterprise (Adeoti et al., 2011; Okorie, 1998; Olomola, 1990). Only 2.7% got loans from commercial banks while 9.2% got loans from cooperative societies to start the business. The remaining relied on their personal savings (57.5%), family members (25%) and friends (6.7%). About 86% of the farmers have been involved in fish farming for not more than 10 years with a mean farm experience of 6.9±6.5 years. Clearly, catfish farming is a recent phenomenon (Aihonsu & Olatingiri, 2012). Although all the respondents had other economic activities they were engaged in aside from fish farming, 55.8% were engaged in the business on part time basis as they saw it as a secondary occupation while the remaining said fish farming was their primary occupation even though they had other businesses. Clearly, fish farming provided opportunities for expanding livelihood choices in the city of Ibadan as fish farmers were also able to engage in other income earning activities after taking care of their fish farms. The push factors for engaging in fish farming as shown in Table 1 were threefold: the increasing demand for fish (57.5%); the high price catfish commands in the market (22.5%) and for subsistence (20%). Clearly, 80% of the fish farmers got involved for commercial reasons.

3.2 Profitability of Cat Fish Farming

The average income from cat fish enterprises in the city of Ibadan was ₦630,049(US$ 3,150.25) (Table 2) while the total average variable cost to enterprise was ₦432,528.75 (US$ 2,162.64) given a gross margin of ₦197,520.25 (US$ 987.60). The average total cost was ₦447,475.96 (US$2,237.38) giving a net income of ₦182, 573.04 (US$912.87). This amounts to a monthly income of about ₦15,214.42 which is less than the national minimum wage of ₦18,000; (US$76.07). Clearly, the high capital investment required and the low net income realizable is a disincentive to attracting young unemployed people. This is partly why in spite of the enormous opportunities in producing fish to meet the gap in supply, the fish subsector has not succeeded in attracting local investors. Only the employed with the necessary financial resources who needs to augment incomes from other economic activities to meet family expenditure are attracted. Table 2 clearly shows that the cost of feed for the fish constitute 79.2% of the total operating cost. Any policy and/or technical measure that substantially reduces the cost of feeding the fish will substantially increase farm income and hence profit. Labour cost constituted only 11.6% of operating cost. The low cost could have resulted from the use of family labour by majority of the fish farmers who were married with fairly large family sizes as shown in Table1.Other cost components were very low .

The financial ratios shows that the expense-structure ratio was 0.03 indicating that for every ₦100 spent on fish farms, ₦3 was incurred on fixed inputs while ₦97 was spent on variable inputs. This suggests that farmers can easily adjust to variations in market conditions as variable costs constitute the largest proportion of farm expenditure. However, it also implies that oscillations in the market price of variable inputs could impact gross margin realizable. Policies that will lead to a reduction in the costs of these inputs, particularly feed will significantly reduce the cost of production and make the enterprise profitable. The benefit - cost ratio of 1.41 suggests that every ₦100 invested in fish farming will yield an additional income of ₦41.

Table 2. Analysis of cost and returns to fish farming

Income/cost Items	Amount (₦)	Percentage in cost category
Revenue	***630,049.00***	
Variable Costs		
Labour	(50,141.67)	11.6
Feed	(342,250.00)	79.2
Fingerlings	(14,250.00)	3.3
Broodstock	(4,593.33)	1.1
Fertilizer	(2,139.17)	0.5
Lime	(758.33)	0.2
Drugs	(5,157.92)	1.2
Transportation	(11,475.00)	2.7
Security	(1,763.33)	0.4
Total variable costs	*(432,528.75)*	
Gross margin	***197,520.25***	
Fixed costs items		
Nets	(2,958.34)	19.8
Plastics	(1,108.42)	7.4
Depreciation	(10,880.45)	72.8
Total fixed cost	*(14,947.21)*	
Total Cost	*(447,475.96)*	
Net revenue	***182,573.04***	
Expense-Structure Ratio		0.03
Benefit-Cost Ratio		1.41

Source: Field Survey, 2014.

3.3 Factors Determining the Profitability of Cat Fish Farms

Equation 3 shows that all the independent variables with the exception of farming experience (X_3) were positively signed in conformity with *a priori* expectations. Farming experience (X_3) was not in conformity with *a priori* expectations as it was negatively signed. Three of the variables, farming experience of respondents (X_3), labour used (X_5) and quantity of feed used (X_6) were also statistically significant.

The experience of respondents in catfish farming (X_3) in contrast to *a priori* expectations was negatively signed although statistically significant. The coefficient of the variable indicates that a unit increase in the experience of catfish farmers will depress income by 13%. This result contradicts findings by Adesiyan and Idowu (2011) and Oluwasola, (2011). Three things could have accounted for this. First, as farmers age, their fish ponds also tend to become older and except remedial actions are taken to keep the fish dams from siltation, productivity could decrease as pond sizes also decrease. Secondly, as farmers age and their children are out of the homes for schools resulting in reduced family labour, the increasing need for labour decreases interest in fish farming. In addition reduced household size which also means a reduction in household expenditure all things being equal could lead to a reduction in efforts put into fish farming. Finally and more critically, when farmers do not find fish farming to be economically rewarding over time, their interest in the enterprise could wane thereby affecting output. The amount of labour used (X_5) was statistically significant with its coefficient indicating that an increase in labour used by a man-day will increase net income by 12.3% (Bamire et al., 2008). This is very important as labour is the most important input in smallholder farm business enterprises. The quantity of feed used (X_6) was also positively associated with net income from catfish farming (Oluwasola & Ajayi, 2013). The coefficient of the variable indicates that a kilogram increase in fish feed used will increase net income by 83.7%.

The adjusted coefficient of determination of R^2 value of 0.915 indicates that about 91.5% of the variability in the net income realized from cat fish farming in the city of Ibadan are determined by the explanatory variables specified in the model.

$$\ln Y = 4.799 + \underset{(0.079)}{0.089\ln X_1} + \underset{0.044)}{0.059\ln X_2} - \underset{(0.072)}{*0.133\ln X_3} + \underset{(0.024)}{0.022\ln X_4} + \underset{(0.050)}{**0.123\ln X_5} + \underset{(0.028)}{***0.837\ln X_6} + \underset{(0.167)}{0.210\ln X_7} \quad (4)$$

R^2 = 0.862

$\bar{R}^2$ = 0.854

F-value = 97.187

N.B. Figures in parenthesis are the standard errors

*significant at 10% level; ** significant at 5% level; ***significant at 1% level

4. Conclusion and Recommendation

The study has shown that catfish enterprises are recent means of livelihood among middle age and highly educated male urban dwellers who although have other means of livelihood, are engaged in catfish farming to augment household income and for subsistence. While supply of fish could enhance the nutrition of households engaged in catfish farming, it could also be targeted by policy to engage unemployed educated youths in the urban centres. Such farmers should be accessed to inputs of feed, credit and water points to enable them engage in the enterprise profitably. The net income realizable for engaging in catfish farming were however too low to attract the unemployed young, able bodied and educated urban dwellers to engage in the enterprise. A major way out of this challenge is to access potential farmers to cheaper feed to enhance the profitability of the enterprise. This could come in two main ways: stemming up research efforts to produce viable fish feed locally as a way out of the high costs of imported feeds or subsidizing fish feed. Nigeria produces only 13.4% of her fishmeal requirements and imports the remaining (Fagbenro & Adebayo, 2005). The major commercial fishmeal processing plant has since closed down because it cannot compete with imported fishmeal. In 2004, the cost of imported fishmeal ranged between US$870 and US$1, 350 per tonne, while the cost of locally produced fishmeal was US$1,500 per tonne (Fagbenro and Adebayo, 2005). Nigeria currently imports about 40,000 metric tonnes (44,444.44 tonnes) valued at US$ 60 million (Global Agricultural Information Network {GAIN}, 2011). Cheaper fish feed will also help retain experienced fish farmers as the net income realized from the enterprise will increase. It will also enhance the quantity as well as the quality of feeds given the fish for increased output and productivity.

The positive but significant association between labour used and net income shows the importance of labour to the production efforts of smallholder farmers. The low cost component of labour in catfish farm enterprises suggests that farmers depended more on family labour which is not sustainable in the long term as family members settle outside the family households. It also suggests that farming was on small scale and oriented mainly for subsistence with marketable surplus exchanged for cash. There is the need for policy measures to attract medium to large scale farmers into the catfish enterprise not only to increase output to bridge the supply gap but also to provide employment for the unemployed urban youths.

References

Adekoya, B. B., & Miller, J. W. (2004). Fish Cage Culture Potential in Nigeria: An Overview. *National Culture, Agricultural Focus, 1*(5), 10-18.

Adeoti, A., Cofie, O., & Idowu, O. (2011). Sustainability of Urban Agriculture in Ibadan, Nigeria: Gender Analysis of Livelihoods. In R. Adeyemo (ed.), *Urban Agriculture, Cities and Climate Change* (pp. 173-178). CuvillierVerlag, Gottingen.

Adesina, A. A., & Zinnah, M. M. (1992). Technology Characteristics, Farmers Perceptions and Adoption Decisions: A Tobit Model Application in Sierra Leone (pp. 297-311). Agricultural Economics, Elsevier Science Publishers, B.V., Amsterdam.

Adesiyan, A. T., & Idowu, E. O. (2011). Urban Agriculture and Household Food Security in Lagos Metropolis, Nigeria. In R. Adeyemo (ed.) *Urban Agriculture, Cities and Climate Change* (pp. 132-137). CuvillierVerlag, Gottingen.

Adewuyi, S. A., Phillip, B. B., Ayinde, I. A., & Akerele, D. (2010). Analysis of Profitability of Fish Farming in Ogun State, Nigeria. *Journal of Human Ecology, 31*(3), 179-184.

Aihonsu, J. O. Y. (2002). Comparative Economic Analysis of Upland and Swamp Rice Production Systems in Ogun State, Nigeria. Unpublished Ph.D Thesis. Department of Agricultural Economics, Obafemi Awolowo Universty, Ile-Ife, Nigeria.

Aihonsu, J. O. Y., & Olatingiri, O. A. (2012). Analysis of Socio-economic Characteristics for Policy Enactment; The Case of Artisanal Fisherfolks in Ibeju-Lekki Local Government Area of Lagos State, Nigeria. In *Refereed Conference Proceedings of the Nigerian Association of Agricultural Economists (NAAE)* pp107-112

Akinola, A. A., & Adeyemo, R. (2008). Adoption and Productivity of Improved Rice Varieties in Osun State, Nigeria. *Ife Journal of Agriculture, 23*(1), 104-116.

Anyawu-Akeredolu, B. (2005). Urban aquaculture in Nigeria. World Aquaculture Society. Annual Report. Retrieved from www.was.org/meetings/abstract data

Ayanwale, A. B., & Osotimehin, K. O. (2001). A Comparative Analysis of Savings Behaviour of Small-scale Farmers and Low Income Industrial Workers in Lagos State of Nigeria. *Ife Research Publication in Geography, 8*, 10-17.

Bamire, A. S., Akinola, A. A., Olubode-Awosola, O. O., & Oyewusi, G. (2008). Comparative Economic Analysis of Crop-Livestock Integration among Smallholder Maize Farmers in the Derived Savanna Zone of Southwestern Nigeria". *Ife Journal of Agriculture, 23*(1), 86-103.

CIA. (2014). World Factbook, 2014.

Eyo, A. A. (1992). Utilization of Freshwater Fish Species in Nigeria. In A. A. Eyo & A. M. Balogun (Eds.), Proceedings of the 10th Annual Conference of the Fisheries Society of Nigeria (pp. 32-38).

Fagbenro, O. A. (2005). Aquaculture in Nigeria: history, status and prospects. A report of FAO World Fish Centre Workshop. Cameroon.

Fagbenro, O. A., & Adebayo, O. T. (2005). A review of the animal and aquafeed in Nigeria. In *A synthesis of the formulated animal and aquafeed industry in Sub-Sahara Africa*. CIFA Occasional Paper No 26 P2C 36.

FAO. (1995). Yearbook Production. Rome: FAO. p. 45.

FAO. (2005b). Small Scale Aquaculture in Sub-Saharan Africa. Revisiting the Aquaculture Target Group Paradigm. Rome: FAO Fisheries.

FAO. (2006b). Fisheries Management in the Federal Republic of Nigeria. Retrieved from www.fao.org/fi/fep/en/NGA/body

Feder, G., Just, R. E.., & Zilberman, D. (1985). Adoption of Agricultural Innovations in Developing Countries: A Survey". *Economic Development and Cultural Change, 33*(2), 255-298. http://dx.doi.org/10.1086/451461

Federal Department of Fisheries (FDF). (2008). Fisheries Statistics of Nigeria. Fourth edition, 1995 – 2007, Nigeria (p. 48).

Fregene, T., Inyang, I., & Awolumote, S. (2011). Operational Attributes of Urban Aquaculture Systems in Ibadan Municipal, Oyo State, Nigeria. In R. Adeyemo (ed.), *Urban Agriculture, Cities and Climate Change*, (pp. 173-178). CuvillierVerlag, Gottingen.

Fregene, T., & Digun-Aweto, T. (2008). Use of Aquaculture Technologies in Osun State, Nigeria. *African Journal of Livestock Extension, 6*, 36-41.

Global Agricultural Information Network (GAIN) (2011) Exporter Guide for Nigeria. (2011). USDA, Foreign Agricultural Service, 24/6/2011.

Greene, W. H. (2005). Econometric Analysis. Pearson Education Publishers, India.

Gujarati, D. N. (2003). Basic Econometrics. Tata McGraw-Hill Publishing Company Limited, India.

Idumah, F. O. (2006). Productivity Differentials among Food Crop Farmers in the Niger Delta. Unpublished Ph.D Thesis, Department of Agricultural Economics, University of Ibadan, Ibadan.

Ifejika, P. I., Ayanda, J. O., & Sule, A. M. (2007). Socio-Economic Variables affecting Aquaculture Production Practices in Borgu Local Government Area of Niger State, Nigeria. *Journal of Agriculture and Social Research, 7*(2), 20- 9. http://dx.doi.org/10.4314/jasr.v7i2.2853

Jagger, P., & Pender, J. (2001). Markets, Marketing and Production Issues for Aquaculture in East Africa: The Case of Uganda. *ICIARM Quart, 24*, 42-51.

Joshi, P. K., & Jha, D. (1992). An Economic Enquiry into the Impact of Soil Alkalinity and Water Logging. *Indian Journal of Agricultural Economics, 47*(2), 195-204.

Kapetsky, J. M. (1981). Some Consideration for the Management of Coastal Lagoon and Estuarine Fisheries. FAO Technical (p. 218).

Komolafe, O. O., & Arawomo, G. A. O. (2007). Reproductive Strategy of *Oreochromisniloticus* in Opa Reservoir, Ile-Ife, Nigeria. *Rev. Biol. Trop., 55*, 595-602. http://dx.doi.org/10.15517/rbt.v55i2.6034

National Bureau of Statistics (NBS). (2012). The Nigeria Poverty Profile. NBS, Abuja.

National Bureau of Statistics (NBS). (2014a). The Nigerian poverty profile 2010, NBS, Abuja.

National Bureau of Statistics (NBS). (2014b). 2012 National baseline youth report, NBS/Federal Ministry of Labour and Productivity, Abuja.

National Population Commission. (2006). Nigeria's National Census, NPC, Abuja.

Oguntoyinbo, J. S. (1994). Climatic Characteristics. In M. O. Filani & F. O. Akintola (Eds.), Ibadan Region (pp. 37-48). Rex Charles Publications, Ibadan.

Okorie, A. (1998). Management of Risks and Defaults in Agricultural Lending. In M. A. Ijere & A. Okorie (Eds.), Readings in Agricultural Finance. Longman, Nig. PLC, Lagos.

Olagunju, F. I., Adesiyan, I. O., & Ezekiel, A. A. (2007). Economic viability of catfish production in Oyo – State. *Nigeria Journal of Human Ecology, 21*(2), 121-124.

Olaoye, O. J, Ashley-Dejo, S. S, Fakoya, E. O, Ikeweinwe, N. B, Alegbeleye, W. O., Ashaolu, F. O., & Adelaja, O. A. (2013). Assessment of Socio-Economic Analysis of Fish Farming in Oyo State, Nigeria. *Global Journal of Science Frontier Research Agriculture and Veterinary, 13*(9), 1-12.

Olayide, S. O., Eweka, J. A., & Bello-Osagie, V. E. (1981). Nigerian Small Farmers: Problems and Prospects in Integrated Rural Development (pp. 1-64). Caxton Press (West Africa) Limited, Ibadan.

Olomola, A. S. (1990). Loan Transaction Cost and Repayment Performance and Small scale Farmers in Ondo State, Nigeria. Unpublished Ph.D. thesis, Department of Agricultural Economics, University of Ibadan, Nigeria.

Oluwasola, O., & Ajayi, D. (2013). Socio-economic and Policy issues Determining Sustainable Fish farming in Nigeria. *International Journal of Livestock Production, 4*(1), 1-8. http://dx.doi.org/ 10.5897/IJLP12.013

Oluwasola, O. (1998). Women in Agriculture in Nigeria. In A. Sesay & A. I. Odebiyi (Eds.), Women in Society and Development (pp. 59-72). Dokun Publishing House, Ibadan, Nigeria.

Oluwasola, O. (2011). Peri-urban farming in Ibadan metropolis. In R. Adeyemo (ed.), *Urban Agriculture, Cities and Climate Change* (pp. 96-103). CuvillierVerlag, Gottingen, Germany.

Pretty, J. N., Morison, J. I. L., & Hine, R. E. (2003). Reducing Food Poverty by Increasing Agriculture Sustainability in Developing Countries. *Agric. Ecosyst. Environ., 95*, 217-234. http://dx.doi.org/10.1016/S0167-8809(02)00087-7

Shimang, G. N. (2005). Fisheries Development in Nigeria: Problems and Prospects. Paper presented at the Homestead Fish Farming Training for Serving and Retired Public Servants, Federal Ministry of Agriculture and Rural Development, Abuja, Nigeria.

This Day. (2015). *Nigeria's unemployment may rise by 2%.* Retrieved May 15, 2015, from www.thisdaylive.com

Tobor, J. G. (1990). The Fish Industry in Nigeria: Status and Potential for Self-sufficiency in Fish Production. *NIOMR Technical, 54*, 34.

UN-HABITAT. (2001). Cities in a Globalizing World: Global Report on Human Settlements". United Nations for Human Settlement (Habitat), Earth scan Publications Ltd., London.

Welcome, R. I. (1979). The Inland Fisheries of Africa. *CIFA Occasional, 7*, 77.

7

Effects of Short-term Tillage of a Long-term No-Till Land on Available N and P in Two Contrasting Soil Types

Miles Dyck[1], Sukhdev. S. Malhi[1,2], Marvin Nyborg & Dick Puurveen[1]

[1] Department of Renewable Resources, University of Alberta, Edmonton, AB, Canada

[2] Previous Address: Agriculture and Agri-Food Canada, Melfort, Saskatchewan, Canada

Correspondence: Miles Dyck, Department of Renewable Resources, University of Alberta, Edmonton, AB, Canada, E-mail: mdyck@ualberta.ca

Abstract

The effects of short-term (4 years) tillage (hereafter called reverse tillage [RT]) of land previously under long-term (29 or 30 years) no-till (NT), with straw management (straw removed [S_{Rem}] and straw retained [S_{Ret}]) and N fertilizer rate (0, 50 and 100 kg N ha^{-1} in S_{Ret}, and 0 kg N ha^{-1} in S_{Rem} plots) were determined in autumn 2011 on ammonium-N, nitrate-N and extractable P in the 0-7.5, 7.5-15 and 15-20 cm soil layers at Breton (Gray Luvisol [Typic Cryoboralf] loam) and Ellerslie (Black Chernozem [Albic Argicryoll] loam), Alberta, Canada. There was no significant effect of RT and straw on ammonium-N, nitrate-N and extractable P in soil. Ammonium-N in soil increased significantly (but small) with N rate in many cases at both sites. Nitrate-N in soil increased with increasing N rate from 0 to 100 kg N ha^{-1} rate at Ellerslie, and up to 50 kg N ha^{-1} rate at Breton. Etractable P in soil decreased markedly with increasing N rate up to 100 kg N ha^{-1} at Breton and up to 50 kg N ha^{-1} at Ellerslie. In summary, increased N fertilizer rates were usually associated with decreased extractable P and increased nitrate-N in soil, but RT and straw had no effect on these nutrients in soil.

Keywords: dynamic organic fractions, N fertilizer, organic C and N, pre-seeding tillage, reverse tillage, soil quality, straw management

1. Introduction

In the Prairie Provinces of Canada, farmers traditionally used tillage for seedbed preparation to facilitate seeding operations, seed germination and weed control. Previous research has shown that tillage, especially when combined with summer fallow, can lead to deterioration of soil quality by diminishing organic C, aggregation, water storing capacity and nutrient supplying power, and can increase the potential for soil erosion (McGill et al., 1988; Lal et al., 1990). In the past several decades, no-tillage (NT, also called zero-tillage, or direct seeding) has gained popularity, because of its potential to enhance storage of water in soil by proper snow management/conservation, improvement in soil fertility/productivity, aggregation/tilth, lower requirement for labour, fuel and machinery, and increase in economic returns compared to conventional tillage (CT) (Selles et al., 1984; Malhi et al., 1988; H. Singh & K. P. Singh, 1994; Larney et al., 1997). However, long-term NT can increase: 1) infestation of perennial grassy weeds, such as quackgrass (Malhi et al., 1988; Campbell et al., 1990; Tessier et al., 1990); 2) stratification of nutrients, especially P, in the surface soil (Grant and Lafond, 1994; Crozier et al., 1999; Lupway et al., 2006; Baan et al., 2009; Jones et al., 2009); 3) emergence of herbicide-resistant perennial weeds (Donald, 1990; Derksen et al., 1993); 4) relatively cool and wet surface soil in spring (Johnson & Lowery, 1985; Jones et al., 1985; Malhi & O'Sullivan, 1990; Wolfe & Eckert, 1999) which can delay seeding and slow crop emergence/early growth, and increase in the potential of nutrient loss in surface water run-off which may cause pollution of water bodies (Ferguson et al., 1996). In addition, in areas where large amounts of crop residue/straw are produced and accumulate on the soil surface, it may be difficult to place the crop seeds properly into the soil, resulting in poor/sporadic germination, especially when proper direct-seeding drills are not available to facilitate seeding.

Therefore, despite the advantages of long-term no-till, producers may wish to introduce tillage occasionally for the purposes of weed, disease and residue control (Irvine et al., 2003). In fact, there is evidence to suggest that "strategic" tillage does not have any long-term negative impacts to crop yields, weed and disease populations or soil quality (Irvine et al., 2003). However, there is still very limited research on the effects of tillage of previously long-term NT soil on crop yield and nutrient uptake and soil fertility on farm fields in Canada (Campbell et., 1988, 1996; Baan et al., 2009), especially in the Parkland region of western Canada where large quantities of crop residue are produced and left on land after harvest. The objectives of this study were to examine the effects of short-term (2 or 3 years) pre-seeding shallow tillage (hereafter called reverse tillage – RT) of land previously under long-term (29 or 30 years) NT on crop yield and nutrient uptake [seed and straw yield, N and P uptake in seed and straw of annual cereals/oilseeds]; soil quality [total organic C (TOC) and N (TON), light fraction organic C (LFOC) and N (LFON) and mineralizable N (N_{min}) in the 0-7.5 and 7.5-15 cm soil layers]; and soil fertility [nitrate-N, ammonium-N and extractable P, in the 0-7.5, 7.5-15 and 15-20 cm soil layers] in two contrasting soil types (a Gray Luvisol soil at Breton and a Black Chernozem [Albic Argicryoll] soil at Ellerslie, Alberta, Canada.). This paper discusses the effects of RT on the amounts of available N and P in soil.

2. Methods

2.1 Location and Experimentation

The field experiments were conducted at Breton (53°07'N, 114°28'W; elevation 830 m) and Ellerslie (53°25'N, 113°33'W; elevation 692 m), Alberta, Canada. The Breton area belongs to ecoregion Peace Lowland/Boreal Transition with a rolling landscape. The soil was an Orthic Gray Luvisol (Typic Cryoboralf), with loam texture, pH of 6.6 and initial total C concentration of 13.75 g C kg^{-1}. The mean annual precipitation of the area is 475 mm and the growing season is from May to August. Approximately 60% of the total precipitation occurs in the growing season. This area has 2356 growing degree days (GDD) at >0 °C and 1335 GDD at >5 °C, 118 days frost free period, mean growing season precipitation 335 mm (range of 182 to 514 mm) and a growing season mean temperature of 14 °C (7 °C to 20 °C). The Ellerslie area belongs to the Aspen Parkland ecological region, which is characterized by a flat glacio-lacustrine landscape. The soil is a Black Chernozem (Albic Argicryoll), with loam texture, pH 6.0 and initial total organic C concentration of 56.45 g C kg^{-1}. The mean annual precipitation of the area is about 450 mm and the growing season is from May to August. Approximately 60% of the total precipitation occurs in the growing season (335 mm with a range of 190 to 440 mm). This area has growing degree days (GDD) of 2419 at >0 °C and GDD of 1402 at >5 °C, a 120 day frost free period, and a mean daily temperature of 14 °C (8 °C to 21 °C) in the growing season.

Table 1. Description of treatments sampled in autumn 2011 at Breton (Gray Luvisol) and Ellerslie (Black Chernozem), Alberta, Canada (experiments established in autumn 1979)

Treatment				
No.	ID	Tillage system	Straw management	Rate of N (kg N ha^{-1})
1a	$NTS_{Rem}0$	No-tillage	Straw removed	0
1b	$RTS_{Rem}0$	Reverse tillage	Straw removed	0
4a	$NTS_{Ret}0$	No-tillage	Straw retained	0
4b	$RTS_{Ret}0$	Reverse tillage	Straw retained	0
3a	$NTS_{Ret}50$	No-tillage	Straw retained	50
3b	$RTS_{Ret}50$	Reverse tillage	Straw retained	50
6a	$NTS_{Ret}100$	No-tillage	Straw retained	100
6b	$RTS_{Ret}100$	Reverse tillage	Straw retained	100

The original experiments with NT and CT with various straw management (removed [S_{Rem}] and retained [S_{Ret}]) and N rate (0, 50 and 100 kg N ha^{-1}) treatments were established in autumn 1979 (with the first growing season in 1980) at the University of Alberta Experimental Farms Breton and Ellerslie, Alberta to examine the influence of fertilizer N-straw management-tillage interactions on crop productivity and soil properties (Malhi et al. 2010, 2011a, 2011b). The treatments were arranged in a randomized complete block design in four replications. Initially, the plots were planted to a continuous barley (*Hordeum vulgare* L.) rotation with two tillage systems (conventional tillage [CT] and zero tillage [ZT]), two straw treatments (straw removed [S_{Rem}] and straw retained [S_{Ret}]) and two N treatments (0 and 56 kg N ha^{-1}) (Nyborg et al., 1995). However in 1991, another N treatment was added (100 kg N ha^{-1}) and the 56 kg N ha^{-1} was reduced to 50 kg N ha^{-1} (Table 1). At the same time, the barley rotation was changed to spring wheat (*Triticum aestivum* L.). In 1997, the rotation was then changed to a barley-wheat-canola (*Brassica napus* L.) rotation. Individual plots were 2.8 m × 6.9 m. Plots under CT were tilled twice, once in the autumn and once in the spring, with a chisel cultivator followed by a coil packer. The ZT plots did not undergo any disturbance, except for seeding drill. The original NT and CT treatments were seeded to various crops in rotations from 1980 to 2010 growing seasons at Breton and from 1980 to 2009 growing seasons at Ellerslie. The NT treatments were split into two (in spring 2010 at Breton and in spring 2009 at Ellerslie) to include: 1) Control NT treatment with no pre-seeding tillage; and 2) Tilled treatment with one tillage using rotary tiller in spring prior to seeding (hereafter called reverse tillage [RT]). The crop was harvested every year from 1980 to 2012 for seed, straw or whole plant yield. In autumn 2011, soil samples were taken from each plot in 8 treatments (Table 1) after crop harvest. The soil samples were then analyzed for available N and P.

2.2 Soil Sampling and Sample Preparation

Soil cores from 8 locations in each plot were collected from the 0-7.5, 7.5-15 and 15-20 cm layers using a 2.4 cm diameter coring tube. Bulk density of the soil was determined by the core method (Culley 1993). The soil samples were air dried at room temperature after removing any coarse roots and easily detectable crop residues, and ground to pass a 2-mm sieve.

2.3 Chemical Analysis

The ground soil samples were analysed for ammonium-N (NH_4-N), nitrate-N (NO_3-N), extractable P (phosphate-P - PO_4-P). Nitrate-N and ammonium-N were extracted using 1:5 soil: 2M KCl solution and their concentrations in extracts determined with a Technicon Autoanalyzer II (Tecnicon Industrial Systems 1973a, 1973b). Phosphorus was extracted using Kelowna extract (Qian et al., 1994) and measured colorimetrically on a Technicon Autoanalyzer (Tecnicon Industrial Systems, 1977).

2.4 Statistical Analysis

The data on ammonium-N, nitrate-N and extractable P were subjected to analysis of variance (ANOVA) using procedures as outlined in SAS (SAS Institute Inc, 2004). Significant ($p \leq 0.05$, or $p \leq 0.10$ when $p \leq 0.05$ not significant) differences between each treatment were determined using LSmeans (Proc GLM, SAS 6.1 for windows). Least significant difference test ($LSD_{0.05}$, or $LSD_{0.10}$ when $LSD_{0.05}$ not significant) for various parameters was used to compare all treatment means, tillage systems (NT vs. RT), straw management (S_{Rem} vs. S_{Ret} for the zero-N treatment) and N fertilizer rates (0, 50 and 100 kg N ha^{-1}).

3. Results

At both sites, there were some significant interaction of tillage, straw and N rate factors on the amount of ammonium-N in soil, but these differences were very small (1 - 4 kg NH_4-N ha^{-1}), and individual factors showed some significant effects on soil ammonium in some depths (Table 2). Averaged over straw and N treatments, ammonium in RT plots was marginally lower ($P<0.10$) than NT for the 0-7.5 cm depth. Averaged over tillage and N rates, ammonium in the S_{ret} plots were significantly lower compared to S_{rem} plots in the 7.5-15 cm depth. The amount of ammonium-N in soil increased significantly (but small) with N rate in many cases at both sites. Ammonium-N in the surface 20 cm ranged from 8.0 to 12.2 kg N ha^{-1} at both sites, suggesting that soil ammonium-N was not very sensitive to fertilizer applications or soil type.

Nitrate-N in different soil layers were relatively low (even after 32 years of N fertilization) at both sites, particularly at Breton where nitrate-N in the 15-20 cm soil layer was extremely low (Table 3). At Ellerslie, however, there were significant straw, tillage, N rate interactions with respect to soil nitrate in all soil layers, and only marginally significant interactions ($P<0.10$) in the 7.5-15cm layer at Breton. The majority of the interaction among treatments at Ellerslie appears to occur in the straw treatments at the 0 N rate. In the zero-N treatment at Ellerslie, compared to NT, nitrate-N in the 0-20 cm soil increased significantly with RT in the S_{Rem} treatment (7.8 kg N ha^{-1} vs. 14.8 kg N ha^{-1}), but there was no effect of RT in the S_{Ret} treatment. When soil nitrate was averaged across all N rates and straw treatments, there was no significant difference in between RT and NT at both sites. Averaged across two tillage systems, S_{Rem} treatments tended to have greater nitrate-N in soil than the S_{Ret} treatment at Ellerslie. At Breton, RT and S_{Ret} treatments had no positive significant effect on the amount of nitrate-N in any soil layer compared to their corresponding treatments. In the 0-20 cm soil, nitrate-N increased with increasing N rate from 8.4 kg N ha^{-1} 0 kg N ha^{-1} to 25.5 kg N ha^{-1} at 100 kg N ha^{-1} rate at Ellerslie. Nitrate-N in soil also increased with N fertilization up to 50 kg N ha^{-1} rate at Breton but the amounts of nitrate-N in soil were very low (i.e., from 2.0 kg N ha^{-1} at 0 kg N ha^{-1} to 3.8 kg N ha^{-1} 50 kg N ha^{-1}).

Significant interactions between tillage, straw and N rate factors were observed with respect to extractable P from the 0-7.5 cm and 0-20 cm layers at Breton and Ellerslie (Table 4). Extractable P in these two layers tended to decrease with increasing N rate and TR. The TR effects are most apparent in the 0-7.5 soil layer and less apparent in 0-20 cm layer Extractable P in the 0-20 cm soil decreased markedly with increasing N rate up to 100 kg N ha^{-1} at Breton (from 83.2 kg P ha^{-1} at 0 kg N ha^{-1} to 52.6 kg P ha^{-1} at 100 kg N ha^{-1}) and up to 50 kg N ha^{-1} at Ellerslie (from 81.5 kg P ha^{-1} at 0 kg N ha^{-1} to 48.6 kg P ha^{-1} at 50 kg N ha^{-1}) (Table 4). With respect to straw, extractable P in the 0-7.5 and 0-20 cm layers was lower in the S_{rem} treatments at Breton, regardless of tillage. Averaged across two tillage systems at zero-N rate, extractable P in the 0-20 cm soil was 83.2 kg P ha^{-1} (S_{ret}) vs. 76.8 kg P ha^{-1} (S_{rem}) at Breton. At Ellerslie, the straw treatments showed a stronger interaction with tillage. For example, in the 0-20 cm soil depth, compared to NT, RT had much greater extractable P in the S_{Rem} treatment at Ellerslie (76.8 kg P ha^{-1} vs. 81.6 kg P ha^{-1}) and much lower extractable P in the S_{Ret} (87.8 kg P ha^{-1} vs. 75.3 kg P ha^{-1}).

Table 2. Effect of long-term tillage, straw and N rate on ammonium-N in soil in autumn 2011 at Breton ((Gray Luvisol soil) and Ellerslie (Black Chernozem soil), Alberta, Canada (experiments established in autumn, 1979)

Treatment[z] (tillage/straw/kg N ha^{-1})	Ammonium-N (kg N ha^{-1}) in soil layers (cm) at Breton				Ammonium-N (kg N ha^{-1}) in soil layers (cm) at Ellerslie			
	0-7.5	7.5-15	15-20	0-20	0-7.5	7.5-15	15-20	0-20
Treatment mean								
$NTS_{Rem}0$	3.4	3.6	2.6	9.6	3.4	3.7	2.4	9.5
$RTS_{Rem}0$	2.9	3.7	2.9	9.5	3.6	3.6	2.7	9.9
$NTS_{Ret}0$	3.3	3.3	2.5	9.1	3.3	3.6	2.4	9.3
$RTS_{Ret}0$	2.7	3.1	2.2	8.0	3.2	3.4	2.3	8.9
$NTS_{Ret}50$	4.1	3.7	2.6	10.4	4.2	4.0	2.8	11.0
$RTS_{Ret}50$	3.3	4.0	2.4	9.7	3.5	4.1	2.5	10.1
$NTS_{Ret}100$	4.5	3.5	2.4	10.4	4.8	5.0	2.4	12.2
$RTS_{Ret}100$	4.1	3.8	2.7	10.6	4.3	4.4	2.5	11.2
$LSD_{0.05}$	0.8	0.9	0.6	1.9	0.7	0.7	0.8	1.5
SEM (Prob)	0.26**	0.31^{ns}	0.20^{ns}	0.63^{ns}	0.24***	0.25**	0.28^{ns}	0.51**
Tillage effect								
NT mean	3.8	3.5	2.5	9.8	3.9	4.0	2.5	10.4
RT mean	3.3	3.7	2.5	9.5	3.7	3.9	2.5	10.1
$LSD_{0.05}$	0.6	0.4	0.3	1.0	0.5	0.5	0.4	1.0
SEM (Prob)	0.20•	0.15^{ns}	0.11^{ns}	0.36^{ns}	0.18^{ns}	0.17^{ns}	0.13^{ns}	0.36^{ns}
Straw effect at 0 N								
S_{Rem}mean	3.2	3.6	2.7	9.5	3.5	3.6	2.6	9.7
S_{Ret} mean	3.0	3.2	2.4	8.6	3.3	3.5	2.3	9.1
$LSD_{0.05}$	0.6	0.2	0.5	1.0	0.6	0.3	0.5	0.9
SEM (Prob)	0.18^{ns}	0.06**	0.16^{ns}	0.30*	0.19^{ns}	0.10^{ns}	0.16^{ns}	0.29^{ns}
N rate effect with S_{Ret}								
0 mean	3.0	3.2	2.4	8.6	3.3	3.5	2.3	9.1
50 mean	3.7	3.8	2.5	10.0	3.8	4.0	2.7	10.5
100 mean	4.3	3.7	2.5	10.5	4.6	4.7	2.5	11.8
$LSD_{0.05}$	0.7	0.6	0.4	1.4	0.5	0.6	0.4	1.1
SEM (Prob)	0.22**	0.20^{ns}	0.14^{ns}	0.45*	0.16***	0.19**	0.14^{ns}	0.36***

[z]NT = No-tillage, RT = Reverse tillage, S_{Rem} = Straw removed, S_{Ret} = Straw retained, 0, 50 and 100 kg N ha^{-1}.

•, *, **, *** and ns refer to significant treatment effects in ANOVA at $P \leq 0.10$, $P \leq 0.05$, $P \leq 0.01$, $P \leq 0.001$ and not significant, respectively.

4. Discussion

Despite some significant statistical interactions, the effects of RT, straw management and N fertilization on soil ammonium-N were small, as also found in our previous studies on these soils (Malhi et al., 2010, 2011b). This suggests that ammonium-N is quickly converted to nitrate soon after ammonium-N is mineralized from the soil organic matter, resulting in little or no build-up of ammonium-N in soil over time under normal soil-climatic conditions.

Table 3. Effect of long-term tillage, straw and N rate on nitrate-N in soil in autumn 2011 at Breton ((Gray Luvisol soil) and Ellerslie (Black Chernozem soil), Alberta, Canada (experiments established in autumn, 1979)

Treatment[z] (tillage/straw/kg N ha^{-1})	Nitrate-N (kg N ha^{-1}) in soil layers (cm) at Breton				Nitrate-N (kg N ha^{-1}) in soil layers (cm) at Ellerslie			
	0-7.5	7.5-15	15-20	0-20	0-7.5	7.5-15	15-20	0-20
Treatment mean								
$NTS_{Rem}0$	1.9	0.6	0.3	2.8	4.3	2.2	1.3	7.8
$RTS_{Rem}0$	1.5	0.9	0.6	3.0	7.6	5.2	2.0	14.8
$NTS_{Ret}0$	0.9	0.5	0.3	1.7	4.9	3.1	1.2	9.2
$RTS_{Ret}0$	1.1	0.7	0.5	2.3	4.0	2.5	1.1	7.6
$NTS_{Ret}50$	1.9	1.0	0.5	3.4	7.6	4.2	1.8	13.6
$RTS_{Ret}50$	2.3	1.4	0.6	4.3	6.2	5.1	2.9	14.2
$NTS_{Ret}100$	1.7	0.8	1.4	3.9	16.8	7.7	3.5	28.0
$RTS_{Ret}100$	1.9	1.2	0.6	3.7	14.1	6.6	2.5	23.2
$LSD_{0.05}$	1.3	0.5	0.8	2.1	5.5	2.8	1.8	9.0
SEM (Prob)	0.42^{ns}	0.17*	0.26^{ns}	0.69^{ns}	1.89***	0.97**	$0.61^{\bullet}$	3.04***
Tillage effect								
NT mean	1.6	0.7	0.6	2.9	8.4	4.3	2.0	14.7
RT mean	1.7	1.0	0.5	3.2	8.0	4.9	2.1	15.0
$LSD_{0.05}$	0.7	0.3	0.4	1.1	4.2	1.9	1.0	6.7
SEM (Prob)	0.22^{ns}	0.10*	0.15^{ns}	0.38^{ns}	1.46^{ns}	0.65^{ns}	0.35^{ns}	2.30^{ns}
Straw effect at 0 N								
S_{Rem}mean	1.7	0.8	0.4	2.9	5.9	3.7	1.7	11.3
S_{Ret} mean	1.0	0.6	0.4	2.0	4.4	2.8	1.2	8.4
$LSD_{0.05}$	0.7	0.3	0.2	0.8	2.8	1.9	0.7	5.3
SEM (Prob)	0.22*	0.10^{ns}	0.07^{ns}	0.24*	0.90^{ns}	0.62^{ns}	0.22^{ns}	1.70^{ns}
N rate effect with S_{Ret}								
0 mean	1.0	0.6	0.4	2.0	4.4	2.8	1.2	8.4
50 mean	2.1	1.2	0.5	3.8	6.9	4.6	2.4	13.9
100 mean	1.8	1.0	1.0	3.8	15.4	7.1	3.0	25.5
$LSD_{0.05}$	0.6	0.4	0.7	1.3	4.4	2.2	1.4	7.0
SEM (Prob)	0.21**	0.13*	0.22^{ns}	0.43*	1.47***	0.72**	0.47*	2.35***

[z]NT = No-tillage, RT = Reverse tillage, S_{Rem} = Straw removed, S_{Ret} = Straw retained, 0, 50 and 100 kg N ha^{-1}.

•, *, **, *** and ns refer to significant treatment effects in ANOVA at $P \leq 0.10$, $P \leq 0.05$, $P \leq 0.01$, $P \leq 0.001$ and not significant, respectively.

Tillage (CT or RT) is expected to increase the amount of nitrate-N soil by increasing mineralization of N from soil organic matter compared to NT (Malhi & Nyborg, 1987; Nyborg & Malhi, 1989). However, in our study, RT and S_{Ret} did not show any significant effect on residual nitrate-N in soil at both sites compared to the corresponding NT and S_{Rem} treatments after 32 years. Similarly, earlier studies in Saskatchewan (Nuttall et al., 1986) and Alberta (Malhi et al., 2006, 2010; Malhi & Lemke, 2007) have also shown no effect of tillage or straw management on residual soil nitrate-N. This lack of difference in residual nitrate-N in soil due to RT or S_{Ret} in this study could be due to uptake of all mineralized nitrate-N by crops in the growing seasons, resulting in little or no additional residual nitrate-N in soil after harvest compared to NT, as evidenced by greater crop yield and N uptake with RT and S_{Ret} than corresponding NT and S_{Rem} treatments (Malhi et al., 2011b, 2015).

Previous studies have shown significant increases in residual nitrate-N in soil after long-term applications of N fertilizer, especially at high N rates (Malhi et al., 1991, 2002, 2009; Guillard et al., 1995). Similarly, in our study, there was an increase in residual nitrate-N in the 0-20 cm soil depth due to N fertilization, but the amounts were relatively small particularly at Breton, even after 32 annual applications of N fertilizer, as also suggested in our previous report (Malhi et al., 2010). There are several possible reasons for these trends. For example, the rate of N in the present study was near or below the rate needed for optimum yield in this soil-climatic region (Mooleki et al., 2010). Also, a portion of the applied N, especially under NT and when straw was retained, may have been immobilized into the soil organic N pool (Malhi et al., 1996). In addition, it is also possible that a portion of the residual soil nitrate-N may have been lost as gaseous N after occasional heavy rainfalls that often occur during summer and/or autumn in this region (Meek et al., 1982; Lessard et al., 1996). The N loss may also have been due to leaching but there was no concrete evidence of leaching because soils in this study were sampled to only 20 cm depth. However, in our previous publication (Malhi et al. 2010) there was an evidence of downward movement of nitrate-N in the soil profile to 90 cm depth of soil sampled in spring 2007. In any case, any nitrate-N leaching below the root zone can be an economic loss to producers, as well as further deep leaching over time can contaminate groundwater and become a potential risk to water quality (Zhang et al., 1996; Vasconcelos et al., 1997; Yuan et al., 2000). The findings suggest the need for deep soil sampling up to 3 or 4 m depth in order to make valid conclusions related to nitrate leaching losses in this on-going study.

Table 4. Effect of long-term tillage, straw and N rate on extractable P in soil in autumn 2011 at Breton ((Gray Luvisol soil) and Ellerslie (Black Chernozem soil), Alberta, Canada (experiments established in autumn, 1979)

Treatment[z] (tillage/straw/kg N ha^{-1})	Extractable P (kg P ha^{-1}) in soil layers (cm) at Breton				Extractable P (kg P ha^{-1}) in soil layers (cm) at Ellerslie			
	0-7.5	7.5-15	15-20	0-20	0-7.5	7.5-15	15-20	0-20
Treatment mean								
$NTS_{Rem}0$	57.4	15.7	2.8	75.9	54.5	18.6	1.7	76.8
$RTS_{Rem}0$	53.0	16.4	8.2	77.6	50.3	28.6	2.7	81.6
$NTS_{Ret}0$	66.9	13.6	3.1	83.6	63.2	21.8	2.8	87.8
$RTS_{Ret}0$	53.9	25.4	3.5	82.8	52.6	20.1	2.6	75.3
$NTS_{Ret}50$	43.7	11.7	3.1	58.5	37.6	12.6	2.3	52.5
$RTS_{Ret}50$	35.7	20.3	4.0	60.0	20.1	14.1	10.7	44.9
$NTS_{Ret}100$	39.4	10.7	2.6	52.7	36.0	14.1	2.4	52.5
$RTS_{Ret}100$	33.4	14.9	4.1	52.4	30.7	16.9	2.5	50.1
$LSD_{0.05}$	14.6	10.7	6.1	19.0	11.9	13.9	9.2	21.5
SEM (Prob)	4.82**	3.53^{ns}	2.00^{ns}	6.28**	4.05***	3.87^{ns}	3.13^{ns}	7.31**
Tillage effect								
NT mean	51.8	12.9	2.9	67.6	47.8	16.8	2.8	67.4
RT mean	44.0	19.3	4.9	68.2	38.4	19.9	4.6	62.9
$LSD_{0.05}$	11.2	5.3	2.8	14.0	11.2	7.1	4.5	15.4
SEM (Prob)	3.81^{ns}	1.78*	0.94^{ns}	4.74^{ns}	$3.86^{\bullet}$	2.44^{ns}	1.54^{ns}	5.32^{ns}
Straw effect at 0								

N								
S_{Rem}mean	55.2	16.1	5.5	76.8	52.4	23.6	3.2	79.2
S_{Ret} mean	60.4	19.5	3.3	83.2	57.9	20.9	2.7	81.5
$LSD_{0.05}$	15.7	9.6	6.5	14.8	8.5	10.9	0.7	14.8
SEM (Prob)	4.81^{ns}	2.98^{ns}	2.00^{ns}	4.52^{ns}	2.72^{ns}	3.53^{ns}	0.23^{ns}	4.76^{ns}
N rate effect with S_{Ret}								
0 mean	60.4	19.5	3.3	83.2	57.9	20.9	2.7	81.5
50 mean	39.7	16.0	3.5	59.2	28.8	13.3	6.5	48.6
100 mean	36.4	12.8	3.4	52.6	33.3	15.5	2.5	51.3
$LSD_{0.05}$	10.4	8.7	1.5	12.5	10.8	8.0	7.5	14.2
SEM (Prob)	3.40***	2.84^{ns}	0.50^{ns}	4.09***	3.64***	2.71^{ns}	2.52^{ns}	4.80***

^{z}NT = No-tillage, RT = Reverse tillage, S_{Rem} = Straw removed, S_{Ret} = Straw retained, 0, 50 and 100 kg N ha^{-1}.

•, *, **, *** and ns refer to significant treatment effects in ANOVA at $P \leq 0.10$, $P \leq 0.05$, $P \leq 0.01$, $P \leq 0.001$ and not significant, respectively.

Previous research under long-term no-tillage (NT) management has shown higher amounts of available P in the surface thin layer (0 to 5 cm or less) under NT than CT (Eckert 1985; Weil et al., 1988; Karlen, 1991; Ismail et al., 1994; Selles et al., 1997; Selles et al., 1999; Lupwayi et al., 2006, Mathers & Nash, 2009, Cade-Menun et al., 2010). Similarly, in our study, the amounts of extractable P in the surface 0-7.5 cm soil layer were greater under NT compared to RT, but the amounts of extractable P in the 7.5-15 and 15-20 cm layers were greater with RT compared to NT. The greater amounts of extractable p in the surface 0-7.5 cm layer under NT than RT were probably because of P "stratification", resulting in accumulation of more extractable P due to P application and from decomposition of crop residues retained on the soil surface under NT, as also suggested by Essington and Howard (2000). It is also possible that increased mixing by tillage under RT may have buried P-rich crop residue and soil into subsoil layers, resulting in increased mineralization of P from crop residue/soil organic matter, thus increasing extractable P content in the subsoil layers. Our findings suggest the importance of NT in maintaining higher concentration of readily plant-available P in soil near the surface where tillage is eliminated, and the need for additional P fertilization under NT because inorganic P at the soil surface under NT may not become fully available to the crop due to its relatively high immobility in the soil profile. Our findings also suggest the economic importance of occasional tillage (i.e., RT) in increasing the availability of extractable P which has accumulated at the soil surface under NT, by moving it into the subsoil layers where crop plant roots can intercept it.

The higher soil extractable P in S_{Ret} treatment than in S_{Rem} treatment in our study was probably due to the return of some P in crop residues in S_{Ret} treatment. This suggests the positive effect of returning crop residue in improving P fertility of soil and thus increasing potential for long-term sustainability of soil productivity, as most soils, especially Black Chernozem, in this region inherently contain insufficient amounts of available P for optimum crop growth and yield. Extractable P of soil in our study decreased significantly with application of N fertilizer, most likely due to the higher crop yield from N application, and subsequently higher P uptake in the seed which was removed from the field, leaving less residual P in soil.

In summary, our findings do not show any negative impact of occasional tillage of long-term NT on soil fertility.

Acknowledgements

The authors thank K Strukoff for technical help and D. Leach for statistical analysis.

References

Baan, C. D., Grevers, M. C. J., & Schoenau, J. J. (2009). Effects of a single cycle of tillage on long-term no-till prairie soils. *Canadian Journal of Soil Science,* 89, 521-530. http://dx.doi.org/10.4141/cjss08041

Cade-Menun, B., Carter, M. R., James, D. C., & Liu, C. W. (2010). Phosphorous forms and chemistry in the soil profile under long-term conservation tillage: A phosphorous-31 nuclear magnetic resonance study.

Journal of Environmental Quality, 39, 1647-1656. http://dx.doi.org/10.2134/jeq2009.0491

Campbell, C. A., McConkey, B. G., Zentner, R. P., Dyck, F. B., Selles, F., & Curtin, D. (1996). Carbon sequestration in Brown Chernozem as affected by tillage and rotation. *Canadian Journal of Soil Science, 75*, 449-458. http://dx.doi.org/10.4141/cjss95-065

Campbell, C. A., Tessier, J. S. J., & Selles, F. (1988). Challenges and limitations to adoption of conservation tillage – soil organic matter, fertility, moisture and soil environment. In: *Land Degradation and Conservation Tillage. Proc. 34th Annual CSSS/AIC Meeting, AIC, Calgary, Alberta, Canada* (pp. 140-185).

Campbell, C. A., Thomas, A. G., Biederbeck, V. O., McConkey, B. G., Selles, F., Spurr, D., & Zentner, R. P. (1998). Converting from no-tillage to pre-seeding tillage: influence on weeds, spring wheat grain yields and N, soil quality. *Soil & Tillage Research, 46*, 175-185. http://dx.doi.org/10.1016/S0167-1987(98)00031-2

Campbell, C. A., Zentner, R. P., Janzen, H. H., & Bowren, K. E. (1990). *Crop rotation studies on the Canadian Prairies*. Canadian Government Publication Centre, Supply and Services Canada, Hull, Quebec, Canada. Publ. No. 184/E, p. 133.

Crozier, C. R., Naderman, G. C., Tucker, M. R., & Sugg, R. E. (1999). Nutrient and pH stratification with conventional and no-till management. *Communications in Soil Science and Plant Analysis, 30*, 65-74. http://dx.doi.org/10.1080/00103629909370184

Culley, J. L. B. (1993). Density and compressibility. In M. R. Carter (Ed.), *Soil sampling and methods of analysis.* Lewis Publishers, Boca Raton, FL, U.S.A.

Derksen, D. A., Lafond, G. P., Thomas, A. G., Loeppky, H. A., & Swanton, C. J. (1993). Impact of agronomic practices on weed communities – tillage systems. *Weed Science, 41*, 409-417.

Donald, W.W. (1990). Primary tillage for Foxtail barley (*Hordeum jubatum* L.) control. *Weed Technology*, 4, 318-321. http://www.jstor.org/stable/3987080

Eckert, D. J. (1985). Effects of reduced tillage on the distribution of soil pH and nutrients in soil profiles. *Journal of Fertilizer Issues, 2*, 86-90.

Essington, M. F., & Howard, D. D. (2000). Phosphorus availability and speciation in long-term no-till and disk-till soil. *Soil Science, 165*, 144-152. http://dx.doi.org/10.1097/00010694-200002000-00005

Ferguson, A. J. D., Pearson, M. J., & Reynolds, C. S. (1996). Eutrofication of natural water and toxic algal blooms. *Issues in Environmental Science and Technology, 5*, 27-41. http://dx.doi.org/10.1039/9781847550088-00027

Grant, C. A., & Lafond, G. P. (1994). The effects of tillage systems and crop rotations on soil chemical properties of a Black Chernozemic soil. *Canadian Journal of Soil Science, 74*, 301-306. http://dx.doi.org/10.4141/cjss94-042

Guillard, K., Griffin, G. F., Allinson, D. W., Yamartino, W. R., Rafey, M. M., & Pietrzyk, W. (1995). Nitrogen utilization of selected cropping systems in the U.S. Northeast II: Soil profile nitrate distribution and accumulation. *Agronomy Journal, 87*, 199-207. http://dx.doi.org/10.2134/agronj1995.00021962008700020011x

Irvine, B., Derkson, D., Moulin, A., McLaren, D., & Grant, C. (2003). Strategic tillage – Evolution or devolution of zero tillage? Retrived from http://www.umanitoba.ca/faculties/afs/MAC_proceedings/2003/pdf/irvine_strategic_tillage.pdf

Ismail, I., Blevins, R. L., & Frye, W. W. (1994). Long-term no-tillage effects on soil properties and continuous corn yields. *Soil Science Society of America Journal, 58*, 193-198. http://dx.doi.org/10.2136/sssaj1994.03615995005800010028x

Johnson, M. D., & Lowery, B. (1985). Effect of three conservation tillage practices on soil temperature and thermal properties. *Soil Science Society of America Journal, 49*, 1547-1552. http://dx.doi.org/10.2136/sssaj1985.03615995004900060043x

Jones, O. R., Allen, R. R., & Unger, P. W. (1990). Tillage systems and equipment for dryland farming. In: R.P. Singh, F. Parr and B.A. Stewart (Eds.), *Dryland Agriculture: Strategies for Sustainability*. Vol. 13. (pp 89-125). Springer-Verlag, New York, NY, U.S.A. http://dx.doi.org/10.1007/978-1-4613-8982-8_5

Jones, C., Miller, P., Engel, R., & Rick, T. (2009). *Strategies to stretch your phosphorous fertilizer dollar* [PowerPoint slides]. Retrieved from http://landresources.montana.edu/soilfertility/documents/PDF/pres/P%20Management_Judith%20Basin%202009.pdf

Karlen, D. L., Berry, E. C., Colvin, T. S., & Kanwar, R. S. (1991). Twelve-year tillage and rotation effects on yields and soil chemical properties in northeast Iowa. *Communications in Soil Science and Plant Analysis, 22*, 1985-2003. http://dx.doi.org/10.1080/00103629109368552

Lal, R., Logan, T. J., & Fausey, N. R. (1990). Long-term tillage effects on a Mollic Ochraqualf in north-west Ohio III. Soil nutrient profile. *Soil & Tillage Research, 15*, 371-382. http://dx.doi.org/10.1016/0167-1987(90)90110-Y

Larney, F. J., Bremer, E., Janzen, H. H., Johnston, A. M., & Lindwall, C. W. (1997). Changes in total, mineralizable and light fraction soil organic matter with cropping and tillage intensities in semiarid southern Alberta, Canada. *Soil & Tillage Research, 42*, 229-240. http://dx.doi.org/10.1016/S0167-1987(97)00011-1

Lupwayi, N. Z., Clayton, G. W., O'Donovan, J. T., Harker, K. N., Turkington, T. K., & Soon, Y. K. (2006). Soil nutrient stratification and uptake by wheat after seven years of conventional and zero tillage in the northern grain belt of Canada. *Canadian Journal of Soil Science, 86*, 767-778. http://dx.doi.org/10.4141/S06-010

Lessard, R., Rochette, P., Gregorich, E. G., Patty, E., & Desjardins, R. L. (1996). N_2O fluxes from manure amended soil under maize. *Journal of Environmental Quality*, 25, 1371-1377. http://dx.doi.org/10.2134/jeq1996.00472425002500060029x

Malhi, S. S., & Lemke R. (2007). Tillage, crop residue and N fertilizer effects on crop yield, nutrient uptake, soil quality and greenhouse gas emissions in the second 4-yr rotation cycle. *Soil & Tillage Research, 96*, 269-283. http://dx.doi.org/10.1016/j.still.2007.06.011

Malhi, S. S., & Nyborg, M. (1987). Influence of tillage on nitrate-N in soil. *Canadex – Tillage/Soil Fertility, December 1987*, Communications Branch, Agriculture Canada, Ottawa, Canada.

Malhi, S. S., & O'Sullivan, P. A. (1990). Soil temperature, moisture and compaction under zero and conventional tillage in central Alberta. *Soil & Tillage Research, 17*, 167-172. http://dx.doi.org/10.1016/0167-1987(90)90014-5

Malhi, S. S., Brandt, S., Ulrich, D., Lemke, R., & Gill, K. S. (2002). Accumulation and distribution of nitrate-N and extractable P in the soil profile under various alternative cropping systems. *Journal of Plant Nutrition, 25*, 2499-2520. http://dx.doi.org/10.1081/PLN-120014709

Malhi, S. S., Brandt, S. A., Lemke, R., Moulin, A. P., & Zentner, R. P. (2009). Effects of input level and crop diversity on soil nitrate-N, extractable P, aggregation, organic C and N, and N and P balance in the Canadian Prairie. *Nutrient Cycling in Agroecosystems, 84*, 1-22. http://dx.doi.org/10.1007/s10705-008-9220-0

Dyck, M., Malhi, S. S., Nyborg, M., & Puurveen, D. (2015). Effects of short-term tillage of a long-term no-till land on crop yield and nutrient uptake in two contrasting soil types. *Sustainable Agriculture Research*, In Review.

Malhi, S. S., Harapiak, J. T., Nyborg, M., & Flore, N. A. (1991). Soil chemical properties after long-term N fertilization of bromegrass: Nitrogen rate. *Communications in Soil Science and Plant Analysis, 22*, 1447-1458. http://dx.doi.org/10.1080/00103629109368505

Malhi, S. S., Lemke, R., Wang, Z. H., & Chhabra, B. S. (2006). Influence of tillage and crop residue management on crop yield, greenhouse gas emissions and soil quality. *Soil & Tillage Research, 90*, 171-183. http://dx.doi.org/10.1016/j.still.2005.09.001

Malhi, S. S., Mumey, G., O'Sullivan, P. A., & Harker, K. N. (1988). An economic comparison of barley production under zero and conventional tillage. *Soil & Tillage Research, 11*, 159-166. http://dx.doi.org/10.1016/0167-1987(88)90023-2

Malhi, S. S., Nyborg, M., Goddard, T., & Puurveen, D. (2010). Long-term tillage, straw and N rate effects on quantity and quality of organic C and N in a Gray Luvisol soil. *Nutrient Cycling in Agroecosystems, 90*, 1-20. http://dx.doi.org/10.1007/s10705-010-9399-8

Malhi, S. S., Nyborg, M., Goddard, T., & Puurveen, D. (2011a). Long-term tillage, straw management and N fertilization effects on quantity and quality of organic C and N in a Black Chernozem soil. *Nutrient Cycling in Agroecosystems, 90*, 227-241. http://dx.doi.org/10.1007/s10705-011-9424-6

Malhi, S. S., Nyborg, M., Goddard, T., & Puurveen, D. (2011b). Long-term tillage, straw and N rate effects on some chemical properties in two contrasting soil types. *Nutrient Cycling in Agroecosystems, 90*, 133-146. http://dx.doi.org/10.1007/s10705-010-9417-x

Malhi, S. S., Nyborg, M., Goddard, T. & Puurveen, D. (2011c). Influence of long-term tillage, straw and N fertilizer management on crop yield, N uptake, and N balance sheet in two contrasting soil types. *Communications in Soil Science and Plant Analysis, 42*, 2548-2560. http://dx.doi.org/10.1080/00103624.2011.609262

Malhi, S. S., Nyborg, M., & Solberg, E. D. (1996). Influence of source, method of placement and simulated rainfall on the recovery of ^{15}N-labelled fertilizers under zero tillage. *Canadian Journal of Soil Science, 76*, 93-100. http://dx.doi.org/10.4141/cjss96-013

Mathers, N. J., & Nash, D. M. (2009). Effects of tillage practices on soil and water phosphorous and nitrogen fractions in a Chromosol at Rutherglen in Victoria, Australia. *Australian Journal of Soil Research, 47*, 46-59. http://dx.doi.org/10.1071/SR08106

McGill, W. B., Dormaar, J. F., & Reinl-Dwyer, E. (1988). New perspectives on soil organic matter quality, quantity, and dynamics on the Canadian Prairies. In *Proc. Canadian Society of Soil Science and Canadian Society of Extension Joint Symposium, Land degradation: Assessment and insight into a western Canadian problem. August 23, 1988, Calgary, Alberta, Canada* (pp. 30-48). Agricultural Institute of Canada.

Meek, B., Graham, L., & Donovan, T. (1982). Long-term effects of manure on soil nitrogen, phosphorus, potassium, sodium, organic matter and water infiltration rate. *Soil Science Society of America Journal, 44*, 1014-1019. http://dx.doi.org/10.2136/sssaj1982.03615995004600050025x

Mooleki, S. P., Malhi, S. S., Lemke, R., Schoenau, J. J., Lafond, G., Brandt, S., … May, W. E. (2010). Effect of nitrogen management, and phosphorus placement on wheat production in Saskatchewan. *Canadian Journal of Plant Science, 90*, 319-337. http://dx.doi.org/10.4141/CJPS09075

Nuttall, W. F., Bowren, K. E., & Campbell, C. A. (1986). Crop residue management practices, and N and P fertilizer effects on crop response and on some soil physical and chemical properties of a Black Chernozem over 25 years in a continuous wheat rotation. *Canadian Journal of Soil Science, 66*, 159-171. http://dx.doi.org/10.4141/cjss86-016

Nyborg, M., & Malhi, S. S. (1989). Effect of zero and conventional tillage on barley yield and nitrate-N content, moisture and temperature of soil in north-central Alberta. *Soil & Tillage Research, 15*, 1-9. http://dx.doi.org/10.1016/0167-1987(89)90059-7

Nyborg, M., Solberg, E. D., Malhi, S. S., & Izaurralde, R. C. (1995). Fertilizer N, crop residue, and tillage after soil C and N content in a decade. In R. Lal, J. Kimble, E. Levine & B. A. Stewart (Eds.), *Advances in Soil Science: Soil management and greenhouse effect.* (pp. 93-99). Lewis Publishers, CRC Press, Boca Raton, FL, U.S.A.

Qian, P., Schoenau, J. J., & Karamanos, R. E. (1994). Simultaneous extraction of phosphorus and potassium with a new soil test: A modified Kelowna extraction. *Communications in Soil Science and Plant Analysis, 25*, 627-635. http://dx.doi.org/10.1080/00103629409369068

SAS Institute. (2004). *SAS product documentation. Version 8*. SAS Inst., Cary, NC, U.S.A. Retrieved from http://support.sas.com/documentation/onlinedoc/index.html (verified 17 July 2009).

Selles, F., Karamanos, R. E., & Bowren, K. E. (1984). Changes in natural 15N abundance of soils associated with tillage practices. *Canadian Journal of Soil Science, 54*, 345-354. http://dx.doi.org/10.4141/cjss84-036

Selles, F., Kochhann, R. A., Denardin, J. E., Zentner, R. P., & Faganello, A. (1997). Distribution of phosphorous fractions in a Brazilian Oxisol under different tillage systems. *Soil & Tillage Research, 44*, 23-34. http://dx.doi.org/10.1016/S0167-1987(97)00026-3

Selles, F., McConkey, B. G., & Campbell, C. A. (1999). Distribution and forms of P under cultivator- and zero-tillage for continuous- and fallow-wheat cropping systems in the semi-arid Canadian Prairies. *Soil and Tillage Research, 51*, 47-59. http://dx.doi.org/10.1016/S0167-1987(99)00027-6

Singh, H., & Singh, K. P. (1994). Nitrogen and phosphorus availability and mineralization in dryland reduced tillage cultivation – effects of residue placement and chemical fertilizer. *Soil Biology & Biochemistry, 26*, 695-702. http://dx.doi.org/10.1016/0038-0717(94)90261-5

Technicon Industrial Systems. (1973a). *Ammonium in water and waste water. Industrial Method No. 90-70W-B. Revised January 1978*. Technicon Industrial Systems, Tarrytown, NY, U.S.A.

Technicon Industrial Systems. (1973b). *Nitrate in water and waste water. Industrial Method No. 100-70W-B. Revised January 1978*. Technicon Industrial Systems, Tarrytown, NY, U.S.A.

Technicon Industrial Systems. (1977). *Industrial/simultaneous determination of nitrogen and/or phosphorus in BD acid digests*. Industrial method no. 334-74W/Bt. Tarrytown, NY, U.S.A.

Tessier, S., Peru, M., Dyck, F. B., Zentner, R. P., & Campbell, C. A. (1990). Conservation tillage for spring wheat production in semiarid Saskatchewan. *Soil & Tillage Research, 18*, 73-79. http://dx.doi.org/10.1016/0167-1987(90)90094-T

Vanderbygaart, A. J., & Kay, B. D. (2004). Persistence of soil organic carbon after plowing a long-term no-till field in southern Ontario, Canada. *Soil Science Society of America Journal, 68*, 1394-1402. http://dx.doi.org/10.2136/sssaj2004.1394

Vasconcelos, E., Cabral, F., & Cordovil, C. M. D. (1997). Effect of solid phase from pig slurry on soil chemical characteristics, nitrate leaching composition and yield of wheat. *Journal of Plant Nutrition, 20*, 939-952. http://dx.doi.org/10.1080/01904169709365307

Weil, R. R., Benedetto, P. W., Sikora, L. J., & Bandel, V. A. (1988). Influence of tillage practices on phosphorus distribution and forms in three Ultisols. *Agronomy Journal, 80*, 503-509. http://dx.doi.org/10.2134/agronj1988.00021962008000030022x

Wolfe, A. M., & Eckert, D. J. (1999). Crop sequence and surface residue effects on the performance of no-till corn grown on a poorly drained soil. *Agronomy Journal, 91*, 363-367. http://dx.doi.org/10.2134/agronj1999.00021962009100030002x

Yang, X. M., & Kay, B. D. (2001). Rotation and tillage effects on soil organic carbon sequestration in a Typic Hapludalf in southern Ontario. *Soil & Tillage Research, 59*, 107-114. http://dx.doi.org/10.1016/S0167-1987(01)00162-3

Yuan, X., Tong, Y., Yang, X., Li, X., & Zhang, F. (2000). Effect of organic manure on soil nitrate accumulation. *Soil Environmental Science, 9*, 197-200.

Zhang, W. L., Tian, Z. X., Zhang, N., & Li, X. O. (1996). Nitrate pollution of ground water in northern China. *Agriculture, Ecosystems & Environment, 59*, 223-231. http://dx.doi.org/10.1016/0167-8809(96)01052-3

Zibilske, L. M., Bradford, J. M., & Smart, J. R. (2002). Conservation tillage induced changes in organic carbon, total nitrogen and available phosphorus in a semi-arid alkaline subtropical soil. *Soil & Tillage Research, 66*, 153-163. http://dx.doi.org/10.1016/S0167-1987(02)00023-5

8

Risk Analysis of Investments In-Milk Collection Centers

Kheiry Hassan M. Ishag[1]

[1] Dhofar Cattle Feed Company, P.O. Box 1220, PC 211, Sultanate of Oman

Correspondence: Kheiry Hassan M. Ishag, Dhofar Cattle Feed Company, P.O. Box 1220, PC 211, Sultanate of Oman. E-mail: kheiryishag@hotmail.com

Abstract

Milk marketing in rural area of Dhofar Region face a lot of difficulties and constrains by individual small scale farmers due to the lack of facilities and access to market. Therefore, farmers reduce their cow and camel milk production and group their animal into three or four groups to be milked in alternative days. Government Authorities decided to establish Milk Collection Centers to facilitate milk marketing and provide raw milk to Dairy industries. A risk analysis for the investment in milk collection centers on rural area of Dhofar Region was conducted in this study. The results showed that all MCC investments had a positive NPV except Shahbi Aseab Center. The study indicates there is a direct relationship between total milk collected, milk price, distant between MCC and Dairy plan and investment profitability. The study revealed an inverse relation between animal population at MCC zone and risk. The probability of achieving returns lower than the opportunity cost was highest for MCC located far from dairy plant which process and market dairy products. Risk premium for four MCC has been calculated relative to Salalah MCC (preferred location) and (Garoun Hirity MCC) was found as preferred MCC location and risk efficiency. In order to make the investment in MCC more attractive, the Government incentives need to be offered to farmers to increase milk production and improve raw milk quality. However, this approach might make investments in bulk milk collection centers feasible. Thus, a recommendable strategy for a successful modernization of the Oman dairy sectors inbound logistics would be to promote an increase in the volume of the milk produced per farm and improve marketing facilities through MCC.

Keywords: Milk Collection Centre MCC, NPV, risk analysis, SERF risk efficiency analysis, Cumulated Distribution Function analysis

1. Introduction

Dairy products are currently one of the main sources of income for a significant share of the Omani farmers at Dhofar Region. In 2014, Oman dairy market is estimated of 167 Million liters per year with annual growth of 6% and only 30% of the demand produced by local dairy plant based at Dhofar Region. Due to lack of facilities and access to market, farmers reduce their cow and camel milk production and group their animal into three or four groups to be milked in alternative days. Government Authorities decided to establish Milk Collection Centers to facilitate milk marketing and provide raw milk to Dairy industries.

Changes in the country's economic environment over the last decade such as market openings and economic stability etc. have made it crucial that the dairy business sector need to be modernized and become more competitive. Operating costs need to be reduced and raw material quality needs to be improved to achieve these goals. The milk collection systems needs to be implemented and being put into operation. Using these MCC systems, milk has to be collected and cooled to 6 ° at each MCC. The raw milk then collected daily by temperature controlled tank trucks to dairy plant for processing and marketing.

One of the probable consequences of the bulk milk collecting system is that it excludes remote areas and scattered small dairy farmers from the dairy business, since their output is considered insufficient to justify investment in MCC. As a contribution to the analysis of this issue, the present work assesses the risks involved in the installation and operation of five MCC at recommended location with appropriate animal population and density. The financial analysis of these investments was performed by (HVA International, 2010). However, this study did not consider risk analysis for this investment.

Volume of raw milk produced and collected is risk-dependent item since output depends on several factors, such as animal population at each location zone, climate & rainy season, costs, and capital investments. Milk production in Dhofar Region is increases during the summer (rainy season) and decreases during the winter (dry season). This seasonal change is due to change in available pasture and fodder crop as more pasture in the summer, less in the winter. Farm investments to expand the herd, improve breeding stock, control diseases, and install milk cooling systems are economic factors that can contribute to increase milk production and assure raw milk quality.

Monte Carlo Simulation models were used in this study to quantify risk and uncertainty in milk collection project at Dhofar Region. The quantitative risk analysis will provide decision and policy makers a means of NPV estimating the probability that the project NPV will fall below zero. The model will also help in improving dairy sector management policy and achieve project objectives simultaneously: sustaining dairy business for food security and preserving the associated natural environment. Danielle D. et al. (2000), use Monte Carlo simulations to estimate profitability of different size of milk cooling tank under risky environment. In this study government subsidy and incentives at five locations were compared by performing dynamic location model.

Using Monte Carlo Simulation dynamic model for project appraisal was addressed by Savvakis C. Savvides in (1994). He argued that this integrated analysis provided a range of outcomes that can reduce the risk of uncertainty and give more reliable results for investor.

Stochastic efficiency with respect to a function (SERF) analysis, which is used in this study, has been used previously to evaluate crop production systems using net return distributions based on empirical data (Barham, Robinson, Richardson, & Rister, 2011; Bryant et al., 2008; Hignight, Watkins, & Anders, 2010; Pendell, Williams, Boyles, Rice, & Nelson, 2007; Ribera, Hons, & Richardson, 2004; Williams, Pachta, Claassen, Roozeboom, & Llewelyn, 2011). SERF provides a more restrictive approach to compare risky alternative technologies by evaluating technology dominance across a wide range of plausible risk preferences.

To sum up, the analysis revealed that, although technologically and financially feasible, cooling tank investments are subject to uncertainties the effects of which can jeopardize efforts to modernize of milk collection system in rural area at Dhofar Region. These issues are discussed in depth in the following analysis.

2. Materials and Methods

Data collection and identification of the equipment needed for MCC were performed through direct contact with suppliers and producers who had already begun their milk collection systems modernization. The choice of tank sizes considered in this study was based on results of a comprehensive diagnostic and survey of the dairy milk production at farm level and dairy industry raw milk requirement.

In order to carry out a financial analysis, cash flows were built for the investment cost and expense and operation cost for each MCC. The main MCC investments were for cooling tanks purchase and the construction of milk receiving area and laboratories and connection of water and electricity.

The operation cost components were consumption of electric energy, consumption of detergents for tank sanitation, maintenance costs, labour cost, and milk transportation cost and investment depreciation. The price offered by dairies industry to farmers for cow and camel is 0.200 RO per Lit plus Government incentives: an additional 0.100 RO per Lit added to the price paid for warm milk for supplying cooled milk. MCC will pay raw milk from farmers by (0.200) and sell it to dairy plant of (0.295). The cost of MCC operation is differing according to distant to dairy plant, milk volume and a reduction of 50% in the typical charge for using the traditional milk collection and transportation system for using the much less costly bulk transport system.

Based on the results from financial analysis, a sensitivity analysis of the investment was carried out, in which some parameters considered in the cash-flow estimation were varied. For each variation, a new net present value (NPV) was calculated for each MCC location. This produced a set of graphs known as Cumulated Distribution Function, which allow the assessment of the degree of uncertainty associated to each MCC investments. Gouse, M. et al. (2009). Assess GM Maize technology amongst smallholders in South Africa. The Cumulative distribution functions (CDFs) are used to compared yield and net returns. The CDFs analysis in this study used to identify the total milk collected revenue and expense items that had the most significant impacts on the financial indicators. The potential risks associated to these items are evaluated in detail in this study.

2.1 Risk Analysis

The financial analysis performed by (HVA International, 2010) considered that the project would be implemented with perfect control of its variables. However, due to uncertainties in the near future, this approach only approximates reality. In practice, project risks should be investigated, defined, and then controlled by policy makers and decision-makers. Risk is defined as the possibility of future loss in predicted return over a certain period of time (Palisade Corporation, 1995; Martins & Assaf NETO, 1992). Current data is imperfect, future data even more imperfect. Every current business decision is associated to a series of hypotheses regarding future events (increase or decrease in raw milk prices, raw milk available at each MCC, climatic conditions, social and political events, etc). Even if prediction techniques were improved, they would not eliminate the uncertainty brought by unforeseen change in various factors affecting profitability. Szekres (1986) states that the decision to take risks depends on the utility function that each individual subjectively attributes to a particular situation. People react to risk with three general types of behavior, risk-avoidance, risk-indifference, or risk-orientation; and the decision-making process is directly related with one of these personal tendencies.

Although sensitivity analysis can point out the need for further study of uncertain variables, by itself it is not sufficient for the analysis of an investment projects risk. Firstly because sensitivity analysis is only a partial analysis, with only one variable considered at a time, it does not permit the negative effect caused by one variable to be counterbalanced by the positive effect from another. Secondly, the mere indication that a project is or is not sensitive to certain variables, though useful, does not satisfy the decision-maker. It is also important to have an idea of the probability of an adverse situation occurring, as well as this situations effect on project results (Noronha, 1987).

@RISK (Palisade Corporation, 2010) software program was used for risk analysis in this study, and Monte Carlo simulation methodology was adopted. This methodology consists of simulating variables previously selected for the cash flow estimation and then calculating financial indicator values (IRR, net present value, etc.). After a number of random interactions, a frequency distribution of the financial indicator considered is generated result.

2.2 Net Present Value

The NPV was used as an evaluation criterion. The net cash flow, calculated by subtracting the cost from the revenue, was discounted by the interest rate to obtain the NPV of the project. If NPV is a function of all both deterministic and stochastic variables, the resulting NPV gets a range of values instead of a single value obtained in a conventional deterministic financial evaluation. NPV is obtained from the below formula.

$$NPV = \sum_{n=0}^{N} \frac{C_n}{(1+r)^n} \quad (1)$$

Where,

ci = the net cash flow in year n (n = 0, 1, 2, ….. n), represented by farm income in this study.

n = the planning period which equals fifteen years in the current analysis.

r = the discount rate.

Monte Carlo simulation is a computational algorithm designed to evaluate the variability or stochastic of the input variables of a model. It can be used to model the effects of key variables on the NPV of a given proposal. The process involves, first, the identification and assessment of the key variables. For each key variable, we fit a probability density function that best describes the range of uncertainty around the expected value. For this purpose, we used survey data at each MCC area and data from MAF Census 2013. The model including these variables is then calculated using randomly-generated input values taken from the underlying probabilistic distribution function. The computer model combines these inputs to generate an estimated outcome value for (NPV). The process is repeated (ten thousand times).

Monte Carlo simulation model is currently regarded as the most powerful technique for cash-flow analysis. It is useful when there are many variables with significant uncertainties. The more complex the project and the more risks and uncertainty that are associated, the more valuable Monte Carlo simulation analysis will be.

2.3 Data collection

Five MCC locations investment were evaluated in this study. This evaluation requires data on the revenue, operating and production costs and costs of capital for each MCC location. The milk collection volume and revenue for each MCC is estimated according to animal population and average milking yield at each location. Data used in this analysis were obtained from farmers face-to-face interviews during filed survey.

Table 1. Camel milk production and feed of survey sample

Group / head	1-50	51-100	100>
Milk production			
Per group/ Lit	22.03	36.25	30.57
Per head/Lit	6.5	7.47	6.47
Animal feed			
Concentrate/group/Kg	100	154	326
Hay/group/Kg	172	245	435
Concentrate/head/Kg	2.88	2.39	2.27
Hay/head/Kg	4.94	3.81	3.12
Average head/ group	35	65	133

Kheiry H. M. Ishag & Abd Al Hamed Al Gasani, 2008.

The above table shows that farmers with big animal group more than 100 camels could not maintain enough cash to feed his animal due to marketing and market access problems. According to animal population and distance from MCC to Dairy plant five MCC locations were recommended. Milk volume, initial unit cost and good quality milk received by dairy plant were also estimated according MCC location.

Table 2. Stochastic variables affect project NPV and at different MCC locations

Variables	Value	Min	Mean	Max
Salalah Milk Collection				
MCC capital cost RO	210 401			
Stating time /year	1			
Initial unit cost RO	0.181	85%	0.190	135%
Initial price RO	0.309	90%	0.305	115%
Total milk collection/Lit/year	803 493	30%	900 000	110%
Early annual demand growth	9.5%	30%	10%	100%
Dairy Plant good milk received %	61%	70%	75%	120%
Garoun Hirity Milk Collection				
MCC capital cost RO	221 332			
Stating time /year	1			
Initial unit cost RO	0.186	65%	0.201	130%
Initial price RO	0.335	90%	0.305	115%
Total milk collection/Lit/year	2 036 639	40%	2 051 890	110%
Early annual demand growth	9.85%	30%	10%	100%
Dairy Plant good milk received %	93%	75%	85%	120%
AL Haq Milk Collection				
MCC capital cost RO	275 034			
Stating time /year	2			
Initial unit cost RO	0.258	85%	0.225	135%
Initial price RO	0.295	90%	0.305	115%
Total milk collection/Lit/year	684 660	40%	1 043 672	110%
Early annual demand growth	8.88%	30%	10%	100%

Dairy Plant good milk received %	87.97%	75%	85%	120%
Tawi Ahater Milk Collection				
MCC capital cost RO	224 375			
Stating time /year	2			
Initial unit cost RO	0.223	90%	0.220	120%
Initial price RO	0.303	90%	0.304	115%
Total milk collection/Lit/year	744 969	30%	750 928	100%
Early annual demand growth	9.86%	30%	10%	100%
Dairy Plant good milk received %	92%	85%	90%	120%
Shahbi Asaeb Milk Collection				
MCC capital cost RO	270 306			
Stating time /year	2			
Initial unit cost RO	0.248	90%	0.250	120%
Initial price RO	0.311	90%	0.304	115%
Total milk collection/Lit/year	644 817	30%	653 510	110%
Early annual demand growth	6%	30%	10%	100%
Dairy Plant good milk received %	93%	85%	90%	120%

Calculated by the Author.

Uncertain variables in simulation mode outlined and estimated in Table 2. Initial unit price, sale price, raw milk collection volume, percentage of good milk received at plant and capital cost of the MCC establishment were the main uncertain variables incorporated in the models. Table 3 shows a recommended MCC locations, animal population and distance from each MCC zone to Dairy plant at Salalah City.

The modeling process began by defining a series of inputs to describe the initial status and behavior of the project and farm system. The underlying behavior of the local milk marketing system was represented using current knowledge and recorded data from survey, MAF and literature. The purpose of qualitative risk analysis in this study is to provide a high level of understanding of risks of the MCC project. Such analysis may increase attention of project management and policy team members to the top risks they need to understand and manage effectively.

Table 3. MCC location, animal population and distance from Dairy plant

MCC location	Distance Km	Cow population	Camel population
Garoun Hirity	30	83 658	36 885
AL Haq City	60	50 489	18 191
Tawi Ahater	62	40 004	9 688
Shahbi Asaeb	90	16 231	9 624
	90	11 304	5 556
Salalah	15	83 658	13 311
	1	88	4 344

Dhofar Governance, MAF Census 2013.

The main risk and uncertainty variables identified in MCC Project were:

- Project capital increase and it is effect on NPV.
- Raw milk availability and it is effect on MCC yield and NPV.

- Raw milk selling price volatility and it is effect on NPV.
- Cost of operation and it is effects on NPV and annual increase in sales price and unit cost.
- Total milk collection volume for year one and thereafter.
- Annual growth of milk collection for each MCC location.
- Percentage of good milk received at Dairy plant each year.
- Discount rate and it is effect on project viability and NPV.

The operation cost estimated according to total raw milk volume and distance between MCC and Dairy plant at Salalah City.

2.4 SERF and Risk efficiency

The analysis of investment decision of MCC needs entails understanding of how investors rank alternative MCC locations with uncertain outcomes, given the stochastic MCC milk yield for each center and the stochastic market price of milk. The economic evaluation of MCC is implemented considering the whole range of net present values (NPVs) and their associated probabilities, along with the relative preferences (utilities) of the decision makers. To assess and compare the economics and the risk efficiency of each MCC location, this study employs stochastic simulation as an unconventional method that incorporates risk in the analysis (Hardaker, J.B., 2004). Stochastic dominance and stochastic efficiency with respect to a function (SDRF and SERF) analyses have a major advantage in that they reduce the set of all possible risky choices to a small group of alternatives. The SERF technique is a novel improved methodology for assessing and ranking risky alternatives but empirical studies using SERF are limited, Especially in agriculture, SDRF and SERF analyses have been used to compare risky alternatives regarding farm production, marketing and financial matters. In this study, SERF analysis has been used to rank MCC locations and to determine risk efficiency of MCC locations, and also to explore the economic viability of MCC investment. However, this analysis will help in identifying best MCC location to start with and Government milk purchase subsidy required for each MCC location.

3. Results and Discussion

The study carried out by (HVA International, 2010), based on estimated cash flows, and showed that investment in MCC Project is only visible under Government price subsidy, since the 8% internal rate of return from that investment would be below the adopted 12% minimum rate of investment attractiveness. The previous study (Danielle et al., 2000), has not evaluated MCC locations risk in their study. However, this study investigated profitability for each recommended MCC location and incorporates risk of price paid to producer and milk collection volume for each MCC location in the analysis. Table 4 shows the result of simulated model for each MCC location and statistics.

The analysis shows that Garoun Hairity Location is highly profitable with RO 1 723 345 NPV and no risk of getting negative NPV and low Coefficient of Variation 0.402%. Shahbi Asaeb location got a negative NPV under different discount rates i.e. (6%-8%-10%-12% and 14%) and high Coefficient of Variation 1.190%. The probability of getting positive NPV for this location is only 20% with 6% discount rate and reduced to 3% at 14% discount rate. Government capital subsidy for this location is highly required to mitigate risk and encourage investors and farmers to participate and benefit from MCC project.

3.1 Cumulated Distribution Function (CDFs) Analysis

The Cumulated Distribution Function (CDFs) analysis performed to illustrate the range and probabilities of NPV for different MCC locations. Figure 1 shows Garoun Hirity location is a superior location and has a high probability of getting positive NPV. However, CDF analysis shows that Al Haq City and Tawi Ahater MCC location are not first-degree or second-degree stochastically dominant as their CDFs lines crossed each other up to cumulative probability of 0.45% is reached and no clear ranking under different ARAC is possible. However, a more improved technics such as stochastic dominance and stochastic efficiency with respect to a function analyses can be used to rank alternative MCC location risk efficiently within specified range of ARAC values.

Table 4. MCC recommended location compared to Salalah location – Statistics for NPVs for each scenario

Models	**Model (1)**	**Model (2)**	**Model (3)**	**Model (4)**	**Model (5)**
Location	**Salalah**	**Recommended MCC Location**			
		Garoun Hairity	**Al Haq City**	**Tawi Ahater**	**Shahbi Asaeb**
Mean	579 727	1 723 345	352 000	261 411	(115 171)
SD	233 264	692 973	275 022	162 332	137 473
CV	0.402%	0.402%	0.780%	0.620%	1.190%
Skewness	0.483	0.5955	0.2470	0.344	0.0183
Kurtosis	3.2598	3.476	3.170	3.061	3.096
Min	30 672	32 957	(754 000)	(189 000)	(756 000)
Max	1 612 876	5 622 000	1 607 209	892 000	392 000
Range	1 643 848	5 654 957	2 361 209	1 081 000	1 148 000

Calculated by the Author.

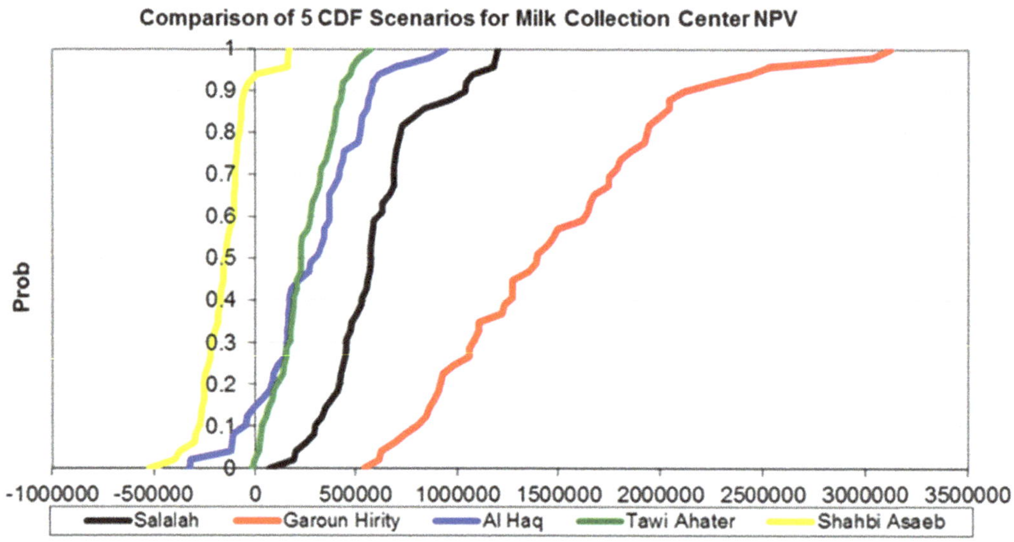

Figure 1. Cumulated Distribution Function CDF analysis of NPV for different MCC locations

3.2 Stochastic Efficiency With Respect to a Function (SERF) Analyses

The stochastic efficiency with respect to a function (SERF) analyses is performed to select and outline the risk efficient MCC location. The Excel Add-In SIMETAR developed by Schumann, Feldman, and Richardson (2011) was used to conduct the SERF analysis based on a negative exponential utility function. Certainty Equivalent graphs were constructed to display and rank NPV across the specified range of ARAC values. The risk premiums were also calculated for each MCC location by subtracting alternative NPV CE values from a preferred MCC location (Salalah NPV) CE values at given ARAC values.

In Figure 2 the SERF method is used to compare five MCC location alternatives simultaneously for a range of ARAC values between (0.00) and (0.0000014). The analysis shows and identifies Garoun Hirity location as a superior location followed by Salalah location and Al Haq City alternatives as the second and third risk-efficient location. The (SERF) analysis indicates that Garoun Hirity is the preferred MCC location for investors over the entire range of risk preferences. While NPV for Garoun Hirity range between 1 472 000 and 1 304 000 risk premiums between RO 749 000 and 722 000 must be paid to investors and farmer risk preference—to persuade them to switch from Shahbi Asaeb to Salalah MCC location the second-most preferred MCC.

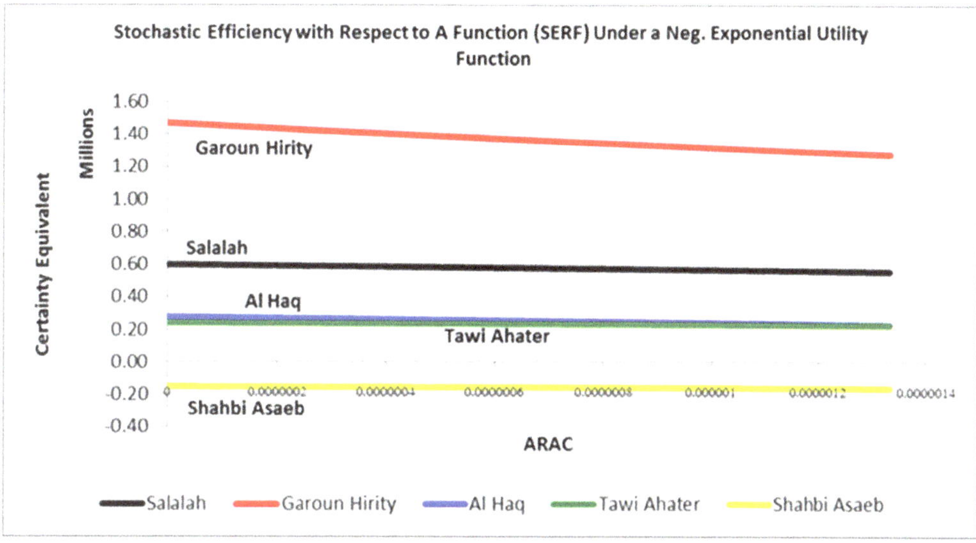

Figure 2. Stochastic Efficiency with Respect to a Function (SERF) of NPV for different MCC locations

The NPV of Garoun Hirity and Salalah location are visible under different Absolute Risk Aversion Coefficient, due to high animal population density and closer distance to Dairy Plant at Salalah City. The raw milk quality expected to be high and milk rejection percentage is low.

Table 5 calculates Certainty Equivalent CE for each MCC location across a range of ARAC. The calculated CEs are displayed on graphs, and the risky alternative with the highest CE at a particular RAC is the most preferred. Rankings five MCC risk alternative performed using SERF, over the range of risk preference, neutral to extremely risk averse, are presented graphically in Figure 2 and numerically in Table 5. Table 5 reveals that under all risk aversion level Shahbi Asaeb location required Government support to operate risk efficiently and encourage investors and farmers to participate in MCC project.

Table 5. Ranking of Risky MCC Alternatives by Risk aversion using CE for NPV (000)

Risk degree	Risk Neutral		Normal Risk		Rather Risk		Extremely Risk	
ARAC	0.00		0.0000004		0.0000008		0.0000014	
Rank	Alternative	CE	Alternative	CE	Alternative	CE	Alternative	CE
1	Garoun H.	1 472	Garoun H.	1 398	Garoun H.	1 334	Garoun H.	1 304
2	Salalah	598	Salalah	584	Salalah	570	Salalah	563
3	Al Haq	278	Al Haq	262	Al Haq	246	Al Haq	238
4	Tawi A.	241	Tawi A.	237	Tawi A.	232	Tawi A.	230
5	Shahbi A.	-150	Shahbi A.	-153	Shahbi A.	-157	Shahbi A.	-159

Calculated by the Author.

3.3 Risk Premium and Willingness to Payment

Risk premiums measure the value to be paid to a Decision Maker of one preferred alternative over a less preferred alternative, and are calculated by subtracting the CE of the less-preferred alternative from the CE of the preferred alternative at each RAC level. Because SERF generates CEs of the Decision Maker's preferences among alternatives MCC location at each risk aversion level, SERF can also estimate the utility-weighted risk premiums between alternatives MCC location. Figure 3 represent the difference between CEs represents what it would take for a Decision Maker to be willing to exchange the preferred location (Salalah) risky alternative for another less-preferred risky alternative. The value of WTP (willingness to payment) is calculated as the difference between the CE for a risky alternative and represents the payment necessary to make the farmers and investors indifferent between the less-preferred alternative and the preferred pilot location alternative at (Salalah) City:

$$\text{WTP} = \text{CE preferred} - \text{CE alternative}$$

The SERF rankings and WTP are used to examine sustainability and analyze of deferent MCC locations. Figure 3 shows how the alternative location were examined in the study and ranked relative to the preferred (Salalah location) at various RACs. Table 6 shows the numerical risk premiums for four risk aversion levels.

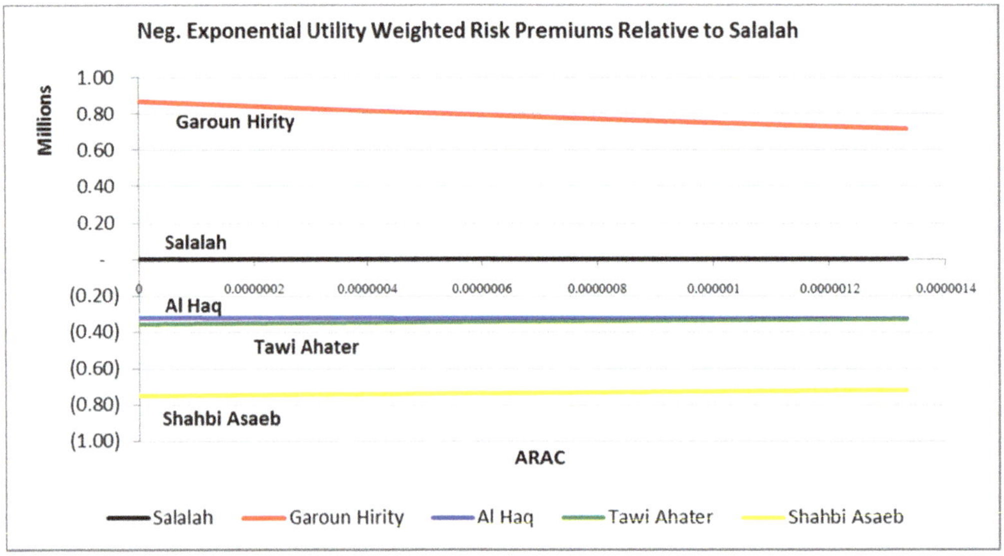

Figure 3. Present Risk Premium for MCC location relative to Salalah location

From Table 6, it is evident that Decision and policy Makers for the risk aversion levels examined have a small risk premium value between the preferred MCC location (Salalah) and the second alternative (Al Haq City) and large risk premium with Shahbi Aseab MCC location. Therefore, a more milk price has to be given to farmers and sufficient capital subsidy has to be given to investors at Shahbi Aseab MCC location to mitigate risk. The study also indicates that Garoun Hirity location is the most preferred location for all ARAC levels as shown in Figure 3. However, at risk neutral level RO 320 000, 357 000 and 749 000 incentives must be paid to Al Haq City, Tawi A. and Shahbi Aseab location respectively to be in the same position of the second-most preferred MCC location at Salalah. In general, the results of the SERF analysis indicate that Garoun hairity MCC location provide a useful risk-management tool for moderately to highly risk-averse farmers and investors.

Price paid to the farmer for milk is a risk-dependent item since it varies according to market conditions and milk composition. The dairy company can offer farmers bonuses and other awards linked to encourage the quality and/or volume of milk supplied. Variation of the price paid to the producer for milk and volume of milk produced on the farm were adopted for risk analysis: Sensitivity analysis found that these were the most unstable parameters of MCC project.

Volume of milk produced is another risk-dependent item since output depends on several factors, such as animal population at each location zone, climate & rainy season, costs, and capital investments. Milk production in Dhofar Region is increases during the summer (rainy season) and decreases during the winter (dry season). This seasonal change is due to change in available pasture: more pasture in the summer, less in the winter. Farm investments to expand the herd, improve breeding stock, control diseases, and install milk cooling systems are economic factors that can contribute to increase milk production and assure raw milk quality.

Table 6. Risk Premium NPV (000) of different MCC locations relative to Salalah location

Risk degree	Risk Neutral		Normal Risk		Rather Risk		Extremely Risk	
ARAC	0.0000		0.0000004		0.0000008		0.0000014	
Rank	Alternative	RP	Alternative	RP	Alternative	RP	Alternative	RP
1	Garoun H.	874	Garoun H.	814	Garoun H.	764	Garoun H.	741
2	Salalah	0	Salalah	0	Salalah	0	Salalah	0
3	Al Haq	-320	Al Haq	-322	Al Haq	-324	Al Haq	-326
4	Tawi A.	-357	Tawi A.	-347	Tawi A.	-338	Tawi A.	-334
5	Shahbi A.	-749	Shahbi A.	-738	Shahbi A.	-727	Shahbi A.	-722

Calculated by the Author.

The risk premium calculation indicates that a range of RO (320 000-326 000) needs to be paid to Al Haq City location to be indifferent with Salalah location. For Tawi Ahater MCC location risk premium needs to be paid will range from RO (357 000-334 000) to be in same position of Salalah Milk Collection location. For Shahbi Asaeb location, the risk premium will range between RO 749 000 to 722 000. However, Salalah Milk Collection is a preferred location to Dairy Plant as it is located near the dairy plant and milk can be transfared to the dairy plant immediately within two hours of milk collection from rural areas. The risk premium analysis also reveals that Certainty Equivalent CE for all recommended MCC locations remain constant across a range of selected ARAC in this research.

3.4 MCC Locations and StopLight Graph Analysis

Stoplight graphs are simple graphical illustrations and show the probability of NPV being greater than a target value (35 000) and less than another target value (15 000) across risky alternatives. Stoplights are quickly interpretable, as they are read much like a traffic stoplight, in this case red is bad, yellow is marginal, and green is good (Richardson, Schumann, & Feldman, 2006).

The probability of a risky alternative generating a net present value less than the lower bound value (15 000) is illustrated by a red region on a bar graph; thus, bad. The probability of an alternative generating a net present value greater than the upper bound value (35 000) is illustrated by a green region; thus, good. The region between the upper and lower bounds is yellow and shows the probability of NPV being between the upper and lower bounds.

The Stoplight graph in Figure 4 illustrates the probability of NPV being less than 15 000 and greater than RO 35 000. Salalah and Garoun Hiriti locations have 100% probabilities of positive NPV, and Al Haq City has 82% whereas Tawi Ahater MCC location has 88% probability of positive NPV. The Stoplight analysis indicates that Shabhi Asaeb location has 92% probability of negative NPV, due to animal population at MCC zone and the long distance to Dairy Plant.

Investment in MCC Project with both the 10 000 and 5 000 liter milk cooling tanks would be feasible and attractive, since their NPV are positive and return in investment (RI) would be significantly above the opportunity cost of capital for all recommended MCC location except Shabhi Asaeb location. Sensitivity analysis, in turn, revealed that an increase of 25% in the incentive paid to farmers for supplying cooled milk (increasing initial price from 0.304 to 0.380) would be sufficient to make the MCC investments attractive, and increase probability of getting positive NPV to 99% at 6% discount rate and to 88% at 14% discount rate for Shahbi Aseab MCC Location.

The result also indicates that the risk of MCCs investment failure can be reduced by selecting MCC location near to Dairy Plant and with large animal population size. The cost of MCC operation and milk volume were the main factors affecting profitability and insure project sustainability as shown by sensitivity analysis.

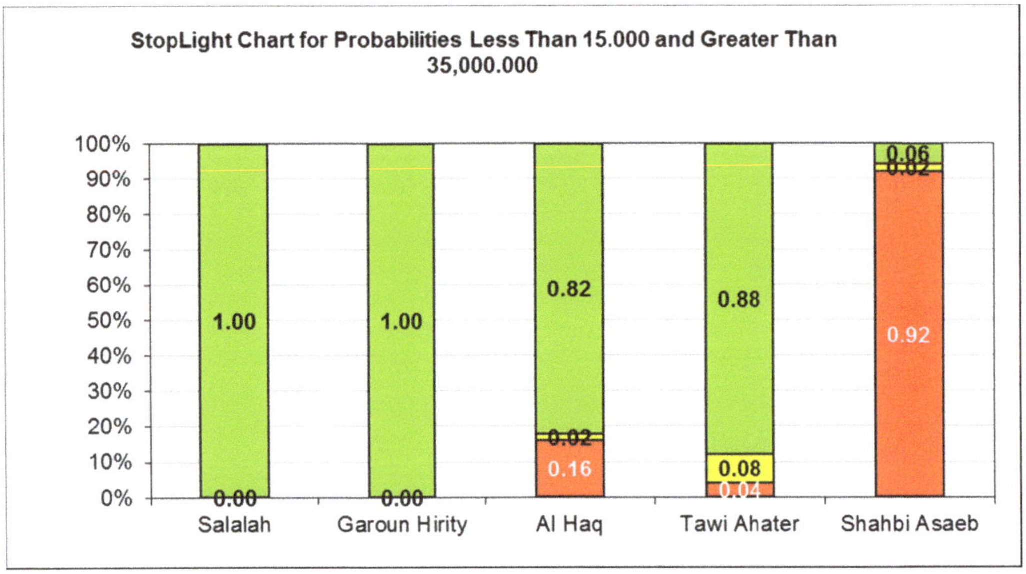

Figure 4. StopLight chart for NPVs of four MCC location compare to Salalah pilot MCC location

4. Conclusions

Investment in different MCCs locations were tested by using dynamic location analysis and incorporating milk volume and milk price risk in the analysis. Risk analyses for the establishing MCCs with 10,000 and 5,000 Lit cooling tanks at four recommended locations were tested and reveal profitability except for Shahhbi Asaeb location. The dynamic risk analysis of this investment shows low probability of getting positive NPV of less than 20% at 6% discount rate and reduced to 3% at 14% discount rate. Incentive price increase of 25% given to MCC at Shahbi Asaeb location can increase probability of getting positive NPV to 99% with 6% discount rate.

The results also indicate that the risk of MCCs investment failure (an increase in profitability of negative NPV) is reduced by the potential for increased milk production at rural area. Conversely, capital requirements will restrain small farmers to invest in upgrading farm facilities and herd improvement investments. This restraint and difficulties faced by small farmers when trying to increase production will probably lead to the exclusion of a significant number of small farmers from the formal dairy product market as the demand for cooled milk delivery by dairy industries grows. As a result, Ministry of Agriculture & Fisheries needs to imposed regulations and implemented herd improvement programs. Moreover, raw milk price incentive needs to be implemented by Government Authorities.

Current social concerns and the fast pace of Oman dairy business modernization suggest that the installation of MCC and cooling tanks will become essential to the survival of dairy plant industries and farmers. It appears that dairy industries will increasingly demand cooled milk transported in isothermal tanks. Therefore, the sustainability strategy recommended for small dairy farmers is to increase milk production by increasing yields through the adoption of herd nutrition and breeding management technologies.

References

Barham, E. H., Robinson, J. R., Richardson, J. W., & Rister, M. E. (2011). Mitigating cotton revenue risk through irrigation, insurance, and hedging. *Journal of Agricultural and Applied Economics, 43*(4), 529-540.

Bryant, K. J., Reeves, J. M., Nichols, R. L., Greene, J. K., Tingle, C. H., & Studebaker, G. E. (2008). Valuing transgenic cotton technologies using a risk/return frame work. *Journal of Agricultural and Applied Economics, 40*(3), 767-775.

Danielle, D. S. A., Carlos, A. B. S., & Sebastião, T. G. (2000). Risk analysis of investments in-farm milk cooling tanks. *Rev. Econ. Sociol. Rural, 41*(2). http://dx.doi.org/10.1590/S0103-20032003000200005

Eschenbach, T. G. (1992). Spiderplots versus tornado diagrams for sensitivity analysis. *Interfaces, 22*(6), 40-46. http://dx.doi.org/10.1287/inte.22.6.40

Gouse, M., Piesse, J., Thirtle, C., & Poulton, C. (2009). Assessing the performance of GM maize amongst smallholders in KwaZulu-Natal, South Africa. *AgBioForum, 12*(1), 78-89.

Hardaker, J. B., Huirne, R. B. M., Anderson, J. R., & Lien, G. (2004b). *Coping with Risk in Agriculture* (2nd Edition). Wallingford: CABI Publishing. http://dx.doi.org/10.1079/9780851998312.0000

Hignight, J. A., Watkins, K. B., & Anders, M. M. (2010). An economic risk analysis of tillage and cropping systems on the Arkansas grand prairie. Paper presented at the Southern Agricultural Economics Association Annual Meeting, Orlando, FL.

Kheiry, H. M. Ishag, & Abd, A. Hamed. A. Gasani. (2008). *Animal Husbandry Survey Report – Dhofar Directorate General of Agriculture & Fisheries*- Ministry of agriculture and fisheries- sultanate of Oman.

Pendell, D. L., Williams, J. R., Boyles, S. B., Rice, C. W., & Nelson, R. G. (2007). Soil carbon sequestration strategies with alternative tillage and nitrogen sources under risk. *Applied Economic Perspectives and Policy, 29*(2), 247-268. http://dx.doi.org/10.1111/j.1467-9353.2007.00341.x

Ribera, L. A., Hons, F. M., & Richardson, J. W. (2004). An economic comparison between conventional and no-tillage farming systems in Burleson County, Texas. *Agronomy Journal, 96*(2), 415-424. http://dx.doi.org/10.2134/agronj2004.0415

Savvakis, C. S. (1994). Risk analysis in investment appraisal. *Project Appraisal Journal, 9*(1), 3-18. http://dx.doi.org/10.1080/02688867.1994.9726923

Schumann, K. D., Feldman, P. A., & Richardson, J. W. (2011). *SIMETAR©: Simulation & econometrics to analyze risk*. College Station, TX: Agricultural and Food Policy Center, Texas A&M University.

Szekeres, S. (1986). Considering uncertainty in project appraisal (p. 60). Washington, USA: International Bank for Reconstruction and Development.

White, J. A. (1998). In K. E. Case, D. B. Pratt & M. H. Agee (Eds.). *Principles of engineering economic analysis*. (4th ed, p. 491), New York, USA: John Wiley & Sons.

Williams, J. R., Pachta, M. J., Claassen, M., Roozeboom, K., & Llewelyn, R. (2011). Wheat stubble to burn or not to burn: An economic analysis. *Western Economics Forum, 10*(1), 32-42.

9

Impact of Energy Price Variability on Global Fertilizer Price: Application of Alternative Volatility Models

Prabuddha Sanyal[1], Leonard A. Malczynski[2] & Paul Kaplan

[1] Cisco Systems Inc., USA

[2] Sandia National Laboratories, USA

Correspondence: Prabuddha Sanyal, 9972 Buckeye St, NW, Albuquerque, New Mexico, USA. E-mail: prabuddha_sanyal@hotmail.com

Abstract

This study evaluates the effects of volatility in crude oil and natural gas prices on fertilizer price variations. Specifically, the study looks at the mean and volatility effects of oil and natural gas prices on both mean and volatility changes in fertilizer prices. Both symmetric models [GARCH (1, 1)] and asymmetric models [GJR (1, 1)] were used to model volatility in fertilizer prices and to evaluate the effects of the volatility over different time periods using Bai-Perron structural break tests. The results show that changes in oil and natural gas prices increased fertilizer prices after the crisis period, during June 2007 to June 2008. Both the ARCH and GARCH had significant effects on fertilizer prices, suggesting that the volatility effects of oil and natural gas prices on fertilizer prices were also significant. Furthermore, the maximum impact of higher energy prices depends on triple superphosphate and diammonium phosphate (DAP) leading to higher production costs and consequent increase in total farm expenditures for crop producers. These higher production costs invariably have a negative effect on farm profitability, thus reducing the investment levels in the farm sector.

Keywords: Oil Price Volatility, Global fertilizer price volatility, Bai-Perron Structural break test, Natural gas price volatility

1. Introduction

1.1 Introduce the Problem

Fertilizers are combinations of nutrients that enable plants to grow. The essential elements of fertilizers are nitrogen, phosphate, and potassium. Almost all nitrogen fertilizer is derived initially from anhydrous ammonia (NH_3), which is made by reacting atmospheric nitrogen and hydrogen from natural gas. The cost of natural gas accounts for 70% to 90% of the production costs of ammonia (Kenkel, 2010). As energy prices increase, the price of ammonia increase as producers competes with other industries for their gas supply. The production of phosphate fertilizer begins with mining rock phosphate. Potassium fertilizers come from mining potash ore deep in the earth. Although there is some potash production in New Mexico and Utah, most potash is imported from Canada. Prices of nitrogen, phosphate, and potassium fertilizers began increasing in 2002 and reached historic highs in mid-2008. During the 12-month period between June 2007 and June 2008, nitrogen, DAP, and potash prices increased by 59%, 359%, and 173% respectively (Huang, 2009).

Over the past decade, owing to higher demand for biofuels and feedstock as well as higher demand for renewable energy, there have been greater linkages between the agriculture and energy markets. The underlying causes of fertilizer price volatility include a shift to a global supply chain, an increase in the production of biofuel, the volatility of prices of raw materials (such as natural gas), and the volatility in exchange rates (Beckman et al., 2013).

This paper investigates the relationship in the volatility of natural gas, crude oil, and global fertilizer prices. As volatility involves financial risks, the results provide useful information for optimum energy use, global agricultural production, and profitability of the farm sector. The remainder of the paper is organized as follows: In the next sub-section, we provide the importance of this problem. This is followed a relevant discussion of the literature. Section 2 develops the method and the hypotheses that are tested. Section 3 discusses the main results of the paper. This is followed by discussion of the results along with the main policy implications of the study.

1.2 Importance of the Problem

Over the past decade, rising energy prices coupled with evolving policies promoting renewable energy and on-farm conservation practices have transformed the relationship between the energy and agriculture sectors (Beckman et al., 2013). Agriculture uses energy both directly in the form of fuel and electricity and indirectly through the use of energy-intensive inputs, such as fertilizers and pesticides. However, over the past decade, record-high energy prices and expanding biofuel production have increased the demand for agricultural commodities and have also increased the demand for biofuels and feedstocks. In this context, several recent studies have examined the relationship between fertilizer markets, commodity markets, and energy markets and how this relationship has evolved over time. No definitive answers can be given on the transmission mechanisms of how one market affects other markets, as the literature is still evolving. Thus, an empirical examination of the linkages between oil and natural gas prices on different fertilizer prices is important. Previous literature has examined only the linkages in mean and volatility effects of oil prices on fertilizer prices. An important determinant of fertilizer prices, namely natural gas price effects is thus missing from the empirical literature. We contribute to the existing studies by looking at the mean and volatility effects of oil and natural gas prices on various fertilizer prices over the period December 1993 to January 2012. In addition, we conduct structural break tests to identify the crisis periods (namely June 2007 to June 2008), where fertilizer price volatility was significant. Finally, we validate our model using ARCH/GARCH models and demonstrate why these models are relevant for understanding the volatility in fertilizer prices.

1.3 An Overview of the Literature

Interest in studying fertilizer price volatility and forecasting begins in the early 1990s. During the mid-1970s and throughout the 1980s, U.S. fertilizer prices were highly volatile. We discuss only the relevant literature for the past five years.

Galbraith (2010) used an error correction model to determine how anhydrous ammonia (AA) and urea prices at different locations in North America adjusted to changes in crop and input prices during two time periods – 2002 to 2005 and 2006 to 2009. Galbraith also considered measures of supply and demand shocks such as natural gas price futures as a measure of supply shock and corn futures as a measure of demand shock. The empirical results suggest that natural gas future prices had relatively more of an impact on nitrogen fertilizer prices during the 2002 to 2005 period. In addition, nitrogen prices tended to adjust more rapidly to increases in natural gas futures prices than decreases. In the 2006 to 2009 period corn futures prices had more of an impact on nitrogen prices. Nitrogen prices tended to respond immediately to decreases in corn futures prices. The nitrogen price responses to increasing corn futures prices were mixed.

Ott (2012) analyzes the drivers behind the recent price volatility of fertilizers and their interplay with energy and food commodity market prices. Three issues are examined: the role of speculations for fertilizer price formation; the interaction among fertilizers, food, and energy prices; and fertilizer price volatility. First, the results showed the presence of speculative behavior in fertilizer markets. However, the speculation on derivative markets was not the cause. Instead the volume traded in the physical markets was the driver behind the volatility in fertilizer markets. It is likely that fertilizer derivative products may have been used as hedging tools and not as speculative ones. Second, the prices of food commodities were driving the fertilizer markets and not vice versa. In addition, higher food prices induced a higher demand for fertilizers, thus increasing food prices to higher levels. Third, the energy sector influenced the increase in fertilizer prices through the input cost channel. For example, rising oil and natural gas prices contributed to a demand for nitrogen nutrients, the production of which depends significantly on energy inputs for production and transport. Fourth, given the oligopolistic fertilizer supply chain and inelastic supply, the upward spiral in fertilizer prices was also due to uncertainty caused by the low levels of inventory. In other words, over the past 15 years, the excess nitrogen supply has disappeared while phosphate and potash supply remained at marginal levels. This implied that no buffer could protect the fertilizer markets when an adverse shock occurred in 2007. Finally, the results show the volatility of energy, food, and fertilizer prices moved closely together when energy prices were increasing. However, when energy prices were declining, food and fertilizer price volatility did not move together, showing that there were asymmetries in the price volatility.

Du and McPhail (2012) examine the co-movement and increasing volatility of ethanol, gasoline, and corn prices over the period, March 2005 to March 2011. A structural change was identified around March 2008 in the pairwise dynamic correlations between the prices using a multivariate GARCH model. A structural vector autoregression model (VAR) model is then estimated on two subsamples, one before and one after the identified change point. Variance decomposition analysis was conducted to examine the channels of variance transmission.

The results of the study show that first, during the earlier period, the responses in one market to price changes in another market were not statistically significant, implying that price variations of individual markets were largely independent and explained by their own shocks. Second, in contrast to the first period, during the second period, all the prices were closely linked due to a strengthened corn-ethanol relationship. Variance decomposition analyses showed that for each market a significant and relatively large share of the price variation could be explained by the price changes in the other two markets. The ethanol shocks had the largest impact on corn prices, suggesting a strong linkage between corn and ethanol markets. The results are robust to the inclusion of the seasonal dummies and various macro and financial indicators.

Chen et al. (2013) empirically evaluate the effect of crude oil prices on global fertilizer prices, both in mean and volatility. Weekly data for the period 2003 to 2008 were used for this analysis. Using descriptive statistics, the results found three time periods with two structural breakpoints in the data for six global fertilizer prices and crude oil prices. Using the Granger causality test, the results showed that most global fertilizer prices were influenced by the crude oil price. The empirical results further demonstrated that crude oil prices had relatively larger effects on global fertilizer prices during the second time period (i.e., after 2008). Thus the sensitivity of fertilizer prices to oil prices increased and became statistically significant. For example, the impact of oil price on the monoammonium phosphate (MAP) price is 1.25% during the first period, 4.9% during the second period, and 6.4% during the third period. This sensitivity may explain why global fertilizer prices reached a peak in 2008, as the crude oil prices reached a high level during the same period.

Various models (including generalized autoregressive conditional heteroskedastic (GARCH), GJR (Note 1), and EGARCH (Note 2)) were selected and estimated on the basis of the Schwartz-Bayes (BIC) criterion and the regularity conditions of the quasi-maximum likelihood estimators to be consistent and asymptotically normal. The main result of the study is that crude oil prices exert a greater impact on fertilizer prices during the latter part of 2008. This result may suggest that as the volatility in global fertilizer prices has increased, vital energy prices and agricultural production are likely to be significantly affected. This may lead to future instability in agricultural commodity prices. Thus it is important to understand the directional relationship between the volatility in energy and global fertilizer prices.

1.4 Hypotheses and Research Design

We specify an ARDL model, since it is likely to explain the short-run dynamics between energy prices and the fertilizer prices.

A general ARDL model for a global fertilizer price can be written as follows:

$$FP_t = \alpha_0 + \sum_{i=1}^{p} \alpha_i FP_{t-i} + \sum_{j=1}^{q} \beta_j \, oilP_{t-j} + \sum_{k=1}^{r} \gamma_k NGP_{t-k} + u_t \quad (1)\ \text{(Note 3)}$$

In Equation (1), the coefficient β_j denotes the effect of the jth period lagged crude oil price on the fertilizer price, which means that the fertilizer price can be predicted by the crude oil price. Similarly the coefficient γ_k denotes the effect of the kth period lagged natural gas price on the fertilizer price. FP denotes the respective fertilizer price series. A test of the null hypothesis that each $\beta_j = 0$ and $\gamma_k = 0$ is a test of Granger non-causality.

All the prices in the model should be stationary series to avoid spurious regression results, whereby the asymptotic standard normal results no longer hold. For this reason, the structural breakpoints of the fertilizer prices are estimated using the Bai-Perron tests as discussed in Section 2.2. If and when the appropriate structural breakpoints are found, the fertilizer price equations will be estimated for different periods.

2. Method

The stationary AR(1)-GARCH(1,1) model is considered for understanding the volatility in global fertilizer price data on the lines of Engle (1982).

2.1 Conditional Mean and Conditional Volatility Models

Let the *AR(k)* process for the fertilizer price series z_t be defined as:

$$z_t = \varphi_1 + \sum_{i=1}^{k} \omega_i z_{t-i} + \varepsilon_t; \ = 1, \ldots\ldots, k; \ t = 1, \ldots\ldots T; \ \varepsilon_t \sim IID(0, \sigma^2); \ |\omega_i| < 1 \quad (2)$$

z_t denotes the fertilizer price, φ_1 is the constant parameter affecting fertilizer prices, z_{t-i} denotes the autoregressive components of fertilizer price series, ω_i represents the autoregressive parameters, and ε_t is the error term. In Equation (2) z_t is assumed conditional on the "immediate past information set" Ω_{t-1}, and thus its conditional mean can be expressed as:

$$E(z_t|\Omega_{t-1}) = \varphi_1 + + \sum_{i=1}^{k} \omega_i z_{t-i} \quad (3)$$

Equation (3) shows that the conditional mean of z_t is time-varying and this is a characteristic of financial time series. Following Engle (2002), the error term is as follows:

$$\varepsilon_t = \mu_t\left(\beta_0 + \sum_{j=1}^{q} \beta_j \varepsilon_{t-j}^2\right) \;;\; j = 1, \dots \dots q \tag{4}$$

Equation (4) is the ARCH(q) model proposed by Engle.

Equivalently, equation (4) may be expressed as:

$$\varepsilon_t^2 = \mu_t^2\left(\beta_0 + \sum_{j=1}^{q} \beta_j \varepsilon_{t-j}^2\right) \tag{5}$$

Taking the expectation of Equation (5) given the relevant information set (π_{t-1}), the conditional variance is given as:

$$var(\varepsilon_t|\pi_{t-1}) = \beta_0 + \sum_{j=1}^{q} \beta_j \varepsilon_{t-j}^2 \text{ since } E(\mu_t^2|\pi_{t-1}) = 1 \tag{6}$$

Equation (6) defines an alternative formulation of the ARCH(q) model, where the $var(\varepsilon_t|\pi_{t-1})$ is a function of the squared error term from past periods ε_{t-j}^2.

In the case of unconditional variance, using the lag operator L, equation (6) becomes:

$$\sigma_t^2 = E(\varepsilon_t^2) = \frac{\beta_0}{1-\beta(L)} \tag{7}$$

where, $\sum_{j=1}^{q} \beta_j \varepsilon_{t-j}^2 = \beta(L)\varepsilon_t^2$ and $\beta(L)$ is the polynomial lag operator $(\beta_1 \mathrm{L} + \beta_2 \mathrm{L}^2 + \cdots + \beta_q \mathrm{L}^q)$

The null hypothesis to be tested based on equation (7) is given by:

$$H_0: \beta_1 = \beta_2 = \cdots = \beta_j = 0$$

This hypothesis is tested using either the *F*-test or nR^2 that follows the Chi-square distribution proposed by Engle (1982). If the null hypothesis is not rejected, then there is no ARCH effect in the model and vice versa.

Engle's ARCH model was extended by Bollerslev (1986), which extends equation (6) by incorporating the lags of the conditional variance equation:

$$\sigma_t^2 = \beta_0 + \sum_{j=1}^{q} \beta_j \varepsilon_{t-j}^2 + \sum_{i=1}^{p} \gamma_i \sigma_{t-i}^2 \tag{8}$$

where, $p \geq 0$; $q > 0$; $\beta_0 > 0$; $\beta_j \geq 0$; $\gamma_i \geq 0$, $j = 1, \dots\dots, q$ and $i = 1, \dots\dots, p$

Equation (8) is the GARCH *(p,q)* model where p and q denote the lagged terms of the conditional variance and the squared error term respectively. The ARCH effect is denoted by $\sum_{j=1}^{q} \beta_j \varepsilon_{t-j}^2$ and the GARCH effect is denoted by $\sum_{i=1}^{p} \gamma_i \sigma_{t-i}^2$.

The asymmetric effects not accounted for in the ARCH and GARCH models is also taken into account. Nelson (1991) proposed the EGARCH *(p,q)* model as follows:

$$\log(\sigma_t^2) = \varphi_2 + [1 - \gamma(\mathrm{L})]^{-1}[1 + \beta(\mathrm{L})]\mathrm{f}\left(\frac{\varepsilon_{t-1}}{\sigma_{t-1}}\right) \tag{9}$$

and

$$\mathrm{f}\left(\frac{\varepsilon_{t-1}}{\sigma_{t-1}}\right) = \alpha\varepsilon_{t-1} + \theta\left(\left|\frac{\varepsilon_{t-1}}{\sigma_{t-1}}\right| - E\left|\frac{\varepsilon_{t-1}}{\sigma_{t-1}}\right|\right) \tag{10}$$

In the EGARCH model, the log of the conditional variance term is a function of the lagged error term. The asymmetric effect is captured by the parameter α in equation (10). Asymmetric effect is present if $\alpha \neq 0$, and there is no asymmetric effect if $\alpha = 0$. The null hypothesis is $\alpha = 0$ and this is tested using the t-statistic.

The asymmetric effect can also be captured using the GJR-GARCH model, which modifies equation (8) as follows:

$$\sigma_t^2 = \beta_0 + \sum_{j=1}^{q} \beta_j \varepsilon_{t-j}^2 + \sum_{i=1}^{p} \gamma_i \sigma_{t-i}^2 + \sum_{j=1}^{q} \delta_j \varepsilon_{t-j}^2 I_{t-j} \tag{11}$$

where, $I_{t-j} = \begin{cases} 1 \; if \; \varepsilon_{t-j} > 0 \\ 0 \; otherwise \end{cases}$.

Thus, there is evidence of asymmetric effect if δ_j<(>)0, which implies that positive (negative) shocks reduce the volatility of z_t by more than negative (positive) shocks of the same magnitude.

2.2 Sequential Bai-Perron Tests for Structural Breaks

Ignoring possible structural breaks can lead to non-rejection of the null hypothesis of non-stationarity, so that the effects of structural breaks may be attributed to the existence of a unit root. For example, Nelson and Plosser (1982) used the Dickey-Fuller unit root test to examine U.S. macroeconomic time series and found the presence of widespread non-stationarity.

In this section a test of the null hypothesis of l breaks against the alternative that an additional break exists on the lines of Bai-Perron is considered. For the model with l breaks, the estimated breakpoints, denoted by $\widehat{t_1}, \ldots\ldots\ldots, \widehat{t_l}$ are obtained by a global minimization of the sum of squared residuals. This strategy proceeds by testing each $(l+1)$ segment for the presence of an additional break by partitioning the time intervals $\widehat{t_1}, \ldots\ldots\ldots, \widehat{t_l}$. In other words, partitioning the sample space into two spaces and then applying the least squares over each of the partitioned sample space can allow us to determine the structural breakpoints in the model.

The test amounts to the application of $(l+1)$ tests of the null hypothesis of no structural change against the alternative hypothesis of a single change. This test is applied to each segment containing the observations $\widehat{t_{i-1}} + 1$ to $\widehat{t_i}$ ($i= 1, \ldots\ldots, l+1$) using the convention that $\widehat{t_0} = 0$ and $\widehat{t_{l+1}} = t$. The model with $(l+1)$ breaks is rejected if the overall minimum value of the sum of squared residuals over all segments where an additional break is included is sufficiently smaller than the sum of squared residuals from the l break model. The break date is the one associated with this overall minimum. This test is defined as:

$$F_t(l+1|l) = \left\{ S_t(\widehat{t_1}, \ldots\ldots\ldots, \widehat{t_l}) - \min_{1 \le i \le l+1} \inf_{\tau \in \omega_{i,\gamma}} S_t(\widehat{t_1}, \ldots\ldots\ldots, \widehat{t_{i-1}}, \tau, \widehat{t_1}, \ldots\ldots\ldots, \widehat{t_l}) \right\} / \widehat{\sigma^2} \quad (12)$$

where,
$$\omega_{i,\gamma} = \left\{ \widehat{t_{i-1}} + (\widehat{t_i} - \widehat{t_{i-1}})\gamma \le \tau \le \widehat{t_i} - (\widehat{t_i} - \widehat{t_{i-1}})\gamma \right\} \quad (13)$$

$\widehat{\sigma^2}$ is a consistent estimate of σ^2 under the null hypothesis. (Note 4)

2.3 Data

Data on energy prices and fertilizer prices were obtained from various sources. The period of data collection was from December 1993 to January 2012. The crude oil spot prices, oil future prices, natural gas spot, and natural gas future prices were obtained from the Energy Information Administration (EIA). The crude oil prices were measured in dollars per barrel, while the natural gas prices were measured in dollars per million British Thermal Units (BTUs). The fertilizer prices were obtained from the World Bank Global Economic Monitor Database and were measured in dollars per metric ton. The potash and nitrogen market data were obtained from the U.S. Department of Agriculture (USDA) National Agricultural Statistics Service (NASS). Both the ARDL model and the GARCH model were used to evaluate the effects of oil, natural gas, and fertilizer prices, and to model the volatility in global fertilizer, oil, and natural gas prices.

2.4 Research Design

Our research design is as follows: In the first stage, we conduct an ARCH Lagrangian multiplier (LM) test to determine the volatility in fertilizer prices. In the second stage, using a Schwartz-Bayes information criterion (BIC), we determine the best model fit. In the final phase, we conduct a post-estimation analysis to validate our GARCH model.

3. Results

In this section, we first provide a few descriptive statistics and graphical analysis to highlight the volatility in oil, natural gas and various fertilizer prices. This is followed by the mean and volatility models of various fertilizer prices using appropriate ARMA-GARCH models. Finally, we validate our model using the null hypothesis of no ARCH-effects in our post-estimation analysis.

3.1 Descriptive Statistics

Table 1. Descriptive statistics of fertilizer prices (Dollars per metric ton): December 1993 to August 2012

Statistics	TS_t	$Urea_t$	PC_t	DAP_t	RP_t
Mean	249.7	198.35	211.16	299.98	76.61
Median	175.31	171	122.5	210.81	43
Maximum	1131.5	770	872.5	1200.63	430
Minimum	122.5	63	103.5	135	33
Std. Deviation	195.17	126.66	172.58	215.96	80.19
Skewness	2.73	1.82	2.08	2.43	2.72
Kurtosis	10.82	7.44	6.73	9.22	10.28
Jarque Bera	826.53	299.68	284.54	567.84	750.94
N	218	218	218	218	218

Source: Computed by the authors.

Table 1 shows the descriptive statistics for various fertilizer prices: Triple superphosphate (TS), urea, potassium chloride (PC), diammonium phosphate (DAP), and rock phosphate (RP), covering the full sample period. There is significant variation in fertilizer prices over the entire sample period as shown by the huge differences between the minimum and maximum values.

Table 2. Correlations of fertilizer prices: December 1993 to August 2012

	TS_t	$Urea_t$	PC_t	DAP_t	RP_t
TS_t		0.81	0.77	0.001	0.91
$Urea_t$	0.81		0.71	0.15	0.69
PC_t	0.77	0.71		-0.17	0.89
DAP_t	0.001	0.15	-0.17		-0.15
RP_t	0.91	0.69	0.89	-0.15	

The correlation coefficient from Table 2 suggests that triple superphosphate (TS) prices are highly correlated with urea, potassium chloride (PC), and rock phosphate (RP) prices. Urea prices are also highly correlated with potassium chloride prices, while potassium chloride prices are highly correlated with rock phosphate prices. Diammonium phosphate (DAP) prices are not correlated with any other fertilizer prices, while rock phosphate prices are highly correlated with TS, urea, and PC prices.

3.3 Graphical Analysis of Fertilizer Price Volatility

In this section, we provide a graphical analysis of oil prices, natural gas prices and fertilizer price volatility for our sample period.

Figure 1. Trend of Real Prices of Crude Oil Spot prices in Oklahoma, Europe and Crude Oil Futures-December 1993 to January 2012

As evident from Figure 1, the correlations between spot oil prices based in Oklahoma, spot oil prices based in Europe and forward prices of oil are highly correlated during the sample period with the correlation coefficient being 0.995 between spot oil prices based in Oklahoma and spot oil prices based in Europe. This may be because the transmission of information in these markets is instantaneous in this market.

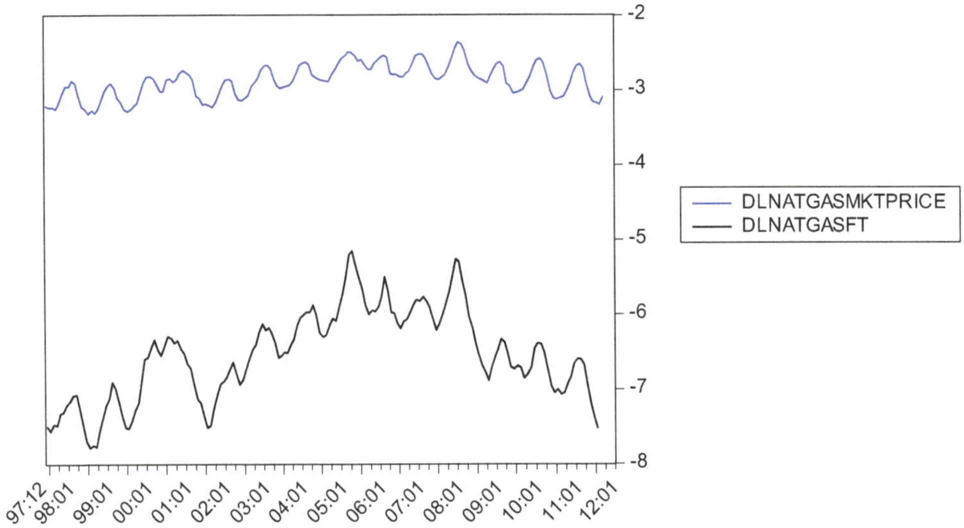

Figure 2. Trend of Real Natural Gas Spot and Future Prices-January 1997 to January 2012

Figure 2 shows the correlations between natural gas market prices and forward prices of natural gas. The correlation coefficient between these two indicators is also quite – 0.77.

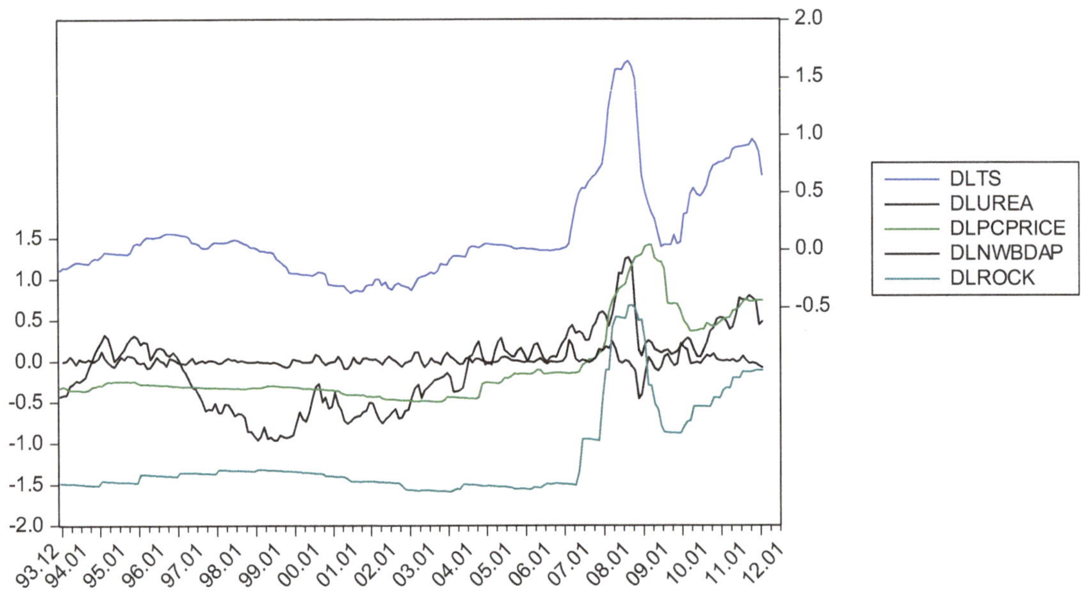

Figure 3. Trend of all Fertilizer Prices-December 1993 to January 2012

Figure 3 shows the price variation of all the fertilizer prices. As evident from this graph, all the fertilizer prices exhibited significant volatility during the crisis period- June 2007 to June 2008, which had implications for agricultural markets.

3.3 Statistics and Data Analysis: Pre-Estimation

The purpose of this section is to determine whether there is presence of ARCH effects in the fertilizer price series. This will then be consistent with large volatility in fertilizer prices over the sample period. Equation (14) specified the models of determining fertilizer prices, where FP stands for fertilizer price.

$$FP_t = \alpha_0 + \sum_{i=1}^{p} \alpha_i \, FP_{t-i} + u_t \; ; \; i = 1, 2, \ldots \ldots, p; \; t = 1, \ldots . T; u_t \sim IID(0, \sigma^2); |\alpha_i| < 1 \qquad (14)$$

Based on Engle's approach, an ARCH- LM test can be performed in two steps: first, estimate equation (14) by OLS and obtain the fitted residuals and second, regress the square of the fitted residuals on a constant and lags of the squared residuals. For example, estimate equation (14) as:

$$\widehat{u_t^2} = \theta_0 + \theta_1 \widehat{u_{t-1}^2 + \cdots \ldots \ldots + \theta_p u_{t-p}^2} + \varepsilon_t \qquad (15)$$

The LM test for the joint null hypothesis is given by: $H_0: \theta_1 = \theta_2 = \cdots = \theta_p = 0$. In other words, there is no ARCH effect present in the model. The usual *F*-test or the test statistic computed by multiplying the number of observations (n) by the coefficient of determination (R^2) obtained from the regression equation of (15) is used. The latter test statistic $n\,R^2 \sim \chi_p^2$, where p is the degree of freedom that equals the number of autoregressive terms in equation (15).

Table 3. ARCH test – Dependent variable: fertilizer prices with different autoregressive parameters and lag length of the dependent variable

Mode	Variables	*p=1*		*p=5*		*p =10*	
		F-test	$n\,R^2$	*F-test*	$n\,R^2$	*F-test*	$n\,R^2$
k= 1	*TS*	40.68***	34.50***	8.20***	35.2***	1.98**	19.02**
	Urea	15.85***	14.89***	9.31***	39.09***	5.089***	42.67***
	PC	1.49	1.494	3.76***	17.73***	2.26**	21.44**
	DAP	51.92***	42.17***	10.55***	43.24***	8.02***	60.11***
	RP	9.40***	9.09***	5.07***	23.25***	4.07***	35.65***
k =2	*TS*	3.97**	3.93**	2.12*	10.39*	2.41***	22.68***
	Urea	0.90	0.91	7.67***	33.24***	4.34***	37.51***
	PC	14.41***	13.62***	7.25***	31.71***	3.90***	34.35***
	DAP	2.93*	2.91*	2.71**	13.10**	3.40***	30.62***
	RP	2.91*	2.90*	4.81***	22.15***	3.26***	29.49***

Source: Computed by the authors.

Notes: Model follows the autoregressive process in equation (1) of order k =1 and 2 respectively and p denotes the lag length for the test statistics based on equation (2).

The asterisks ***, **, and * denote 1%, 5%, and 10% levels of significance respectively.

Table 3 shows the test-statistics for the existence of ARCH effects in the variables. All the fertilizer price series show evidence of ARCH effects as evidenced by the *F*-test and $n\,R^2$ up to 10 lags for the triple superphosphate series. All the test statistics are statistically significant at 1% and 5% respectively, thus rejecting the "no ARCH" hypothesis.

3.4 Sequential Bai-Perron Test Results for Structural Breaks

The sequential Bai-Perron multiple breakpoint tests were conducted for each of the fertilizer price series. The break-test options are given for a trimming value of 0.15 with the maximum number of breaks being 5. The results are reported in Table 4.

Table 4. Bai-Perron multiple breakpoint test for fertilizer price series based on *l+1* vs. *l* sequentially determined breaks

Fertilizer prices	Break tests	Scaled F-Statistic	Critical Value	Break Dates**
TS	0 vs 1*	101.37	8.58	1999 M09 and 2007 M05[a]
	1 vs 2*	199.15	10.13	
	2 vs 3*	73.41	11.14	
	3 vs 4*	57.09	11.83	
Urea	0 vs 1*	158.44	8.58	1997 M04 and 2006 M12[b]
	1 vs 2*	216.80	10.13	
	2 vs 3*	111.03	11.14	
	3 vs 4*	46.38	11.83	
PC	0 vs 1*	61.59	8.58	2005 M06
	1 vs 2*	71.25	10.13	
DAP	0 vs 1*	132.52	8.58	1999 M07 and 2007 M03[c]
	1 vs 2*	206.59	10.13	
	2 vs 3*	90.9	11.14	
RP	0 vs 1*	30.22	8.58	2007 M12
	1 vs 2*	18.14	10.13	
	2 vs 3*	30.84	11.14	
	3 vs 4*	30.43	11.83	
	4 vs 5*	26.99	12.25	

Source: Computed by the authors.

Table 4, column 5 identifies the break dates for each of the fertilizer price series. Most of these break dates correspond to the current fertilizer price volatility during the period June 2007 to June 2008. Additional break dates were also identified for each of the series but are not reported because the Bai-Perron multiple breakpoint tests the null of constant parameters against the alternative of two (unknown) breaks. If these breakpoints are denotes as $(\widehat{T_1}, \widehat{T_2})$, the Wald test for non-constancy of the parameters can be constructed. The least squares estimates $(\widehat{T_1}, \widehat{T_2})$ jointly minimize the sum squared errors. For 2 breaks, these estimates require $O(n^2)$ regressions which are cumbersome but feasible. On the other hand, for 3 breaks, these estimates require $O(n^3)$ regressions, which are not feasible. Thus, the two breakpoints of the null of constant parameters is used against the alternative of two unknown breaks.

3.5 Main Results

The empirical estimates for the alternative volatility models for the four fertilizer price series are given in Tables 5 through 8 for different time periods based on the structural break tests. Suitable models for triple superphosphate prices are ARMA(1,1) and GARCH (2,2) with threshold effects present for the entire sample period. For the period 2007 month six to 2012 month one, the appropriate model is ARMA(1,0) and GJR(1,1) (Note 5). The short-run persistence of shocks in the two periods are 0.469 and 0.934 respectively, while the long-run persistence of shocks are 0.899 and 0.954 respectively. These empirical outcomes suggest that the mean increase in triple superphosphate prices during the period 2007 month six to 2012 month one may be attributed to higher mean prices of crude oil and natural gas prices. Thus, from a policy perspective, it is important to understand the linkages between mean and volatility in energy prices and mean and volatility in fertilizer prices.

Table 5. Mean and volatility in triple superphosphate prices for Entire Sample Period (1993-2012) and for various Sub-Samples

Period	**1993 M12 to 2012 M01**	**2007 M06 to 2012 M01**
Series	ARMA(1,1)	ARMA(1,0)
(TS)	GARCH (2, 2) [a,b] with threshold effects present	GJR (1,1)
α_0	3.32***	2.27***
	(0.28)	(0.856)
Log Oil spot price (-1)	0.13***	0.446***
	(0.05)	(0.16)
Log Oil Spot price (-2)	0.60***	0.302*
	(0.04)	(0.128)
Log of Natural Gas price (-1)	-0.29***	0.223**
	(0.07)	(0.105)
AR(1)	0.86***	0.881***
	(0.015)	(0.053)
MA (1)	0.47***	
	(0.06)	
β_0	.001**	0.004**
	(.0008)	(0.002)
β_1	0.146***	0.721**
	(0.054)	(0.287)
β_2	0.215	
	(0.14)	
γ_1	0.54	0.025
	(0.46)	(0.211)
γ_2	-0.11	
	(0.28)	
δ_1	.215	0.426
	(0.14)	(0.59)
Short-run persistence[a]	0.469	0.934
Long-run persistence[b]	0.899	0.954
Second order stationarity condition[c]	0.791	
BIC	-2.28	-1.26

The asterisks ***, ** and * denote significance at 1%, 5% and 10% respectively.

[a] In case of GARCH (2,2) with threshold effects present, short-run persistence is defined as $(\sum\beta_i + \delta_1/2)$.

[b] In case of GARCH (2,2) with threshold effects present, long-run persistence is defined as $(\sum\beta_i + \sum\gamma_i + \delta_1/2)$.

[c] The second order stationarity condition is given by $\sum\beta_i + \sum\gamma_i < 1$.

Oil prices are the spot prices based on Oklahoma spot prices.

Table 6. Mean and volatility in urea prices for Entire Sample Period (1993-2012) and for various Sub-Samples

Period	**1993 M12 to 2012 M01**	**1993 M12 to 1997 M04**	**1997 M05 to 2006 M11**	**2006 M12 to 2012 M01**
Series (TS)	ARMA(1,1) GJR(1,1)	ARMA(1, 1) GJR(1,1)	ARMA(1,1) GJR(1,1)	ARMA(1,1) GJR(1,1)
Mean Equation				
α_0	6.64*** (1.39)	4.78*** (0.79)	1.95*** (0.22)	2.77*** (0.81)
Log Oil spot price (-1)	-0.157** (0.07)	0.15 (0.12)	0.602*** (0.079)	0.53*** (0.16)
Log of Natural Gas price (-1)	0.043 (0.07)	-0.08 (0.12)	0.346** (0.138)	0.29* (0.15)
AR(1)	0.98*** (0.01)	0.86*** (0.05)	0.595*** (0.079)	0.66*** (0.09)
MA (1)	0.35*** (0.07)	0.66** (0.07)	0.452*** (0.059)	0.44*** (0.14)
Variance Equation				
β_0	0.0005** (0.0002)	0.002*** (7.39E-06)	0.007*** (2.56E-06)	0.0004 (0.0005)
β_1	0.20** (0.102)	-0.24 (0.16)	-0.036 (0.25)	0.19 (0.11)
β_2				
γ_1	0.89*** (0.05)	0.55*** (0.23)	0.358*** (0.134)	0.96*** (0.12)
γ_2				
δ_1	-0.25** (0.114)	0.34 (0.37)	-.093 (0.237)	-0.34*** (0.11)
Short-run persistence[a]	0.075	-0.07	-0.081	0.02
Long-run persistence[b]	0.965	0.48	0.277	0.98
BIC	-1.876	-2.09	-1.439	-1.033

In case of GJR (1, 1) model, short-run persistence is defined as (β_1+ $\delta_1/2$).

In case of GJR (1, 1) model, long-run persistence is defined as (β_1+γ_1+ $\delta_1/2$).

The asterisks ***, ** and * denote significance at 1%, 5% and 10% respectively.

The BIC criterion is used to determine the optimal lag length in the variance equation.

** There was significant presence of autocorrelation in the residuals for the 1993 month 12 to 2012 month one and we were unable to correct for this using first difference specification.

For the potassium chloride price series, no single model appropriately describes the mean and volatility of the prices. The second order stationarity conditions were also not satisfied for each time period and thus we do not report these results

Table 7. Mean and volatility in DAP prices for Entire Sample Period (1993-2012) and for various Sub-Samples

Period	**1993 M12 to 2012 M01**	**1993 M12 to 1999 M06**	**1999 M07 to 2007 M02**	**2007 M04 to 2012 M01**
Series (TS)	ARMA(1,1) GARCH(2,1)	ARMA (1,1) GARCH (2,1) with threshold effects present	ARMA(2, 2) GARCH (2,1)	ARMA(1,1) GARCH (2,1)
Mean Equation				
α_0	4.62*** (0.13)	5.33*** (0.21)	4.33*** (0.71)	2.31*** (0.65)
Log Oil spot price (-1)	0.144*** (0.02)	-0.06 (0.038)	0.076*** (0.019)	0.36*** (0.09)
Log Oil Spot price (-2)		0.09** (0.04)		0.334*** (0.04)
Log of Natural Gas price (-1)	0.211*** (0.05)	-0.05 (0.044)	0.104*** (0.039)	0.352*** (0.128)
AR(1)	0.96*** (0.01)	0.76*** (0.09)	0.599*** (0.177)	0.89*** (0.07)
AR (2)			0.379** (0.173)	
MA(1)	0.60*** (0.08)	0.47** (0.07)	1.38*** (0.13)	0.73*** (0.06)
MA (2)			0.559*** (0.102)	
Variance Equation				
β_0	0.002*** (9.45E-06)	0.002 (0.0004)	0.001*** (0.0003)	0.0007*** (2.36E-05)
β_1	0.09 (0.13)	-0.07 (0.06)	0.877** (0.463)	-0.08 (0.092)
β_2	-0.08* (0.05)	0.06 (0.069)	0.64* (0.35)	0.166* (0.09)
γ_1	0.36*** (0.125)	0.64 (0.54)	-0.871*** (0.067)	0.75*** (0.13)
γ_2				
δ_1		0.085 (0.22)		
Second order stationarity condition[c]	0.37	0.88	0.646	0.836
BIC	-2.823	-3.731	-3.773	-1.898

The asterisks ***, ** and * denote significance at 1%, 5% and 10% respectively.

[c] The second order stationarity condition is given by $\sum\beta_i + \sum\gamma_i < 1$.

The BIC criterion is used to determine the optimal lag length in the variance equation.

Table 8. Mean and volatility in rock phosphate prices for Entire Sample Period (1993-2012) and for various Sub-Samples

Period	**1993 M12 to 2012 M01**	**1993 M12 to 2007 M11**	**2007 M12 to 2012 M01**
Series (TS)	ARMA(2,2) GARCH(3,1)	ARMA(2,2) GARCH(3,1)	ARMA (1,1) GARCH (2,2) with presence of threshold effects
Mean Equation			
α_0	3.6*** (0.07)	3.55*** (0.112)	2.70*** (0.51)
Log Oil spot price (-1)	0.037*** (0.01)	0.044*** (0.017)	0.51*** (0.09)
Log of Natural Gas price (-1)	0.016 (0.02)	0.089*** (0.019)	0.126* (0.07)
AR(1)	0.87*** (0.03)	0.45 (1.17)	0.90*** (0.03)
AR (2)		0.50 (1.14)	
MA(1)	0.28*** (0.04)	0.82 (1.06)	-0.05 (0.05)
MA (2)		0.18 (0.26)	
Variance Equation			
β_0	0.0002*** (3.87E-05)	0.0004*** (2.03E-05)	7.53E-05 (0.0002)
β_1	0.39** (0.179)	0.46 (0.62)	0.01 (0.07)
β_2	0.05 (0.09)	0.05 (0.26)	0.52** (0.26)
β_3	0.04 (0.06)	-0.05 (0.04)	
β_4			
γ_1	-0.32 (0.23)	0.19 (0.27)	0.19 (0.13)
γ_2			0.21** (0.10)
γ_3			
γ_4			
δ_1			-0.04 (0.07)
Second order stationarity condition[c]	0.16	0.653	0.93
BIC	-5.25	-4.33	-1.03

The asterisks ***, ** and ** denote significance at 1%, 5% and 10% respectively.

[c] The second order stationarity condition is given by $\sum\beta_i + \sum\gamma_i < 1$.

The BIC criterion is used to determine the optimal lag length in the variance equation.

For the urea price series, an appropriate model is ARMA (1, 1) and GJR (1, 1) for the different time periods. The short-run persistence of shocks for the periods 1993 month twelve to 2012 month one, 1993 month twelve to 1997 month four, 1997 month five to 2006 month eleven, and 2006 month twelve to 2012 month one are respectively 0.075, -0.07, -0.08, and 0.02, while the long-run persistence of shocks are 0.965, 0.48, 0.28, and 0.98 respectively. These empirical estimates imply that short-run shocks do not persist, while the long-run shocks continue to persist. This may suggest that policy makers can look into the long-run relationship between oil price, natural gas price, and urea prices and find the equilibrium relationship between these prices. (Note 6)

For the DAP price series, the ARMA (1, 1) and GARCH (2, 1) describe the mean and volatility of the price series for the entire sample. Again, the second order stationarity conditions are satisfied for each period. It is interesting to note that the effects of lagged crude oil spot prices and natural gas prices is significant during the period 2007 month four to 2012 month one and has a much higher mean effect on DAP price series than other time periods. This again suggests higher and significant effects of the mean of crude oil prices and natural gas prices on DAP price series after the crisis period.

For the rock phosphate price series the ARMA (2, 2) and GARCH (3, 1) describe the mean and volatility equations appropriately. There is strong mean effects of lagged crude oil price and lagged natural gas prices on rock phosphate prices during the period 2007 month twelve to 2012 month one. The ARCH and GARCH effects are only significant during the period 2007 month twelve to 2012 month one.

Overall, the results suggest that appropriate volatility models for all the time periods can only be described for the urea and DAP price series, but not for other fertilizer price series. The results also suggest significant ARCH and GARCH effects for all the fertilizer prices during the period January 2007 to January 2012. This suggests that the lagged conditional variance and the squared error terms have significant effects on the conditional variance equation. Also the mean effects of lagged crude oil prices and lagged natural gas prices considerably have positive and significant impact on the fertilizer price series. This result corroborates with other studies that suggest higher and significant effects of the mean of crude oil prices and natural gas prices on DAP price series after the crisis period. (Note 7)

3.6 Post-Estimation and Validation

This study also conducted some post-estimation analyses to determine if the volatility models captured these effects. The post-estimation ARCH test is carried out using both the *F*-test and chi-square distributed $n R^2$ test. The results are presented in Table 9 and do not reject the null hypothesis of no ARCH effects. Most of the values are not statistically significant at the conventional level of significance. Thus, this study further validates the theoretical literature that ARCH/GARCH models are suitable for dealing with volatility in fertilizer prices.

4. Discussion

This paper examined the effects of crude oil price volatility and natural gas price volatility on fertilizer price volatility. This relationship is important as agriculture demands and supplies both energy inputs and energy outputs. To model the volatility in fertilizer prices, both symmetric models GARCH (1, 1) and asymmetric models GJR (1, 1) were considered. One interesting innovation of this study was to evaluate the effects of volatility over different time periods using Bai-Perron structural break tests. Results showed that the mean changes in oil prices and natural gas prices contributed to the increase in the mean of various fertilizer prices after the crisis period i.e. from June 2007 to June 2008. In addition, both the ARCH and GARCH effects were significant after the crisis period, suggesting that the volatility effects of oil and natural gas prices on fertilizer prices were also significant. The out-of-sample forecasts of the effect of higher energy prices on fertilizer prices were also considered. The results showed that, depending on the scenarios postulated, the maximum impact of higher energy prices fall on triple superphosphate and diammonium phosphate. The implication of this result is that higher production costs will add significantly to total expenditures for crop producers. These higher production costs, coupled with lower output prices, are likely to have a negative effect on farm profitability, thus reducing the investment levels in the farm sector.

Table 9. Post-estimation analysis of presence of ARCH effects of various fertilizer prices for the entire sample (1993-2012) and for various Sub-Samples

Fertilizer price	Period	*p=1*		*p=5*		*p =10*	
		F-test	$n\,R^2$	*F-test*	$n\,R^2$	*F-test*	$n\,R^2$
TS	1993 M12 to 2012 M01	0.048	0.048	0.33	1.70	2.49***	23.38***
	1993 M12 to 1999 M09	0.008	0.008	0.20	1.13	0.26	3.05
	1999 M10 to 2007 M05	0.172	0.176	0.171	0.909	0.14	1.59
	2007 M06 to 2012 M01	0.41	0.43	0.57	3.05	0.59	6.64
Urea	1993 M12 to 2012 M01	6.54**	6.40**	1.51	7.53	0.87	8.86
	1993 M12 to 1997 M04	0.34	0.36	0.46	2.61	0.59	7.22
	1997 M05 to 2006 M11	0.101	0.103	0.89	4.54	0.69	7.24
	2006 M12 to 2012 M01	1.52	1.53	0.88	4.53	1.40	13.28
PC	1993 M12 to 2012 M01	2.42	2.42	1.19	5.99	0.82	8.33
	1993 M12 to 1997 M04	0.081	0.085	0.58	3.23	0.54	6.72
	1997 M05 to 2006 M11	0.016	0.017	0.06	0.33	0.15	1.70
	2006 M12 to 2012 M01	0.31	0.32	0.22	1.21	0.38	4.47
DAP	1993 M12 to 2012 M01	0.07	0.07	1.16	5.80	3.38***	30.43***
	1993 M12 to 1999 M06	0.053	0.055	0.24	1.34	1.38	13.16
	1999 M07 to 2007 M02	2.58	2.57	1.11	5.62	1.09	10.96
	2007 M04 to 2012 M01	0.041	0.043	0.30	1.64	0.63	7.03
RP	1993 M12 to 2012 M01	0.002	0.002	0.05	0.28	3.46***	31.06***
	1993 M12 to 2007 M11	3.32*	3.21*	0.68	3.49	0.38	4.00
	2007 M12 to 2012 M01	1.16	1.18	0.43	2.39	0.81	8.73

Source: Computed by the authors (2015).

The asterisks ***, **, and * denote 1%, 5%, and 10% levels of significance respectively.

p denotes the lag length for the test statistics based on equation (15).

References

Bai, J., & Perron, P. (1998). Estimating and Testing Linear Models with Multiple Structural Changes. *Econometrica, 66*(1), 47-78. http://dx.doi.org/10.2307/2998540

Beckman, J., Borchers, A., & Jones, C. A. (2013). *Agriculture's Supply and Demand for Energy and Energy Products*, EIB-112, Economic Research Service, USDA, May 2013.

Bollerslev, T. (1986), Generalized Autoregresive Conditional Heteroscedasticity, *Journal of Econometrics, 31*, 307-327. http://dx.doi.org/10.1016/0304-4076(86)90063-1

Chen, P. Y., Chang, C. L., Chen, C. C., & McAleer, M. (2013). Modelling the Effects of Oil Prices on Global Fertilizer Prices and Volatility. *Journal of Risk and Financial Management, 5*, 78-114. http://dx.doi.org/10.3390/jrfm5010078

Du, X., & McPhail, L. L. (2012). Inside the Black Box: the Price Linkage and Transmission between Energy and Agricultural Markets. *The Energy Journal, 33*(2), 171-194. http://dx.doi.org/10.5547/01956574.33.2.8

Engle, R. (1982). Autoregressive Conditional Heteroskedasticity with Estimates of the Variance of United Kingdom Inflation. *Econometrica, 50*, 987-1007. http://dx.doi.org/10.2307/1912773

Engle, R. (2002). New Frontiers for ARCH Models. *Journal of Applied Econometrics, 17*, 425-446. http://dx.doi.org/10.1002/jae.683

Galbraith, C. (2010). An Examination of Factors Influencing Fertilizer Price Adjustment, *Paper presented at the*

Agricultural and Applied Economics Association 2010 AAEA, CAES and WAEA Joint Annual Meeting, Denver, Colorado, July 25-27, 2010.

Huang, W-yuan (2009), *Factors Contributing to the Recent Increase in U.S. Fertilizer Prices, 2002-08,* Report No. AR-33, Economic Research Service, USDA, February 2009.

Kenkel, P. (2010). *Causes of Fertilizer Price Volatility*, Division of Agricultural Sciences and Natural Resources, Oklahoma State University. Retrieved from http://agecon.okstate.edu/coops/files/agec-261.pdf

Nelson, C. R., & Plosser, C. I. (1982). Trends and Random Walks in Macroeconomic Time Series, *Journal of Monetary Economics, 10*, 139-162. http://dx.doi.org/10.1016/0304-3932(82)90012-5

Nelson, D. (1991). Conditional Heteroskedasticity in Asset Returns: A New Approach. *Econometrica, 59*, 347-370. http://dx.doi.org/10.2307/2938260

Ott, H. (2012), *Fertilizer markets and their interplay with commodity and food prices*, JRC Scientific and Policy Reports, European Commission. Retrieved from http://www.jrc.ec.europa.eu/

Phillips, P. C. B., &Perron, P. (1988). Testing for a Unit Root in Time Series Regression. *Biometrika, 75*, 335-346. http://dx.doi.org/10.1093/biomet/75.2.335

Salisu, A. A., & Fasanya, I. O. (2012). Comparative Performance of Volatility Models for Oil Price, *International Journal of Energy Economics and Policy, 2*(3), 167-183.

Notes

Note 1. The GJR model is a GARCH variant that includes leverage terms for modeling asymmetric volatility clustering. In the GJR formulation, large negative changes are more likely to be clustered than positive changes. The GJR model is named for Glosten, Jagannathan, and Runkle (1993).

Note 2. The exponential GARCH (EGARCH) model is a GARCH variant that models the logarithm of the conditional variance process. In addition to modeling the logarithm, the EGARCH model has additional leverage terms to capture asymmetry in volatility clustering. The EGARCH(P,Q) model has P GARCH coefficients associated with lagged log variance terms, Q ARCH coefficients associated with the magnitude of lagged standardized innovations, and Q leverage coefficients associated with signed, lagged standardized innovations. We considered the EGARCH model as an additional specification in this study. However, this model did not perform well with respect to the BIC criterion. Thus, we do not report these results.

Note 3. All the variables in the above equation were converted to natural logarithms so that second-order moment conditions were satisfied.

Note 4. Bai and Perron (1998) show that for the test to have better power, $\widehat{\sigma^2}$ should also be consistent under the alternative hypothesis. The advantage of the above test is that it also allows the results to carry through for different distributions across segments for the regressors and the errors.

Note 5. We conducted regression models for the period 1993 month twelve to 1999 month nine and 1999 month ten to 2007 month five using ARMA(1,1) GJR(1,1) and ARMA(1,1) GARCH(2,2). As the residuals displayed significant autocorrelation owing to the short-sample size, we could not correct for this using first difference specifications. Thus, these results are not reported and may be obtained from the authors upon request.

Note 6. Determining the equilibrium relationship between energy and fertilizer prices and the substitution between crops and biofuels in a production function framework however is not trivial as it requires the formulation of full microeconomic model that depicts these relationships. Such an exercise is beyond the scope of this paper.

Note 7. For example, Chen et al. (2013) finds that the mean effect of oil price on global fertilizer prices was positive and significant after the crisis period. However, the authors did not incorporate the mean effects of natural gas prices on global fertilizer prices in their model.

10

What Can Be Learned about the Adaptation Process of Farming Systems to Climate Dynamics Using Crop Models?

Sandro L. Schlindwein[1], Frank Eulenstein[2], Marcos Lana[2], Stefan Sieber[3], Jean-Philippe Boulanger[4], Edgardo Guevara[5], Santiago Meira[5], Elvira Gentile[6] & Michelle Bonatti[3]

[1] Federal University of Santa Catarina, Florianópolis, Brazil

[2] The Leibniz Centre for Agricultural Landscape Research, Institute of Land Use Systems, Müncheberg, Germany

[3] The Leibniz Centre for Agricultural Landscape Research, Institute of Socio-Economics, Müncheberg, Germany

[4] ECOCLIMASOL Climate Risk Management Solutions, Buenos Aires, Argentina

[5] INTA – National Institute of Agricultural Technology, Pergamino, Argentina

[6] Department of Geography, University of Buenos Aires, Buenos Aires, Argentina

Correspondence: Sandro L. Schlindwein, Federal University of Santa Catarina, CCA – ENR, CP. 476, 88040-900 Florianópolis (SC), Brazil. E-mail: sandro.schlindwein@ufsc.br

Abstract

The objective of this paper is to reflect and discuss how the use of crop models by aware practitioners might trigger learning of how to think and act differently about the adaptation process of farming systems to climate dynamics. The development of adaptation strategies is discussed from the perspective of contrasting metaphors, since the metaphors in use have distinctive practical implications regarding how crop models might be used for adaptation purposes. Further, in this paper it is pointed out that adaptation should be understood as the result of a learning process and therefore the use of crop models for adaptation purposes must be transformed. Instead of seeing them only as tools to secure yield of cropping systems under a changing climate they must be conceived as components of learning systems for adaptation of whole farming systems.

Keywords: adaptation, climate change, crop models, learning

1. Introduction

The scientific evidence of human induced global warming and climate change is very strong. Increasing is also society´s awareness about these phenomena and their projected impacts. Less agreement is found however, on the extension of these impacts on ecosystems and society, and how to tackle them properly. As has been discussed by Ison (2010), this is because climate change is among those situations that should be considered as "wicked problems", or "wicked situations", since it is a 'pressing and highly complex policy issue involving multiple causal factors and high levels of disagreement about the nature of the problem and the best way to tackle it'.

Two main general approaches have been adopted so far to tackle global warming and climate change and its impacts: mitigation and adaptation. Simply stated, mitigation measures involve the control and reduction of the emission of greenhouse gases like CO_2, or enhancing the sinks of these gases to keep future warming as low as possible. Mitigation will not be discussed in further detail in this paper, and the interested reader is referred to the vast literature already existing on this issue (as in Fischer, Shah, Tubiello, & van Velhuizen, 2005; Tubiello & Fischer, 2007; Ortiz et al., 2008). Adaptation, on the other hand, is related to changes carried out in socio-ecological systems (as in farming systems) in response to current and expected impacts of climate change aiming to moderate harm or exploit beneficial opportunities (based on Moser & Ekstrom, 2010).

Agriculture is the most climate dependent human activity, and the extent to which farming systems will be affected by climate dynamics (climate change and variability) might vary across crops and regions and adaptation scenarios, as the IPCC projections have shown (Intergovernmental Panel on Climate Change [IPCC], 2014). Accordingly, also the adaptation measures will be different from site to site, and as outlined by Howden et al. (2007) the immense diversity of farming systems and practices will offer a large array of possible adaptation options. Nevertheless, all these site specific adaptation options share some common assumptions, as for example that adaptation should be based on the reduction of the vulnerability of the considered farming systems to climate threats, keeping the risks of negative impacts as low as possible. Therefore, adaptation refers to an ongoing, never ending local decision making process rather than to a final state or condition of a system of interest resulting from the adoption of some global policies or initiatives. This 'local' perspective on adaptation has also been claimed by Smit, Burton, Klein and Wandel (2000), for whom "adaptation depends fundamentally on the characteristics of the system of interest, including its sensitivities and vulnerabilities".

To develop adaptation measures for cropping systems, some of the impacts of climate dynamics on them have been evaluated with the use of crop models (Ewert, 2004; Challinor, Wheeler, Craufurd, Slingo, & Grimes, 2004; Challinor, Ewert, Arnold, Simleton, & Fraser, 2009), which are models that reproduce the growth and development of a certain crop (more details on item 3). With the aid of such models, it is possible, for example, to project and evaluate the effects of increasing temperatures and radiation on crop biophysical responses, development and yield. Crop models have been originally developed for application at the field (local) level, but their application at larger scales as for climate change assessments has become a common practice (Adam, van Bussel, Leffelaar, van Keulen, & Ewert, 2011). Therefore, crop models have been seen as a tool to support the process of developing adaptation strategies of cropping systems to climate dynamics because they allow to evaluate its effects on crop development and yield. To put it in other terms, crop models have been adopted to improve the understanding of the impacts of climate dynamics on crop development and yield, and therefore adopted to assist the development of cropping systems adapted to a changing and variable climate.

However, accepting the claim presented at the beginning of this paper that climate change characterizes a "wicked problem situation", an approach to develop adaptation measures based on the adoption of single tools as crop models raises a number of key issues, particularly if they can deal effectively with such a situation considering its complexity. Therefore, for the purpose of this paper, a very important issue is how crop models could be integrated into learning systems for adaptation, and what can be learned about the adaptation process with their use.

The objective of this paper is to reflect and discuss how the use of crop models by aware practitioners might not only enact the design and implementation of cropping systems adapted to a changing climate, but whether they might also trigger learning of how to think and act differently about the adaptation process of whole farming systems to improve the wicked situations within which these systems are embedded.

2. How Should We Understand Adaptation to Climate Change?

For the IPCC in its Fifth Assessment Report (AR5) "adaptation [to climate change] is the process of adjustment to actual or expected climate and its effects, seeking in human systems to moderate harm or exploit beneficial opportunities" (IPCC, 2014). In the IPCC's Fourth Assessment Report, Easterling et al. (2007) have discussed this issue considering its implications for food, fiber and forest products. They assumed that adaptation means both the actions of adjusting practices, processes and capital in response to the actuality or threat of climate change as well as changes in the decision environment, such as social and institutional structures, and altered technical options that can affect the potential or capacity for these actions to be realized. It follows that adaptation might be understood as a vulnerability reducing process relying not only on the adoption of "hard" measures based on the engineering of existing biophysical systems of interest (as cropping systems), but also as a process relying on "soft" measures related to social, political and historical factors of the context within which these systems are embedded. At the end, adaptation certainly relies on a combination of "hard" and "soft" measures aiming to reduce the vulnerability of the considered systems of interest.

Although in recent years there have been notable developments regarding knowledge and practice on adaptation to climate change and variability, what we can notice so far, as in Adger et al. (2007), is that despite its complexity most of these adaptation developments still consist on the proposition of a set of practices to be potentially adopted everywhere, following the same path adopted in case of mitigation (as almost always has been the case of global mitigation initiatives).

According to Collins and Ison (2009), etymologically adaptation means 'fitted or suited' and to adapt is 'to fit' or 'make suitable'. To discuss the notion of adaptation, these authors explore two metaphors that for the design of adaptation strategies have far reaching practical implications, in science as well as in policy and in intervention. For them, one of the metaphors is that of 'adaptation as fitting into', that corresponds to the most widespread understanding of the process of adapting to a given situation. Interpreted according to this metaphor, adaptation unfolds based on the assumption that both the situation and what is to be adapted are known in advance. They are predetermined. It seems reasonable to assume that most of the already proposed practices on adaptation of agriculture to climate dynamics are based on this understanding, and a significant implication of it is that the same set of adaptation practices recommended for a given situation could be transferred to different contexts. A good example here of this type of adaptation practice in farming systems is the introduction of crops cultivars more tolerant to droughts, or the adoption of agronomic practices to minimize droughts effects.

The other organizing metaphor discussed by Collins and Ison (2009) is that of 'adaptation as co-evolution'. This metaphor is based on the understanding that adaptation results from a process of mutual interaction between the system of interest (a farming system) and its environment. So, in human systems of interest adaptation can be seen as a process of learning and development. For Collins and Ison (2009) the 'adaptation as co-evolution' metaphor draws attention to the systemic concept of 'emergence', because 'adaptation' might be seen as an emergent property of a co-evolutionary process. From this perspective on adaptation follows that current farming systems are adapted to their biophysical environments as a result of co-evolutionary processes in which humans have learned how to manage them under circumstances of high (also climatic) complexity and uncertainty. Landraces might be considered the typical result of this kind of adaptation process. The persistence of Agriculture in human history and its ubiquity might be seen therefore as an emergent property of this relational dynamics between humans and their changing biosphere.

A further practical implication of adaptation conceptualized as co-evolution, is that adaptation practices must be seen as context or site-specific, and they need to be designed taking into consideration local site features. From this perspective, adaptation would be the result of recursive interactions between the system of interest and its environment over time. For Ison (2010) to understand adaptation as co-evolution seems to be the only way forward for managing in a climate changing world, and he makes the point that this requires an effective form of praxis – a systems practice.

However, conceptualizing a process like adaptation to climate dynamics in terms of metaphors may not only reveal some important features about the nature of this process, but also conceal or hide others. As has been warned by Collins and Ison (2009), climate change adaptation seen as the result of a co-evolutionary process cannot be restricted to studies of climate change *sensu stricto*, but as a systemic praxis it will have to consider all domains of human concern.

Furthermore, there are certainly other possible metaphors underlying the expression 'adaptation to climate change' as even Collins and Ison (2009) discuss. In his book Ison (2010) discusses briefly another metaphor for adaptation which is 'adaptation for'. This metaphor relies on the perspective that in a climate changing world it is necessary to build a discourse and praxis of hope to allow the construction of new horizons of knowing and doing. Therefore, conceptualizing the adaptation process of farming systems in these terms might implicate a radical change of perspective. To give a concrete example, in a study published by Assad et al. (2008) the impacts of climate change on the geography of Brazilian agriculture are discussed, showing that it could be profoundly modified, displacing crops from one region to another, with significant (negative) socio-economic impacts, demanding major adaptations efforts in order to keep it as it is currently. Although this could well be the case, this perspective (typically based on the "adaptation as fitting into" metaphor) simply ignores all the new opportunities and possibilities that this change in the geography of agricultural production bears up, triggering structural changes and new forms of practice in whole regions, opening up new spaces for learning. This is also a good example of how it is important to develop awareness of the metaphor in use, because it is framing the features of the adaptation process in place.

2.1 Adaptation to Climate Change as a Learning Process

The conceptualization of adaptation as 'co-evolution', as a learning and development process, has some distinct practical implications. Accepting this conceptualization of adaptation, implies to accept and to consider that since it is not possible to completely know the situation in the long-term, it is also not possible to know in advance what is to be adapted. As has been briefly described, adaptation will result from the process of mutual, recursive interactions between the changing climate situation and the considered systems of interest (as for example, a farming system). This recurrent process of adapting to changing climate situations, based on the transformation of the systems of interest as well as of their environments, might be thus understood as the result of a learning process.

Learning itself is a very complex process, and for the purpose of this paper it is enough to assume that learning emerges in the process of engaging with a situation with the purpose of improving it. It means that learning implies change of some form and in some degree. So, learning might emerge when in the process of engaging with a climate change situation a change of understandings leads to changes on practices. But here it must be stressed that this understanding on learning is much different than that of O'Neill (2008) since based on a different epistemology. In his discussion about learning in a climate change context this author assumed an instrumental objectivist epistemology and defined learning as the acquisition of new information that leads to changes (to reductions?) in uncertainty. However, considering what has been discussed so far regarding the metaphors of adaptation in use, it might be expected that learning will also happen when moving from the understanding of adaptation based on the fitting into metaphor to the co-evolution metaphor, because this move from one understanding to another will enact new forms of practice, and not because it has any noticeable impact on the existent level of uncertainty. Therefore, according to the perspective espoused in this paper, the aim is not to change the level of uncertainty, but to change the way we engage with it.

Ultimately, the key issue regarding adaptation to climate change goes much beyond than identifying a set of practices or measures fitted to a given situation, although this could be necessary in some moment along the adaptation process. Rather, the key issue is how to set up an adaptation process based on learning and development in a situation characterized by uncertainty, exploring the opportunities that the conceptualization of adaptation as 'co-evolution' provides for the design of learning systems for concerted social action towards adaptation to a dynamic climate.

Therefore, this learning perspective on adaptation to climate dynamics has important implications considering how crop models have been and could be used for adaptation purposes. The challenge is how to use them as part of an integrated approach for effectively engaging stakeholders in the adaptation process, creating awareness of the multitude of aspects involved in this process, and in the whole situation.

3. Crop Models

In crop modeling, a large variety of models have been developed since the 1960s. Crop models have many current and potential uses for answering questions in research, crop management, and policy (Boote, Jones, & Pickering, 1996), as for example in many of the questions related to the impacts of climate change and variability on cropping systems. Briefly defined, crop models, or crop simulators, are computer programs that mimic the growth and development of crops. Data of weather, soil, and crop management are processed to predict crop ontogenesis, phenology, maturity date, yield, fertilizer´s efficiency and other important elements of crop production. The calculations performed in crop models are based on the existing knowledge of the physics, physiology and ecology of crop responses to the environment (United States Department of Agriculture [USDA], 2007). There exists a large literature on crop models (Sirotenko, 2001; Batchelor, Basso, & Paz, 2002; Lobell & Burke, 2010), and details about their features will not be discussed here. A good overview on them, particularly on the DSSAT (Decision Support System for Agrotechnology Transfer) package, that incorporates models of different crops with software that facilitates the evaluation and application of crop models for different purposes, can be read in Jones et al. (2003).

The adoption of crop models in many different agricultural situations and for many different purposes is not new. For Boote, Jones and Pickering (1996), a number of crop growth models have been adopted to evaluate the consequences of global climate change, as increasing temperatures and altered rainfall patterns on crop yields. However, for these authors, the efforts to evaluate the effects of a changing climate on yields adopting crop models should be placed with more emphasis on year-to-year fluctuations caused by weather variability. As has been claimed by Hansen and Jones (2000) dynamic, process-level crop models are playing an increasing role in translating information about climate variability into prediction and recommendations at a range of scales, tailored to the needs of agricultural decision makers. Crop models have been also adopted to evaluate agricultural production risks across a wide range of climatic conditions. Despite the high uncertainty of this sort of analysis, particularly for large time scales, they have been adopted as a tool to support strategic decision making in crop production and agricultural land use planning.

For Jones et al. (2003) in the development of DSSAT the systems approach provided "a framework to conduct research and to understand how the system and its components function". Although the "systemic" nature of crop models have been mentioned by these authors, it is necessary to address briefly what is commonly understood, in conceptual terms, as "the systems approach" in the context of crop modeling, as well as its implications for practice. It is clear from the statement of Jones et al. (2003) that they assume the existence of well-defined "cropping systems", existing as discrete entities with distinguishable boundaries over which crop models can be developed and the resulting models will be run. This stance is typical of a first-order systems approach. For Russell and Ison (2000), the first-order tradition, on which this systems approach is based, is characterized by concerned intervention, the definition of clear goals, the 'naming' of the problem, and the proposal of a rational solution. Accordingly, this systems approach should allow the understanding not only of the functioning of the system, but also allow the prediction of its behavior under given conditions, as well as how to manage or to control it (Jones et al., 2003). Furthermore, DSSAT, which is currently the most used package for running crop models for different purposes, is based on the assumption that knowledge (in form of production technology) can be transferred, an epistemological stance also typically linked to first-order traditions.

The epistemological assumptions behind crop model building is an important aspect that must be considered when using crop models for managing or intervening in "messy" situations or "wicked problems" like climate change. "Messy" or "wicked" situations are highly complex, characterized by uncertainty, interdependencies and which can be observed and interpreted from multiple perspectives. As a consequence, if systemic failures are to be avoided, any intended intervention on such situations can no longer be undertaken by means of prediction and control. These situations simply don't follow deterministic rules. Rather, human intervention on such situations must be based on the acceptance that they are intrinsically unpredictable. It means that for effectively managing such situations of complexity it is necessary to espouse an epistemological stance different than first-order traditions making possible the enactment of new forms of practice. This must also include the way crop models have been used in the development of adaptation strategies. But how to transform crop models into learning devices for raising systemic awareness about the situation?

4. Transforming the Use of Crop Models: From Tools to Optimize Crop Yields to Components of Learning Systems for Adaptation

For Mingers and Brocklesby (1999) a tool is an artifact, often a computer software, that can be used in performing a particular technique, or a specific activity. The particular technique or specific activity that crop models can perform is, for example, to identify the growth factors that contribute most to the achievement of a desired yield level of a cropping system, optimizing it. Thinking in terms of the metaphors discussed so far, this use of crop models as a tool for optimizing purposes can be characterized according to the 'fitting into' metaphor, since it is based on the assumption that the situation (in this case restricted to the climate itself or to some of its variables) as well as what is to be adapted (a cropping system) are known in advance. Then, crop models are used to fit a cropping system into the given environment, setting the practices to be adopted to achieve the desired goal: a certain optimized yield level.

The issue of concern here, however, is not if this "fitting process" is effective or not for the design of cropping systems eventually adapted to a changing climate, but if from this modality of use of crop models learning (of what type? to what extent?) about the adaptation process of a whole farming system to a dynamic climate can emerge. It must be clear at this point that although learning about adaptation to climate dynamics has not been an explicit objective of crop models, their use also does not preclude it at all. However, the point being made here is that depending on the perspective driving the use of crop models for adaptive practices learning can be constrained instead of being enabled. Consequently, to adopt crop models only as a tool to perform simulations to maximize crop yield seems not to be very helpful to improve or to transform the wicked situations in which these cropping systems are embedded, despite their value in assisting decision making in agricultural land use at local levels (Note 1).

To move forward "beyond the dominance of empiricist prediction" it is necessary to depart from the view of crop models only as tools to fix cropping systems under a changing climate. Just to give an example of how to avoid a recipe-like use of crop models, they could be regarded as part of a set of guidelines to design adaptation strategies, becoming part of a methodology in the sense suggested by Ison (2010). As claimed by this author, a methodology involves the conscious braiding together of theory and practice in a given situation, characterizing a context specific enactment. Therefore, instead of using crop models under a changing climate to deterministically and systematically intervene in a cropping system, changing sowing dates, varieties, level of fertilizer input, or anything else to optimize crop yields avoiding a decrease in economic return, based on a non-deterministic and systemic approach crop models are seen as just one among many other tools mobilized to explore novel possibilities of maintaining viable farming systems. This implies a move from the relative certainty of a model prediction (the empiricist tradition) to the uncertain possibilities of an ongoing adaptation process, or the move from the cropping system to the situation of concern. It is from this way of putting crop models into practice in a changing climate situation that practitioners may take responsibility for learning from the process. In this case it is expected that aware practitioners (any stakeholder involved in the process) would be "able to judge what is appropriate for a given context in terms of managing a process" (Ison, 2010) – the adaptation – instead of managing simply an output – the crop yield, since not always optimizing crop yields might be the best option if the purpose is to improve the whole situation and to avoid undesirable unintended consequences.

Furthermore, the uncertainty of future climate constrains the possibility of developing or securing in advance a set of practices to be adopted in an unknown situation. Instead of being informed by the "adaptation as fitting into" metaphor, adaptation to climate change must be informed by the "co-evolution metaphor", as has been discussed previously. It means that the tool-like use of crop models for adaptation purposes must be transformed: they need to be integrated into learning systems for adaptation.

But what is a learning system? From a first-order perspective, the design of a learning system comprises the combination of elements and processes interconnected as subsystems as well as the specification of some boundary conditions – to define what is part of the system, what is not – for the purpose of learning (Blackmore, 2005; Ison, 2010). However if we move to a second-order understanding, adopting a systemic design practice perspective, instead of speaking about the design or the modelling of a 'learning system', it is the situation of concern that will be considered "as if it were a learning system". As discussed by Ison (2010), the implication of this perspective is that a 'learning system' can only ever be said to exist after its enactment, that is, on reflection (Note 2). But if learning as suggested by Kolb (1984) is the creation of knowledge through the transformation of experience, then it is the experience of using crop models that needs to be transformed in order to create knowledge (taken to be a set of practices) about adaptation. Transforming the way crop models have been used is an important step to create the circumstances for effectively engaging stakeholders into the adaptation process, making the improvement of wicked situations like climate change possible. In this way crop models will also contribute to build capacity for social collectives and individuals to respond to a dynamic climate, and stakeholders will move "from a goal-oriented thinking towards thinking in terms of learning".

The key issue regarding the use of crop models to design adaptation strategies to climate dynamics of farming systems can now be re-formulated in the following terms: what can be learned about the situation in case crop models were used as if they were part of a learning system to design adaptation strategies? Based on this learning perspective crop models will be understood not only as a set of rules or a computer program to simulate the development of crops to optimize yields, but as devices taking part in a process of designing systems of interest (farming systems) to tackle a variable and changing climate; they might become part of an epistemological device that opens up new learning possibilities about human-induced climate change, its far reaching implications on ecosystems and society and how to tackle them.

5. Final Remarks

The process of adapting agricultural land use to a dynamic climate and the farmer's adaptation capacity should not be reduced to the capacity to adopt or adjust certain practices assumed to be more suitable in future climatic conditions. In fact, as has been discussed in this article, the adaptation process should be understood as resulting from the permanent capacity to learn with a changing climate transforming wicked situations for the better. To achieve this, the process of developing adaptation strategies should be conceptualized in terms of the 'adaptation as co-evolution' metaphor with learning itself being its emergent property.

This demands however that practitioners be aware of their epistemological choices when designing adaptation strategies to climate dynamics in agricultural systems, for which they might mobilize different resources, among them crop models. So, for example, to frame human-induced climate change as a "wicked situation" is a matter of choice or preference that, however, calls for special forms of practices in the adaptation process. This epistemological awareness might transform the "hard" perspective on using crop models to a "soft", learning based intervention on a situation. Crop models will be seen now as devices to facilitate learning about the whole situation rather than to optimize yields. In analogy to what has been discussed so far, it might be expected that learning will also happen when moving from the tool-like adoption of crop models to the view of crop models as components of learning systems for adaptation, because this move from one understanding to another will enact new forms of practice.

It is certainly possible to learn something about climate change adopting crop models as if they were tools to design adaptation strategies in agricultural land use systems. But this adaptation approach is no longer enough. Agricultural land use is a complex human activity system immersed in an environment of extremely high uncertainty about future climate. Under such circumstances crop models alone can't cope with all the complexity; they can't absorb all the variety present in such situations. Therefore, crop models must be part of an integrated approach to adaptation, enabling new forms of practice for managing the complexity of the decision making process driving adaptation of agricultural land use. By accepting this shift in perspective regarding the use of crop models for adaptation purposes, it is up to the crop modelers to engage in processes leading to the integration of crop models into learning systems for adaptation, enabling the emergence of learning and awareness about the whole situation.

In terms of adaptation to climate dynamics in agricultural land use, it only can make sense to use crop models in case they do not close options for future adaptability to change. In the “messy”, “wicked” situation we find ourselves, any designed adaptation strategy must avoid ‘a narrowing of possibilities on pathways towards effective action’. If we assume that there is no right or wrong answer in using crop models to design adaptation strategies, we must agree that it is necessary to evaluate the context where crop models can be used for effective action.

Regarding effective action, it must also be considered that although decision support systems as DSSAT have been used to facilitate the formulation of adaptation strategies to climate dynamics, the resulting model based farming systems not necessarily are transformed into farming practices, even in conditions of high climatic risk. So, for example, McCown (2012) reports about a program of research that has been conducted to see if, and under what conditions, computer simulations of farming scenarios can be valued by farmers. This research program was a response to the researcher´s growing awareness of persistent low interest in software of decision support systems among farmers, at least in Australian dryland cropping. Under these climatic conditions decision makers ‘rely heavily on intuitive judgment underpinned by experience’ and not necessarily on results of decision support systems, which might remain an enigma for them (McCown, 2012).

Furthermore, it is important to emphasize that the “problem” of climatic change of anthropogenic causes can’t be reduced to technological or economic aspects, as remembered by Orr (2009), and that for managing this situation it is necessary to learn to think and act differently, as has been claimed by Ison (2010). It is based on this process of learning to think and act differently that should emerge the capacity to adapt to climate change, and not as the result of identifying or adopting practices better fitted to a given climatic situation. This clearly points to the need of adopting a different research approach if learning is to be fostered, a research approach which is also oriented towards practice facilitating action which is systemically desirable and/or culturally feasible. A good example of such a research approach is ‘systemic inquiry’ understood as an institutional form and process designed for engaging with uncertainty and complexity (http://www.open.ac.uk/blogs/govan/?page_id=23 accessed 30 september 2015) like situations of climate change adaptation. As systemic inquiry progresses it involves the enactment of a process of learning amongst those who already have, or through participation build, a stake in an issue of concern (Ison, 2002; SLIM, 2004), as might be the case of farmers and researchers committed to improve whole farming systems under a dynamic climate. Therefore, paraphrasing Bawden (2010), to the challenge of climate change it is necessary to give a learning response.

Acknowledgments

The first author wishes to express his gratitude to FAPESC of the state government of Santa Catarina, Brazil, and to the German Academic Exchange Service (DAAD) for financial support during his permanence at ZALF in January/February 2011, when this work has been initiated. He also expresses his gratitude to the Leibniz-Centre for Agricultural Landscape Research (ZALF) and to its Institute of Land Use Systems for providing the necessary resources and for hospitality during his stay on several occasions. The research leading to these results also has received funding from the European Community's Seventh Framework Programme (FP7/2007-2013) under Grant Agreement Nº 212492: CLARIS LPB. A Europe-South America Network for Climate Change Assessment and Impact Studies in La Plata Basin.

References

Adam, M., van Bussel, L. G. J., Leffelaar, P. A., van Keulen, H., & Ewert, F. (2011). Effects of modeling detail on simulated potential crop yields under a wide range of climatic conditions. *Ecological Modelling, 222*(1), 131-143. http://dx.doi.org/10.1016/j.ecolmodel.2010.09.001

Adger, W. N., Agrawala, S., Mirza, M. M. Q., Conde, C., O’Brien, K., Pulhin, J., … Takahashi, K. (2007). Assessment of adaptation practices, options, constraints and capacity. In M. L. Parry, O. F. Canziani, J. P. Palutikof, P. J. van der Linden, & C. E. Hanson (Eds.), *Climate Change 2007: Impacts, Adaptation and Vulnerability. Contribution of Working Group II to the Fourth Assessment Report of the Intergovernmental Panel on Climate Change* (pp. 717-743). Cambridge, UK: Cambridge University Press.

Assad, E. D., Pinto, H. S., Zullo Junior, J., Marin, F. R., Pellegrino, G. Q., Evangelista, S. R., & Otavian, A. F. (2008). *Aquecimento global e a nova geografia da produção agrícola no Brasil*. Brasília, DF: Embaixada Britânica.

Basso, B., Ritchie, J. T., Pierce, F. J., Braga, R. P., & Jones, J. W. (2001). Spatial validation of crop models for precision agriculture. *Agricultural Systems, 68*(2), 97-112. http://dx.doi.org/10.1016/S0308-521X(00)00063-9

Batchelor, W. D., Basso, B., & Paz, J. O. (2002). Examples of strategies to analyze spatial and temporal yield variability using crop models. *European Journal of Agronomy, 18*(1-2), 141-158. http://dx.doi.org/10.1016/S1161-0301(02)00101-6

Bawden, R. (2010). The community challenge: the learning response. In C. Blackmore (Ed.), *Social Learning Systems and Communities of Practice* (pp. 39-56). London: Springer

Blackmore, C. (2005). Learning to appreciate learning systems for environmental decision making – a "work in progress" perspective. *Systems Research & Behavioral Science, 22*(4), 329-341. http://dx.doi.org/10.1002/sres.697

Boote, K. J., Jones, J. W., & Pickering, N. B. (1996). Potential uses and limitations of crop models. *Agronomy Journal, 88*(5), 704-716. http://dx.doi.org/10.2134/agronj1996.00021962008800050005x

Challinor, A. J., Wheeler, T. R., Craufurd, P. Q., Slingo, J. M., & Grimes, D. I. F. (2004). Design and optimisation of a large area process-based model for annual crops. *Agricultural and Forest Meteorology, 124*(1-2), 99-120. http://dx.doi.org/10.1016/j.agrformet.2004.01.002

Challinor, A. J., Ewert, F., Arnold, S., Simelton, E., & Fraser, E. (2009). Crops and climate change: progress, trends, and challenges in simulating impacts and informing adaptation. *Journal of Experimental Botany, 60*(10), 2775-2789. http://dx.doi.org/10.1093/jxb/erp062

Collins, K., & Ison, R. (2009). Editorial: living with environmental change: adaptation as social learning. *Environmental Policy and Governance, 19*(6), 351-357. http://dx.doi.org/10.1002/eet.520

Easterling, W. E., Aggarwal, P. K., Batima, P., Brander, K. M., Erda, L., Howden, S. M., … Tubiello, F. N. (2007). Food, fibre and forest products. In M. L. Parry, O. F. Canziani, J. P. Palutikof, P. J. van der Linden, & C. E. Hanson (Eds.), *Climate Change 2007: Impacts, Adaptation and Vulnerability. Contribution of Working Group II to the Fourth Assessment Report of the Intergovernmental Panel on Climate Change* (pp. 273-313). Cambridge, UK: Cambridge University Press.

Ewert, F. (2004). *Modelling Changes in Global Regionalized Food Production Systems Under Changing Climate and Consequences for Food Security and Environment – Development of an Approach for Improved Crop Modelling within IMAGE.* Wageningen/Bilthoven: Plant Production Systems Group, Department of Plant Sciences, Wageningen University & Netherlands Environmental Assessment Agency (MNP), National Institute for Public Health and Environment (RIVM).

Fischer, G., Shah, M., Tubiello, F. N., & van Velhuizen, H. (2005). Socio-economic and climate change impacts on agriculture: An integrated assessment, 1990-2080. *Philosophical Transactions of the Royal Society B: Biological Sciences, 360*(1463), 2067-2083. http://dx.doi.org/10.1098/rstb.2005.1744

Hansen, J. W., & Jones, J. W. (2000). Scaling-up crop models for climate variability applications. *Agricultural Systems, 65*(1), 43-72. http://dx.doi.org/10.1016/S0308-521X(00)00025-1

Howden, S. M., Soussana, J. F., Tubiello, F. N., Chhetri, N., Dunlop, M., & Meinke, H. (2007). Adapting agriculture to climate change. *Proceedings of the National Academy of Sciences, 104*(50), 19691-19696. http://dx.doi.org/10.1073/pnas.0701890104

Intergovernmental Panel on Climate Change. (2014). Summary for policymakers. In C. B. Field, V. R. Barros, D. J. Dokken, K. J. Mach, M. D. Mastrandrea, T. E. Bilir, M. Chatterjee, K. L. Ebi, Y. O. Estrada, R. C. Genova, B. Girma, E. S. Kissel, A. N. Levy, S. MacCracken, P. R. Mastrandrea, and L. L. White (Eds.), *Climate Change 2014: Impacts, Adaptation, and Vulnerability. Part A: Global and Sectoral Aspects. Contribution of Working Group II to the Fifth Assessment Report of the Intergovernmental Panel on Climate Change* (pp. 1-32). Cambridge, UK and New York, NY, USA: Cambridge University Press.

Ison, R. (2002). Some reflections on a knowledge transfer strategy: a systemic inquiry. In A. Cristovão and L. O. Zorini (Eds.), *Farming and rural systems research and extension: Proceedings of the Fifth IFSA European Symposium* (pp. 647-656). Florence, Italy.

Ison, R. (2010). *Systems Practice: How to Act in a Climate-Change World.* London: Springer.

Jones, J. W., Hoogenboom, G., Poster, C. H., Boote, K. J., Batchelor, W. D., Hunt, L. A., … Ritchie, J. T. (2003). The DSSAT cropping system model. *European Journal of Agronomy, 18*(3-4), 235-265. http://dx.doi.org/10.1016/S1161-0301(02)00107-7

Kolb, D. A. (1984). *Experiential Learning: Experience as the Source of Learning and Development.* New Jersey: Prentice Hall.

Lobell, D. B., & Burke, M. B. (2010). On the use of statistical models to predict crop yield responses to climate change. *Agricultural and Forest Meteorology, 150*(11), 1443-1452. http://dx.doi.org/10.1016/j.agrformet.2010.07.008

McCown, R. L. (2012). A cognitive systems framework to inform delivery of analytic support for farmer´s intuitive management under seasonal climatic variability. *Agricultural Systems, 105*(1), 7-20. http://dx.doi.org/10.1016/j.agsy.2011.08.005

Mingers, J., & Brocklesby, J. (1997). Multimethodology: towards a framework for mixing methodologies. *Omega – The International Journal of Management Science, 25*(5), 489-509. http://dx.doi.org/10.1016/S0305-0483(97)00018-2

Moser, S., & Ekstrom, J. (2010). A framework to diagnose barriers to climate change adaptation. *Proceedings of the National Academy of Sciences, 107*(51), 22026-33031. http://dx.doi.org/10.1073/pnas.1007887107

Oliver, Y. M., Robertson, M. J., & Wong, M. T. F. (2010). Integrating farmer knowledge, precision agriculture tools, and crop simulation modelling to evaluate management options for poor-performing patches in cropping fields. *European Journal of Agronomy, 32*(1), 40-50. http://dx.doi.org/10.1016/j.eja.2009.05.002

O'Neill, B. (2008). Learning and climate change: an introduction and overview. *Climatic Change, 89*, 1-6. http://dx.doi.org/10.1007/s10584-008-9443-8

Orr, D. S. (2009). *Down to the Wire. Confronting Climate Collapse.* New York: Oxford University Press.

Ortiz, R., Sayre, K. D., Govaerts, B., Gupta R., Subbarao, G.V., Ban, T., … Reynolds, M. (2008). Climate change: Can wheat beat the heat? *Agriculture, Ecosystems and Environment, 126*(1-2), 46-58. http://dx.doi.org/10.1016/j.agee.2008.01.019

Russell, D. B., & Ison, R. L. (2000). The research-development relationship in rural communities: an opportunity for contextual science. In R. Ison & D. Russell. *Agricultural Extension and Rural Development: Breaking Out of Traditions. A Second-Order Systems Perspective* (pp. 10-31). Cambridge, UK: Cambridge University Press.

Sirotenko, O. D. (2001). Crop modeling: Advances and problems. *Agronomy Journal, 93*(3), 650-653. http://dx.doi.org/ 10.2134/agronj2001.933650-ax

SLIM. (2004). *SLIM framework: social learning as a policy approach for sustainable use of water.* Retrieved September 29, 2015, from https://sites.google.com/site/slimsociallearningforiwm/system/app/pages/search?scope=search-site&q=slim+framework

Smit, B., Burton, I., Klein, R. J. T., & Wandel, J. (2000). An anatomy of adaptation to climate change and variability. *Climatic Change, 45*, 223-251. http://dx.doi.org/10.1023/A:1005661622966

Tubiello, F. N., & Fischer, G. (2007). Reducing climate change impacts on agriculture: Global and regional effects of mitigation, 2000-2080. *Technological Forecasting and Social Change, 74*(7), 1030-1056. http://dx.doi.org/10.1016/j.techfore.2006.05.027

United States Department of Agriculture. (2007). *What are crop simulation models?* Retrieved February 1, 2011, from http://www.ars.usda.gov/main/docs.htm?docid=2890

Notes

Note 1. This might be the case, for instance, in the context of precision farming. In such circumstances, crop models have been used in combination with other tools for yield map prediction, increasing the availability of information on spatial variation to allow zone-specific management prescriptions (Basso, Ritchie, Pierce, Braga & Jones, 2001; Oliver, Robertson & Wong, 2010). However, also this kind of adoption of crop models to improve decision making doesn´t break out of optimizing traditions since the ultimate objective is to maximize yields.

Note 2. Since it is not the purpose of this paper to discuss in more detail these different perspectives on learning systems and its implications, the interested reader is referred to Ison (2010) for further details.

11

Micro-Dosing of Lime, Phosphorus and Nitrogen Fertilizers Effect on Maize Performance on an Acid Soil in Kenya

P. O. Kisinyo[1], P. A. Opala[2], V. Palapala[1], S. O. Gudu[1], C. O. Othieno[3] & E. Ouma[1]

[1] Rongo University College, P. O. Box 103-40404, Rongo, Kenya

[2] Maseno University, Private bag, Maseno, Kenya

[3] University of Eldoret, P. O. Box 1125-30100, Eldoret, Kenya

Correspondence: P. O. Kisinyo, Rongo University College, Rongo, Kenya.

E-mail: kisinyopeter@yahoo.com; dean-agriculture@ruc.ac.ke

Abstract

High cost of inorganic fertilizers and lime has precluded their use by smallholder farmers to remedy the problem of soil acidity and infertility in Kenya. To address the problem, we tested a precision technique referred to as micro-dosing, which involves application of small, affordable quantities of inorganic inputs on an acid soil in Busia County, Kenya. Experimental treatments were N-fertilizer (0 and 37.5 kg N ha^{-1}), P-fertilizer (0 and 13 kg P ha^{-1}) and lime (0, 0.77 and 1.55 tons lime ha^{-1}). 37.5 kg N and 13 kg P ha^{-1} are 50% of the recommended fertilizer rates for maize production in Kenya while 0.77 and 1.55 tons lime ha^{-1} are 25 and 50% of the actual requirement. Soil chemical changes, maize grain yield and nutrient recovery were determined. Lime and P-fertilizer significantly affected only the top-soil pH, Ca, Mg and available P, while the effects of N-fertilizer were evident on both top- and sub-soil N likely due to its faster mobility than P and lime. Grain P-fertilizer recovery efficiencies were 14 and 16-27% due to 13 kg P and 13 kg P + 0.77-1.55 tons lime ha^{-1}, respectively. N-fertilizer recovery efficiencies were 37 and 42-45% due to 37.5 kg N and 37.5 kg N + 0.77-1.55 tons lime ha^{-1}, respectively. Fertilizers applied to supply 37.5 kg N, 13 kg P and 0.77-1.55 tons lime ha^{-1} increased grain yield above the control by 134, 39 and 12-22%, respectively, therefore micro-dosing of these inputs can increase maize production on Kenyan acid soils.

Keywords: fertilizer N and P, lime, western Kenya, smallholder

1. Introduction

Agricultural production in the high rainfall areas of Kenya is constrained by soil acidity and soil fertility depletion (Kanyanjua et al., 2002). The yields of maize, which is the staple food in these regions, are very low, averaging less than 1 ton ha^{-1} against a potential of 4 tons ha^{-1}. To mitigate these constraints, many soil fertility management technologies have been attempted. These ranges from sole use of inorganic fertilizers to sole use of organic inputs (Sanginga & Woomer, 2009; Okalebo et al., 2006) but these approaches have several drawbacks which have hindered their adoption by smallholder farmers in Kenya. In particular, use of organic inputs is labour intensive, they are of often low in nutrient content and not available in sufficient quantities on smallholder farms. Inorganic fertilizers, on the other hand, are expensive and most smallholder farmers cannot afford them (Sanchez et al., 1997). However, use of fertilizers remains the key to increasing food productivity in Africa (Quinones et al., 1997).

The main problem with most of the current fertilizer management recommendations is that they target maximization of yields or profits without consideration of the agricultural risks and resource constraints faced by many smallholder households (Twomlow et al., 2010). There has been therefore growing interest to find alternative interventions that promote efficient and sustainable use of fertilizer tailored to smallholder farmers' socio-economic conditions (Nziguheba, 2007). One such technology is the use of precision technique referred to as micro-dosing, which involves the application of small, affordable quantities of inorganic inputs and seed at the time of planting or as top dressing three to four weeks after emergence (ICRISAT, 2008). For example, while the recommended N and P fertilizer rates for maize production in western Kenya are 75 kg N and 26 kg P ha^{-1} (Kenya Agricultural Research Institute (KARI), 1994), other research however suggests that farmers can increase their average yield by 50-100% by applying as little as 10 kg N ha^{-1} (Twomlow et al., 2010) therefore improving the prospects of adopting the use of fertilizers. In addition, fertilizer use efficiency could be improved by liming the acid soils to eliminate Al toxicity (Kisinyo et al., 2014), thereby resulting in a reduction in the amount of fertilizer required to achieve a target yield. Much of the work on micro-dosing of fertilizers has been conducted in semi-arid regions of Zimbabwe using mainly N fertilizers. There is need therefore to extend these evaluations in other environments, such as western Kenya, where rainfall is high and soil acidity in addition to N and P deficiencies limit crop production, to test the robustness of this technology. The objectives of this study were therefore to determine effect of micro-dosing of lime, P and N fertilizer on: i) soil chemical properties and ii) maize performance on an acid soil in Busia County, Kenya.

2. Materials and Methods

2.1 Site Description

The study was carried out in Busia County, 1300 m asl, (0°19'N & 34°12'E). The average annual rainfall is 1600 mm which is distributed over two main cropping seasons, the long rainy season which is from March to July and the short rainy season from September to December (Jaetzold & Schmidt, 1983). The initial soil properties are presented in Table 1. The soils were moderately acidic, low in organic C (< 2%), N, P and exchangeable bases (Ca, Mg and K) but high in Al saturation (> 20%) (Landon, 1991). The soil type at the site has been classified as an Orthic Ferralsol (FAO, 1988).

Table 1. Initial soil chemical and physical characteristics of the study site

Parameter	value
pH – H_2O (soil: water; 1:2.5)	5.44
Bicarbonate extractable P (mg kg^{-1})	2.44
Total N (%)	0.09
Nitrate (mg kg^{-1})	1.57
Organic C (%)	1.66
Ca (cmol kg^{-1})	3.12
Mg (cmol kg^{-1})	1.81
K (cmol kg^{-1})	0.41
Exchangeable Al (cmol kg^{-1})	1.63
Al saturation (%)	23.4
ECEC (cmol kg^{-1})	6.97
Specific gravity	2.44
Sand (%)	54
Silt (%)	30
Clay (%)	26
Soil Texture	Sandy clay loam
Soil Classification	Orthic Ferralsol

2.2 Experimental Design, Layout and Management

The study was conducted over two seasons; long and short rains in the 2008. The experiment was a 2 × 2 × 3 split-split plot laid out in randomized complete block design with N as the main plot, P as the sub-plot and lime as sub-sub plot. Two N rates, 0 representing a practice where farmers apply no fertilizer and 37.5 kg N ha^{-1}, which is half the recommended N rate for the area (KARI, 1994), were used in the main plot. P was also applied at 0 and 13 P kg ha^{-1} for similar reasons as N. Liming material containing 21% calcium oxide was applied at 0, 0.77 and 1.55 $tons^{-1}$ ha in the sub-sub plots. The lime requirement of soil was calculated using the equation proposed by Cochrane et al. (1980). The equation aims at reducing the % Al saturation to a level that is commensurate with crop Al tolerance using the following formula:-

$$\text{Ca cmol kg}^{-1}\text{ soil} = 1.5[\text{Al} - \text{RAS}\ (\text{Al} + \text{Ca} + \text{Mg})/100] \quad (1)$$

Where Al = cmol kg^{-1} soil in the original exchange complex, RAS = required % Al saturation, Ca = cmol kg^{-1} soil in the original exchange complex and Mg = cmol kg^{-1} soil in the original exchange complex. The approximate lime requirements (tons $CaCO_3$ cmol kg^{-1}) was calculated by multiplying the Ca cmol kg^{-1} by the soil specific gravity.

A RAS value of 20% was used since most maize germplasm grown in Kenya are sensitive to >20% Al saturation (Ligeyo, 2007; Kisinyo et al., 2013). Thus, the calculated actual lime requirement was 3.09 tons ha^{-1} while 0.77 tons ha^{-1} and 1.55 tons ha^{-1} represent 25% and 50% of the actual lime requirement, respectively.

Plots measuring 3.5 m by 3 m were demarcated after ploughing to the appropriate tilth in the year 2008 long rain season. Each plot was separated from the next by a spacing of 1 m. Lime was applied only once in the long rains season by evenly broadcast and thoroughly mixing it with the soil 30 days prior to planting. This was to allow for adequate time for it to react with the soil. Fertilizers phosphate as triple superphosphate and N as calcium ammonium nitrate were spot applied into the planting holes and thoroughly mixed with soil at the time of planting in each season. Nitrogen was split-applied with 30% applied at planting and the rest as a top-dress at six weeks later. Two maize seeds (hybrid 513) were planted per hill at a spacing of 75 cm by 25 cm and were thinned to one plant per hill two weeks post-emergence. Maize was managed using the recommended agronomic practices for the area and harvested at physiological maturity.

2.3 Determination of Crop Response to Lime, P and N Fertilizer Applications

The grain yield was computed using the following formulae:

$$\text{Yield per plot} = (\text{Total fresh weigh} \times \text{Sample dry weight}) / \text{Sample fresh weigh} \quad (2)$$

$$\text{Yield (kg ha}^{-1}) = (\text{Yield per plot} \times 10{,}000 \text{ m}^2)/\text{Effective area (m}^2) \quad (3)$$

where effective area is part of the plot harvested which is less the guard row and plants at the end of each row.

Plant fertilizer recovery efficiency was calculated using the following equation by Dobermman (2005):

$$\text{Nutrient recovery efficiency (\%)} = (NT - No)/\text{T} \times 100 \quad (4)$$

where NT = nutrient uptake in treated plots (kg ha^{-1}), No = nutrient uptake in control plots (kg ha^{-1}) and T = rate of the nutrient applied (kg ha^{-1}).

2.4 Soil Sampling and Analyses

Nine sub - soil samples were taken at the experimental site in March 2008 with a soil auger at the 0-20, 20 - 40 and 40 - 60 cm soil depths in a zig - zag manner prior to treatment application. Samples from each depth were thoroughly mixed and about 1.0 kg composite sample taken to the laboratory for determination of the initial soil properties at the site. The samples were subsequently taken from each plot at appropriate intervals at the same depths to monitor changes in selected soil chemical properties due to treatment applications. The soil samples were air-dried and the ones taken before treatment applications were analysed for texture, specific gravity, pH (1: 2.5; soil: water), bicarbonate extractable P, exchangeable bases (Ca^{2+}, Mg^{2+}, K^+ and Na^+), acidity (Al^{3+} and H^+), organic carbon (%C) and total N (%N). The samples taken after treatment applications were analyzed for pH, available P, nitrates, exchangeable Ca^{2+}, Mg^{2+} and Al^{3+}. Soil specific gravity was calculated by determining the weight of solid particles per weight of equal volume of water. The grain samples were analysed for P and N contents. Detailed laboratory procedures for specific gravity determination are outlined by Smith (1981) while procedures for soil chemical and plant analyses are described by Okalebo et al. (2002).

2.5 Statistical Analysis of Data

All the maize yields data were subjected to analysis of variance (ANOVA) with the split-split plot design using General Statistics (Genstat, 2010). Means were separated using pooled standard error of difference of means (s.e.d) whenever treatment effects were significant at $p \leq 0.05$).

3. Results and Discussion

3.1 Effects of Lime on Soil pH, Ca, Mg and Al levels

Figure 1 shows the effect of lime on soil pH. Lime at the rates of 0.77 and 1.55 tons ha^{-1} took about 60 and 126 days to increase soil pH to the highest pH peaks of about 5.7 and 6.0, respectively. Lower lime rate (25% of the actual lime requirement) increased soil pH to highest peak in relatively shorter time and thereafter the pH began to decline earlier compared to where half of the requirement was applied. This was likely because Ca^{2+} ions in lower lime rate was depleted very fast, resulting in increase in soil acidity. Similar results were reported by Sanchez (1976) who found that the residual effect of lime depends on the amount of Ca^{2+} and/or Mg^{2+} ions still remaining in the liming material at any given time. In a Kenyan acid soil, 2 tons lime ha^{-1}, representing 50% of the actual lime requirement kept soil pH above 5.5 for about 2 years (Kisinyo et al., 2014). The exchangeable Ca^{2+} and Mg^{2+} increased with increasing lime rates (Figures 2 and 3) while the exchangeable Al^{3+} decreased with increasing rates of lime.

Lime had no significant effect on soil pH, exchangeable Ca^{2+} and Al^{3+} in the sub-soils (20-40 and 40-60 cm depths) (Figures 1- 4). This is attributable to the slow solubility of lime and hence low mobility within the soil (Yorst & Ares, 2007). In a similar study, very little changes in soil pH, exchangeable Ca^{2+} and Al^{3+} were reported in the sub-soil (12 -85 cm depth), $2^1/_2$ years after lime application at rates between 1.5 – 6.0 tons $CaCO_3$ ha^{-1} (Arya, 1990). Surface application of lime is not likely therefore to correct sub-surface soil acidity.

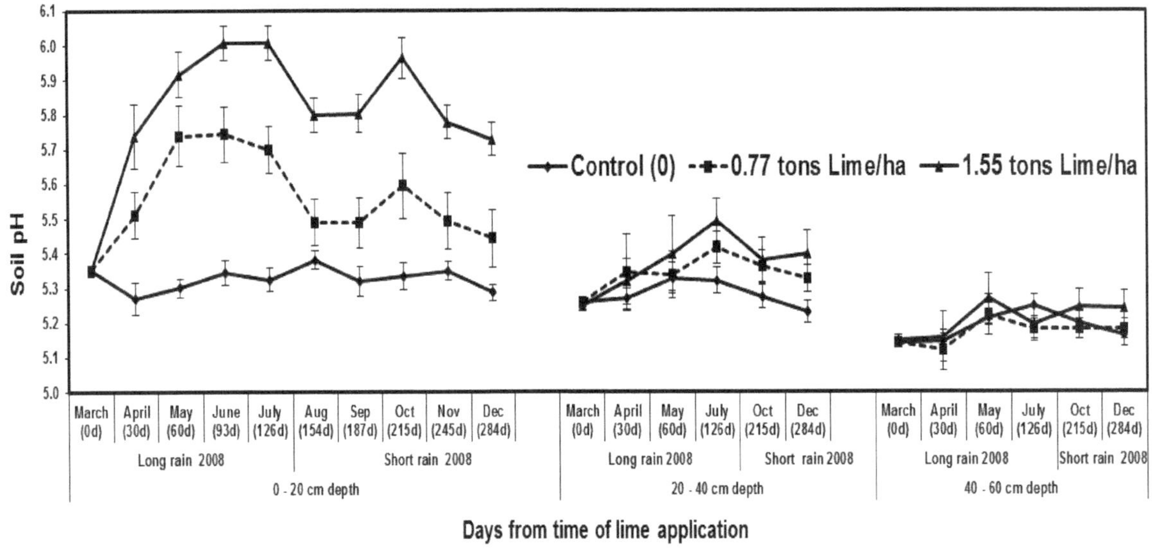

Figure 1. Effect of lime on soil pH during the cropping period; *d=days from time of lime application and error bars indicates standard error of means (SEM)*

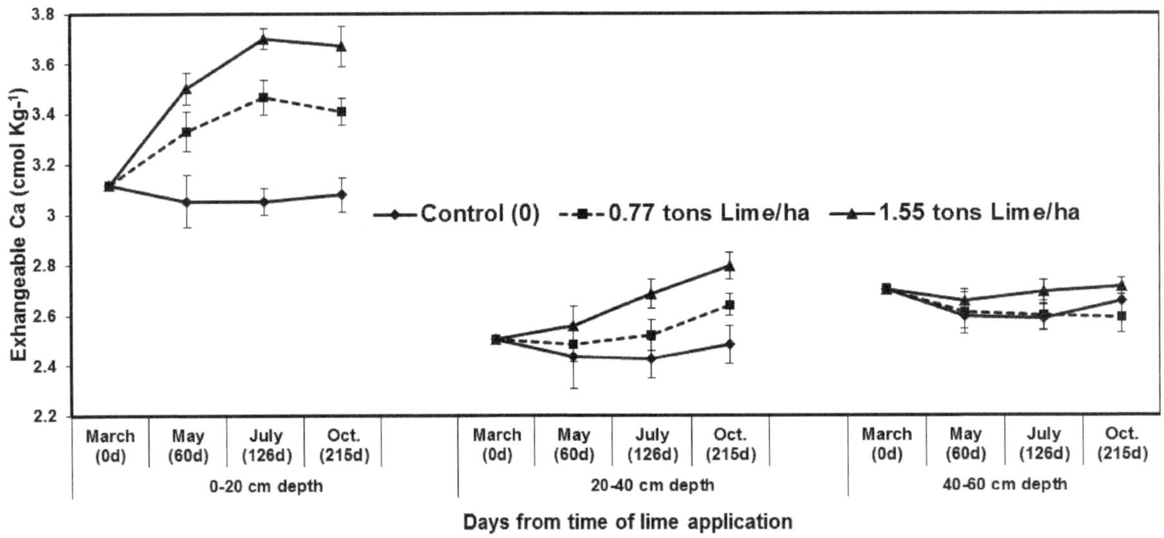

Figure 2. Effect of lime on exchangeable calcium during the cropping period

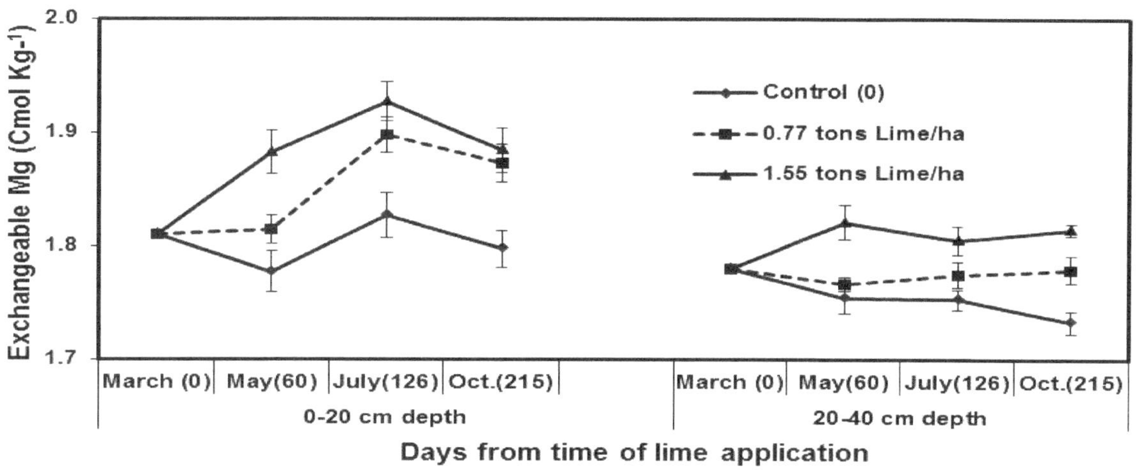

Figure 3. Effect of lime on exchangeable magnesium during the cropping period

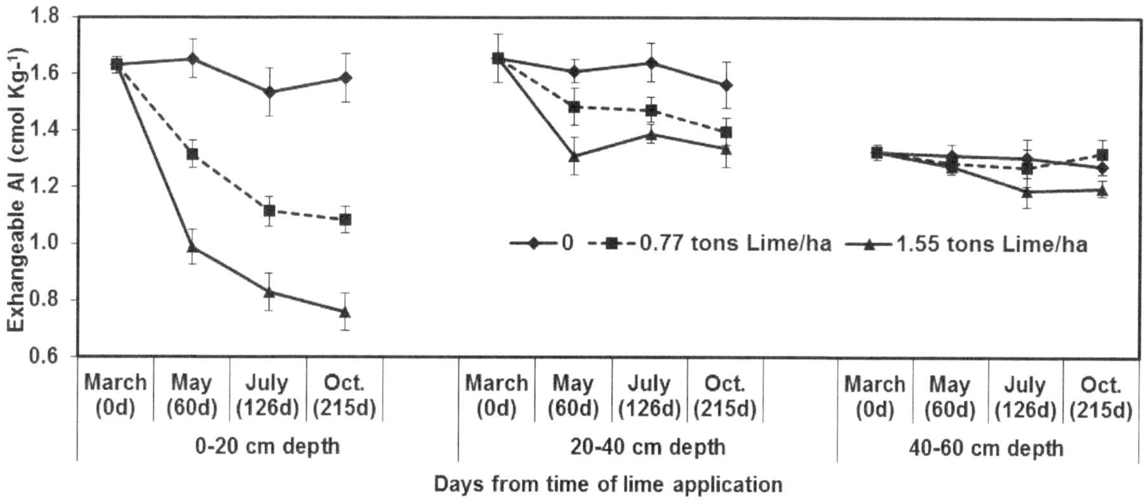

Figure 4. Effect of lime on exchangeable aluminium during the cropping period

3.2 Effects of P Fertilizer and Lime on Soil Available P

Figure 5 shows the effects of lime and P fertilizer on soil available P at the 0-20 cm depth. Generally, both P fertilizer and lime had significant effect ($p \leq 0.05$) on available P. A combined application of P fertilizer and lime increased soil available P more than the application of either of them alone. The available P reached its highest peak in about 30 days (in May) from date of its application (in April) and thereafter there was a general declining trend during the cropping season. However, in the month of October there was another peak due to a second P fertilizer application in mid-August, during the short rain planting season. The increase in soil available P as a result of liming observed in this study is consistent with others studies (Tisdale et al., 1990; van Straaten, 2002) which attributed it to reduced P sorption. Many other studies have investigated the effects of lime application on P retention and extractability, but consistent improvements in the availability of soil P have not been obtained (Mansell et al., 1984; Holford et al., 1994; Curtin & Syers, 2001). Liming is, however, often successfully applied to remedy other soil constraints such as soil acidity as was demonstrated in the present study.

Figure 6 shows the effects of P fertilizer on soil available P down the soil profile. Soil available P generally decreased down the soil profile because of the immobile nature of P. In a similar study, no significant changes in soil available P below the layer of P fertilizer incorporation in acid soils were reported (Arya, 1990). From the present study, it is evident that surface application of lime and P fertilizer applications may not correct P deficiency on acid soils.

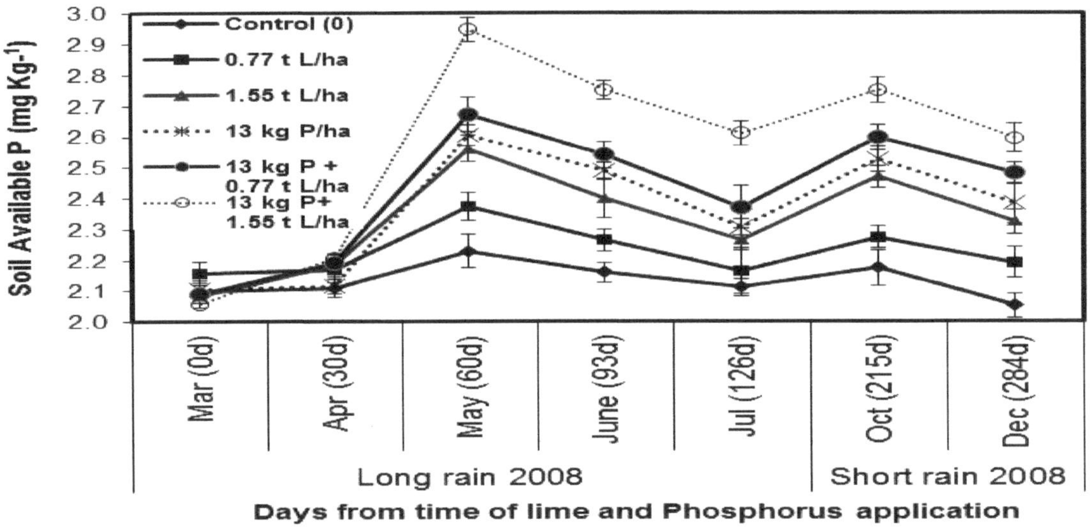

Figure 5. Effect of P fertilizer and lime on soil available P at the 0 - 20 cm soil depth during the cropping period

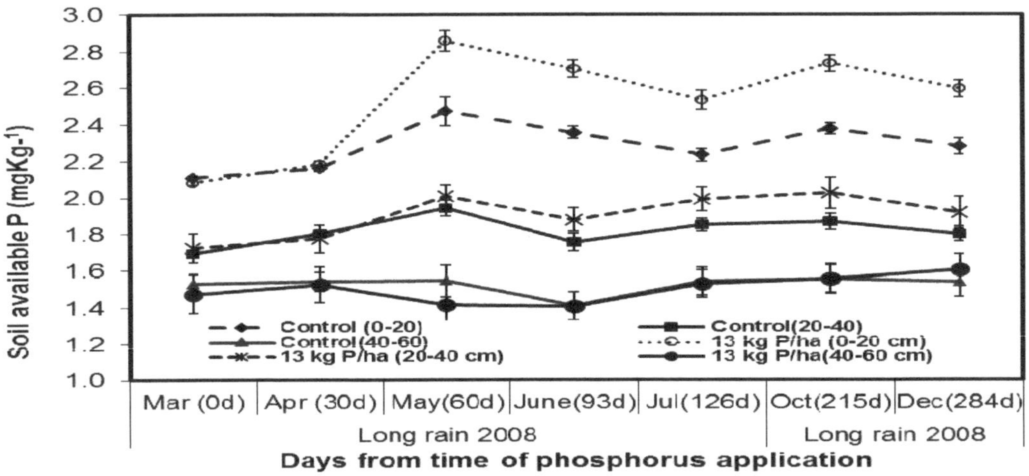

Figure 6. Effect of P fertilizer and lime on soil profile available P during the cropping period

3.3 Effect of N Fertilizer on Soil Nitrate Levels

Soil nitrate levels generally increased at the onset of rains i.e. from March to May even on plots which did not receive N fertilizer application (Figure 7). This was followed with a decline in plots which did not receive N fertilizer throughout up to July when the crop was harvested. A similar trend of increase was observed during the short rains due to N fertilizer applications. Increased nitrate levels observed on all plots at the onset of rains was likely due to the so called 'Birch effect' attributed to rapid mineralization of killed microbial biomass and labile organic matter released on drying and also to rapid nitrification of NH_4^+-N accumulated during the dry season (Birch, 1958; Seneviratne & Wild, 1985). The effect of N fertilizer on nitrate levels down the soil profile was more pronounced than that of lime on soil pH, exchangeable Al^{3+}, Ca^{2+} and Mg^{2+} and also that of P fertilizer on soil available P reported above in Figures 1 to 6. This was because unlike phosphate, Ca^{2+} and Mg^{2+} ions, nitrate ions are very mobile within the soil and are therefore easily leached by rainfall water down the soil profile (Giller et al., 1997).

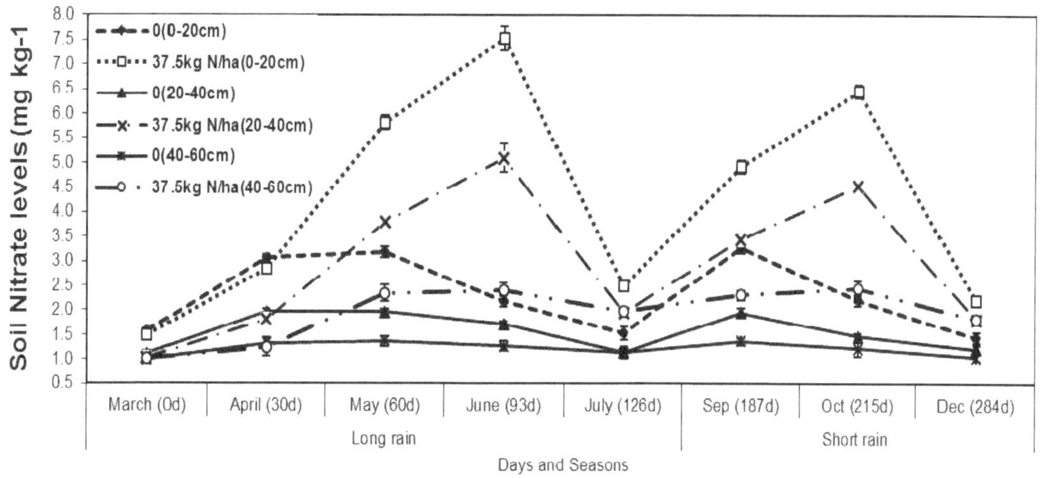

Figure 7. Effect of N fertilizer on soil profile nitrate levels

3.4 Effect of Lime on Phosphorus and Nitrogen Fertilizers Recovery Efficiency

Table 2 shows the effect of lime on grain P and N fertilizer recovery efficiencies. Lime increased both grain P and N fertilizers recovery efficiencies, which increased with increase in rates of lime. The mean grain P fertilizer efficiencies were about 14, 16 and 27 % due to 13 kg P, 13 kg P + 0.77 tons lime and 13 kg P + 1.55 tons lime ha^{-1}, respectively. The average annual N fertilizer recoveries were about 37, 42, and 45% due to 37.5 kg N, 37.5 kg N + 0.77 tons lime and 37.5 kg N + 1.55 tons lime ha^{-1}, respectively. In acid tropical soils, only about 10-20% of the applied P fertilizer is normally recovered by plants in the year of application due to high P sorption by Al and Fe sesquioxides (Keerthisinghe, 2001). P fertilizer recoveries of between 18 to 31% have been reported on Nigerian acid soils (Fofana et al., 2007). Thus, P fertilizer recoveries in this study are within the range normally reported in tropical acid soils. Nitrogen fertilizer recovery efficiencies were also within the range of 30-70% normally recovered by crops (Bock, 1984). Micro-dosing lime increased P and N fertilizers recovery efficiencies primarily by restoring soil pH to levels that are favourable for plant growth and microbial activity. It also eliminated toxic elements and ensured the availability of essential elements. Under such optimal growth conditions crops are able to absorb and use fertilizers more efficiently (Adams & Martin, 1984). Therefore, it is evident from this study that higher N and P fertilizer recovery efficiencies in acid soils can be achieved through micro-dosing lime.

Table 2. Effect of lime on the grain P and N fertilizers recovery efficiency (%)

Phosphorus fertilizer	recovery efficiency			Nitrogen fertilizer	recovery efficiency		
Treatments	LR	SR	Mean	Treatments	LR	SR	Mean
13 kg P ha^{-1}	13.6	14.2	13.9	37.5kg N ha^{-1}	37.6	36.2	36.9
13 kg P ha^{-1} +0.77 tons lime ha^{-1}	15.7	16.4	16.1	37.5kg N ha^{-1}+ 0.77 tons lime ha^{-1}	42.7	40.9	41.8
13 kg P ha^{-1}+1.55 tons lime ha^{-1}	26.3	26.9	26.6	37.5kg N ha^{-1} + 1.55 tons lime ha^{-1}	46.4	43.9	45.2
Mean	18.5	19.2		*Mean*	42.2	40.3	

3.5 Effects of N, P Fertilizers and Lime on Maize Grain Yield

The maize grain yield data is presented in Table 3. The highest mean annual grain yield increase above the control of 134% was obtained with the application of 37.5 kg ha^{-1} N, followed by 13 kg P ha^{-1} (39%) and lime (12-22%). The increased maize grain yield due to lime, N and P fertilizers application reported in this study confirm work by several authors that soil N, P deficiencies and acidity limit maize production in these soils. In similar studies in Kenyan highland acid soil, 50% of the recommended N fertilizer rate and 25-50% of the actual lime requirement increased maize grain yield by about 106 and 13-27%, respectively (Kisinyo, 2011) while Opala et al. (2007) observed significant responses to application of only 6 kg P ha^{-1}. Similarly, in Zimbabwe, 33-50% of the recommended N fertilizer rate increased maize grain yield and adoption of fertilizer use by SHF

(ICRISAT, 2008). Combinations of any two or three of the inputs i.e. lime, P and N, had more effect on maize yield than any one of them applied alone likely due to synergism.

Table 3. Effects of N, P fertilizers and lime on maize grain yield (tons ha^{-1})

Treatments		Long rain					Short rain				
N kg ha^{-1}	P kg ha^{-1}	Control	0.77 t lime ha^{-1}	1.55 t lime ha^{-1}	Mean for P	Mean for N	Control	0.77 t lime ha^{-1}	1.55 t lime ha^{-1}	Mean for P	Mean for N
0	0	0.474	0.595	0.641	0.570	0.857	0.569	0.614	0.730	0.638	0.905
	13	0.950	1.171	1.309	1.143		1.007	1.179	1.329	1.171	
	0	1.667	1.874	1.973	1.838	2.013	1.763	1.876	2.017	1.885	2.112
37.5	13	2.009	2.181	2.372	2.187		2.139	2.333	2.544	2.338	
Mean Lime		1.355	1.275	1.455			1.369	1.500	1.655		
CV%		3.7					3.2				
$s.e.d_{(0.05)}\ N$		0.016					0.015				
$s.e.d_{(0.05)}\ P$		0.020					0.005				
$s.e.d_{(0.05)}\ lime$		0.021					0.012				
$s.e.d_{(0.05)}\ N*P$		0.010					0.016				
$s.e.d_{(0.05)}\ N*Lime$		ns					ns				
$s.e.d_{(0.05)}\ P*Lime$		0.024					ns				
$s.e.d_{(0.05)}\ N*P*Lime$		0.031					0.026				

t = tons & ns= not significant.

4. Conclusion

Lime increased soil pH, exchangeable Ca^{2+} and Mg^{2+} and reduced exchangeable Al^{3+} which resulted to increased soil available P. Both lime and P fertilizer significantly increased soil available P of the top but not sub-soil due to their slow mobility. Surface applications of lime and P fertilizer cannot therefore adequately correct sub-soil acidity and P deficiency. However, the effect of N fertilizer on soil nitrate levels was significant in the sub-soil because nitrate ions are very mobile within the soil profile and are therefore easily leached by rain water. Micro-dosing of lime increased recovery efficiency of N and P fertilizers and therefore, has the potential to increase nutrient recovery necessary for high crop yields. Application of small amounts of lime and fertilizers P and N significantly increased maize grain yield suggesting that micro-dosing has the potential to increase maize production on Kenyan N and P-deficient acid soils. However, there may be need for further economic evaluation because farmers make their decisions on whether or not to adopt a technology based on economic considerations.

Acknowledgements

We are grateful to The McKnight Foundation for funding the research work. We are also very thankful to the farmer, Mr. Peter Omitto for allowing us to conduct the research on his farm.

References

Adams, F., & Martin, J. B. (1984). Liming effects on nitrogen use and efficiency. In R. O. Hauck (Ed.), *Nitrogen in Crop Production* (pp. 416-424). American Society of Agronomy, Madison.

Arya, L. M. (1990). *Properties and Processes in upland acid soils in Sumatra and their management for crop production*. Winrock Intl. Inst. of Agri. Dev., Raleigh.

Birch, H. F. (1958). The effect of soil drying on humus decomposition and nitrogen availability. *Plant Soil, 10*, 9-31. http://dx.doi.org/10.1007/BF01343734

Bock, B. R. (1984). Efficient use of nitrogen in cropping systems. In R. O. Hauck (Ed.), *Nitrogen in Crop Production* (pp. 272-292). American Society of Agronomy, Madison.

Cochrane, T. T., Salinas, J. G., & Sanchez, P. A. (1980). An equation for liming acid mineral soils to compensate crop aluminium tolerance. *Tropical Agriculture (Tinidad), 57*, 33-40.

Curtin, D., & Syers, J. K. (2001). Lime-induced changes in indices of soil phosphate availability. *Soil Science Society of America Journal, 65*, 147-152. http://dx.doi.org/10.2136/sssaj2001.651147x

Dobermman, A. (2005). Nitrogen use efficacy-state of the art. IFA International Workshop on Enhanced Efficiency Fertilizers, Frankfurt, Germany, June 28-30. Retrieved Spetember 11, 2010 from http://soilfertility.unl.edu/Matrials%20to%20included/nitrogen%20use520state%20of%20the%20art_Dobermman.pdf

FAO. (1988). Revised legend of the FAO-UNESCO soil map of the world. Food and Agriculture Organization of the United Nations, Rome.

Genstat. (2010). *The GenStat Teaching Edition.* GenStat Release 7.22 TE, Copyright 2008, VSN International Ltd.

Giller, K. E., George, C., Ehaliatis, C., Adams, E., Sakala, W. D., & Mafongoya, P. L. (1997). Building soil nitrogen capital in Africa. In P. J. Buresh et al. (Eds.), *Replenishing soil fertility in Africa* (pp. 151-192). SSSA special publication no. 51. Madison. Wisconsin.

Fofana, B., Wopereis M. C. S., Bationo, A., Breman, H., & Mando, A. (2007). Millet nutrient use efficiency as affected by natural soil fertility, mineral fertilizer use and rainfall in the West African Sahel. *Nutrient Cycling in Agroecosystems, 81*, 25-36. http://dx.doi.org/10.1007/s10705-007-9146-y

Holford, I. C. R. (1983). Effects of lime on phosphate sorption characteristics and soluble phosphate of fifteen acid soils. *Australian Journal of Soil Science, 21*, 333-342. http://dx.doi.org/10.1071/SR9830333

ICRISAT. (2008). *Micro Doses, Mega Benefits. Promoting Fertilizer Use in Semi – Arid Zimbabwe.* A manual for extension staff. FAO, Rome.

Jaetzold, R., & Schmidt, H. (1983). F*arm management handbook of Kenya, Vol. IIA Western Kenya and Vol. II B (Central Kenya):* Natural conditions and farm management information. Ministry of Agriculture/GAT Nairobi and GTZ/Ashburn.

Kanyanjua, S. M., Ireri, L., Wambua, S., & Nandwa, S. M. (2002). Acid soils in Kenya: Constraints and remedial options. *KARI Technical Note No.11.*

KARI. (1994). Fertilizer use recommendations. Vol. 1-22. *Fertilizer Use Recommendations Project.* Kenya Agricultural Research Institute, Nairobi Kenya.

Keerthisinghe, G., Zapata, F., Chalk, P., & Hocking P. (2001). Integrated approach for improved P nutrition of plants in tropical acid Soils. In W. J. Horst et al. (Eds), *Plant nutrition – Food security and sustainability of agro-ecosystems* (pp. 974-975). Netherlands: Springer.

Kisinyo, P. O., Othieno, C. O., Gudu, S. O., Okalebo, J. R., Opala, P. A, Maghanga, J. K., … Ogola, B. O. (2013). Phosphorus Sorption and Lime Requirements of Maize Growing Acids Soil of Kenya. *Sustainable Agriculture Research, 2*, 116-123. http://dx.doi.org/10.5539/sar.v2n2p116

Kisinyo. P. O. (2011). *Constraints of soil acidity and nutrient depletion on maize (Zea mays L.) production in Kenya.* Ph.D. Thesis. Moi University.

Kisinyo, P. O., Othieno, C. O., Gudu, S. O., Okalebo, J. R., Opala, Ng'etich W. K., Nyambati, R. O., … Opile, W. R. (2014). Immediate and Residual Effects of Lime and Phosphorus Fertilizer on Soil Acidity and Maize Production in Western Kenya. *Experimental Agriculture,* 50, 128-143. http://dx.doi.org/10.1017/S0014479713000318

Landon, J. R. (1991). *Booker Tropical Soils Manual: A handbook for soil survey and agricultural land evaluation in the tropics and subtropics* (pp. 133-137). New York: John Wiley & Sons Inc.

Ligeyo, D. O. (2007). *Genetic analysis of maize (Zea mays L.) tolerance to aluminium toxicity and low phosphorus stress and development of synthetics for use in acid soils of western Kenya.* Ph. D. Thesis. Moi University.

Mansell, G. P., Pringle, R. M., Edmeades, D. C., & Shannon, P. W. (1984). Effects of lime on pasture production on soils in the North Island of New Zealand. 3. Interaction of lime with phosphorus. New Zeal. *Journal of Agriculture Research, 27*, 363-369. http://dx.doi.org/10.1080/00288233.1984.10430637

Nziguheba, G. (2007). Overcoming phosphorus deficiency in soil1511s of Eastern Africa: recent advances and challenges. In A. Bationo et al. (Eds.), *Advances in integrated soil fertility management in sub-Saharan Africa: challenges and opportunities* (pp. 49-160). New York: Springer.

http://dx.doi.org/10.1007/978-1-4020-5760-1_13

Okalebo, J. R., Gathua, K. W., & Woomer, P. L. (2002). *Laboratory methods of soil analysis: A working manual.* (2nd ed.). TSBR-CIAT and SACRED Africa, Nairobi, Kenya.

Okalebo, J. R., Othieno, C. O., Woomer, P. L., Karanja, N. K., Sesmoka, J. R. M., Bekunda, M. A., … Mukhwana, E. J. (2006). Available technologies to replenish soil fertility in East Africa. *Nutrient Cycling in Agroecosystems, 76*, 153-170. http://dx.doi.org/10.1007/s10705-005-7126-7

Opala, P. A., Jama, B. J., Othieno, C. O., & Okalebo, J. R. (2007). Effect of phosphate fertilizer application methods and nitrogen on maize yields in western Kenya. An agronomic and economic evaluation. *Experimental Agriculture, 43*, 477-487. http://dx.doi.org/10.1017/S0014479707005315

Quinones, M. A., Borlaug, N. E., & Dowswell, C. R. (1997). A fertilizer-based green revolution for Africa.. In P. J. Buresh et al. (Eds.), *Replenishing soil fertility in Africa* (pp. 81-95). SSSA special publication no. 51. Madison. Wisconsin.

Sanchez, P. A. (1976). *Properties and management of soils in the tropics*. New York, USA: John Willey and Sons, Inc.

Sanchez, P. A., Shepherd, K. D., Soule, M. J., Place, F. M., Buresh, P. J., Izac, A-M. N., … Woomer, P. L. (1997). Soil fertility replenishment in Africa: An investment in Natural resource capital. In R. J. Buresh , P. A. Sanchez & F. Calhoun (Eds.), *Replenishing soil fertility in Africa.* (pp. 1-46). SSSA special publication no. 51. Madison.

Sanginga, N., & Woomer, P. L. (2009). *Integrated Soil Fertility Management in Africa: Principles, Practices and Development Processes*. Tropical Soil Biology and Fertility Institute of the International Centre for Tropical Agriculture, p. 263. Nairobi, Kenya.

Seneviratne, R., & Wild, A. (1985). Effect of mild drying on the mineralization of soil nitrogen. *Plant Soil, 84*, 175-179. http://dx.doi.org/10.1007/BF02143181

Smith, M. J. (1981). *Soil Mechanics. Longman Scientific and Technical*, (4th ed.). England, UK.

Tisdale, S. L., Nelson, W. L., & Beaton. J. D. (1990). *Soil fertility and fertilizers*. New York: Macmillan.

Twomlow, S., Rohrbach, D., Dimes, J., Rusike, J., Mupangwa, W., Ncube B., … Maphosa, P. (2010). Micro dosing as a pathway to Africa's Green Revolution: Evidence from broad scale on-farm trials. *Nutrient. Cycling in Agroecosystems, 88*, 3-15. http://dx.doi.org/10.1007/s10705-008-9200-4

van Straaten, P. (2007). *Agrogeology: the use of rocks for crops*. Enviroquest Limited, 352 River Road, Cambridge, Ontario N3C 2B7 Canada, p. 440.

Yorst, R. S., & Ares, A. (2007). Phosphorus and lime requirement of tree crops in tropical acid soils: A review. *Journal of Tropical Forest Science, 19*, 176-185.

12

Growth in Total Factor Productivity in the Egyptian Agriculture Sector: Growth Accounting and Econometric Assessments of Sources of Growth

Boubaker Dhehibi[1], Ali Ahmed Ibrahim Ali El-Shahat[2], Aymen Frija[1] & Aden-Aw Hassan[1]

[1] Social, Economic and Policy Research Program (SEPRP), International Center for Agricultural Research in the Dry Areas (ICARDA), P.O. Box 950764, Amman 11195, Jordan

[2] Department of Agricultural Economics, Faculty of Agriculture, Zagazig University, Egypt

Correspondence: Boubaker Dhehibi, Social, Economic and Policy Research Program (SEPRP), International Center for Agricultural Research in the Dry Areas (ICARDA), P.O. Box 950764, Amman 11195, Jordan. E-mail: B.Dhehibi@cgiar.org

Abstract

This research aims to assess the Total Factor Productivity (TFP) of the whole agricultural sector in Egypt for the period 1961-2012 using Törnqvist index calculations. Particularly, it aims to investigate: (1) the most important factors explaining the TFP growth in the Egyptian agriculture (2) estimating changes in technical efficiency and technical change and determining the magnitude of their contribution to the overall TFP growth, and lastly, (4) based on these findings, providing policy implication and recommendations that allows enhancing and sustaining future growth of agricultural production in Egypt.

The currently analysis provided relevant results which might help us understanding the structural trend of the Egyptian agricultural sector, and understanding the most significant variables affecting this trend. Such results will have important policy implications for promoting further growth in the Egyptian agricultural sector. The empirical findings showed that rural development variables were found to significantly and negatively affect agricultural productivity. This demonstrates that agricultural activity is still a marginalized activity which is linked to low levels of income and is a source of employment for low productive labor. Moreover, a negative significant effect of the infrastructure variable on the productivity gains of the agricultural sector in Egypt was found which might indicates a form of low integration of farmers within large neighboring markets. These findings highlighted the decisions makers to take a deeper look at their rural infrastructure strategy, knowing that it may affect the productivity of the agricultural sector as whole.

Keywords: total factor productivity, agricultural growth, Tornqvist index, growth determinants, Egypt

1. Introduction

Growth of agricultural productivity is considered as backbone of sustainable growth poverty reduction in developing countries such Egypt. Trends of economic development of such countries show that differences in poverty reduction rates over the past decades have been closely related to differences in agricultural performance – particularly the rate of agricultural productivity growth.

Despite decades of investment in new agricultural technologies and rural development, hunger and poverty continue to plague large areas of Egypt. While increasing agricultural productivity perhaps remains the single most important determinant of economic growth and poverty reduction, serious doubts are emerging as to whether agricultural productivity in Egypt can be further increased and how. The role of policies, investments, resources constraints, and other related-factors in increasing agricultural productivity in Egypt is also questionable.

Growth of agricultural productivity is a prerequisite for achieving sustainable growth and significant reduction of poverty in developing countries. Development economists view productivity growth in the agricultural sector as critically important for growth of agricultural production and balance of food supply and demand. Trends of economic development in the Middle East and North African (MENA) countries show that differences in poverty reduction rates over the past decades have been closely related to differences in agricultural performance – particularly the rate of agricultural productivity growth

Despite decades of investment in new agricultural technology and rural development, hunger and poverty continue to plague large areas of the MENA region. While increasing agricultural productivity perhaps remains the single most important determinant of economic growth and poverty reduction in the rural areas of MENA countries, serious doubts are emerging as to whether agricultural productivity can be further increased and how.

Particularly, development experts need greater understanding of the links between agricultural productivity, employment implications and poverty – to what extent will farm productivity growth be labor-saving or employment generating Moreover, future agricultural growth in the region is likely to be severely constrained by limited water resources, population growth (especially youth) and the extreme weather volatility in addition to the political instability.

The objective of the current research paper is to provide an overview of the historical TFP growth of the Egyptian agricultural sector and to analyze its main determinants. The TFP calculation for Egypt is expected shed light on technical change and economic growth of this sector. The final aim of this research will be to provide guidance for future investments for the development of this sector in Egypt. For this purpose, a global dataset about agricultural inputs and outputs (values and volumes) in Egypt was constructed for the period 1961-2012. This data was used to calculate annual Törnqvist productivity indexes for the mentioned period, reflecting the TFP growth (change) of the Egyptian agricultural sector. Another set of variables was used to assess their relationship to the calculated Törnqvist indexes, through a log-linear econometric regression.

The remaining of the paper is divided into 5 sections. The next section presents a general overview of the Egyptian agriculture; the third and fourth sections present a theoretical discussion of TFP growth drivers and the methodology used for this research, respectively. Results are presented and discussed in the fifth section. A last section concludes.

2. Overview of the Agricultural Sector and Its TFP Growth in Egypt

The Egyptian economy has traditionally relied heavily on the agricultural sector for food, fiber and other products. The agricultural sector provides the livelihood for about 55% of the inhabitants and employment for about 34% of the total labor force. In addition, agriculture contributes with about 20% to the gross domestic product (GDP) and with about 20% of the total exports value. Crops production contributes with an average of 61.19 % to the agricultural GDP while livestock contributes with 30.87 % and fisheries with 7.93 %. The demand for agricultural products is increasing due to population growth and the need for more export earnings. Agriculture in Egypt has witnessed significant developments over the last two decades with direct effects on the role of the agricultural sector in national income formation and exports. Such developments have been also marked by farmers' delivery as related to the cropping structure, applied technology, levels of income, and farmers' response to market changes.

Agricultural growth rates in Egypt have been widely differed from one period to another due to the effect of the general economic conditions and to the different investment and development strategies implemented during the last decades. The 1981/1982 and 1986/1987 periods were at the forefront of the periods that achieved a higher growth of the agricultural GDP, estimated at an annual rate of 3%. The 1987/1988 and 1991/1992 periods were at the lower edge, with annual growth rate of around 2%. In the last decade there has been some improvement of this growth rates, reaching 3.3% in 2006/2007. At the beginning of the sixth development plan, 2007/2008 – 2011/2012, the projected rate of growth was estimated at 3.6%.A quick analysis of these figures shows that the structural changes experienced by the agricultural sector in Egypt during the 1990s were positively affecting the growth rates.

For a long period of time, Egyptian exports remained confined to cotton, rice, onion and citrus. Since mid-1970s, exports have witnessed a drastic change: diversification of the exported commodities, expanded export markets, and increased export returns. Total annual export value reached 1230 Million USD during the period 2005–2007, with the aforementioned four crops constituting some 50% of the total export value, while other crops such as vegetables, fruits, medicinal and aromatic plants are constituting the remaining share. On the other hand, export markets have not been confined to Arab and European markets but have been expanded to include African and South Eastern Asian markets, among others.

In spite of the fact that rates of agro-industrialization are still below expectations, the last two decades have witnessed a significant development in this regard. Many agro-industrial units using the latest techniques have been established, meeting world quality standards and capable of accessing several foreign markets. It is noteworthy that processing industry has not been limited to the direct processing of agricultural products but has also been expanded to include several inputs such as packaging material, fertilizers, pesticides, irrigation supplies, as well as other main agricultural inputs. These achievements have led to a significant increase of processed agricultural products. Agricultural activities' income levels in Egypt have been remarkably increasing during the last two decades, but remain low in average. Net returns per feddan (Note 1), have increased from an average of 684 EGP per year during 1980 – 1985 to about 1046 EGP per year in 2007.

A literature review of the TFP growth in the Egyptian agricultural sector reveals that empirical studies were focusing on analyzing TFP of specific agricultural commodities (Rodriguez & Elasraag 2015; Abou-Saad, 2012; Bahloul, 1999). Rodriguez and Elasraag (2015), decomposed the productivity growth of Egyptian cotton production using stochastic frontier approach, and were tracking the changes of the TFP composition over the period 1990–2008. They mainly showed that TFP of this production system does not increase in Egypt. According to their results, this is mainly due to the negative average contributions of changes in allocative efficiency and in the technical progress. Abou-Saad (2012) used Cobb-Douglass and *translog* function models for single-output multi-input stochastic production frontiers and a *translog* multi-output multi-input stochastic input distance function in order to measure the TFPs of four major crops in Egypt: wheat, maize, rice, and cotton. His both approaches provided similar results indicating high level of technical efficiency (TE) in grain production. Productivity growth was very low but the high TE indicates that such a growth needs to come from new technologies. The author also considers that there is a need to invest in research and development in order to reverse a continuous decline in technological progress (TP). Bahloul (1999) assessed the TFP of the overall agricultural sector in Egypt between 1975 and 1992, using the Törnqvist –Theil index. His results indicate an average TFP growth rate of 0.016% for the studied period. The results also indicate that the annual average wages affect the TFP growth negatively, while the number of agricultural workers and the number of tractors are among the factors which are positively and significantly affecting this growth rate.

3. Theoretical Discussion: Determinants of Agricultural Productivity Growth

Several factors have been identified in the specialized literature as sources of productivity change in the agricultural sectors. These are mainly related to the research and development, extension, education, infrastructure, government programs and policies, technology transfer and foreign R&D spillovers, health, structural change and resource reallocation, terms of trade, etc. The productivity measures itself do not provide any information about the separate role/effect of each of these factors. However, complementary econometric regressions can relate the calculated productivity changes to some of these variables in order to quantify the extent of their effect on the agricultural sector performances. We hereby provide a brief literature overview about the role and importance of some of these TFP growth determinants.

Agricultural Research & Development (R&D): Investments in R&D are in fact considered as main determinant of productivity growth in the agricultural sector. In fact, farmers are benefiting from new technologies, varieties, and breeding practices which are mostly developed through research channels and promoted through development ones. Moreover, agricultural research is not only required to increase the agricultural productivity, but also to prevent productivity fallings due to degradation and other overexploitation problems. In the most developed countries, investments in agricultural research and development can be divided into public and private investments, with clear dominance of private investments. However, this figure is completely different in the less developed countries, where mainly the public sector is investing in agricultural R&D. Another important R&D factor in the developing countries is the foreign R&D spillovers. Extension is also considered as part of R&D expenditures, which is actually due to its importance to enhance the scope and intensity of adoption of new technical innovations provided by different research systems and technology transfer channels. Agricultural extension systems aim to reduce the time lag between development of new technologies and their adoption. The sooner the benefits from research are received by farmers and consumers, the higher will be the rate of return to that research expenditure. This time lag between the development of the technology and its adoption by farmers is actually not only depending on the extension services performances but is also function of farmer's skills and level of education. For this reason, some variables reflecting farmer's levels of education and skills were also introduced in our analysis.

Education and Human Capital: Farmer's education is determining their cognitive capacity, ability to take rational decisions, openness for new learning and technologies, search for information, and managing available resources at their farm level. For this reason, an agricultural sector where most of the farmers are having high education level will certainly perform differently from and agricultural sector dominated by lowly skilled and educated farmers. Thus, education is an investment in "human capital" analogous to a farmer's investment in physical capital. Education also hastens the rate of development of new technologies by training scientists.

Another, though less obvious, effect of education is to help consumers better evaluate the potential risks posed by new products and technologies. The potential benefits of a new technology may not be realized if consumers do not appreciate it and purchase the final products.

Investments in agricultural infrastructure: Investment in public agricultural capital, (physical infrastructure in particular) accounts for the largest share of agricultural public budgets in many countries. The role of infrastructure is to expand the productive capacity by increasing resources and enhancing the productivity of private investments (Munnell, 1992). Many studies have found a significant positive relationship between investments in infrastructure and agricultural productivity (Gopinath & Roe, 1996; Yee et al., 2000). The most obvious example of how public investment in infrastructure might affect agricultural productivity is through investment in public transportation and in irrigation infrastructure. As an example, an improved highway system can allow for better market integration of farmers and can reduce costs of acquiring production inputs and of transporting outputs to market.

Level of trade openness in the agricultural sector: Terms of trade are defined as the export-import unit values ratio. In the literature, Agricultural exports expose the producers in a country to an international competitiveness which spurs efficient production technologies. Besides, agricultural imports are a sign of a problematic agricultural sector. An increase in terms of trade reduces inefficiency and consequently increases TFP. This implies that any increase of the export unit value (or equivalent any decrease of the import unit value) enhances TFP. A number of studies in the literature state that favorable agricultural terms of trade is a strategic necessity for enhancing technology adoption as well as mobilization of higher investment levels in transforming agriculture (Dantwala, 1976; De Janvry & Subbarao, 1986). An alternate body of opinion claims that non-price factors (mainly technology, infrastructure, research and extension) are more significant for sustainable agricultural growth in world economies where prices are used as a policy instrument for obtaining a desirable allocation of resources. Terms of trade are important source of information for policy-makers. Changes in inter-sectoral terms of trade cause redistribution of income not only among sectors but also among income classes. Such redistributive flows of income affect the saving, investment, and marketing capacities of farmers.

Other variables: Some other variables are also identified in the specialized literature as determining the agricultural productivity growth. In this study, we also considered variables such as: (i) the balanced territorial development reflected by the share of rural population (in the overall Egyptian population), (ii) the share of agricultural employment (iii) importance of the main strategic crop (wheat) in the total harvested areas, etc. Full definition of these variables as well as the proxies used to integrate them into our modeling of the TFP determinants will be presented in section 4.3 below.

4. Methodological Framework

4.1 TFP Definition and Conceptualization

The two most commonly used measures of productivity are single (partial) factor productivity (SFP) and total (multi) factor productivity (TFP). When multiple heterogeneous inputs are used in the production process, aggregation of these inputs may require the use of price indices. This implies that productivity can be affected by both changes in relative prices of inputs and by the input use per unit of output (Kathuria et al., 2011). TFP measures account for the use of a number of inputs in the production process and are therefore more suitable for performance measurement and comparison across firms and for a given firm over time (Coelli et al., 2005). In this context, TFP can be defined as a ratio of aggregate outputs relative to aggregate inputs used. This aggregation of inputs and outputs raises the problems of index number (i.e. How can we aggregate inputs and outputs without biasing our calculation?).

Three different views exist on what TFP means (Lipsey & Carlaw, 2002). The first conventional opinion considers TFP as the measure of the rate of technical change (see for example, Law, 2000; Krugman, 1996; Young, 1992 among others). The second view (Jorgensen & Griliches, 1967) considers that TFP only measures free lunches of technical change, which are mainly associated with externalities and scale effects. The third view is highly skeptical whether TFP measures anything useful (Metcalfe, 1987; Griliches, 1995). Kathuria et al. (2011) provides the following possibilities on what TFP growth means in literature:

TFP Growth = Output growth – Input growth

= Technical/Technological change/Progress

= Embodied (or endogenous) technical change + Disembodied (exogenous) technical change

= Changes in technical efficiency + Technological progress

Among these definitions, the later authors mention that the first one is the most commonly used. As per definition, TFP growth incorporates all the residual factors after accounting for input growth, and has also been hailed as an "index of ignorance" (Abramovitz, 1956).

4.2 TFP Measurements Methods

4.2.1 Frontier vs Non-Frontier Approaches for TFP Measurement

Measurement of TFP can be done using non-frontier and/or frontier approaches. Each of these approaches is further divided into parametric and non-parametric techniques. Non frontier approaches include growth accounting methods (or non-parametric index-based methods) and econometric parametric approaches. Frontier approaches include the non-parametric Malmquist index methodology and the stochastic production frontier method.

In frontier approach, the objective is to estimate the best obtainable positions based on the estimation of a bounding function, given inputs and prices levels. For example, a cost frontier traces the minimum attainable cost given input prices and output while a "production frontier" traces the set of maximum attainable output for a given set of inputs and technology. This approach is different from the parametric non frontier approaches where an average function is often estimated by the ordinary least square regression as the line of best fit through the sample data (Kathuria et al., 2011). Moreover, the frontier approaches identify the role of technical efficiency in overall firm performances while non-frontier approaches assume that firms are technically efficient. This difference results in different interpretation of TFP growth estimated from both approaches. TFP growth as obtained from frontier approach consists of two components: (i) outward shifts of the production function resulting from technological progress, and (ii) technical efficiency enhancement related to the movements towards the production frontier. On the other hand, the non-frontier approach only considers technological progress as a measure of TFP growth.

Both frontier and non-frontier approaches can be estimated through parametric and non-parametric techniques. Parametric estimations need the specification of a functional form for the frontier and parameters are estimated through econometric techniques using sample data and outputs. One important implication of this issue is that the accuracy of the derived estimates is sensitive to the specified functional form. In contrast, this latter point is the strength of the non-parametric methods (such as data envelopment analysis DEA, or other mathematical programming methods), which are parameters free and does not assume any functional forms. However, one shortcoming of the latter non parametric approaches is that no direct statistical tests can be carried out to validate the estimates.

4.2.2 Indexes for TFP Measurement: The Törnqvist Index

A common feature of the TFP index number is that the empirical estimation of different TFP indexes is based on different weighting methods of inputs and outputs. In most empirical studies, the Divisia, Solow, and the Törnqvist indexes are frequently used. Among index number methods, Törnqvist-Theil Index, which is an approximation to Divisia Index, was used in this study to construct aggregate output and aggregate input indexes. Explanation on theoretical properties and issues in measurement of the productivity through the Törnqvist Index can be found in Diewert (1980) and Coelli et al. (2005). The Törnqvist outputs, inputs and TFP indexes can be expressed under the logarithmic form as follows:

Output index:

$$\mathrm{Ln}\left(\frac{Q_t}{Q_{t-1}}\right) = \frac{1}{2}\sum_j (R_{j,t} + R_{j,t-1})\mathrm{Ln}\left(\frac{Q_{j,t}}{Q_{j,t-1}}\right) \quad (1)$$

Input index:

$$\mathrm{Ln}\left(\frac{X_t}{X_{t-1}}\right) = \frac{1}{2}\sum_i (S_{i,t} + S_{i,t-1})\mathrm{Ln}\left(\frac{X_{i,t}}{X_{i,t-1}}\right) \quad (2)$$

TFP index:

$$\mathrm{Ln}\left(\frac{TFP_t}{TFP_{t-1}}\right) = \mathrm{Ln}\left(\frac{Q_t}{Q_{t-1}}\right) - \mathrm{Ln}\left(\frac{X_t}{X_{t-1}}\right) \quad (3)$$

Where; $R_{j,t}$ is the share of output (j) in total revenue in time (t); $Q_{j,t}$ is the output (j) in time (t); $S_{i,t}$ is the share of input (i) in total input cost, and $X_{i,t}$ is the input (i) in time (t).

Compared to other methods for TFP calculation, the advantage of the Törnqvist index is related to its capacity to decompose TFP growth into outputs and inputs growth indexes. Moreover, the Törnqvist index also consider the inputs and outputs values (prices) which is not the case for other non-parametric indexes such as the Malmquist index.

4.2.3 Factors and Drivers Determining TFP Growth

At this level of the paper, the econometric estimation of the relationship between TFP growth and different factors has been conducted. The most important factors supposed to have an effect on the TFP growth of the Egyptian agricultural sector have been included. In a stylized form, the following regression model (with expected signs in parentheses) has been used:

$$TFP = f(BTD, IIC1, RR, TO, INF) \tag{4}$$

Where:

TFP = Annual Total Factor Productivity values of the Egyptian agricultural sector;

BTD (+) =Balanced Territorial Development Indicators expressed by the Rural GDP per capita (Current LE/Capita/Year);

IIC1 (+) = Index of Innovation Invention Capital- expressed by the number of scientist year;

RR (+) =Resources Reallocation expressed by the "agricultural employment share (%);

TO (+) =Trade Openness expressed by the following formula: (Import + export)/total production (%);

INF (+) =Infrastructure Investments, expressed by the "road density" (km /km2 agricultural land) - 1000 Km.

The log-linear form was considered as functional form for the Equation (4). The log-linear form allows for estimating coefficients that can be directly interpreted as elasticities. In addition, as pointed out in the pioneering work by Jud and Hyman (1974), Equation (4) contains a weak residual variance relative to other functional forms for the same data set and adjusts the data better than the linear specification for both forecasted parameter signs and statistical significance.

The standard Ordinary Linear Squared (OLS) method, if applied to non-stationary data series, can produce spurious regression. The OLS regression can give high R^2, low Durbin-Watson (DW) statistics, and significant t-values of the estimated coefficients, suggesting a significant relationship between dependent and explanatory variables, when in fact they are completely unrelated. Conventionally, the factors explaining TFP have been studied by expressing variables in logarithmic form. This is similar to the first differencing of variables in time series analysis. Provided the original series are integrated of order 1, as is normally the case, expressing the variables in logarithmic terms ensures a stationary data series and means that the OLS method can safely and directly be used (Hendry, 1995).

4.3 Data Description and Statistical Analysis

The data used in this study was collected from different sources. FAO's annual time series (from 1961 to 2012) of Egyptian agricultural (including crops and livestock) production, total cultivated area, total labor employed in the agricultural sector, machinery deployed in the agricultural activities, animal capital, and fertilizer consumption are used to construct the Törnqvist index. The FAO data is complemented, when needed, with data from national statistical agencies, especially when such alternative data is more accurate or up to date. Data from FAO was continuously cross checked with the available (sometimes limited) data from national sources. The following Table 1 presents the list of the variables used in the empirical model (TFP growth measurement). Therefore, the TFP determinant variables where primarily collected from national sources, while some of them were also collected from international databases, as it is for the UNDP-human development index, and the road density.

Summary statistics of the variables (average inputs values) used in the empirical model, including mean, minimum and maximum values and standard deviations and the inefficiency variables are summarized in Table 2. The figures from this table shows that average labor value (1961-2011) is the highest among other inputs. Labor input is followed by the natural resources value (residual), animal feed, capital stock, and fertilizers, respectively.

Table 1. List of Inputs and outputs variables used for the Egyptian agricultural TFP calculation

Variables	Description	Source
Inputs/outputs variables		
Total Agricultural Output (in value)	This variable is representing the total annul value (in current local currency) of the agricultural production in Jordan. This variable is also disaggregated into crops and livestock outputs.	FAO Database
Labor (in quantity and value)	Labor is an agricultural input. This variable is representing the annual quantity (number of active persons) and value (in current local currency: LCU (Note 2)) of labor used in the agricultural sector. The total value is calculated based on average wages in Egypt.	Ministry of Planning (MOP)
Seeds (in quantity and value)	This variable is describing the aggregated annual quantities (in tons) and values (LCU) of all crops seeds used in the agricultural sector. This is not taking into consideration the seedlings and saplings.	FAO database
Machinery (in quantity and value)	The machinery variable is reflected by the annual number and value (LCU) of new tractors in use. Other machineries were not considered in this input vector.	FAO database
Pesticides (in quantity and value)	Pesticide variable is describing the overall annual quantities (in Tons) and value (LCU) of pesticides and other treatment products used in the agricultural sector for different cropping activities.	FAO database
Animal Feed (in quantity and value)	The feed variable is describing the annual quantity (in tons) and value (LCU) of animal feeds used for the livestock activity in Jordan.	FAO database
Capital Stock (in quantity and value)	Capital stock is an important variable representing the annual value (LCU) of fixed inputs (land, live head of livestock, tree plants, livestock infrastructure, etc.). For the quantity of this variable, we used an aggregated normalized index regrouping the annual values of the most important among these fixed capital assets.	FAO database
Natural Resources (water/land) (in quantity and value)	Natural resource input vector is regrouping the rest of inputs which are hard to account for. It is usually including land and water resources used for the agricultural activities. The value of this input corresponds to the residual difference between the overall agricultural output value and the value of all the previous input vectors.	FAO Database + CAPMAS

Source: Own elaboration (2014).

Table 2. Descriptive statistics of the input and output variables, calculated over the period 1961-2011

Inputs and outputs	Unit	T	Minimum	Maximum	Mean	S. Deviation
Quantity of Labor	Thousand Workers	52	3600.0	6965.0	4928.4	958.8
Labor Value	Million LE	52	99.0	13165.9	3717.8	3903.6
Fertilizers Quantities	1000 Tons	52	238.9	1472.4	844.1	403.4
Fertilizers Values	Million LE	52	4.7	16604.7	2749.1	4690.6
Total Seeds Quantities	1000 Tons	52	402.8	1079.7	676.2	180.8
Value (Current million LE)	Million LE	52	7.5	4554.8	1258.7	1435.1
Number of Tractors		52	12837.0	103188.0	55149.3	33431.3
Depreciated Tract Value	Million LE	52	0.1	56.9	14.9	18.1
Pesticides Quantities	Tones	52	1859.3	13213.0	7075.1	3246.8
Pesticides Values	Million LE	52	4.7	1398.7	161.3	297.3
Total Animal Feed Quantities	1000 Tons	52	27310.7	71372.	48643.8	12285.3
Total Animal Feed Value	Million LE	52	238.8	24082.8	3996.3	5605.3
Capital Stock Quantity		52	300.0	2517.5	924.1	674.7
Capital Stock Value	Million LE	52	15.3	8242.6	2059.3	2510.4
Total Agricultural Output	Million LE	52	753.0	188832.0	36089.3	49878.6

Source: Own elaboration (2014) based on FAOSTAT (2013) and other sources.

5. Results and Discussion

5.1 Agricultural TFP Growth of Egypt

The trends of the output, inputs and TFP indexes are presented in Figure 1. The three indexes trends are going up slowly and smooth during the 1966 – 1982 period. The dominance of the irrigated agriculture in Egypt might explain the steady levels of agricultural outputs, inputs use and productivity. Irrigation creates in fact lower dependence to climate volatility and makes the levels of outputs less volatile among years. The privatization and economic liberalization of the agriculture sector have been started since 1981. The structural adjustment plan (SAP) of Egypt also started in the early 90s. The estimation of the Equations (1), (2) and (3) for the Egyptian agricultural sector are presented in the following Figure 1 and Table 3.

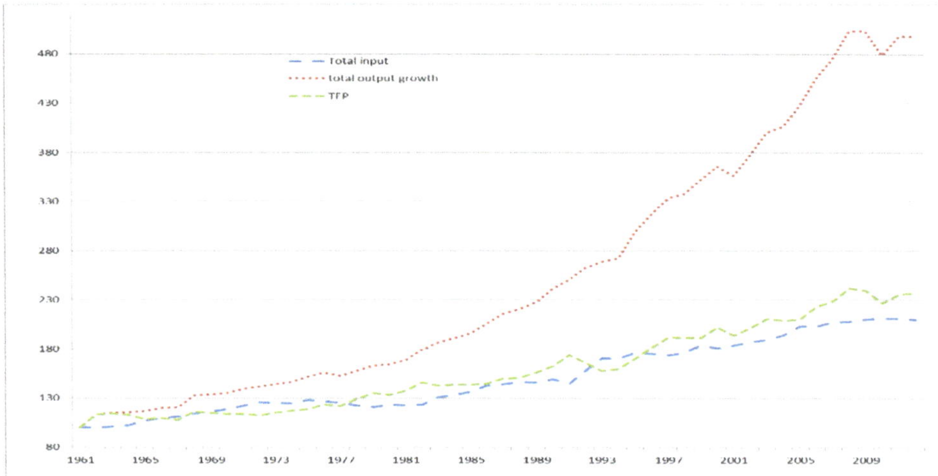

Source: Elaboration by the authors (2014)

Figure 1. Trends of the output, inputs, and TFP indexes calculated (base 100 for 1966)

Furthermore, our results for the Egyptian case show that trends of agricultural, crop and livestock output values have been increased faster since 1999-2000. The trends of labor, fertilizers, capital stock and seeds values have been strongly increased since 1990. The crops revenue shares in the agricultural revenue were fluctuating during 1961 – 2011. It decreased from 69.4% in 1961, to 61.6% in 1982, to 55.6% in 1984 and then increased to 71.5% in 1992. The shares of livestock revenue in total agricultural revenue were also fluctuating during the same period. It increased from 30.7% in 1961, to 38.4% in 1982, to 44.4% in 1984 and then decreased to 28.6% in 1992. These fluctuations justify the variability on the annual growth rates of the selected agricultural inputs and outputs.

Table 3. Normalized (base 100 for 1962) Values of Output, Input and TFP Indexes for the Egyptian agricultural sector calculated based on the Tornqvist-Theil method

	Normalized Output Index	Normalized Input Index	Normalized TFP Index
1962	100	100	100
1970	120.31	108.11	13.55
1975	144.37	125.07	19.29
1980	158.51	123.57	34.94
1985	184.17	129.21	54.95
1990	222.33	145.56	76.77
1995	270.29	163.66	106.63
2000	341.09	177.92	163.17
2005	393.48	191.52	201.96
2010	482.80	208.05	274.74
2011	498.16	211.21	286.95

Source: Elaboration by the authors (2014).

The annual growth rates of the studied inputs and outputs variables are ranging between 0.7% (e.g., natural resource quantity) and 18.2% (e.g., fertilizers values). The increase of agricultural output resulted from an increased use of traditional inputs: These were mainly cultivated areas and growth in TFP. On average, modern inputs (fertilizers and machinery) contributed little to the agricultural output growth and the difference between growth in output and the sum of total contributions by factor inputs and TFP is nearly equal to growth in efficiency, which on average made the lowest contribution to growth in output.

5.2 Factors Determining TFP Growth

Concerning the TFP determinant, Equation (4) was estimated using the SPSS statistical package. As mentioned in the methodological section, the log linear form of the equation was considered in this econometric estimation. Results of the estimation reveal some significant variables affecting the productivity growth of the agricultural sector in Egypt. Table 4 presents the results of this estimation. The estimated relationship between the TFP and the TFP determinants and the estimated parameters are not significant. This is due mainly to the obvious fluctuations in the values of the TFP determinants during the studied period.

Table 4. TFPG determinants in the Egyptian agricultural sector (1980–2011)

	Dependent variable $LnTFP_t$		
Parameters	**Estimated coefficients**	**t-ratios**	**p-value**
Constant	-0.12	-0.17	0.86
$LBTD_t$ (Balanced Territorial Development Indicators)	-0.07^{**}	-1.35	0.18
$LIICI_t$ (Index of Innovation Invention Capital- IIC - # scientists-year)	0.09	0.36	0.71
LRR_t (Resources Reallocation: Agricultural employment share)	0.005	0.01	0.99
LTO_t (Trade Openness)	0.04	0.65	0.51
$LINF_t$ (Infrastructure)	-0.05^{**}	-1.60	0.12
T	33		
R^2	0.41		
F-statistic	0.81 ($p<0.51$)		
Log likelihood	66.65		

Source: Author's calculation based on coefficient estimates of the linear regression model (2014).
*Note: ** Significant at 10% level.*

The indicators for the estimation performance are quite satisfying. The R^2 is equal to 0.41, showing that 41% of the TFP variations in Egypt, over the period of analysis, are explained by the regressed variables considered in our analysis. Concerning TFP determinants for the agricultural sector in Egypt, many important issues can be raised: First, rural development variables were found to significantly and negatively affect agricultural productivity. Put differently, when the rural GDP per capita increases, the agricultural productivity growth of the agricultural sector decreases. This demonstrates that agricultural activity is still a marginalized activity which is linked to low levels of income and is a source of employment for low productive labor. This type of structural problem cannot be handled solely within the framework of an agricultural development strategy but implies a wider vision of integrated rural development where agriculture is developed in parallel/synergy with other economic sectors.

A second issue related to TFP determinants is the negative significant effect of the infrastructure variable on the productivity gains of the agricultural sector in Egypt. Based on specialized literature, if the coefficient of this variable was negative, this might indicate a form of low integration of farmers within large neighboring markets. However, the positive sign of this variable could indicate the high level of fragmentation of agricultural lands due to the development of more roads and unpaved rural roads (Dhehibi et al., *2014*). It is again important that policy makers take a deeper look at their rural infrastructure strategy, knowing that it may affect the productivity of the agricultural sector as whole.

6. Concluding Remarks and Policy Implications

The aim of this study was to calculate the TFP growth of the agricultural sector in Egypt and to assess its determinants. The Törnqvist index methodology was used to calculate the TFP growth and was complemented by an econometric regression of the obtained TFP scores on a set of potentially explicative variables.

Results indicated that the trends of agricultural, crop and livestock output values have been increasing rapidly in Egypt since 1999-2000, while trends of labor, fertilizers, capital stock and seeds uses have been dramatically increasing since 1990. The crops revenue shares in the agricultural revenue have been quite fluctuated during 1961 – 2011. It's decreased from 69.4% in 1961, to 61.6% in 1982, to 55.6% in 1984 and then increased to 71.5% in 1992. In addition the livestock revenue shares in agricultural revenue have been relatively fluctuated during the same period. It increased from 30.7% in 1961, to 38.4% in 1982, to 44.4% in 1984 and then decreased to 28.6% in 1992.Results also show that the Törnqvist-Thiel Indexes for agriculture, crops and livestock have been gradually increased during 1961 – 2011, with respective normalized values of 114, and 498, in 1962 and 2011.

Furthermore, results of the analysis of the TFP growth determinants in Egypt shows the existence of major structural problems within the agricultural sector. In fact, the level of rural GDP per capita was surprisingly found to be negatively related to the TFP growth, indicating that the agricultural activity remains marginalized and synonym of low income levels. This also means that capital accumulation in the agricultural sector will be low, thus affecting the investment levels. Investments in infrastructures (in our case roads density used as proxy) was also found to be negatively correlated to the TFP growth, indicating that these investments have not been targeting the rural areas where most of the agricultural activities are located.

Increased productivity together with the sources of growth implies that the Egyptian government's effort on formulating and implementing a comprehensive strategy that strengthens public and private sectors linkages in helping to facilitate adoption of new technologies and improved management of resources that will contribute to the TFP growth of agriculture; such a strategy is more likely to yield better results in terms of employment, income generation and hence poverty alleviation. Thus, the measurement of productivity and the analysis of agricultural sector performance in this country is an essential steps in a systematic study of the relationship between agricultural-sector productivity, growth in per capita income and real GDP and the incidence of poverty.

Acknowledgment

The authors thank the 'Agricultural Productivity with an Emphasis on Water Constraints in the Middle East and North Africa (MENA)' project sponsored by the Economic Research Service (ERS) - United States Department of Agriculture (USDA) for funding this research.

References

Abou-Saad, I. M. (2012). Total factor productivity change in Egyptian agriculture: A governorate level analysis using stochastic frontiers; PhD dissertation, University of Connecticut, USA.

Abramovitz, M. (1956). Resources and output trends in the U.S. since 1870. *American Economic Review, 46*(2), 5-23.

Bahloul, A. Q. (1999). Total Factor Productivity and Sources of Long-Term Growth in Egyptian Agriculture Sector. *Egyptian Review of Agricultural Economics, 9*(2).

Coelli, T. J., Rao, D. S. P., O'Donnell, C. J., & Battese, G. E. (2005). An Introduction to Efficiency and Productivity Analysis (2nd ed.). Springer, New York.

Dantwala, M. L. (1976). Agricultural policy since independence. *Indian Journal of Agricultural Economics, 31*(4), 762-782.

De Janvry, A., & Subbarao, K. (1986). Agricultural Price Policy and Income Distribution in India, Oxford University Press, Delhi.

Dhehibi, B., Frija, A., Telleria, R., & Aw Hassen, A. (2014). The Effect of Trade Liberalization on the Sustainability of Agricultural Sectors in Egypt and Tunisia: A New Framework Based on TFP Growth Structure. International Conference: Building Sustainable Agriculture for Food Security in the Euro-Mediterranean Area: Challenges and Policy Options, 20-21 November, 2014 - Rabat, Morocco.

Diewert, W. E. (1980). Capital and theory of productivity measurement. *The American Economic Review, 70*(2), 260-67.

FAOSTAT. (2013). *FAO Database*. Retrieved from http://faostat3.fao.org/home/E

Gopinath, M., & Roe, T. L. (1996). Sources of growth in US GDP and economy-wide linkages to the agricultural sector. *Journal of Agricultural and Resource Economics, 21*(2), 325-340.

Griliches, Z. (1995). R&D and productivity: econometric results and measurement issues. In P. Stone-man (Ed.), *Handbook of the Economics of Innovation and Technological Change* (pp. 52-89). Basil Blackwell, Oxford, UK and Cambridge, Mass.

Hendry, D. F. (1995). *Dynamic Econometrics*. Oxford University Press, Oxford. http://dx.doi.org/10.1093/0198283164.001.0001

Jorgenson, D. W., & Griliches, Z. (1967). The explanation of productivity change. *Review of Economic Studies, 34*(3), 249-283. http://dx.doi.org/10.2307/2296675

Jud, D., & Hyman, J. (1974). International demand for Latin American tourism. *Growth and Change, 5*(1), 25-31. http://dx.doi.org/10.1111/j.1468-2257.1974.tb00278.x

Kathuria, V., Raj, R. S. N., & Sen, K. (2013). Productivity measurement in Indian manufacturing: a comparison of alternative methods. *Journal of Quantitative Economics, 11*(1/2), 148-179.

Krugman, P. (1996). *The Myth of Asia's Miracle: Pop Internationalism*. MIT Press, Cambridge.

Law, M. T. (2000). *Productivity and Economic Performance: An Overview of the Issues*. The Fraser Institute, Vancouver, BC.

Lipsey, R. G., & Carlaw, K. I. (2002). *The Measurement of Technological Change*. Unpublished.

Metcalfe, S. (1987). Technical change. In J. Eatwell, et al. (Eds.), *A Dictionary of Economics*. The New Palgrave, McMillan, London.

Munnell, A. H. (1992). Policy watch: infrastructure investment and economic growth. *Journal of Economic Perspectives, 6*(4), 189-198. http://dx.doi.org/10.1257/jep.6.4.189

Rodriguez, X., & Elasraag Yahia, H. (2015). Assessing the Total Factor Productivity of Cotton Production in Egypt. *PLoS One, 10*(1), e0116085. http://dx.doi.org/10.1371/journal.pone.0116085

Yee, J., Huffman, W., Ahearn, M., & Newton, D. (2000). Sources of agricultural productivity growth at the state level, NC-208 – Meeting on Agricultural Productivity: Data, Methods, and Measures, 9–10 March, Washington, DC.

Young, A. (1992). A tale of two cities: factor accumulation and technical Change, NBER Macroeconomic Annual, MIT Press, Cambridge.

Notes

Note 1. 1 feddan=0.42 hectares.

Note 2. 1 Egyptian pound = 0.13 US$.

13

Development of Four-Column Data Storage Model for Data-Manipulation of Greenhouse Gases and Soil Properties

Bence Matyas[1,2], György Matyas[3], Mihaly Szendrei[4], Ankit Singla[2,5], Yuhua Kong[6], Janos Katai[1], Agnes Zsuposne Olah[1] & Kazuyuki Inubushi[2]

[1] Centre for Agricultural and Applied Economic Sciences, University of Debrecen, Debrecen, Hungary

[2] Graduate School of Horticulture, Bioresource Science, Chiba University, Matsudo, Chiba, Japan

[3] Oracle APEX hosting, Individual entrepreneur, Hungary

[4] Prec Cast Ltd., Sátoraljaúhely, Hungary

[5] Department of Microbiology and Fermentation Technology, Sam Higginbottom Institute of Agriculture, Technology and Sciences, Allahabad, Uttar Pradesh, India

[6] College of Forestry, Henan Agricultural University, Henan, China

[*] Correspondence: Kazuyuki Inubushi, Graduate School of Horticulture, Bioresource Science, Chiba University, Matsudo, Chiba, Japan. E-mail: inubushi@faculty.chiba-u.jp

Abstract

Studies on greenhouse gas emissions and climate change require huge amounts of data. Generally, these data can be inducted directly from measuring devices or from different systems like geographical information systems and other specific systems, which are used for experimental data storage and management. Database technology has come a long way over the years, however even with these advances in technology and the availability of supercomputers processing large amounts of data from different data storage methods is still a challenge. In the present study, one physical data table was generated for data-manipulation and data storage using a web-developed platform. Data storage with four-column usage assures the possibility for continuous refinement during the research processes. It is because of the determination of new variables after the entity identification process.

Keywords: Data-manipulation, non-relational data storage, four columns data storage, greenhouse gases, physical data table

1. Introduction

Apart from the water vapour, carbon dioxide (CO_2), methane (CH_4) and nitrous oxide (N_2O) are important greenhouse gases (GHG) contributing 60, 20 and 6% towards global warming, respectively(IPCC, 2007).The global temperatures have increased 0.8°C over the past century and are predicted to increase another 1.1-6.4 degrees over the next century (Peters et al., 2013).The concentrations of CO_2, CH_4 and N_2O in the atmosphere has increased from pre-industrial levels of 280 ppm, 715 and 270 ppb to about 379 ppm, 1,774 and 319 ppb, respectively, in 2005 (IPCC, 2007). Although the concentration of CH_4 and N_2O is much lesser than CO_2, both gases have 25 and 298 times higher global warming potential than CO_2, respectively (IPCC, 2007). Ravishankara et al.(2009) reported that N_2O can also lead to depletion of the stratospheric ozone layer. The emissions of CO_2 resulting from the soil respiration and vegetation are the principal sources from which this gas enters into the atmosphere (Raich & Schlesinger, 1992). Rice is considered as one of the most important anthropogenic sources that accounts for 10-15% of the global CH_4 emission to the atmosphere (Cheng et al., 2008). Agricultural soil is a major source of N_2O (Hayakawa et al., 2009; Singla et al., 2013, 2014a; Singla & Inubushi, 2015).Several reports are available in the published literature which studied the emissions of these GHG under various soil types, cropping, irrigation and fertilizer management (Cabrera et al., 1994; Glatzel et al., 2004; Kong et al., 2013; Singla & Inubushi, 2014a, b; Singla et al., 2014b). It will be useful and time-saving if such large amount of data could be stored in one project with one platform usage. The development of a data-manipulation model could meet such requirement and provide a useful platform for data storage. Generally, data mining projects consist of three essential tasks: data collection, data preparation, and data modelling (Pyle,

1999; Westerman, 2001). The generation of such models also ensures possibilities for storing different data in one data table with one platform usage (Imhoff et al., 2003). Significant benefits of such system models manifest when many different types and huge amounts of data are required to be compiled in one project. Users could be able to input data as usual data tables (relational) while using data models. The integrated mechanisms of such models ensure that the system stores input data into one physical data table (non-virtual). The objective of the present study was to develop input correlations between soil attributes, their GHG production potential, and geographical information system (GIS) data of sampling parameters under one platform.

2. Method

2.1 Data Sites Used in the Study

The seven sampling sites from central Japan (34° 49'~36°36' N, 139°00'~54' E) and eight sites from eastern Hungary (47°55'~48°12' N, 21°23'~67' E) were selected for data storage. The physical parameters of soil sampling sites in both countries were taken as: land use, latitude/longitude, altitude, soil orders (FAO taxonomy), annual mean temperature (°C), annual mean precipitation (mm), and the soil moisture (%). The chemical and microbial properties of soils from both countries were taken as: pH (H_2O), EC (d S m^{-1}), total C {mg g^{-1}dry soil(ds)}, total N (mg g^{-1}ds), C/N ratio, microbial biomass C (MBC) (μg g^{-1} ds), microbial biomass N (MBN) (μg g^{-1} ds), soluble organic C(SOC) (μg g^{-1} ds), nitrate-N(NO_3^--N) (μg g^{-1} ds), ammonium-N(NH_4^+-N) (μg g^{-1} ds), nitrite-N(NO_2^--N) (μg g^{-1} ds), and MBC/ total C (%). The concentrations of CO_2, N_2O, and CH_4 were measured using gas chromatographs (GC-14B, Shimadzu, Japan) equipped with a thermal conductivity detector, electron capture detector and flame ionization detector, respectively (Singla & Inubushi, 2013). Equations in multiple regression models for the relationship between GHG production/consumption potential and the soil properties were stored in the database (Table 1). All the statistical analyses were completed using SPSS Statistics 20 (IBM, New York, USA).

Table 1. Multiple regression models for the relationship between the cumulative greenhouse gas (GHG) production/consumption (Y) and the soil properties (X)

Country	#	Equations	Adjusted R^2	p
Japan (n = 7)	#1	Y1 = 1130.0 X1 – 32.0	0.860	0.002
	#2	Y1 = 4.3 X4 + 298.2	0.666	0.016
	#3	Y1 = 0.435 X3 + 296.9	0.548	0.035
	#4	Y3 = 31.48 X10 – 355.8	0.510	0.043
Hungary (n = 8)	#5	Y1 = 1750.3 X2 – 574.0	0.942	< 0.001
	#6	Y1 = 291.0 X7 – 321.1	0.699	0.006
	#7	Y1 = 5.663 X3 – 301	0.877	< 0.001
	#8	Y2 = 32.7 X2 – 7.4	0.936	< 0.001
	#9	Y2 = 0.102 X3 – 1.9	0.803	0.002
	#10	Y2 = 5.1 X7 – 2.0	0.605	0.014
	#11	Y2 = 0.018 Y1 + 3.4	0.958	< 0.001
	#12	Y3 = 6.262 X10 – 44.98	0.609	0.014
	#13	Y3 = 3.3 X8 – 31.2	0.477	0.035
Japan and Hugary (n = 15)	#14	Y1 = 108.9 X7 + 55.01	0.408	0.006
	#15	Y2 = 13.64X2 + 0.867	0.255	0.032
	#16	Y3 = 1.137 X6 – 0.169 X5 – 22 X9 + 26.2	0.811	< 0.001

Y1: cumulative carbon dioxide (CO_2) production; Y2: cumulative nitrous oxide (N_2O) production; Y3: cumulative methane (CH_4) consumption; X1: nitrite-nitrogen (NO_2^--N); X2: ammonium nitrogen (NH_4^+-N); X3: microbial biomass carbon (MBC); X4: microbial biomass nitrogen (MBN); X5: soluble organic carbon (SOC); X6: total carbon (TC); X7: total nitrogen (TN); X8: soil water content; X9: the ratio of MBC to TC; X10: soil carbon-nitrogen (C:N) ratio. #1 to #13 were analyzed by single regression method. #14, #15, and #16 were analyzed by stepwise regression method.

2.2 Database

Oracle Application Express (Apex) was chosen for geographical and experimental data storage. Apex provided opportunity to easy access because users could share works (by hierarchical levels) in same web platform without any software installation. Data of both countries were stored in Apex data tables at the first stage. The structure of entity identification was determined with key usage (Figure 1). The data tables which were used under GHG work space are available at web-link: http://apex.oracle.com/pls/apex/f?p=4500:1000:8427548650555

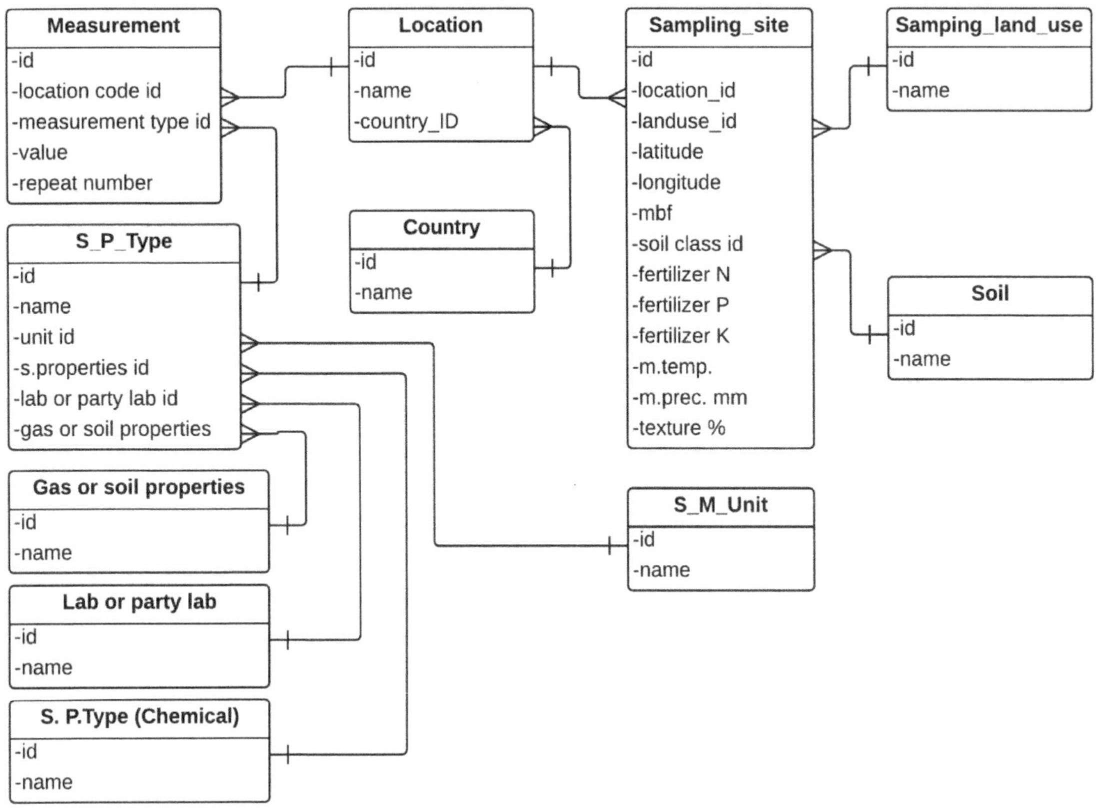

Figure 1. Measurement, Country, Location, Sampling_site, Sampling_land_use, Soil,S_P_Type, S_M_Unit, P. Type, Lab or party, Gas or soil properties tables were created to store data with APEX platform. ID columns were used to create relationships between these data tables. Number values could be entry with columns of ID, value, repeat number, latitude, longitude, altitude (mBf), ferti N, ferti P, ferti K, m.temp., m.prec., texture usage. Text could be entry with columns of name. Boolean options could be chosen with columns of lab or party, and gas or soil properties

2.3 Data-Manipulation Model

The developed data storage model called Joker Tao (JT) was used for storing data of different systems like data of GIS and input experimental data in one data table. After the data entry process in Apex data tables, data of CO_2, N_2O and CH_4 production/consumption potentials and the soil properties under different land-use types in central Japan and eastern Hungary (Kong et al., 2013) were stored in one physical data table with four column usage (Figure 2). The data lines arose manually or due to events (prompts from the model). Different data with attributes and correlations were stored in uniform storage format. Each of the attributes was key indexed. A multitude of huge amounts of data lines generated a qualitative leap in this system model. When data lines were created, it was done automatically which could be logical. Attributes of data storage, data maintenance and properties function both as cognitive and feature. At the same time, the above explanatory properties monitor and enforce the feedback environment. The most elementary entity of a set is an object, which is determined by three factors: attribute of describing uniqueness, attribute of specified and characterized entity, and value of attribute of specified and characterized entity (Table 2).

Table 2. Set theory of entity identification with the developed four columns data storage system model

Baseline			**A U B (union)**			**A ∩ B (intersection)**		
ID	Attribute	value	ID	Attribute	value	ID	attribute	Value
1	Cardinality	1	1	Cardinality	1	4	cardinality	1
1	set A	Yes	1	set A	yes	4	of set B	yes
2	Cardinality	1	2	Cardinality	1	4	of set A	yes
2	set A	Yes	2	set A	yes	**A\B**		
3	Cardinality	1	3	Cardinality	1	ID	attribute	Value
3	set A	Yes	3	set A	yes	1	cardinality	1
4	Cardinality	1	4	Cardinality	1	1	of set A	Yes
4	set B	Yes	4	set B	yes	2	cardinality	1
4	set A	Yes	4	set A	yes	2	of set A	Yes
23	Cardinality	1	23	Cardinality	1	3	cardinality	1
23	set B	Yes	23	set B	yes	3	of set A	Yes

The record of this set theory shows how identification process was completed. Data of Baseline and Union main columns are same, because the event of base consist A and B sets determination. Cardinality was used for storing number data type as value. Intersection and minus procedure were used to represent query of dataset. Relationship between sets of A and B was determined how the model interprets the input data, because the possibility was guaranteed all data could be entity and attributes at the stage of data entry.

3. Results

Structure of data storage was determined with attributes of conceptual-functional name, functional logic and physical functionality. Tables of sampling country, sampling land use, sampling location and sampling sites were created to input attributes of sampling. Tables of measurements, measurement unit, types of measurement and types of soil properties were created to input measured data (Figure 1).

The developed integrated mechanism ensures that all input data could be stored in one physical data table with four columns usage. The process of insertion from data of multi-tables into one physical data table was determined with a Java code given below:

```
public static String getEntityName() throws Exception
{
Connection conn = broker.getConnection();
PreparedStatementpstmt = conn.prepareStatement("select * from joker tao");
ResultSetrs = pstmt.executeQuery();
inti = 0;
while (rs.next()) {
i++;
}
System.out.println("Number of records: " + i);
broker.freeConnection(conn);
return "";
}
```

The developed system model operates whichever nature, real (data from physical data table) or virtual (data from virtual data tables) objects. The structure of data storing was determined vertically and not horizontally as similar to Menon (2005). Actual query determines how the system interprets the data, because every stored data could be entity and attribute in this model. Identifier of Group_ID, Uniq_ID, Field_ID, Array_Index, Seek_Value,

Field_Value were created to separate attributes of different entities:

```
public static void insertJokerRow(Integer GROUP_ID, Integer UNIQ_ID, Integer FIELD_ID, Integer ARRAY_INDEX, String SEEK_VALUE, String FIELD_VALUE) throws Exception {
if (GROUP_ID == null) pstmt.setNull(1, 2); else pstmt.setInt(1, GROUP_ID.intValue());
if (UNIQ_ID == null) pstmt.setNull(2, 2); else pstmt.setInt(2, UNIQ_ID.intValue());
if (FIELD_ID == null) pstmt.setNull(3, 2); else pstmt.setInt(3, FIELD_ID.intValue());
if (ARRAY_INDEX == null) pstmt.setNull(4, 2); else pstmt.setInt(4, ARRAY_INDEX.intValue());
if (SEEK_VALUE == null) pstmt.setNull(5, 12); else pstmt.setString(5, SEEK_VALUE);
if (FIELD_VALUE == null) pstmt.setNull(6, 12); else pstmt.setString(6, FIELD_VALUE);
pstmt.execute();
  }
```

Data.txt file was created to complete the insertion process. The method reads input data from this text file and inserts them into one data table. Identifier of Group_ID, Uniq_ID, Array_Index, and Field_Value were used for entity identification:

```
public static void readFile() throws Exception
  {
    File f = new File("data.txt");
BufferedReaderbr = new BufferedReader(new FileReader(f));
while (br.ready()) {
        String line = br.readLine();
int GROUP_ID = Integer.parseInt(line.substring(0, 10));
int UNIQ_ID = Integer.parseInt(line.substring(11, 21));
int ARRAY_INDEX = Integer.parseInt(line.substring(22, 32));
        String FIELD_VALUE = line.length() > 32 ? line.substring(33, line.length()) : "";
insertJokerRow(Integer.valueOf(GROUP_ID), Integer.valueOf(UNIQ_ID), null, Integer.valueOf(ARRAY_INDEX), null, FIELD_VALUE);
    }
br.close();
}
```

Thanks to the above Java code the all data tables were converted to one physical data table. Concept of data storage was determined with columns of ID, Name, Attribute and Value (Table 3).

Table 3. Relationship determination (variable of Genux_proximus_idwas used to hierarchical entity identification. Value of 1 in the fourth row means that the entity which was determined with number 2 ID records are in relation to entity which were determined with value of 100 ID records)

ID	ATTRIBUTE	INDEX	Value	explanation
1	1	1	100	group membership
1	2	1	Genus proximus_id	
1	13	1	74	
2	1	1	100	
2	2	1	name	
2	13	1	71	71=alpha
3	1	1	100	
3	2	1	Sub_group _ID	sub group membership
3	13	1	74	code
4	1	1	100	Basic attribute

Associated attributes of this system model means that the system could initiate the search with a parallel processing mode to complete queries. Every single A and B search path is further engraved (marked) and path number increases.

$$\text{Path number} = \text{path number} + 1.$$

The model logs if path number is a large amount. The more frequent searches are priorities. The mentioned searching mode is completed with large amounts of data rows and cognitions usage. One data table was used to identify attributes (columns), entities (rows) and relationships between stored data. Relationships between stored data were determined with variables of ID and genus_proximus_idusage in four columns physical data table. Every record is a different entity in a relational data storage models (Menon, 2005), but a record with same ID values is the entity in this system model (Figure 2) like in some specific non-relational database systems (Imhoff et al., 2003; Sadalage & Fowler, 2013).The developed model operates data by the same operation (proceeding) which leaves the data type out of consideration.

4. Discussion

Differently, the relational data storage models with multi-table using (Imhoff et al., 2003), only one physical data table was used for data-manipulation. In this case, there is no need to create relationships between data tables or use horizontal column expanding, which could slow down the time of the query. More data rows were used in this model than in relational multi-level database systems, but path number usage could shorten the searching line during the time in which query is completed. Data storing with four-column usage assures the possibility for continuous refinement during the researching processes, because new variables could be determined after the entities identification process. Data attributes, soil sampling parameters, values of measurements and values of GHG production/consumption potentials were stored in one physical data table which could retrieve these data using integrated mechanisms in one platform. It offers a tool and opportunity for the researchers and developers to coordinate parallel teamwork and continuous data entry processes.

We created a customized code table, which allows any data whether entity, attribute, data connection or formula, to each be stored and managed under one physical data table. JT also handles relational data models differently; it allows data lines with the same ID values to identify a single entity. Similarly, non-relational models are also unique in JT; the inputted data models are not stored in an unstructured concept, making it easier to manage a huge amount of data without creating several applications. When compared to NoSQL models the speed of the JT system lies in its use of vertical data expansion and the elimination of the sequential search approach, which slows queries. JT is self-learning as all equations can be stored within the code table while maintaining the characteristics of the balance method similar to data stored in the physical data table.

The present study showed that the development of four columns data storage model using one physical table to store data from multi tables will be useful for research and development projects in any discipline.

Acknowledgements

This study was financially supported by the Hungarian-Japanese intergovernmental S&T cooperation program for the year 2009–2011. The corresponding author is thankful to György Mátyás, the idea owner of JT. The first version of JT was developed (R. number: INNO-1-2008-0015 MFB- 00897/2008) in an EU application. The authors are also thankful to the Call Tec Consulting Ltd. which was the first company in Hungary to obtain highest Oracle ranking.

References

Cabrera, M. L., Chiang, S. C., Merka, W. C., Pancorbo, O. C., & Thompson, S. (1994). Nitrous oxide and carbon dioxide emissions from pelletized and non pelletized poultry litter incorporated into soil. *Plant and Soil, 163*,189-195. http://dx.doi.org/10.1007/BF00007967

Cheng, W., Sakai, H., Hartley, A., Yagi, K., & Hasegawa, T. (2008). Increased night temperature reduces the stimulatory effect of elevated carbon dioxide concentration on methane emission from rice paddy soil. *Global Change Biology, 14*, 644-656. http://dx.doi.org/10.1111/j.1365-2486.2007.01532.x

Glatzel, S., Basiliko, N., & Moore, T. (2004). Carbon dioxide and methane production potentials of peats from natural, harvested and restored sites, Eastern Quebec, Canada. *Wetlands, 24*, 261-267. http://dx.doi.org/10.1672/0277-5212(2004)024[0261:CDAMPP]2.0.CO;2

Hayakawa, A., Akiyama, H., Sudo, S., & Yagi, K. (2009). N_2O and NO emissions from an Andisol field as influenced by pelleted poultry manure. *Soil Biology and Biochemistry, 41*, 521-529. http://dx.doi.org/10.1016/j.soilbio.2008.12.011

Imhoff, C., Galemm, N., & Geiger, J. G. (2003). Mastering data warehouse design: relational and dimensional techniques. Indianapolis, Indiana: Wiley Publishing, Inc..

IPCC. (2007). Changes in atmospheric constituents and in radiative forcing. In Climate change, the physical science basis, contribution of working group I to the fourth assessment report of the intergovernmental panel on climate change, Cambridge University Press, Cambridge, UK.

Kong, Y., Nagano, H., Katai, J., Vago, I., Olah, A. Z., Yashima, M., & Inubushi, K. (2013). CO_2, N_2O and CH_4 production/consumption potentials of soils under different land-use types in central Japan and eastern Hungary. *Soil Science and Plant Nutrition, 59*,455-462. http://dx.doi.org/10.1080/00380768.2013.775005

Menon, R. M. (2005). *Expert Oracle JDBC Programming*. New York: Springer-Verlag.

Peters, G. P., Andrew, R. A., Boden, T., Canadell, J. G., Ciais, P., Quéré, C. L., ... Wilson, C. (2013). The challenge to keep global warming below 2°C. *Nature Climate Change, 3*, 4-6. http://dx.doi.org/10.1038/nclimate1783

Pyle, D. (1999). *Data preparation for data mining*. San Francisco, CA: Morgan Kaufmann,

Raich, J. W., & Schlesinger, W. H. (1992). The global carbon dioxide flux in soil respiration and its relationship to vegetation and climate. *Tellus, 44*(B), 81-99. http://dx.doi.org/10.1034/j.1600-0889.1992.t01-1-00001.x

Ravishankara, A. R., Daniel, J. S., & Portmann, R. W. (2009). Nitrous oxide (N_2O): the dominant ozone-depleting substance emitted in the 21st century. *Science, 326*, 123-125. http://dx.doi.org/10.1126/science.1176985

Sadalage, P. J., & Fowler, M. (2013). NoSQL Distilled: A brief guide to the emerging world of polyglot persistence. Addison-Wesley: Pearson Education.

Singla, A., Dubey, S. K., Iwasa, H., & Inubushi, K. (2013). Nitrous oxide flux from Komatsuna (*Brassica rapa*) vegetated soil: A comparison between biogas digested liquid and chemical fertilizer. *Biology and Fertility of Soils, 49*, 971-976. http://dx.doi.org/10.1007/s00374-013-0796-4

Singla, A., & Inubushi, K. (2013). CO_2, CH_4 and N_2O production potential of paddy soil after biogas byproducts application under waterlogged condition. *International Journal of Agriculture Environment and Biotechnology, 6*, 233-239.

Singla, A., Dubey, S. K., Singh, A., & Inubushi, K. (2014b). Effect of biogas digested slurry-based biochar on methane flux and methanogenic archaeal diversity in paddy soil. *Agriculture Ecosystems and Environment, 197*, 278-287. http://dx.doi.org/10.1016/j.agee.2014.08.010

Singla, A., & Inubushi, K. (2014a). Effect of biochar on CH_4 and N_2O emission from soils vegetated with paddy. *Paddy and Water Environment, 12*, 239-243. http://dx.doi.org/10.1007/s00374-014-0950-7

Singla, A., & Inubushi, K. (2014b). Effect of biogas digested liquid on CH_4 and N_2O flux in paddy ecosystem. *Journal of Integrative Agriculture, 13*, 635-640. http://dx.doi.org/10.1016/S2095-3119(13)60721-2

Singla, A., Sakata, R., Hanazawa, S., & Inubushi, K. (2014a). Effect of biogas digested slurry based-biochar and digested liquid on N_2O, CO_2 flux and crop yield for three continuous cropping cycles of komatsuna (*Brassica rapa* var. perviridis). *Biology and Fertility of Soils, 50*, 1201-1209. http://dx.doi.org/10.1007/s00374-014-0950-7

Singla, A., & Inubushi, K. (2015). Effect of slag-type fertilizer on N_2O flux from komatsuna vegetated soil and CH_4 flux from paddy vegetated soil. *Paddy and Water Environment, 13*, 43-50. http://dx.doi.org/10.1007/s10333-013-0405-z

Westerman, P. (2001). Data warehousing: using the Wal-Mart model. (1972). Ethical standards of psychologists. Washington, DC: American Psychological Association. http://dx.doi.org/10.1016/b978-155860684-5/50001-6

14

A Comparison between Conventional Tilled and Bicropped Winter Wheat Grown in Monoculture over Four Years

J. M. Finnan[1], J. I. Burke[2] & T. M. Thomas[1]

[1] Oak Park Research Centre, Carlow, Republic of Ireland

[2] School of Agriculture and Food Science, University College Dublin, Belfield, Dublin

Correspondence: J. M. Finnan, Teagasc Crops Research Centre, Oak Park, Carlow, Republic of Ireland. E-mail: john.finnan@teagasc.ie

Abstract

A four year experiment was conducted at a site in the south-east of Ireland in which medium and high input conventional winter wheat production systems were compared to no input and low input systems in which winter wheat was direct drilled into an understory of white clover. Whole crop and grain yields from all systems were strongly related to external input levels, yields from bicropped treatments were poor. Nitrogen uptake and grain yields from the conventional treatments declined during the course of the study whereas nitrogen uptake and yields from bicropped treatments were more stable. Fertiliser N application significantly depressed biological production efficiency and altered biomass partitioning. The proportion of biomass partitioned to the stem decreased with fertiliser N, differences between treatments persisted until final harvest. Although the clover sward was still present in the fourth year, this component of the bicrop was gradually replaced by weeds as the experiment progressed in spite of several attempts to control weeds. It is suggested that further research is needed to identify a clover management strategy which ensures the persistence of the white clover sward and allows it to enrich soil fertility in such a way as to be of benefit to the accompanying wheat crop.

Keywords: Winter wheat, bicropping, nitrogen uptake, harvest index, biological production efficiency, nitrogen utilization efficiency

1. Introduction

It has long been established that increased production and forage quality will result when legumes are grown along with grasses in pastures (Haystead & Marriott, 1978; Simpson, 1965; Whitney, 1977). In addition to increased production, a concurrent increase in soil nitrogen (N) occurs and annual amounts in excess of 100 Kg N ha^{-1} can be accumulated (Clement & Williams, 1967). Legumes are often included as a component in crop rotations because of their beneficial effects on soil fertility. Succeeding crops benefit from enhanced availability of N after residual legume shoots and roots have decomposed. This source of organic N can be made available to succeeding crops after ploughing and arable cropping and its availablitity reduces the requirement for fertiliser N (Clement & Williams, 1967). Additionally, the inclusion of a legume in a rotation leads to increases in the levels of soil organic matter as well as to improvements in soil structure (Ladd et al., 1983). The practise of crop rotation as well as interest in the use of legumes as sources of N declined in developed regions of the world when cheap forms of inorganic N fertiliser became available (Groya & Sheaffer, 1985). In contrast, mixed cropping and intercropping systems including legumes have been and continue to be used in countries where N fertilizers are difficult to obtain (Remison, 1978; Martin & Snaydon, 1982).

The achievement of high grain yields in winter wheat crops is dependent on the availability of nitrogen (N), an essential element involved in all of the plants metabolic processes. Substantial increases in grain yields have been made possible by the availability of N fertiliser (Bell et al., 1995). However, N is costly to produce and is easily dissipated in the environment if applied incorrectly, a sizable portion of applied N is lost by leaching and denitrification (Huber et al., 1977). Thus, both environmental and economic considerations have driven a quest to reduce inputs of N fertiliser both by developing management practises more responsive to the physiological needs of crops and by increasing the efficiency with which absorbed nitrogen is used to produce grain. Cereal-legume intercropping systems offer one possible way of minimising inputs of fertiliser N by utilizing the legume component to improve soil fertility (Ofori & Stern, 1987). Brophy et al. (1987) demonstrated that N transfer was possible between a legume and an associated non-legume. Simultaneous seeding of legumes and small grains has provided mixed results (Stewart et al., 1980; Brandt et al., 1989). However, the development of direct drills has enabled cereal crops to be drilled into an already established clover understory. Williams and Hayes (1991) showed that spring cereals could be direct drilled into a clover understory once clover growth had been suppressed. Jones (1992) drilled spring barley, spring oats and winter barley into an understory of white clover and found that whole crop silage and grain yields were similar to those obtained by traditional methods. Jones and Clements (1993) successfully grew winter wheat in a white clover understory over several years although yields were low. White and Scott (1991) examined the effects of small grain species, living mulch species and topdress N in a two year study in which winter cereals were direct drilled into living mulches. Of the living mulches tested, red and white clovers showed the greatest potential to produce dry matter with little interference to the cereals and appeared to contribute nitrogen to rye. Schmidt et al. (2003) showed that the presence of a clover understory had a beneficial effect on earthworm populations, previous work had shown that earthworms increased the biomass and N uptake of wheat (Schmidt & Curry, 1999).

Previous experiments had shown that cereals could be successfully drilled into a clover understory, that the clover understory could survive several cereal crops and that the yield of cereal crops drilled into a clover understory was comparable to those of conventionally sown cereal crops. In this study winter wheat-clover bicropping sytems which received either no inputs of N fertiliser and plant protection chemicals or low inputs of these additives were compared to medium and high input commercially used systems in which winter wheat was sown into tilled soil. The objective of the study was to examine the possibility of a low input production system for winter wheat which would utilise biologically fixed N and which would not need high inputs of fertiliser N or plant protection chemicals.

2. Materials and Methods

The field investigations were carried out in one experimental field at Oak Park Research Centre, Carlow, Republic of Ireland, Latitude $52^0 51' 12''$N and Longitude $6^0 54' 15''$W. The soil is a free draining medium-heavy textured limestone soil with 3% organic carbon content. The field, which had been intensively cropped for over 20 years, was left fallow during 1993 following a crop of spring barley harvested in August 1992. All plots received 37 Kg P/ha and 74 Kg K/ha during each year of the study. Additionally, the experimental area received 70 Kg K/ha in March 1995 as soil analysis had indicated the level of this element to be particularly low.

2.1 Treatments

The trial layout was a randomised complete block with four replications, each plot measuring 60 metres long and 10 metres wide. The same plots were used for each treatment in all four years of the study. The four treatments were replicated in each block. These were; winter wheat drilled into a white clover sward, no fertiliser nitrogen, no plant protection chemicals (Bicropped, zero); winter wheat drilled into a white clover sward, 50 Kg N/ha, low input of plant protection chemicals (Bicropped, low); winter wheat sown into a conventional ploughed seedbed, 160 Kg N/ha, low input of plant protection chemicals (Conventional, medium); winter wheat sown into a conventional ploughed seed bed, 200 Kg N/ha, high input of plant protection chemicals (Conventional, high). Further details are provided in Table 1.

Table 1. Details of treatments

Treatment	Bicropped, Zero	Bicropped, Low	Conventional, medium	Conventional, high
Nitrogen	0 Kg/ha	50 Kg/ha	160 Kg/ha	200 Kg/ha
Growth Regulator			Half Rate Growth Regulator	Full Rate Growth Regulator
Herbicide			Half Rate Autumn Herbicide	Full Rate Autumn Herbicide
Insecticide			Full Rate pre-harvest Insecticide	Full Rate Autumn Insecticide Full Rate pre-harvest Insecticide
Fungicide		2 * Half Rate Fungicide applications	2 * Half Rate Fungicide applications	3 * Full Rate Fungicide applications

Towards the end of June 1996, a large number of aphids were found on bicropped plots. A selective insecticide (perimicarb) was used to control this problem. Grass weeds (*Poa annua*, *Poa trivialis*, *Agrostis stolinifera*, *Agrostis tenuis*) became increasingly problematical in the bicropped plots as the experiment progressed and a number of attempts were made to eradicate this problem. Avadex granules (active ingredient tri-allate) were applied in November 1996, bicropped crops were sprayed with Fusilade (active ingredient fluazifop-P-butyl) in September 1997 and Topik (active ingredient clodinafop-propargyl) was sprayed on bicropped treatments in Febuary 1998. In March 1998, it was found necessary to spray bicropped plots with Benazolin to control *Galium aparine*. It was still found necessary to spray bicropped plots with glyphosate in August 1998 to facilitate harvesting in spite of all attempts to control weeds.

The period after sowing was very wet in October 1997 and slug numbers increased dramatically. Consequently, methaldehyde pellets were broadcast on all plots on 17th October, methiocarb pellets were broadcast on 23rd October.

2.2 Clover Establishment

White clover (cv. Donna) was sown into the plots of the bicropped treatments on 28th April 1994. The clover was drilled at a seeding rate of 9.0 Kg/ha in rows 15 cms apart. A herbicide mixture containing MCPB, MCPA and benazolin was sprayed on 14th June to control annual broad leaved weeds. The clover was cut, wilted for three days and baled as silage during September 1994. Clover regrowth was grazed by sheep.

2.3 Wheat Establishment

In each year of the experiment, clover was cut and removed from the bicropped plots before wheat was sown. Wheat was sown into the bicropped plots using a Hunter rotaseeder drill at a row spacing of 22.8 cms. This drill uses a set of powered rotating blades to till a shallow 5 cm wide strip of soil. It was found necessary to drag a heavy chain behind the drill to ensure sufficient coverage of the seed.

Conventional treatments were ploughed and tilled before wheat was sown using a seed drill at a row spacing of 12 cms. All treatments were sown at a seeding rate of 200 Kg/ha in the 1994/95 and 1995/96 growing seasons. All treatments were sown at a seeding rate of 170 Kg/ha in the 1996/97 and 1997/98 growing seasons. Sowing dates are given in Table 2.

Table 2. Dates of Sowing and harvest

	Sowing Date	Whole Crop Harvest Date	Final Harvest Date
1994/95	11/10/94	10/7/95	4/8/95
1995/96	18/10/95	23/7/96	16/8/96
1996/97	9-10/10/96	15/7/97	15/8/97
1997/98	8-9/10/97	-	31/8/98

2.4 Cereal Seedling Establishment

The establishment of wheat plants was determined during the spring after sowing. In this assessment, two adjacent 0.5m row lengths were dug up at three random locations per plot and the number of plants were counted.

2.5 Measurements of Soil N

On 25th November 1998, a soil corer was used to take soil samples from each plot. Four soil cores were taken from each plot, each core was divided into a 0-15cm fraction as well as a 15-30 cm fraction. Ammonium nitrogen and nitrate nitrogen was determined in both fractions.

2.6 Intermediate Harvests

Twenty plants were sampled at random from each plot at 2-3 week intervals from spring until final harvest in 1995, 1996 and 1997. The roots were removed before the plants were divided into leaves, stems and ears. The individual plant parts were dried to constant weight at 70 degrees centigrade. The dried plant parts were weighed before being ground. Percentage N was determined on the ground material using a LECO FP-228 Nitrogen determinator (LECO Corporation, St Joseph, Michigan, U.S.A.). Biological production efficiency was determined at each of the individual harvests as accumulated biomass/ accumulated nitrogen.

2.7 Whole Crop Silage

Whole crop silage was harvested in 1995, 1996 and in 1997. A swath (1.25m*10m) was cut from each plot approximately one month before final harvest when the wheat was at dough stage. The total freshweight of the freshly harvested material was determined. Sub-samples were taken to determine botanical composition and dry matter content.

2.8 Final Harvest

Three 0.5 m^2 quadrats were harvested from each plot immediately before final harvest. The roots were removed from the plants before the ears were seperated, from the straw and threshed into grain and chaff, all plant parts were then dried. Harvest index was calculated as grain weight divided by (grain weight+weight of chaff+weight of straw). The percentage of nitrogen in the grain was determined using a LECO FP-228 Nitrogen determinator (LECO Corporation, St Joseph, Michigan, U.S.A.). Nitrogen harvest index was determined as grain nitrogen / total plant N determined at the last intermediate harvest . Nitrogen utilization efficiency (Moll et al., 1982) was calculated as grain produced per unit of whole plant nitrogen. The total uptake of N in 1998 was determined using these three quadrats of plants harvested immediately before final harvest. After drying, the individual plant parts were ground before percentage N was determined using a LECO N determinator.

Grain yield was quantified by harvesting two 60m strips from each plot using a plot combine. A sub-sample of the grain harvested from each plot was taken for analysis of percentage moisture. After harvest, straw was baled and removed from each plot.

2.9 Statistics

In the case of soil nutrient levels and plant establishment, two way analysis of variance was used to determine differences between treatments. For all other parameters, multiway analysis of variance was used to determine the effect of treatment, time and to test for interaction. Data was analysed using the GLM procedure of the Minitab statistical package (Minitab Inc., 3081 Enterprise Drive, PA 16801-3008).

3. Results

3.1 Soil Nutrient Status

The results of measurements of soil mineral N content are shown in Table 4. Treatments had no effect on soil ammonium nitrogen. Levels of nitrate in the 15-30 cm fraction were higher in bicropped treatments although the effect was not statistically significant. There were significantly higher levels ($p<0.05$) of nitrate in bicropped treatments at 15-30 cms depth.

3.2 Establishment of Wheat Plants

Plant establishment was lower in the bicropped treatments in comparison to the conventional treatments in all four years of the study (Table 3). Both conventional treatments had significantly more plants per m^2 ($p>0.01$) compared to both bicropped treatments in 1995. There were no significant differences between treatments in 1996 although conventional treatments did have more plants/m^2 compared to bicropped treatments. In 1997, the high input conventional treatment had significantly ($p<0.05$ - $p<0.01$) more plants/m^2 compared to both bicropped treatments and the low input conventional treatment had significantly more plants/m^2 ($p<0.05$) compared to the low input bicropped treatment. The low input conventional treatment had significantly more plants/m^2 ($p>0.05$) compared to the high input bicropped treatment in 1998. There were no significant differences in plant establishment between the two bicropped treatments or between the two conventional treatments.

Table 3. Wheat plant establishment 1994-1998

	8/3/95	3/3/96	2/4/97	20/2/98
Bicropped, zero	348.1	332.1	301.6	245.6
Bicropped, low	344.8	327.0	323.4	207.8
Conventional, medium	434.0	373.7	357.7	284.9
Conventional, high	451.4	368.9	380.6	250.9
s.e.d (9 d.f)	23.28	21.7	23.6	21.5

Table 4. Soil measurements of nitrate and ammonium nitrogen (mg/Kg dry soil)

	0-15 cms nitrate nitrogen	15-30 cms nitrate nitrogen	0-15 cms ammonium nitrogen	15-30 cms ammonium nitrogen
Bicropped, zero	7.7	7.3	5.9	5.0
Bicropped, low	7.7	7.9	5.1	4.9
Conventional, medium	4.3	3.8	5.9	4.6
Conventional, High	4.2	3.9	6.4	4.7
s.e.d (9 d.f.)	3.1	1.3	0.6	0.3

Table 5. Treatment effect on winter wheat yield, nitrogen accumulation, biomass and nitrogen distribution and treatment effects on whole crop silage. *,**,***,**** significant at the 0.05, 0.01,0.001,0.0001 level respectively, n.s. not significant

Year	Treatment	Grain Yield (t/ha @ 15%m)	Ears/m2	Grains/ear	1000 gw (g)	Wheat Biomass (t/ha)	Harvest Index	Nitrogen Uptake (Kg/ha)	Nitrogen Utilization efficiency	Whole Crop Silage Yield (t/ha D.M)
1995	Bicropped, Zero	1.6	275.4	13.8	41.2	8.4	21.2	64.2	20.8	9.3
	Bicropped, Low	3.1	282.4	24.0	44.2	12.4	24.5	89.5	30.0	13.0
	Conventional, Medium	8.6	453.4	40.2	46.1	16.9	34.0	170.0	43.6	20.7
	Conventional, High	10.1	540.0	39.2	46.4	20.7	39.2	205.2	43.1	21.9
1996	Bicropped, Zero	1.5	289.3	14.7	36.8	6.1	28.2	43.5	26.7	7.6
	Bicropped, Low	2.4	316.6	20.4	39.0	9.7	30.7	64.7	31.3	11.3
	Conventional, Medium	8.2	496.0	43.4	39.6	20.0	41.5	182.0	39.1	15.7
	Conventional, High	9.8	571.3	43.4	40.8	22.8	45.0	240.0	34.5	17.9
1997	Bicropped, Zero	1.6	260.3	16.8	40.3	6.7	29.0	59.3	22.6	8.9
	Bicropped, Low	2.6	252.6	26.8	42.1	11.0	33.5	102.2	22.2	10.3
	Conventional, Medium	6.0	389.0	45.7	35.4	17.2	37.7	175.1	29.7	12.9
	Conventional, High	7.3	411.7	49.1	38.8	18.1	40.2	203.6	30.8	15.1
1998	Bicropped, Zero	1.5	296.6	13.0	40.9	6.5	29.2	55.3	23.1	
	Bicropped, Low	2.5	328.6	19.4	42.1	9.7	31.5	86.6	24.5	
	Conventional, Medium	6.3	406.3	39.9	39.9	15.6	38.2	162.5	32.9	
	Conventional, High	7.4	402.3	44.6	44.6	15.9	40.2	194.1	32.7	
	Year (3 d.f.)	****	****	****	****	****	****	n.s.	****	**** (2 d.f.)
	Treat (3 d.f.)	****	****	****	****	****	****	****	****	**** (3 d.f.)
	Year * Treat (9 d.f.)	****	****	n.s.	****	****	****	**	n.s.	**** (6 d.f.)

3.3 Whole Crop Silage

Whole crop silage yields were closely related to input levels, there were statistically significant treatment effect during the experiment (p<0.0001, Table 5). Whole crop silage yield from each treatment declined during the three years (significant time effect p<0.0001, Figure 1). The rate of decrease in silage yield from year to year was greatest in the conventional treatments, there was a statistically significant interaction between year and treatment (p<0.0001). The rate of decline in whole crop silage yields was ameliorated in bicropped treatments as weed yields increased from year to year almost compensating for the decrease in wheat and clover yields.

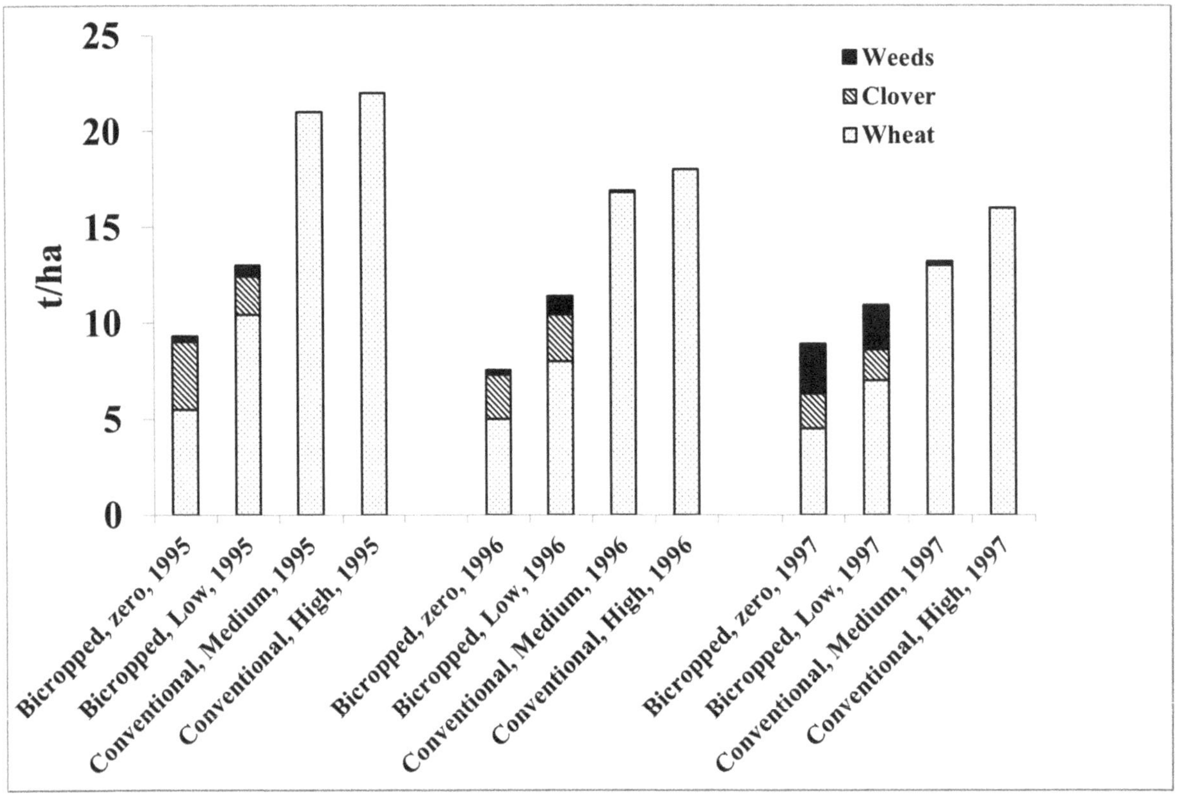

Figure 1. Trends in whole crop silage harvest 1995-1997. Yields (t/ha dry matter) of weeds, clover, winter wheat

The clover understory was not threatened to any great extent by either biotic orabiotic factors during the course of the study. Although the clover sward was still present during the fourth year, it's extent had been dramatically reduced due to competition from weeds (Figure 1). In effect, the clover sward was gradually replaced during the course of the study by a crop of weeds. Several attempts were made to control the spread of weeds by different chemical means. These attempts were not successful. The most problematic weeds were grass weeds (*Poa annua*, *Poa trivialis*, *Agrostis stolinifera*, *Agrostis tenuis*) and *Galium aparine*.

3.4 Wheat Yield

Grain yields and total biomass yields from wheat were strongly related to input levels during the experiment (p<0.0001, Table 5). Yields of all treatments declined during the course of the study (time effect p<0.0001). There was also a significant time by treatment interaction (p<0.0001), wheat yields from conventional treatments declined during the course of the study whereas yields from bicropped treatments were more stable. The high input bicropped treatment outyielded the low input bicropped treatment in all four years of the study.

3.5 Nitrogen Uptake

Nitrogen uptake in 1995, 1996 and 1997 is shown in Figure 2 as an index of nitrogen nutrition. The addition of fertiliser nitrogen resulted in dramatic increases in nitrogen uptake by the wheat plants. The total amount of nitrogen accumulated by each treatment prior to final harvest in each of the four years of the study is shown in Table 5. Nitrogen uptake increased with input level (p<0.0001) but there was a significant (p<0.01) interaction between time and treatment during the course of the study. Nitrogen uptake by the conventional treatments decreased during the course of the study whereas nitrogen uptake by the bicropped treatments was more stable.

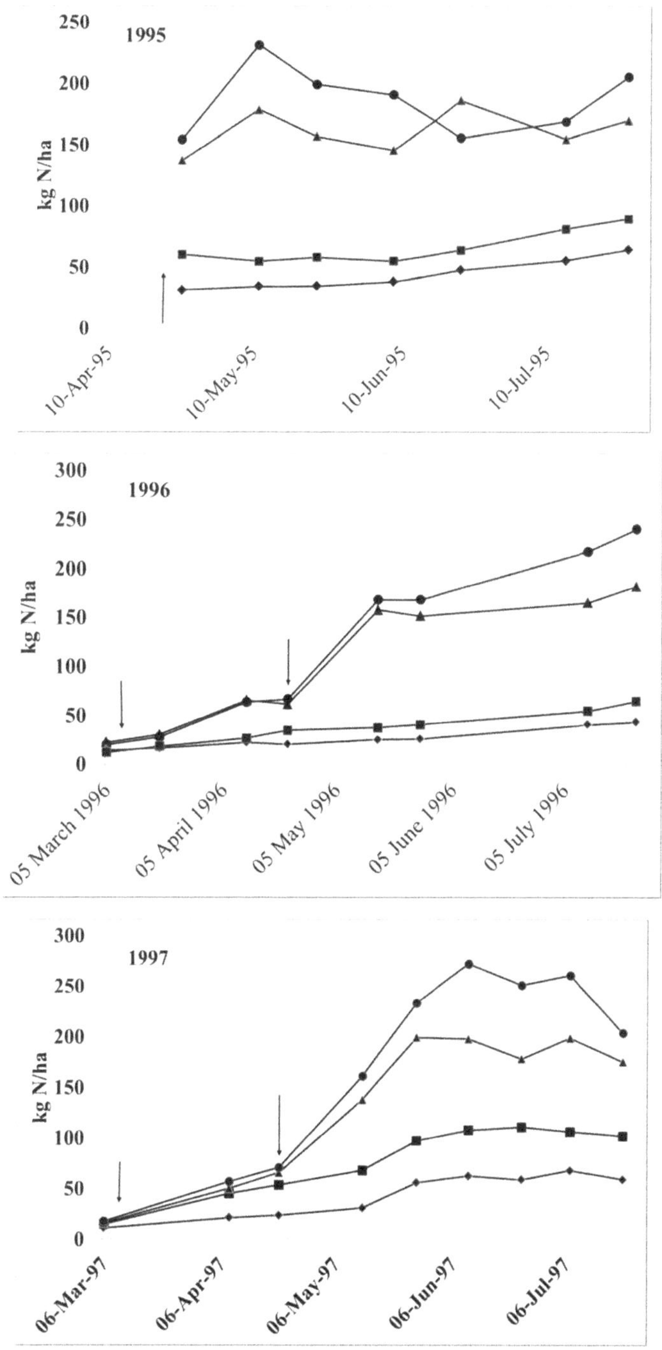

Figure 2. Nitrogen accumulation by winter wheat in the 1995, 1996 and 1997 growing seasons for the bicropped zero (diamonds), bicropped low (squares), conventional medium (triangles) and conventional high (circles) treatments. Arrows show the timings of fertilizer applications

3.6 Partitioning of Biomass and Nitrogen

Nitrogen application altered biomass partitioning. Less biomass was concentrated in the stem and more in the leaf as nitrogen application increased (Figures 3 and 4). Supply of fertiliser nitrogen suppressed biological production efficiency (BPE), the biomass produced per unit of nitrogen absorbed (Figure 4). The reduction in

BPE was proportional to the level of fertiliser applied. Application of the first split of nitrogen (50 Kg N for Bicropped low, 70 Kg N for conventional treatments) resulted in an initial reduction in BPE in 1996 and 1997. The second split of nitrogen (90Kg N for conventional medium, 130 Kg N for conventional high) resulted in a further suppression of BPE proportional to the level of nitrogen applied. The first intermediate harvest in 1995 took place after the second split of nitrogen had been applied, differences in BPE were already apparent at this stage, the parameter decreasing with quantity of applied nitrogen. Biological production efficiencies at the end of each growing season are shown in Table 5. Biological production efficiencies decreased with time in all treatments ($p<0.0001$) and there was a statistically significant treatment effect ($p<0.0001$).

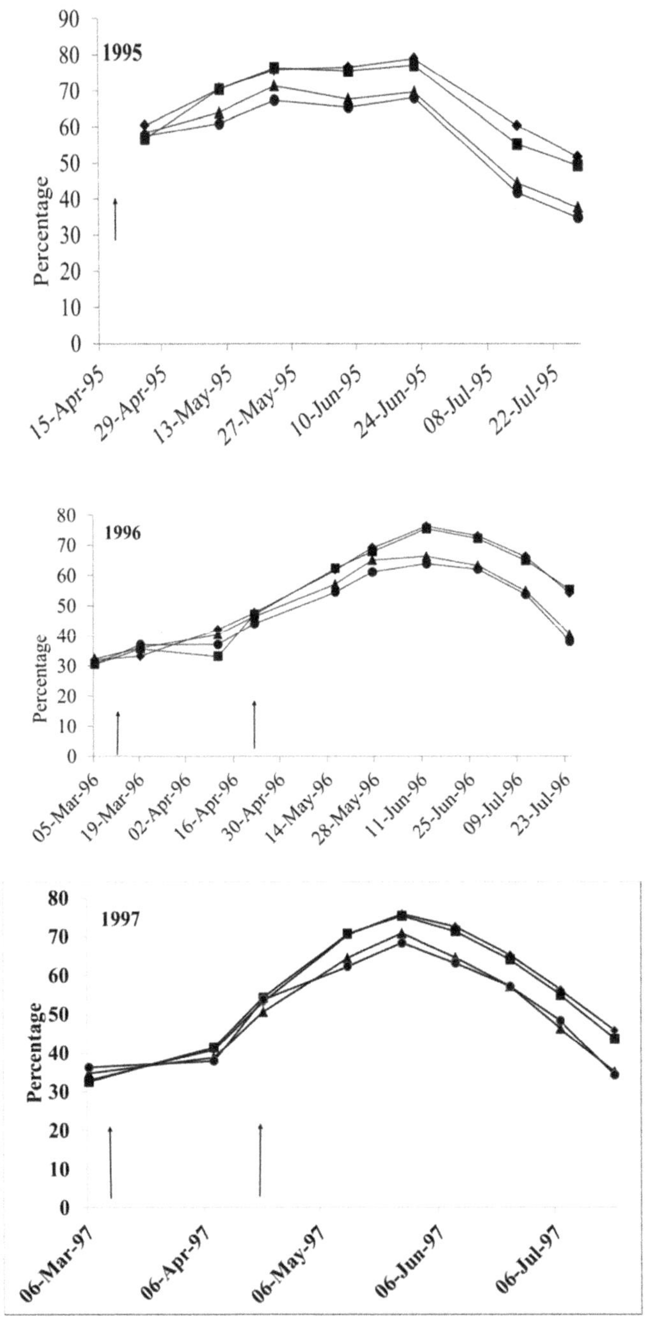

Figure 3. Winter wheat stem dry matter as a proportion of total dry matter. Trends 1995-1997 for the bicropped zero (diamonds), bicropped low (squares), conventional medium (triangles) and conventional high (circles) treatments. Arrows show the timings of fertilizer applications

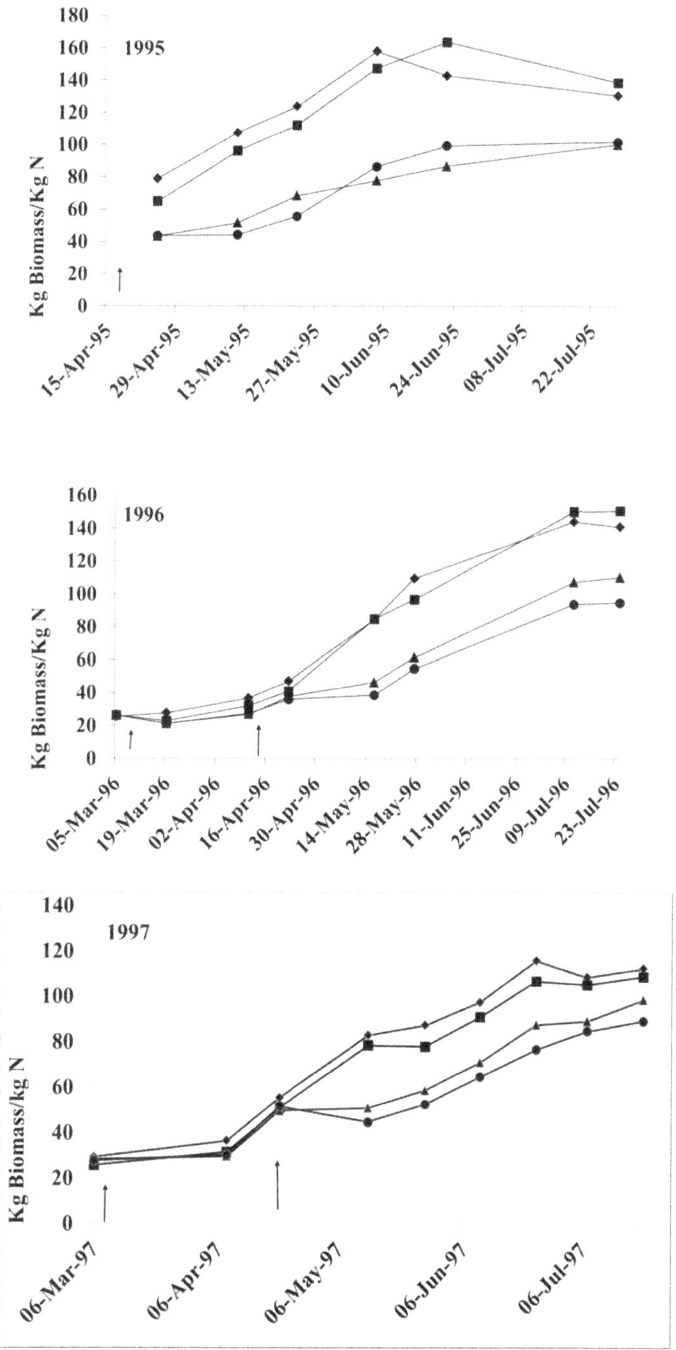

Figure 4. Trends in winter wheat biological production efficiency 1995-1997. Biological production efficiency represents biomass formed per unit of nitrogen. Data presented for the bicropped zero (diamonds), bicropped low (squares), conventional medium (triangles) and conventional high (circles) treatments. Arrows show the timings of fertilizer applications

Both harvest index and nitrogen harvest index increased with input level. Bicropped treatments were less efficient at allocating biomass and absorbed nitrogen to the growing grains (Table 5). The effect of treatment was highly significant on both harvest index and nitrogen harvest index ($p<0.0001$). Harvest index and nitrogen harvest index increased in both bicopped treatments from year to year during the course of the experiment. However, this trend was not evident in the conventional treatments. Bicropped treatments were less efficient at converting absorbed N into grain yield throughout the experiment (Table 5), conventional treatments had higher nitrogen utilization efficiencies (NUE) in each of the four years of the study. The effect of treatment on NUE was statistically significant ($p<0.0001$).

4. Discussion

It is possible to establish successive winter wheat crops in a white clover understory although plant establishment was lower in direct drilled treatments. Jones (1992) also found reduced seedling emergence when cereal crops were direct drilled into a clover understory. Cornish and Lymbery (1987) have attributed reduced emergence in direct drilled crops to a combination of uneven sowing depth and to rougher seedbeds. In this study, the reduced plant establishment was at least partially attributable to the type of direct drill used to sow the bicropped wheat. Unless conditions were completely dry the powered rotating blades did not provide enough soil to cover the seeds sufficiently leaving seeds vulnerable to predation by birds. Additionally, the rotating blades had the effect of compacting the soil under the newly sown seeds providing a difficult environment for establishment to take place.

Nitrogen uptake by the bicropped winter wheat crops was low despite being more stable than that of conventional treatments which declined during the course of the study. Higher levels of nitrate nitrogen were found in the soils of bicropped treatments, a similar result was reported by Jones and Clements (1993). However, levels of N uptake in all treatments were similar to those reported in conventionally sown winter wheat nitrogen trials suggesting that the contribution from biologically fixed nitrogen was low (Powlson et al., 1986, Ortiz-Monasterio et al., 1997). Although this experiment was not a nitrogen trial *per se*, differences between treatments were largely a function of the level of nitrogen fertilization.

Differences in biomass accumulation between the two methods of cultivation were less than differences in grain yield because of differences in harvest index between treatments. Low harvest index was a contributory factor to the poor grain yields obtained from the bicropped treatments. Harvest index increased with the level of applied nitrogen irrespective of the method of cultivation. Spiertz and Ellen (1978) reported an increase in harvest index with nitrogen supply, Dreccer et al. (2000) reported a similar result. However, most studies have reported that harvest index decreases with applied N (Donald and Hamblin, 1976; Pearman et al., 1978; Ortiz-Monasterio et al., 1997; Le Gouis et al., 2000, Garner & Dyke, 1969, Delogu et al., 1998). It is possible that high plant population density in the widely spaced rows of the bicropped treatments may have contributed to the substantial differences in harvest index between the two methods of cultivation. Puckridge and Donald (1967) reported a decrease in harvest index with increasing population density in wheat crops grown in Australia. Nitrogen harvest index also fell with increasing N, in contrast to the results of Spiertz and Ellen (1978), Le Gouis et al. (2000) and Dreccer et al. (2000). Poor harvest index in bicropped treatments ensured that nitrogen utilization efficiency was also low in these treatments and indeed the latter parameter followed the same trend as harvest index, increasing with added nitrogen. In contrast, a decrease in nitrogen utilization efficiency with added nitrogen has been reported by several authors (Sowers et al., 1994; Dhugga & Waines, 1989, Ortiz-Monasterio et al., 1997, Delogu et al., 1998).

The addition of fertiliser N suppressed the efficiency of biological production and altered the partitioning of biomass so that more assimilate was partitioned into the leaf and less to the stem. Low biological production efficiencies at high levels of nitrogen have also been reported by Ortiz-Monasterio et al. (1997) and Le Gouis et al. (2000). Thus, the bicropped treatments with their low levels of added N were more efficient at converting absorbed N into biomass. However, although efficient at producing biomass, the bicropped treatments stored a higher percentage of their biomass in the stem compared to their conventional contempories. This pattern of biomass allocation continued up to final harvest. Bicropped treatments appeared unable to use the extra stem reserves to produce leaf laminae for photassimilate production or to use the same reserves for grain filling. The importance of stem reserves for grain filling particularly under conditions when current assimilation cannot meet the demands of developing grains has been demonstrated (Austin et al. 1977, 1980). In this experiment the efficiency of assimilate partitioning was a function of nitrogen supply and appears to have been determined around the time of the second split of N.

Subsequent to this experiment, other studies have also reported lower grain yields in wheat-clover bicropping systems compared to conventional systems (Sarunaite et al., 2006; Thorsted et al., 2006a). Yield reductions have been attributed to the presence of clover, the negative effects of clover being stronger than the positive effects (Thorsted et al., 2006b). Thorsted et al. (2006a) attributed such yield reductions to interspecific competition for light and N during vegetative growth and for soil water during grain filling. Several studies have reported that grain yields in cereal/wheat bicropping systems can be substantially increased by legume defoliation by either herbicide (Bergkvist, 2003) or by the use of mechanical treatments (Thorsted et al., 2002; Thorsted et al., 2006c). Bergkvist (2003) concluded that competition from clover needs to be minimised when wheat is at the tillering stage in order to maximise grain yield while Thorsted et al., (2006c) reported that the largest yield increases were obtained when mechanical weed control was applied during stem elongation. Mechanical treatment increased N transfer between clover and wheat, possibly as a result of the fact that legume defoliation results in root and nodule death. However, the above studies in which wheat yields in wheat-clover bicropping systems were stimulated by legume defoliation were relatively short in duration (2 years) and the effects of such treatments on the longer term persistence of clover swards is, as yet, unknown.

The foremost practical problem encountered in this experiment was the control of weeds in bicropped treatments. Weed control is typically more problematical when crops are established by minimum cultivation compared to conventional tillage. Mouldboard ploughing in the latter method minimises the number of weed seeds that can germinate. Christian and Bacon (1990) reported that direct drilling increased populations of annual meadow grass (*Poa annua*) and rough stalked meadow grass (*Poa trivialis*). Ball et al. (1994) found that direct drilled plots of winter barley became infested with soft brome (*Bromus mollis* L.). Experiments in England have demonstrated that grass weeds can pose a significant threat when winter cereals are grown continuously using minimum cultivation (Wilson et al., 1989). There were several attempts in this study to control weed infestation using several herbicides, none of these attempts were successful. Weed control options were restricted by the need to choose a herbicide which had no detrimental effect on clover. The absence of suitable herbicides delayed the introduction of direct drilling for many years (Cannell and Hawes, 1994). Similarly, at present a clover-wheat intercrop is not sustainable over a number of years due to the lack of a suitable herbicide to control weeds in the clover component of the bicrop.

Higher levels of nitrate nitrogen found in the soil of bicropped treatments were not sufficient to be of benefit to the nitrogen economy of the bicropped wheat crops which had insufficient nitrogen available to them at the appropriate time to generate a sink for grain yield. Reducing the need for fertiliser nitrogen by providing wheat with a supply of biological nitrogen is a noble aspiration. However, further research is needed to identify a legume management strategy that could provide wheat with sufficient nitrogen at the appropriate time.

Acknowledgements

The co-operation and assistance of Peter Tiernan, Ivan Mitchell and Pat Prior is gratefully acknowledged. Additionally, the authors wish to thank the Commission of the European Communities for financial support.

References

Austin, R. B., Edrich, J. A., Ford, M. A. & Blackwell, R. D. (1977). The fate of the dry matter, carbohydrates and 14C lost from the leaves and stems of wheat during grain filling. *Annals of Botany, 41*, 1309-1321.

Austin, R. B., Morgan, C. L., Ford, M. A., & Blackwell, R. D. (1980). Contributions to grain yields from pre-anthesis assimilation in tall and dwarf barley phenotypes in two contrasting seasons. *Annals of Botany, 45*, 309-319.

Ball, B. C., Lang, R. W., Robertson, E. A. G., & Franklin, M. F. (1994). Crop performance and soil conditions on imperfectly drained loams after 20-25 years of conventional tillage or direct drilling. *Soil and Tillage Research, 31*, 97-118. http://dx.doi.org/10.1016/0167-1987(94)90074-4

Bell, M. A., Fischer, R. A., & Byerlee, S. K. (1995). Genetic and agronomic contributions to yield gains: a case study for wheat. *Field and Crops Research, 44*, 55-65. http://dx.doi.org/10.1016/0378-4290(95)00049-6

Bergkvist, G. (2003). Effect of white clover and nitrogen availability on the grain yield of winter wheat in a three season intercropping system. *Acta Agriculturae Scandanavica, Section B, Soil and Plant Science, 53*, 97-109. http://dx.doi.org/10.1080/09064710310011953

Brandt, J. E., Hons, F. M., & Haby, V. A. (1989). Effects of subterranean clover interseeding on grain yield, yield components and nitrogen content of soft red winter wheat. *Journal of Production Agriculture, 2*(4), 347-351. http://dx.doi.org/10.2134/jpa1989.0347

Brophy, L. S., Heichel, G. H., & Russelle, M. P. (1987). Nitrogen transfer from forage legumes to grass in a systematic planting design. *Crop Science, 27*, 753-758. http://dx.doi.org/10.2135/cropsci1987.0011183X002700040030x

Cannell, R. Q., & Hawes, J. D. (1994). Trends in tillage practises in relation to sustainable crop production with special reference to temperate climates. *Soil and Tillage Research, 30*, 245-282. http://dx.doi.org/10.1016/0167-1987(94)90007-8

Christian, D., & Bacon, E. T. B. (1990). A long term comparison of ploughing, tine cultivation and direct drilling on the growth and yield of winter cereals and oilseed rape on clayey and silty soils. *Soil and Tillage Research, 18*, 311-331. http://dx.doi.org/10.1016/0167-1987(90)90117-V

Clement, C. R., & Williams, T. E. (1967). Leys and organic matter. II. The accumulation of nitrogen in soils under different leys. *Journal of Agricultural Science, Cambridge, 69*, 133-138. http://dx.doi.org/10.1017/S0021859600016543

Cornish, P. S., & Lymbery, J. R. (1987). Reduced early growth of direct drilled wheat in southern New South Wales: causes and consequences. *Australian Journal of Experimental Agriculture, 27*, 869-880. http://dx.doi.org/10.1071/EA9870869

Delogu, G., Cattivelli, L., Pecchioni, N., De Falcis, D., Maggiore, T., & Stanca, M. (1998). Uptake and agronomic efficiency of nitrogen in winter barley and winter wheat. *European Journal of Agronomy, 9*, 11-20. http://dx.doi.org/10.1016/S1161-0301(98)00019-7

Dhugga, K. S., & Waines, J. G. (1989). Analysis of nitrogen accumulation and use in bread and durum wheat. *Crop Science, 29*, 1232-1239. http://dx.doi.org/10.2135/cropsci1989.0011183X002900050029x

Donald, C. M., & Hamblin, J. (1976). The biological yield and harvest index of cereals as agronomic and plant breeding criteria. *Advances in Agrononomy, 28*, 361-405. http://dx.doi.org/10.1016/S0065-2113(08)60559-3

Dreccer, M. F., Schapendonk, A. H. C. M., Slafer, G. A., & Rabbinge, R. (2000). Comparative response of wheat and oilseed rape to nitrogen supply: absorption and utilisation efficiency of radiation and nitrogen during the reproductive stages determining yield. *Plant and Soil, 220*, 189-205. http://dx.doi.org/10.1023/A:1004757124939

Garner, H. V., & Dyke, G. V. (1969). The Broadbalk Yields. *Report of Rothamsted Experimental Station 1968, 2*, 26-49.

Groya, F. L., & Sheaffer, C. C. (1985). Nitrogen from forage legumes: harvest and tillage effects. *Agronomy Journal, 77*, 105-109. http://dx.doi.org/10.2134/agronj1985.00021962007700010025x

Haysteed, A., & Marriott, C. (1978). Fixation and transfer of nitrogen in a clover-grass sward under hill conditions. *Annals of Applied Biology, 88*, 453-457. http://dx.doi.org/10.1111/j.1744-7348.1978.tb00740.x

Huber, D. M., Warren, H. L., Nelson, D. W., & Tsai, C. Y. (1977). Nitrification inhibitors- new tools for food production. *Bio-Science, 27*, 523-529. http://dx.doi.org/10.2307/1297812

Jones, L. (1992). Preliminary trials using a white clover (*Trifolium repens* L.) understory to supply the nitrogen requirements of a cereal crop. *Grass and Forage Science, 47*, 366-374. http://dx.doi.org/10.1111/j.1365-2494.1992.tb02282.x

Jones, L., & Clements, R. O. (1993). Development of a low input system for growing wheat (*Triticum vulgare*) in a permanent understory of white clover (*Trifolium repens*). *Annals of Applied Biology, 123*, 109-119. http://dx.doi.org/10.1111/j.1744-7348.1993.tb04078.x

Ladd, J. N., Amato, M., Jackson, R. B., & Butler, J. H. A. (1983). Utilization by wheat crops of nitrogen from legume residues decomposing in soils in the field. *Soil Biology and Biochemistry, 15*, 231-238. http://dx.doi.org/10.1016/0038-0717(83)90064-0

Le Gouis, J., Beghin, D., Heumez, E., & Pluchard, P. (2000). Genetic differences for nitrogen uptake and nitrogen utilisation efficiencies in winter wheat. *European. Journal of Agronomy, 12*, 163-173. http://dx.doi.org/10.1016/S1161-0301(00)00045-9

Martin, M. P. L. D., & Snaydon, R. W. (1982). Intercropping barley and beans. 1. Effects of planting pattern. *Experimental Agriculture, 18*, 139-148. http://dx.doi.org/10.1017/S0014479700013612

Moll, R. H., Kamprath, E. J., & Jackson, W. A. (1982). Analysis and interpretation of factors which contribute to efficiency of nitrogen utilization. *Agronomy Journal, 74*, 562-564. http://dx.doi.org/10.2134/agronj1982.00021962007400030037x

Ofori, F., & Stern, W. R. (1987). Cereal–legume intercropping systems. *Advances in Agronomy, 41*, 41-90. http://dx.doi.org/10.1016/S0065-2113(08)60802-0

Olesen, J. E., Berntsen, J., Hansen, E. M., Petersen, B. M., & Petersen, J. (2002). Crop nitrogen demand and canopy area expansion in winter wheat during vegetative growth. *European Journal of Agronomy, 16*, 279-294. http://dx.doi.org/10.1016/S1161-0301(01)00134-4

Ortiz-Monasterio, J. I., Sayre, K. D., Rajaram, S., & McMahon, M. (1997). Genetic progress in wheat yield and nitrogen use efficiency under four nitrogen rates. *Crop Science, 37*, 898-904. http://dx.doi.org/10.2135/cropsci1997.0011183X003700030033x

Pearman, I, Thomas, S. M., & Thorne, G. N. (1978). Effect of nitrogen fertilizer on growth and yield of semi-dwarf and tall varieties of winter wheat. *Journal of Agricultural Science, Cambridge, 91*, 31-45. http://dx.doi.org/10.1017/S0021859600056604

Powlson, D. S., Pruden, G., Johnston, A. E., & Jenkinson, D. S. (1986). The nitrogen cycle in the Broadbalk wheat experiment: recovery and losses of 15N – labelled fertilizer applied in spring and inputs of nitrogen from the atmosphere. *Journal of Agricultural Science, Cambridge, 107*, 591-609. http://dx.doi.org/10.1017/S0021859600069768

Puckridge, D. W., & Donald, C. M. (1967). Competition between wheat plants sown at a wide range of densities. *Australian Journal of Agricultural Research, 18*, 193-211. http://dx.doi.org/10.1071/AR9670193

Remsion, S. U. (1978). Neighbour effects between maize and cowpea at various levels of N and P. *Experimental Agriculture, 14*, 205-212. http://dx.doi.org/10.1017/S001447970000870X

Sarunaite, L., Kadziuliene, Z., & Kadziulis, L. (2008) The impact of legumes on cereals in a crop rotation and legume-cereal bi-cropping. Biodiversity and animal feed: future challenges for grassland production. Proceedings of the 22nd General Meeting of the European Grassland Federation, Uppsala, Sweden, 9-12 June 2008. *Grassland Science in Europe, 13*, 625-627.

Schmidt, O., & Curry, J. P. (1999). Effects of earthworms on biomass production, nitrogen allocation and nitrogen transfer in wheat-clover intercropping model systems. *Plant and Soil, 214*(1-2), 187-198. http://dx.doi.org/10.1023/A:1004723914623

Schmidt, O., Clements, R. O., & Donaldson, G. (2003). Why do cereal-legume intercrops support large earthworm populations. *Applied Soil Ecology, 22*(2), 181-190. http://dx.doi.org/10.1016/S0929-1393(02)00131-2

Simpson, J. R. (1965). The transfer of nitrogen from pasture legumes to an associated grass under several systems of management in pot culture. *Australian Journal of Agricultural Research, 16*, 915-926. http://dx.doi.org/10.1071/AR9650915

Sowers, K. E., Pan, W. L., Miller, B. C., & Smith, J. L. (1994). Nitrogen use efficiency of split nitrogen applications in soft white winter wheat. *Agronomy Journal, 86*, 942-948. http://dx.doi.org/10.2134/agronj1994.00021962008600060004x

Spiertz, J. H. J., & Ellen, J. (1978). Effects of nitrogen on crop development and grain growth of winter wheat in relation to assimilation and utilization of assimilates and nutrients. *Netherlands Journal of Agricultural Science, 26*, 210-231.

Stewart, R. H., Lynch, K. W., & White, E. M. (1980). The effect of growing clover cultivars in association with barley cultivars upon grain yield of the barley crop in the year of sowing and the subsequent year. *Journal of Agricultural Science, Cambridge, 95*, 715-720. http://dx.doi.org/10.1017/S0021859600088110

Thorsted, M. D., Olesen, J. E., & Koefoed. (2002). Effects of white clover cultivars on biomass and yield in oat/clover intercrops. *Journal of Agricultural Science, Cambridge, 138*, 261-267. http://dx.doi.org/10.1017/s0021859602002010

Thorsted, M. D., Olesen, J. E., & Weiner, J. (2006a). Width of wheat rows and clover strips influence grain yield in winter wheat/white clover intercropping. *Field Crops Research, 95*, 280-290. http://dx.doi.org/10.1016/j.fcr.2005.04.001

Thorsted, M. D., Weiner, J., & Olesen, J. E. (2006b). Above- and below-ground competition between intercropped winter wheat *Triticum aestivum* and white clover *Trifolium repens*. *Journal of Applied Ecology, 43*, 237-245. http://dx.doi.org/10.1111/j.1365-2664.2006.01131.x

Thorsted, M. D., Olsen, J. E., & Weiner, J. (2006c). Mechanical control of clover improves nitrogen supply and growth of wheat in winter wheat/white clover intercropping. *European Journal of Agronomy, 24*, 149-155. http://dx.doi.org/10.1016/j.eja.2005.07.004

White, J. G., & Scot, T. W. (1991). Effects of perennial forage-legume living mulches on no-till winter wheat and rye. *Field Crops Research, 28*, 135-148. http://dx.doi.org/10.1016/0378-4290(91)90079-B

Whitney, A.S. (1977). Contribution of forage legumes to the nitrogen economy of mixed swards. A review of relevant research. In: Ayanaba, A., Dart, P.J. (Eds.), Biological nitrogen fixation in farming systems in the tropics, 89-114. John Wiley, New York.

Williams, E. D., & Hayes, E. D. (1991). Growing spring cereals in a white clover (*Trifolium repens*) crop. *Journal of Agricultural Science, Cambridge, 117*, 23-37. http://dx.doi.org/10.1017/S002185960007893X

Wilson, B. J., Moss, S. R., & Wright, K. J. (1989). Long term studies of weed populations in winter wheat as affected by straw disposal, tillage and herbicide use. Proceedings, Brighton Crop Protection Conference-Weeds, 1989 BCPC Publications, Croydon, UK, pp3A-6, 131-136.

15

Development of Corn Stover Biofuel: Impacts on Corn and Soybean Markets and Crop Rotation

Farzad Taheripour[1], Julie Fiegel[2] & Wallace E. Tyner[1]

[1] Department of Agricultural Economics, Purdue University, West Lafayette, Indian, USA

[2] HYLA Mobile, USA

Correspondence: Farzad Taheripour, Department of Agricultural Economics, Purdue University, 403 West State St., West Lafayette, IN 47907-2056, USA.

Abstract

What would be the impacts of a viable market for corn stover? A partial equilibrium model and a linear programing model were used to determine to what extent the existence of a viable market for corn stover would affect the traditional corn-soybean crop rotation in the US. We find that with government support production of biofuel from corn stover could significantly increase. That boosts profitability of farming corn in combination with harvesting corn stover versus soybeans. We show that if corn stover is demanded for biofuel, then a major shift will be observed in crop rotations in the US.

Keywords: soybeans, corn, corn stover, cellulosic biofuel, land use, corn-soybean rotation, partial equilibrium

1. Introduction

1.1 Importance

First generation biofuels and their impacts on: 1) greenhouse gas emissions, 2) oil imports, and 3) markets for agricultural commodities and food prices have been examined from different angles in recent years. These studies showed that first generation biofuels, which are produced from food crops will not be able to replace a large portion of oil-based liquid fuels, because their rapid expansion could cause adverse impacts on food supply (Abbott, Hurt, & Tyner, 2008; Abbott, Hurt, & Tyner, 2011; Trostle, 2008; Zilberman et al., 2013) and/or induce major unintended land use changes which in turn will lead to increases in greenhouse gas emissions (Hertel, 2010; Tyner & Taheripour, 2012). Instead, second-generation biofuels produced from forest or crop residues such as wood chips, corn stover, or wheat straw offer an alternative to first generation biofuels. These residue-based biofuels could have minor impacts on both food prices and land use change. In addition, they could make a major contribution in greenhouse gas emission reduction targets if removed in a sustainable manner (English et al., 2013). Another advantage for second-generation biofuels produced from agricultural residues is that they provide a potential new source of income for farmers (Thompson & Tyner, 2013).

If second generation biofuels became economically viable and a massive volume of biofuels are produced from agricultural residues, it then can have major impacts on agricultural commodity markets. Corn stover is a major and abundant feedstock in the USA, which is expected to be used in biofuel production, if the technology becomes economically viable. Converting corn stover to biofuels, if a significant market develops, can affect profitability of corn production versus other crops produced in the USA, in particular. Thompson and Tyner suggest that if a viable corn stover market existed, it could have a large impact on farmers' crop rotations and land allocation decisions. However, that research was done at the farm level with a given set of crop prices and ignored interactions between farm and market level variables. The purpose of this paper is to determine to what extent producing biofuels from corn stover can affect demands for and supplies of corn and soybeans and their market prices. In addition, it examines to what extent producing biofuels from corn stover could affect switching from corn-soybean rotation to continuous corn.

To determine how a corn stover market would impact corn and soybean markets, a partial equilibrium model was developed which links corn, soybeans, corn stover, ethanol, dried distiller grains, and gasoline markets at the USA national level and determines market-clearing prices at different crude oil prices. The resulting prices were used in a linear programming model to determine how farmers would allocate their land.

1.2 Background and Outline

Corn and soybeans represented about 65% of USA harvested areas in 2012. The harvested area under these two crops has grown from 58.6 million hectares in 2000 to 66.1 million hectares in 2012 (Figure 1). These crops are traditionally produced in a corn-soybean rotation to gain agronomic benefits such as nitrogen fixing and breaking pest cycles. However, as shown in Figure 1 the share of corn in total harvested areas of these two crops has increased from 50% in 2000 to about 54% in 2012. The recent increase in the share of corn in total harvested area of corn and soybeans is mainly due to the expansion in corn demand for ethanol production, which changed the soybean to corn price ratio from 2.54 in 2000 to 2 in 2012, in favor of corn.

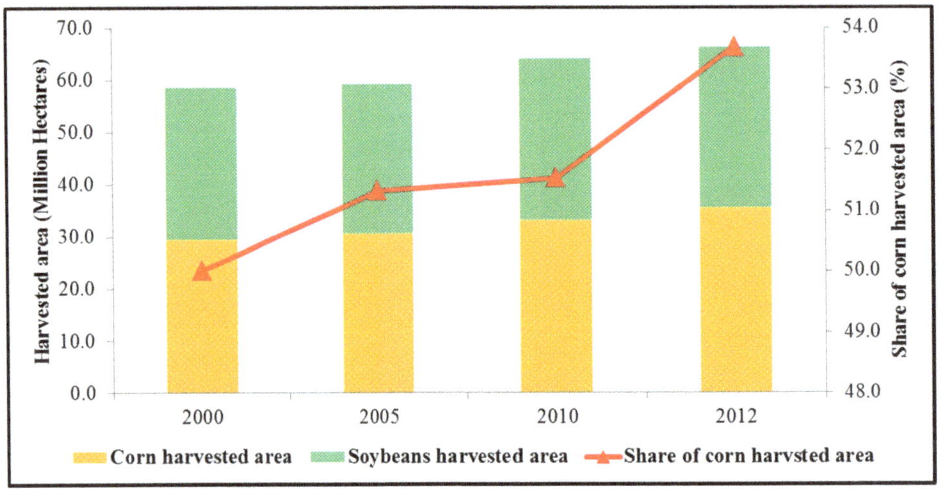

Figure 1. Corn and Soybean harvested areas, 2000-12

Due to higher demand for corn in recent years some farmers, in particular in the Midwest, are shifting their planting pattern from a corn-soybean rotation to a corn-corn or corn-corn-soybean rotation as shown in Figure 2. This figure compares changes in the planted areas of corn and soybeans in the 12 largest corn-soybean producing states for the time period of 2000 to 2102. As shown in Figure 2 the changes in corn planted areas were larger than the corresponding changes for soybeans everywhere, except for North Dakota which experienced slightly larger expansion in soybeans. States such as Illinois, Iowa, and Minnesota experienced large increases in planted area of corn with decreases in soybeans, implying more continuous corn. South Dakota and Nebraska had large increases in corn planted area with much smaller increases in soybeans, again representing more continuous corn. Other states presented below the 45 degree line in Figure 1 also had larger changes the corn planted areas versus soybean planted areas.

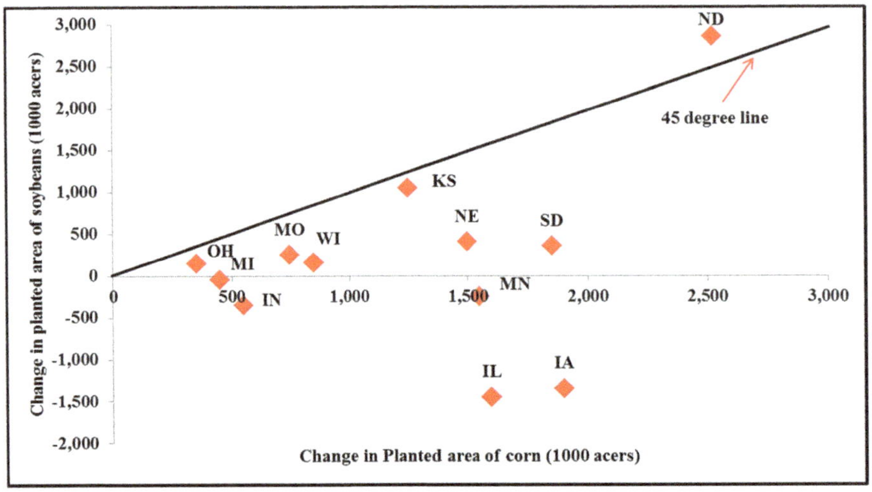

Figure 2. Changes in corn and soybean planted areas in 12 largest corn-soybeans producer states, 2000-12

Converting corn stover to biofuel could change relative profitability of corn and soybeans in favor of corn, when corn stover price is higher than its production costs. Production costs of corn stover include fertilizer costs to maintain productivity of land after stover removal and collection, storage, and transportation costs (Thompson & Tyner, 2013). These authors have concluded that if a viable corn stover market existed, it could have a large influence on farmer's crop rotation in favor of corn-corn rotation. However, they ignored the fact that relative prices of corn and soybeans could be different in the presence of a viable market for corn stover.

Consider a case where converting corn stover to biofuel is profitable, either due to market forces or government supports. In this case, farmers who produce corn and soybeans will bring profitability of corn stover collection into account. When corn stover production is a profitable operation for a farmer, in each planting period he/she will compare gains from a joint production of corn and corn stover with profits from soybean production at given prices (Note 1). In this case, if the joint profits from corn production and corn stover is higher than the profit from soybean production, then corn will be produced. If a large group of farmers decides to follow this choice, then the corn-corn rotation will increase among farmers, which will lead to an increase in corn production and a reduction in soybean production, other factors being constant. This could affect relative prices of corn and soybeans at the market level.

In this paper we first develop a partial equilibrium model to examine impacts of converting corn stover to ethanol on markets for ethanol, gasoline, corn, corn stover, and soybeans at the USA aggregation level. The model developed in this paper is based on the model developed and used by (Tyner & Taheripour, 2008). Then we will feed the results of this model into the Purdue Crop/Livestock Linear Programing (PCLP) model (Doster et al., 2008) to examine farmers' land allocation behavior in the presence of a viable market for corn stover. We test the sensitivity of the land allocation process at the farm level with respect to changes in key economic factors.

2. Method and Implemented Models

2.1 Partial Equilibrium Model

The partial equilibrium model developed in this paper is an extended version of the model developed by and used in several other articles (Taheripour & Tyner, 2008; Tyner & Taheripour, 2007; Tyner, Taheripour, & Perkis 2010; Tyner, Taheripour, & Hurt August, 2012). The original model follows an integrated partial equilibrium modeling structure which displays linkages among crude oil, gasoline, ethanol, and corn markets. The model captures the demand and supply sides of the corn, corn ethanol, dried distiller grains, and gasoline markets. The demand side of the corn market consists of three major corn users: foreign users (q_{cxd}), domestic uses for food and feed (q_{cdd}), and the corn ethanol industry (q_{ced}). The foreign and domestic demands for food and feed are modeled using constant price elasticity functional forms. The corn demand for the ethanol industry is $q_{ced} = y.q_{se}$. Here, y shows the corn-ethanol conversion factor and q_{se} is quantity of ethanol supply. Hence, total demand for corn is: $q_{cd} = q_{cxd} + q_{cdd} + q_{ced}$.

On the supply side, a constant return to scale Cobb-Douglas production function is used to model the supply side of the corn market. In this production function, capital, labor, land, and a composite input (which represent fertilizer, pesticides, seeds, energy and other items) are used to produce corn. In this function all inputs, except the composite input, are constant in the short-run. This production function is used to determine the supply for corn, q_{cs}.

In this model, in the short-run the demand for liquid fuel, g_{gd}, only responds to its own price using a constant price elasticity functional form. However, in the long-run, demand can grow with income and population. The supply side of the fuel market is comprised of gasoline producers and ethanol producers. Gasoline supply is produced from crude oil. The supply of gasoline, q_{gos}, is a function of its price and the price of crude oil. It follows a constant elasticity functional form as well. Ethanol is produced from corn. The supply of ethanol, q_{es}, is a function of its own price and the price of corn following a constant elasticity functional form as well. In this model it is assumed that every gallon of ethanol is presumed to contain 70% of the energy of a gallon of gasoline. Hence, total supply of gasoline equivalent is: $q_{gs} = q_{gos} + 0.7*q_{es}$.

Distiller's Dried Grains with Solubles (DDGS) is a co-product of the ethanol industry. DDGS can be used as a substitute for corn and to some extent soybean meal in the livestock industry and also alleviate impacts of ethanol production on the corn market. DDGS can increase profitability of ethanol industry as well. As a substitute for corn, DDGS covers a portion of corn demand: $q_{DDGS} = \Upsilon.q_{ced}$. Here q_{DDGS} is the quantity of DDGS produced and Υ is the corn-DDGS conversion factor. The model evaluates profitability of the ethanol industry and assumes that this industry will expand/contract until profits reach zero. Given this assumption and according to the predetermined supply and demand elasticities, the model determines equilibrium prices and their corresponding quantities for corn, ethanol, and gasoline and other endogenous variables for given exogenous variables defined in the model (such as crude oil price).

Several new components are introduced into this partial equilibrium model to handle new markets for soybeans, corn stover, and a biofuel produced from corn stover. The first component added to the model is a drop-in biofuel named bio-gasoline. This drop-in biofuel can be produced from corn stover with the following supply function: $q_{gstov} = A_{estov}(p_{pg})^{gstov}(p_{stov})^{-gstov}$. The energy content of bio-gasoline is assumed to be equal to the energy content of gasoline. In the presence of bio-gasoline, total supply of gasoline will be equal to: $q_{gs} = q_{gos} + 0.7*q_{es} + q_{gstov}$. Similar to ethanol industry, the bio-gasoline industry will expand until profits reach zero. Profits per gallon of bio-gasoline are estimated by: $\pi s = p_g - cap_{stov} - var_{stov}$. Here, cap_{stov} and var_{stov} are capital and variable costs per gallon of bio-gasoline. At equilibrium $\pi s=0$.

In the new model it is assumed that the capital costs of producing corn ethanol is zero when markets operate below the existing production capacity. However, if an expansion in capacity is required the model takes into account the required capital costs.

The supply of and demand for corn stover are added to the model as well. The supply of stover is presented by: $q_{stov} = A_{stov}(p_{stov}/c_{stov})^{estov}$. Here q_{stov} and p_{stov} represent supply and market price of stover, A_{stov} is the constant term, c_{stov} stands for production costs of stover per metric ton (including all costs items such as collection costs, costs to maintain land productivity, and transportation costs), and *estov* indicates price elasticity of supply. The production costs are divided into two segments of fixed and variable costs. The variable costs are assumed to be sensitive to changes in crude oil price to cover impacts of changes in crude oil price on collection and transportation costs of stover. The corn stover total production costs follows an increasing trend from \$68 per ton at \$60 per gallon of crude oil to about \$100 per ton at \$160 crude oil price. The demand for corn stover is determined by bio-gasoline production with the following equation: $qd_{stov} = q_{gstov}\ (V_{stov})$. Here V_{stov} is the conversion rate of corn stover to bio-gasoline. For details about the cost structure of the bio-gasoline industry and corn stover activity see Fiegel (Fiegel, 2012).

A market for soybeans is also added to the model. In this market, the demand for soybeans is defined as: $qd_{soy}=A_{soy}(1/(p_{soy})^{esoy})$. Here qd_{soy} and p_{soy} represent demand for soybeans and its price, A_{soy} shows the constant term of the demand function, and *esoy* indicates the own price elasticity of demand for soybeans. The model determines the supply of soybeans (qs_{soy}) using the allocated land to this product (l_{soy}) from the following equation: $qs^{soy}=l_{soy}.yield_{soy}$. Here, $yield_{soy}$ represents soybean yield. The model assumes that total supply of land (l_{tot}) for corn and soybean production is fixed in the short-run. Hence, it determines areas under soybean production using the following relationship: $l_{soy}=l_{tot}-(q_{cs}/yield_{corn})$.

Finally, the model imposes a zero profit condition to allocate land between corn and soybeans. Indeed the model assumes that farmers maximize their profit when they allocate their land between corn and soybeans. It is assumed that at equilibrium: $\pi_{soy}=\pi_{corn}+\pi_{stov}$.

The revised model is calibrated using data obtained for 2010 for the US economy. For details see Fiegel (Fiegel, 2012). Then the model is solved for several alternative crude oil prices (an exogenous key variable of the model) ranging from \$60 per barrel to \$160 per barrel. For each crude oil price two alternative policies are tested. The first policy assumes that the government will not support converting corn stover to biofuel. The second policy assumes that the government will pay a fixed subsidy of \$1.01 per gallon of bio-gasoline. Note that in both cases mentioned here we assumed that the government pays no subsidy for corn ethanol.

2.2 Farm Level Model

To examine impacts of having a viable market for corn stover on crop rotation at a farm level the PCLP model which was originally developed at Purdue University is used and modified. PCLP is a linear programming model which helps determine profit-maximizing decisions for a given farm according to its possible crop activities, its existing resources, and according to current prices of commodities and input costs. The PCLP model takes specific data such as land, labor, capital, crop yields, crop prices, and detailed input costs and determines activities which maximize farmer's profits. To adapt this model a new activity called stover collection is added to this model. This activity covers all steps and their corresponding costs which are required to collect and sell corn stover to a bio-refinery at current prices. These steps and their corresponding costs are defined in detail in (Fiegel, 2012; Thompson & Tyner, 2013). Then the model is solved for a group of farmers who participated in the Top Crop Farmer Workshops in 2007-2010 under two alternative cases of with and without corn stover activity to find out their optimal choices under these two different cases. To tune the PCLP model with market conditions in the presence of corn stover activity, prices obtained from the partial equilibrium model are used. Then the sensitivity of the results with respect to changes in the assumptions and parameters behind the corn stover activity are examined.

3. Results

3.1 Impacts at Market Level

The simulation results cover a wide range of cases for many variables under all alternative scenarios developed for this paper. In what follows, we only present a selection of key results which are important for our analyses. Two key variables are outputs of corn ethanol and stover bio-gasoline under alternative cases. The simulation results are presented in Table 1.

Table 1. Key simulation results obtained from partial equilibrium model

Description	Base 2010	Crude oil scenarios					
Crude oil price ($/barrel)	76.69	60.00	80.00	100.00	120.00	140.00	160.00
Case I: With no subsidy							
Ethanol supply (billion gallons)	13.23	7.00	12.06	13.71	14.25	16.70	18.77
Bio-gasoline supply (billion gallons)	0.00	0.01	0.11	0.56	1.60	2.55	3.49
Corn supply	12.45	11.19	12.21	12.53	12.64	13.16	13.61
Available corn stover (million tons)	348.52	313.22	341.74	350.80	354.04	368.54	381.01
Removal rate of corn stover (percent)	0.00	0.04	0.51	2.66	7.51	11.51	15.28
Soybeans supply (billion bushels)	3.33	3.69	3.40	3.31	3.27	3.13	3.00
Corn harvested area (million acres)	81.40	73.16	79.82	81.93	82.69	86.08	88.99
Soybeans harvested area (million acres)	76.60	84.84	78.18	76.07	75.31	71.92	69.01
Corn price ($/bushel)	5.18	4.22	5.00	5.32	5.39	5.79	6.13
Corn stover price $/ton)	0.00	26.14	47.00	68.81	88.69	101.61	112.75
Soybeans price ($/bushel)	11.30	9.21	10.85	11.46	11.69	12.82	13.92
Soybean to corn price ratio	2.18	2.18	2.17	2.15	2.17	2.21	2.27
Case II: With $1.01 subsidy							
Ethanol supply (billion gallons)	13.23	6.34	11.88	14.31	15.65	18.20	20.24
Bio-gasoline supply (billion gallons)	0.00	2.32	4.17	6.33	8.44	9.67	10.67
Corn supply	12.45	11.07	12.27	12.89	13.31	13.87	14.32
Avilable corn stover (million tons)	348.52	309.93	343.66	360.86	372.67	388.41	400.86
Removal rate of corn stover (percent)	0.00	12.47	20.20	29.25	37.74	41.48	44.37
Soybeans supply (billion bushels)	3.33	3.72	3.38	3.20	3.08	2.92	2.80
Corn harvested area (million acres)	81.40	72.39	80.27	84.28	87.04	90.72	93.63
Soybeans harvested area (million acres)	76.60	85.61	77.73	73.72	70.96	67.28	64.37
Corn price ($/bushel)	5.18	4.10	4.66	4.70	4.49	4.75	4.99
Corn stover price $/ton)	0.00	83.11	98.11	111.76	123.75	132.68	140.97
Soybeans price ($/bushel)	11.30	9.05	10.97	12.20	13.17	14.65	16.00
Soybean to corn price ratio	2.18	2.21	2.35	2.60	2.93	3.08	3.21

Table 1 shows that the supply of bio-gasoline will be very limited at low levels of crude oil price in particular when the government does not support bio-gasoline production. However, with a bio-gasoline subsidy, the market will produce significant amounts of bio-gasoline especially at medium and higher crude prices. A similar impact would result from a mandate such as the Renewable Fuel Standard. For example, when the crude oil price is around $100 per barrel, bio-gasoline production will be about 0.56 billion gallons when no subsidy is paid, and it will be about 6.33 billion gallons with subsidy. The simulation results indicate that the supply of corn ethanol increases as the crude oil price rises. The results also indicate that the bio-gasoline subsidy has a positive impact on supply of corn ethanol as well. For example, when the crude oil price is about $100 per barrel, supply of corn ethanol is about 13.71 billion gallons when there is no bio-gasoline subsidy. However, with bio-gasoline subsidy the supply of corn ethanol is about 14.31 billion gallons for the same crude oil price of $100 per barrel. This is due to the positive impact of bio-gasoline subsidy on corn supply and its negative impact on corn price.

In general, corn supply increases as crude oil goes up. However, the expansion in corn supply is faster in the presence of the bio-gasoline subsidy. With bio-gasoline subsidy, corn production and converting corn stover to biofuel is more attractive. For example, when the crude oil price is about $100 per barrel, corn supply is about 12.53 billion bushels with zero bio-gasoline subsidy and 12.89 billion bushels with subsidy. Under such conditions, farmers collect and supply a portion of available corn stover. For example, at the crude oil price of $100 per barrel, the available corn stover is about 360.86 million tons with the bio-gasoline subsidy, but farmers only collect about 29% of available corn stover.

While bio-gasoline production elevates demand for and supply of corn, it negatively affects supply of soybeans because corn and soybeans compete for limited land, in particular in the short run. The simulation results indicate that supply of soybeans decreases when crude oil and bio-gasoline production goes up, under both cases of with and without bio-gasoline subsidy. The supply of soybeans goes down faster in the presence of bio-gasoline subsidy. Reduction in supply of soybeans increases its market price and its relative price compared to corn price. Figure 3 highlights changes in the soybean to corn price ratio when crude oil price is changing, for both cases of with and without bio-gasoline subsidy. The relative price remains around 2.2 as crude oil price increases when the government does not support production of bio-gasoline. However, it grows to 3.2 as the crude oil price approaches $160 per barrel in the presence of bio-gasoline subsidy. Of course, this is only a short run result, and we would expect to see other crop adjustments (not in the PE model) in the real world.

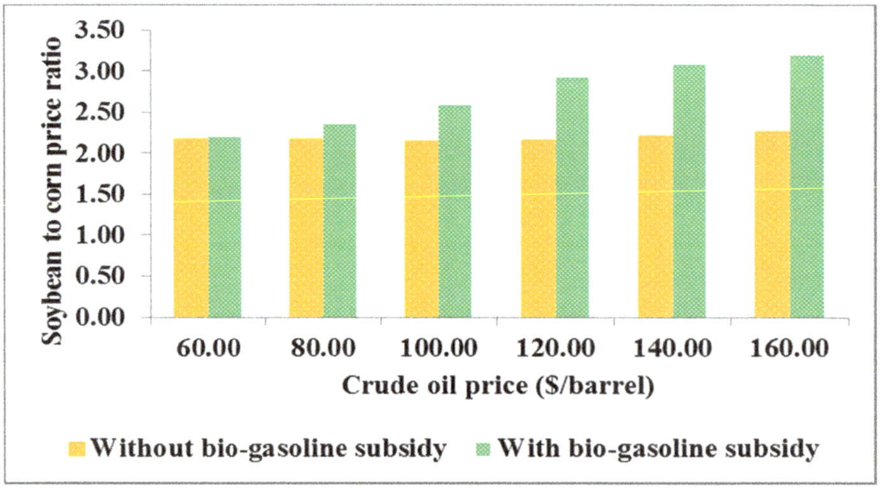

Figure 3. Projected soybeans to corn price ratio with and without bio-gasoline subsidy

From the partial equilibrium analyses provided in this section we can conclude that: 1) with no bio-gasoline subsidy a limited amount of corn stover will be converted to biofuel even at a very high crude oil price level; 2) The bio-gasoline subsidy could significantly boost production of this biofuel in particular at medium and higher crude oil prices; 3) under scenarios developed here, no more than 45% of available corn stover will be removed for biofuel production; 4) converting corn stover to bio-gasoline boosts corn production, increases corn-corn rotation, and decreases supply of soybeans; and 5) converting corn stover to bio-gasoline changes corn to soybean price ratio in favor of soybeans, at least in the very short term.

In our partial equilibrium analyses we examined market behavior for a wide range of crude oil prices. To assess impacts of converting corn stover to bio-gasoline on farmers' behavior at farm level we use the simulation results related to crude price of $100 per barrel with the bio-gasoline subsidy. We selected this case because bio-gasoline will not be produced without the government subsidy, and the crude oil price was expected to be around $100 per barrel in the near future (U.S. Department of Energy 2012).

3.2 Impacts at Farm Level

To examine impacts of a viable market for corn stover at a farm level we tuned the PCLP model with market clearing prices obtained from the PE model as described above. We then made the following experiments for each farm to assess its response with respect to changes in key economic variables:

i. Base case, no stover removal,

ii. Base case, with stover removal with tuned market prices in the presence of bio-gasoline subsidy under status quo assumptions on tillage costs, soybean and corn yields, cost associated with corn stover activity, and harvesting technology,

iii. No saving in tillage costs,

iv. Change in yield due to change in rotation, known as *yield drag*,

v. Reduction in corn stover price

vi. Change in corn stover harvesting technologies,

vii. Impacts of new harvesting technology but no savings in tillage costs.

The first case provides a status quo situation where there is no market for corn stover. The second case evaluates impacts at farm level in the presence of a market for corn stover. In this case it is assumed that: 1) corn-corn rotation with stover removal reduces tillage costs by $25 per acre; 2) corn-corn rotation does not affect corn and soybean yields; 3) corn stover farm price and corn stover delivery price are about $85.40 per ton, and $111.80 per ton, respectively; and 4) farmers will use the rake and bale system to remove corn stover. In cases *iii* to *vii* these basic assumptions are relaxed. In case *iii* it is assumed that corn-corn rotation with stover removal does not reduce tillage costs. Case *iv* takes into account the fact that corn-corn rotation can affect yield. In case *v* the farm level and delivery prices used in the base case are reduced by 20%. Case *vi* investigates impacts of using a new technology to remove corn stover. This technology is known as the corn-rower. Compared to the rake and bale system, this technology eliminates the need to rake the stover after harvesting corn and in general reduces the costs of stover activity (Fiegel, 2012). Finally, the last case assumes that farmers will use the corn-rower technology to harvest corn stover, but this technology does not reduce tillage costs.

While the above experiments are tested for individual participant farmers, in what follows we report the results obtained from the pool of participant farmers. The key simulation results are shown in Table 2.

Table 2. Land allocation patterns for alternative scenarios examined using PCLP model at farm level (figures are shares in total acreage in percent except otherwise noted)

Description	Case i	Case ii	Case iii	Case iv	Case v	Case vi	Case vii
Corn-soybean acreage with no stover removal	42.0	3.6	3.6	3.6	4.3	3.6	3.6
Corn-corn acreage with no stover removal	14.7	1.3	1.3	0.8	1.3	1.3	1.3
Corn-soybean acreage with stover removal	0.0	12.2	27.9	13.8	21.5	18.2	27.4
Corn-corn acreage with stover removal	0.0	66.0	37.7	64.2	43.2	53.9	30.5
Soybeans acreage	42.0	15.8	28.5	16.4	28.4	21.8	33.8
Other uses	1.3	1.2	1.0	1.1	1.3	1.2	3.4
Total acreage participating in workshop	100.0	100.0	100.0	100.0	100.0	100.0	100.0
Stover harvest acres	0.0	78.2	65.6	78.0	64.8	72.1	57.9
Stover removal rate (tons/acres)	0	1.18	1.01	1.15	0.99	1.93	1.57

Table 2 shows that in the base case when there is no stover removal activity, participant farmers allocate their land mainly to corn-soybean rotation or only prefer to produce soybeans. In this case farmers only allocate about 14.7% of their land to the corn-corn rotation. However, when corn stover removal is introduced as a new option under the base case scenario farmers allocate about 66% of their land to the corn-corn rotation and remove stover from their land. In this scenario about 15.8% of available land will remain in corn-soybean rotation. In this case corn stover will be removed from 78.2% of available land at a rate of 1.18 tons per acre. This shows that if corn stover is demanded for biofuel production, then a major shift will be observed in corn rotation. The rest of the scenarios show sensitivity of these results with respect to the assumption behind the base case. The results of case *iii* indicate that when corn-corn rotation does not reduce tillage cost, then only 37.7% of available land will be used in corn-corn rotation with corn stover removal. Corn-soybean rotation with stover removal and soybeans will have significant shares (27.9% and 28.5, respectively) in this case. The results obtained from case *iv* emphasize that yield drag does not affect the results of the base case significantly. Case *v* which assumes stover price is 20% lower than the base case provides results similar to case *iii* which projects higher tillage costs for the case of corn-corn rotation with stover removal. The next case, which examines the impacts of a lower cost of stover removal technology, indicates a relatively high share of 54% for corn-corn rotation with stover removal. Finally, the last case indicates that if the corn-rower technology does not reduce tillage coasts, we are then essentially back to a situation similar to the case *iii* where we removed tillage costs savings.

4. Conclusions

With a viable corn stover market and stover at a farm price of $85.40/ton, the large majority of farmers in the PCLP data set found it profitable to harvest stover in the base case. The sensitivity cases demonstrated how farmer behavior might change under different assumptions. We recognize, of course, that every farm is different with different soil types, management, equipment, etc. Clearly, PCLP captures some but not all of these differences.

Figure 4 displays the allocation of acres for each scenario. The figure shows that most farms chose to profit from continuous corn rotation with stover removal. Only in the cases of reduced tillage savings and decrease in stover price does corn-soybean with stover removal and soybean production increase. These results may be interpreted as meaning that if a viable stover market developed, we could see an increase in corn-corn rotation in the Midwest. We might see more soybeans grown in other areas. In fact the data from recent years shows this trend already beginning to happen.

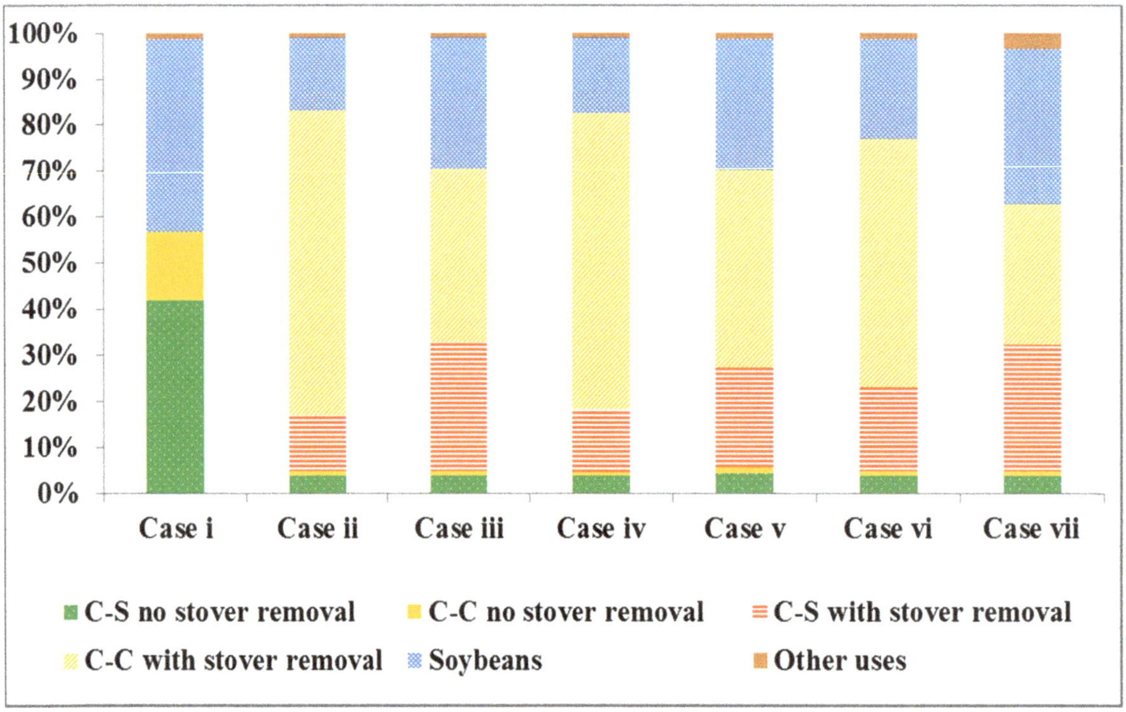

Figure 4. Land allocations for alternative scenarios

Table 2 displays the results for the base case and the five sensitivity cases. Twenty-three of 25 representative Midwest farms harvest stover. Twenty-three farms also would harvest stover for all given sensitivity cases except with the reduced stover price, where 21 found it profitable with the new price.

Acknowledgments

The authors are grateful to the Indiana Corn Soybean Alliance, who provided partial funding for this research.

References

Abbott, P., Hurt, C., & Tyner, W. E. (2008). *What's Driving Food Prices?*: Farm Foundation Issue Report.

Abbott, P., Hurt, C., & Tyner, W. E. (2011). *What's Driving Food Prices in 2011?* In *Issue Report* Farm Foundation.

Doster, D. H., Dobbins, C. L., Patrick, G. F., Miller, W. A., Preckel, P. V., Valentin, L., & Erickson, B. (2008). *Purdue PC-LP Farm Plan B-21 Crop Input Form*. Edited by Dept. of Agricultural Economics. West Lafayette, Indiana: Purdue University.

English, A., Tyner, W. E., Sesmero, J., Owens, P., & Muth, D. (2013). Environmental tradeoffs of stover removal and erosion in Indiana. *Biofuels, Bioproducts, and Biorefining, 7*(4), 78-88.

Fiegel, J. L. (2012). *Development of Viable Corn Stover Market: Impacts on Corn and Soybean Markets*, Department of Agricultural Economics, Purdue University, West Lafayette, IN.

Hertel, T., Golub, A., Jones, A., O'hare, M., Plevin, R., & Kammen, D. (2010). Effects of US Maize Ethanol on Global Land Use and Greenhouse Gas Emissions: Estimating Market-mediated Responses. *Bioscience, 60*(3), 223-231. http://dx.doi.org/10.1525/bop.2010.60.3.8

Taheripour, F., & Tyner, W. E. (2008). Ethanol Policy Analysis - What Have We Learned So Far? *Choices, 23*(3), 6-11.

Thompson, J., & Tyner, W. E. (2011). *Corn Stover for Bioenergy Production: Cost Estimates and Farmer Supply Response*, RE-3-W. Edited by Purdue Cooperative Extension Service. West Lafayette, IN: Purdue University.

Thompson, J., & Tyner, W. E. (2013). Corn Stover for Bioenergy Production: Cost Estimates and Farmer Supply Response. *Biomass and Bioenergy, 62*, 166-173. http://dx.doi.org/10.1016/j.biombioe.2013.12.020

Trostle, R. (2008). *Global Agricultural Supply and Demand Factors Contributing to the Recent Increase in Food Commodity Prices*. Edited by U.S. Department of Agriculture Economic Research Service. Washington, D.C.

Tyner, W. E., & Taheripour, F. (2007). Renewable Energy Policy Alternatives for the Future. *American Journal of Agricultural Economics, 89*(5), 1303-1310. http://dx.doi.org/10.1111/j.1467-8276.2007.01101.x

Tyner, W. E., & Taheripour, F. (2008). Policy Options for Integrated Energy and Agricultural Markets. *Review of Agricultural Economics, 30*(3), 387-396. http://dx.doi.org/10.1111/j.1467-9353.2008.00412.x

Tyner, W. E., & Taheripour, F. (2012). Land Use Changes and Consequent CO2 Emissions due to US Corn Ethanol Production. *Encyclopedia of Biodiversity* (2nd Ed.).

Tyner, W. E., Taheripour, F., & Hurt, C. (2012). *Potential Impacts of a Partial Waiver of the Ethanol Blending Rules*. Farm Foundation.

Tyner, W. E., Taheripour, F., & Perkis D. (2010). Comparison of Fixed Versus Variable Biofuels Incentives. *Energy Policy, 38*, 5530-5540. http://dx.doi.org/10.1016/j.enpol.2010.04.052

U.S. Department of Energy. (2012). *Annual Energy Outlook 2013* - Early Release. Washington, D.C.

Zilberman, D., Hochman, G., Rajagopal, D., Sexton, R., & Timilsina. G. R. (2013). The Impact of Biofuels on Commodity Food Prices: Assessment of Findings. *American Journal of Agricultural Economics, 95*(2), 275-281. http://dx.doi.org/10.1093/ajae/aas037

Notes

Note 1. In this paper it is assumed that corn and corn stover production is a single process and the farmer makes the decision at planting time. Of course a farmer can decide to produce corn at the planting time and later on he/she can decide whether to collect the corn stover or not.

16

Exploiting Cropping Management to Improve Agricultural Water Use Efficiency in the Drylands of Eastern Uganda

Robert Mulebeke[1], Geofrey Kironchi[2] & Moses M. Tenywa[3]

[1] Department of Agriculture, Kyambogo University, Kampala, Uganda

[2] Department of Land Resource Management and Agricultural Technology, University of Nairobi, Nairobi, Kenya

[3] Department of Agricultural Production, Makerere University, Kampala, Uganda

Correspondences: Robert Mulebeke, Department of Agriculture, Kyambogo University, Kampala, Uganda. E-mail: rmulebeke@yahoo.co.uk

Abstract

A remarkable challenge lies in maximizing agricultural water productivity, particularly in the drought prone regions of sub Saharan Africa. It is hypothesized that water use efficiency (WUE) can be increased by selection of appropriate cropping management systems. This study seeks to establish the effects of cropping management on water use efficiency in cassava-sorghum cropping systems in the drylands of eastern Uganda. A randomised complete block design (RCBD) consisting of six treatments: sole cassava, sole sorghum, sole cowpea, cassava + sorghum, cassava + cowpea, and sorghum + cowpea, replicated three times were used. Two tillage practices; mouldboard ploughing (Mb) and, ripping (Rp) were used to assess the effect of tillage. WUE (kg ha^{-1} mm^{-1}) was calculated as a ratio of yield (kg ha^{-1}) to evapotranspiration (ET) (mm). ET was estimated using the soil water balance. WUE varied significantly (α= 0.05) between cropping systems with the highest observed in cassava (34.38 kg ha^{-1} mm^{-1}) while the lowest was 3.76 kg ha^{-1} mm^{-1} for sorghum. WUE did not differ appreciably in both Mb and Rp tillage practices. Farmers growing sole cassava could use either of the tillage practices. The best yield was recorded in cassava + cowpea cropping system under Mb ploughing and sole sorghum under Rp gave the poorest combined yield (1,676 kg ha^{-1}).

Keywords: evapotranspiration, soil moisture storage, tillage, water use efficiency

1. Introduction

Agricultural water productivity is associated with the failure to meet the Millennium Development Goals (MDGs), especially where rainfall is limited and most times with erratic distribution in sub-Saharan Africa. Farmers have been exposed to water management strategies albeit in isolation at plant, plot, and farm level (Cooper et al., 2008), but have not measured up to the challenge. Now, there is a paradigm shift among nations and international institutions towards integrated water resource management (IWRM) (Tollan, 2002; Schulze, 2004), where policy-makers are now focusing on demand management, carrying capacity of the natural environment (Snellen & Schrevel, 2004), while linking water resources directly to development initiatives (UN-HABITAT, 2003). Among the biophysical measures driving this paradigm shift is the campaign to: 1) increase water use efficiency and land productivity, 2) continue efforts to explore ways to ''*grow more food with fewer drops*'' under sustainable conditions through research and development, capacity building and spread of technology (ICID, 2005).

A major challenge in the drought-prone regions of sub Saharan Africa is to balance the factors controlling crop water use and soil moisture availability (Rockström, 2003; Boko et al., 2007). Water use efficiency (WUE) is envisaged to provide a solution to this challenge by enhancing the partitioning of the available rainfall through manipulation of tillage and cropping systems. Well aware that efficiency is an input-output relationship, agricultural production undergoes several biophysical processes which most often dictate the final product (Hsiao et al., 2007) therefore, ability to relate an appropriate biophysical process to the harvestable crop product is a must in improving agricultural water use efficiency. Further still, the socioeconomic implications of delivering, using, managing, or buying water have exacerbated the impacts of supplying water for agricultural production, making farmers in the drylands more vulnerable.

Suggestions to improve and sustain food production in the drylands can be through choice of practices and cropping systems that demand less nutrients, labour, and capital for the production process (Andales et al., 2006) but also use water more efficiently. Reviews of WUE methods reported by Nielsen et al. (2005) show that use of crop residues on the soil surface increased WUE of corn and sunflower by 28 and 17% respectively due to change from conventional tillage to no-till in Kansas, USA while, WUE of winter wheat at Akron, Colombia increased from 6.9 to 7.5 and to 8.4 kg ha^{-1} mm^{-1} in a wheat-fallow-conventional till, wheat-fallow no-till, and wheat-corn-fallow no-till respectively. On the contrary, crop residues have several competing users in eastern Uganda like: dry season feed for livestock, thatching material for houses and, are easily damaged by termites, therefore use of crop residues for soil surface cover may not be feasible.

Soil moisture fluxes, a function of precipitation, runoff, infiltration, and evapotranspiration, determine crop water use (Allen et al., 1998; Angus & Herwaarden, 2001; Nielsen et al., 2005) however, water use can be regulated through choice of seedbed management practices that make soil moisture more available for uptake by crops. It is also true that water use is dependent on the inherent physiology and growth habit of the plant vis-à-vis the environmental stresses (Nielson et al., 2005; Makoi & Ndakidemi, 2010), notwithstanding the fact that, the prevailing microclimate can be engineered through choice of crops and cropping systems. Indeed, the process worthy exploiting in this soil-plant-atmosphere relationship is how rainwater is partitioned into water used by the crop and/or stored within the root zone ("green" water) under the different tillage and cropping management practices. In this case, WUE would refer to a ratio of crop yield to water used (yield/evapotranspiration) yet crop water use is subject to how farmers manage the crop in question.

In the drylands of eastern Uganda cassava (*Manihot esculanta*), sorghum (*Sorghum bicolor*) and cowpea (*Vigna unguiculata*) are dominant staple crops in most smallholder households. These crops are cultivated as pure stands in a crop rotation plan or by intercropping (Adipala et al., 1997; Ainembabazi et al., 2005; Otim-Nape et al., 2005; Whyte & Kyadondo, 2006). The common practice is to till the soil using animal power (Tenywa et al., 1999), usually up to a plough depth of 20 cm therefore moisture storage and consequently root penetration is limited to this volume. For purposes of this study, crop combinations and field management strategies used in growing these crops defines the cassava-sorghum based cropping system. Additionally, rainfall in this area was estimated to be partitioned thus: 93.2% ET, 3.8% runoff (Wasige et al., 2004), however, ET of up to 200% was recorded in a sorghum field indicating a severe negative water balance (Kizito, 2004). Therefore, the possibility of translating this magnitude of ET (93 to 200% of rainfall received) into yields is great, especially when the proportion of evaporation to ET is reduced and a steady water supply to the crop is ensured through improved soil moisture storage.

Crop and seedbed manipulations that positively impact on the soil water balance were suggested (Benli et al., 2006; Lipiec et al., 2006; Turner, 2004) and have been popularized as 'conservation agriculture' (FAO, 2007, Verhulst, et al., 2010).Conservation practices are based on promotion of biophysical processes above and below the ground. These soil and crop manipulations could target reducing evaporative losses from plant and soil surfaces while, increasing storage in the root zone and transpiration (Gicheru et al., 2004; D'Haene et al., 2008), with the resultant quantity being improved WUE. Agronomists have defined water use efficiency as a ratio of harvestable yield to water consumed by the crop (Turner, 2004; Passioura, 2006; Rijsbern & Manning, 2006; Bouman, 2007; Hsiao et al., 2007) in a simplified expression;

$$WUE(\text{kg ha}^{-1}\text{mm}^{-1}) = \text{Crop yield (kg ha}^{-1}) / \text{Water supply (mm)} \quad (1)$$

In this study WUE is defined as the ratio of harvestable (marketable) crop yield to evapotranspiration (water used by the crop) on the understanding that, the maximum rate of evapotranspiration, usually a function of available water in the soil, determines water uptake by a crop (Abbate et al., 2004; Ritzema, 2006; Bodner et al., 2007) and consequently the total yield. Water supply and water used by the crop are related and can be derived from the water balance components. Since it was not possible to precisely account for the loss of productive water through evaporation, the collective term of evapotranspiration was adopted to refer to water used by the crop.

Taking advantage of the knowledge on water balance parameters, seedbed management practices and manipulation of cassava, sorghum, cowpea cropping cycles to manage water resources in rainfed agricultural ecosystems is still a challenge. This study therefore, aimed at establishing the effect of tillage and cropping systems on water use efficiency in cassava-sorghum cropping systems in the drylands of eastern Uganda.

2. Materials and Methods

2.1 Study Site Description

The study was located in eastern Uganda (34° 0' E and 1° 40' N) in the Usuk sandy farm-grasslands agro-ecological zone (Wortman & Eledu, 1999) found in the greater Teso farming system. The Teso farming system lies at approximately 1036 and 1127 m above sea level and is predominantly an agro-pastoral system (Parsons, 1960). The area covers the administrative districts of Amuria, Katakwi, Kumi, and Soroti in eastern Uganda (Appendix 1).

2.1.1 Climate

The area experiences bimodal rainfall where the long rain season is usually from mid-March to June and the short rain season is from August to November. There is a short dry spell between the two rain seasons i.e. mid-June to mid-July, however areas bordering northeast experience earlier dry seasons. The seasonal mean precipitation ranges between 650 and 900 mm, but up to 25-30% of the precipitation is received outside the annual growing period (Kayizzi et al., 2007).

The mean annual maxima and minima temperature is 31.3 °C with an annual mean of 24 °C. Extreme temperatures of approximately 35^0C are common in the month of February. Potential ET is higher than precipitation for most part of the year (Table 1). Relative humidity ranges from 66% to 83% at 0900hours East African time and 35% to 57% at 1500hours, thereby reducing chances of rainfall. This condition is expected to support crop growth but the seasonal growing period is reduced to 72-120 days (Komutunga & Musiitwa, 2001).

Table 1. Long-term mean monthly rainfall, evapotranspiration and temperature for Soroti, eastern Uganda

	Jan	*Feb*	*Mar*	*Apr*	*May*	*Jun*	*Jul*	*Aug*	*Sep*	*Oct*	*Nov*	*Dec*
Rainfall (mm)	20.1	47	72.9	182.9	192	125	125	184.9	142	116.1	72.5	38
ET (mm)	121.7	118.9	129	106.7	90.2	90.8	86.2	92.7	92.7	106	108.2	111
Temp (°C)	25.5	25.9	25.9	24.4	23.5	23.1	22.5	22.6	23.3	24.1	24.5	24

Source: Major land resources areas of Uganda, Yost and Eswaran, (1990) p. 96.

2.1.2 Soils

The soils of the Teso region originate from rocks of the precambrian age basement complex comprising mainly granites, mignalites, gneiss, schists and quartzites. These rocks give rise to four major soil series Serere and Amuria catenas, Metu complex and Usuk series that are mainly of the ferralitic type (sandy sediments and sandy loams). The dominant soil type in the area is *Petroferric Haplustox*. Sandy soils stand out in the area (Ssali, 2000) are well drained, friable and characterized by low water holding capacity and low organic matter levels which intensify the water deficit challenge. The soil physical and chemical properties at the experimental site are described in Table 2 below.

2.1.3 Land Use and Livelihood Strategies

Land use and livelihoods are inextricably linked (ecologically, economically, socially, and politically) that any change in former dramatically impacts on the latter. The eastern Uganda population is predominantly agro-pastoral basically producing at subsistence level (Whyte & Kyaddondo, 2006). Now, most households in the Teso region derive their livelihood on increasingly small land holdings ranging between 0.5 to 4 ha (national average = 0.4 to 3ha) per household (Okwi et al., 2006), hence forcing intensive production systems and/or seeking non-farm income in order to ensure food self sufficiency. Crop production is reported to be declining (NARO, 2002), despite the use of nutrient inputs and pest management strategies (Kayizzi et al., 2007), a limitation thought to arise from water stress conditions. Cassava is a key root crop (NARO, 2002) serves both as a major source of calorie intake and a 'new' cash crop to over 60% of the households in eastern Uganda (Otim-Nape et al., 2005) while, sorghum is ranked 2nd most important cereal in eastern Uganda. These major crops are grown in root crop/cereal/legume mixtures (Otim-Nape & Zziwa, 1990) like, the sorghum-cowpea intercrop (Adipala et al., 1997) or crop rotations beginning with cassava (Otim-Nape et al., 2005)mainly for pest management, soil fertility improvement and income aspects yet, these crop mixtures have a potential to exploit in integrated water resource management practices. Therefore, conditioning the already known crop, tillage and crop management practice to improve water use efficiency is thought to be a major contribution to land use and livelihood security in the cassava-sorghum based cropping systems in eastern Uganda.

Table 2. Soil Physical and chemical characteristics at the study site

	Profile depth	
	0-10 cm	*20-40 cm*
Clay (%)	4	4
Silt (%)	6	28
Sand (%)	90	68
OM (%)	2.72	1.01
Textural class (FAO)	Sandy	Clay loam
Bulk density (g cm^{-3})	1.63	1.54
K_{sat} (cm hr^{-1})	5.8	7.6
pH (H_2O)	5.9	5.2
CEC (me/100 g)	9.2	10.2
N (%)	0.12	0.1
P (me/100 g)	2.15	1.25
K (me/100 g)	0.2	0.21
Na (me/100 g)	0.05	0.09
Ca (me/100 g)	6.21	4.01
Mg (me/100 g)	2.11	1.21

2.2 Experiment Design

The study was set in a randomised complete block design (RCBD). The treatments consisted of six cropping systems, i.e. i) sole cassava, ii) sole sorghum, iii) sole cowpea, iv) cassava + sorghum, v) cassava + cowpea, vi) sorghum + cowpea and replicated three times. Two tillage (seed bed management) practices, i.e. mouldboard ploughing and ripping using ox-drawn equipment were practiced. Each experimental unit was 10 × 7 m separated by a 1m gap. Replicates were located on a similar slope profile along a defined transect in order to minimise variations due to lateral soil moisture flux and soil fertility gradient. The experiments were conducted in three growing seasons. Cassava was planted in the first rains of 2010 (late March), using a 20 ± 5 cm cutting with a spacing of 1 m × 1 m, giving a total plant population of 10,000 plants ha^{-1}. Harvesting was done in April 2011. Unlike the cassava crop, sorghum and cowpea were planted three times i.e. late march 2010, August 2010 and April 2011.

2.3 Determination of Evapotranspiration

The soil water balance was used to derive ET (Abbate et al., 2004; Benli et al., 2006; Bodner et al., 2007). When micro-meteorological data was not available to run the original Penman–Monteith water balance model (Van Vosselen et al., 2005), some parameters representing the classical law of mass conservation were used. Precipitation (P) was the only source of moisture into the soil system because there was no irrigation (I = 0) and there was no groundwater within reach of the root zone (CP = 0). Subsurface flow into (SF_i) and out (SF_o) of the soil was negligible due to the small slope gradient (< 3%) and the surface conditions could not allow runoff to take place (RO = 0). Therefore the soil water balance equation is re-arranged to give the expression;

$$P = ET + DP + \Delta S \tag{2}$$

where; *ET* is evapotranspiration,

DP is loss to the soil layers below the root zone, and

ΔS is the amount stored in the soil at a given period.

The basic soil-plant-atmosphere relationship being explained here is the balance between rainfall received (*P*), water retained in the root zone (ΔS), water percolating beyond the root zone (*DP*) and, water used by the crop (*ET*). We notice here that two basic parameters; *P* and ΔS are measured and *DP* is estimated. But in principle, *DP* only occurs when the upper horizons are saturated and can be estimated based on Darcy's law (Ritzema, 2006) using the equation;

$$DP = -k\frac{\partial H}{\partial Z} \tag{3}$$

where; k is hydraulic conductivity (mm day^{-1})

$\frac{\partial H}{\partial Z}$ is the hydraulic gradient

However literature indicates that in dryland agricultural ecosystems, where the slope gradient is almost flat deep drainage is negligible. In particular, a water balance model for this catchment indicated that only 8.3% of the rainfall received was stored (Wasige et al., 2004), hence a minimal volume is expected to go to deep percolation. Therefore, to estimate *ET* throughout the life cycle of a crop, we need to determine soil moisture content at planting time and at crop maturity. Angus and van Herwaarden (2001) used the expression;

$$ET = P + \Delta S \tag{4}$$

where, ΔS represents soil water content at sowing time minus soil water content at maturity.

2.4 Crop Yield Determination

2.4.1 Cassava

The main yield components used was fresh and dry storage root weight (Ntawuruhunga et al., 2001). Measurements were done on five plants from each plot randomly chosen during harvest of cassava in the different treatments. Total fresh weight was taken after uprooting the plant and weighed in the field using a commercial weighing scale (25 kg ±10 g), and then marketable storage roots were removed and weighed separately. For the measurements done in the laboratory, one storage root was randomly taken from each of the five plants selected in each plot. This was used to obtain the average diameter, length and fresh weight of the storage root. The storage root was prepared for drying by manually removing the peel (corky periderm and cortex) using a kitchen knife, then the starchy flesh was sliced to thin chips to ease sun drying. In order to ensure uniform level of drying, the materials were placed in an oven at 60 °C for 24 hours before taking the dry weight measurements. The dry weight of the storage roots was taken using a precision laboratory scale (1000 g ± 0.001 g) and the values used to derive the yield per hectare.

2.4.2 Cowpea

The yield parameters taken for cowpea include; number of pods per plant, number of seeds per pod, weight per hectare. A one square meter area was randomly chosen in each plot then the number of plants counted and the average number of pods per plant determined. Mature pods were harvested, air dried and hand threshed. Grain weight was taken after all seeds were oven dried at 60 °C for 24 hours.

2.4.3 Sorghum

The sorghum variety grown has one head (panicle) per plant. Ten plants were randomly selected from the mid rows per plot from which panicles were cut using a hand knife. Heads from each plot were sun dried, then threshed by abrasion using light pressure from a piece of wood. Traditionally, in eastern Uganda sorghum and millet are threshed by beating harvested heads with a dry smooth piece of wood after sun drying.

2.5 Determination of Water Use Efficiency

The yield obtained is dependent on the water used (evapotranspiration) throughout the growing period, which defines WUE. Therefore after substituting water supply with *ET* in Equation 1, we obtain a working expression following similar arguments by Moitra et al. (1996), Xu and Hsiao, (2004), and Payero et al. (2008).

$$WUE(\text{kg ha}^{-1}\text{mm}^{-1}) = \text{Crop yield (kg ha}^{-1})/\text{ET (mm)} \tag{5}$$

2.6 Statistical Data Analysis

Data were analysed using GenStat software. Analysis of variance (two way ANOVA) was performed to identify differences between treatments and means separated using LSD at α=0.05. However, the cowpea and sorghum crop was grown for three seasons hence, a variation due to season introduced. The first two seasons had three replicates and the third season was replicated four times, so the ANOVA for unbalanced designs was adopted where season and replication were blocked as nuisance factors. However, some further analysis would require introduction of season as a factor.

3. Results and Discussion

3.1 Effect of Cropping Management Practice on Evapotranspiration

Evapotranspiration (*ET*) was calculated following the water balance model. Since, precipitation (P) cannot vary in all treatments, and the change in soil moisture content (ΔSM) was known then, ET as a sum of P and ΔSM, was dependent on soil moisture storage, represented by the soil water content at sowing time minus soil water content at maturity and the ability of the crop to extract moisture from the soil.

ET varied significantly between cropping systems (F pr = <0.001) but, did not show any difference with respect to tillage practice (F pr = 0.105) at 95% confidence level. On the contrary, the interaction between tillage practices and cropping systems presented significant differences (F pr = <0.001). The sole cassava (526.5 mm) and cassava + cowpea (524.0 mm) cropping pattern recorded higher ET under Rp than Mb while, the sorghum + cowpea intercrop performed slightly better in Mb (526.5 mm) than Rp (519.3 mm) ploughed plots. The ET for sole cowpea cropping system was similar and highest in both tillage practices (Figure 1).The rainfall data used here were average records of two non-recording rain gauges, one located adjacent to the experiment site and the other at the Soroti Flying School Meteorological Department located about 3 km from the experimental site. The amount of rainfall received during the growing seasons was 728.4, 350.0 and 477.0 mm for season I (March – July 2010), season II (August – November 2010) and season III (April – July 2011) respectively. The ΔSM was derived from soil moisture content observations made through the growing period at two soil profile depths (0-10 and 20-40 cm).

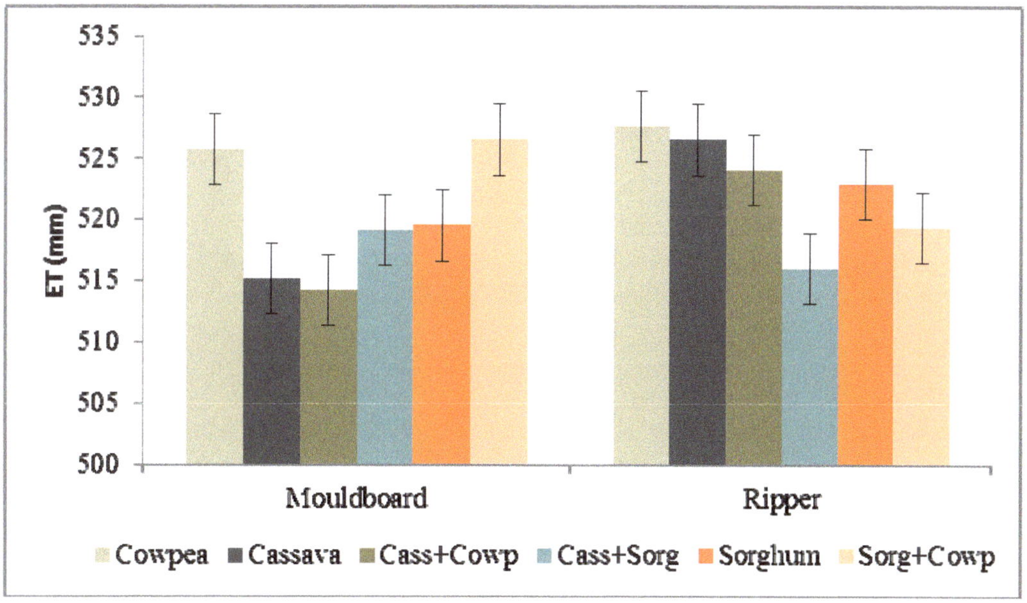

Figure 1. Effect of cropping management on evapotranspiration in the cassava-sorghum cropping system

There was a significant difference in ΔSM between, cropping systems, tillage x cropping system interaction, seasons (F pr = < 0.001) and tillage practice (F pr = 0.044) at 95% confidence level.

When effect of seasons was removed and each season analysed independently, it was observed that tillage practice in season I (F pr = 0.065), tillage practice x cropping system interaction in season II (F pr = 0.165) were not significantly different at 95% confidence level. In season III all treatments were significant (F pr = <0.001).Therefore for the calculation of the water balance, the accumulated means for the three seasons were used.

There was more water extracted from the soil system during the growing period as indicated by the negative sign on the values where cassava was grown as a sole crop or as an intercrop (Table 3).

Table 3. Effect of cropping management on soil moisture storage

Tillage Practice	*Cropping system*	*ΔSM(mm)*
Mouldboard	Sole Cassava	-4.215*d*
	Sole Cowpea	1.031*ba*
	Sole Sorghum	-2.054*c*
	Cassava + Cowpea	-4.736*d*
	Cassava + Sorghum	-2.275*c*
	Sorghum + Cowpea	1.410*b*
Ripper	Sole Cassava	1.441*b*
	Sole Cowpea	1.962*bc*
	Sole Sorghum	-0.390*a*
	Cassava + Cowpea	0.175*a*
	Cassava + Sorghum	-3.832*d*
	Sorghum + Cowpea	-2.154*c*

Note: *Different letters against the values indicate significant differences at* $\alpha = 0.05$.

The high negative values exhibited by sole cassava, cassava + cowpea, cassava + sorghum, and sole sorghum cropping systems under mouldboard ploughing compared to the positive and relatively low values under ripper ploughing show that the crops under ripping are less likely to show signs of water stress during the growing period. It was noted that the ripper ploughing practice could hold more water in the 20-40 cm horizon which could then be made available to deep rooted crops like cassava and cowpea.

3.2 Effect of Cropping Management on Yield of Cassava, Cowpea, and Sorghum

The cassava crop was grown between March 2010 and April 2011 (one season) while, the cowpea and sorghum crop was grown for three seasons (March – July 2010; August – November 2010; April – July 2011). Generally, the yield of cassava cowpea, and sorghum varied between the two tillage practices and the cropping systems (Table 4). On introduction of the effect of seasons only intercrops involving sorghum varied significantly (F pr = 0.001) at 95% confidence level.

Much as the marketable yield of cassava can be both in terms of fresh weight and dry weight, for this study dry weight was used in order to harmonise the crop yield records for all crops. The yield of cassava differed significantly between cropping systems but was similar as a sole crop under mouldboard and ripper ploughing practices. The cassava + cowpea intercrop in the mouldboard ploughed plot recorded the highest cassava yield (20,023 kg ha^{-1}) compared to other treatments.

Cowpea yield was significantly different in both tillage (F pr = 0.013) and cropping systems (F pr = <0.001) but, the interaction between tillage and cropping systems were non-significant (F pr = 0.102). Additionally, variation due to seasons was not significant (F pr 0.161) at 95% confidence level. The yield of sorghum was significantly different between tillage practices (F pr = 0.009), cropping systems (F pr = 0.004) and season (F pr = 0.001) were recorded.

The traditional practice of growing cassava as a sole crop under mouldboard ploughing gave similar yields like ploughing using a ripper. However, the high yield recorded when cassava was intercropped with cowpea (20,023 kg ha^{-1}) under mouldboard ploughing is evidence that these crops complemented each other in resource use. Agronomists and soil fertility scholars have advanced the contribution of legumes in soil fertility enhancement in mixed cropping systems however, efficient use of soil moisture provides the missing link in enhancing productivity in these cropping systems. Additionally, the cowpea yield in the cassava + cowpea intercrop was still better than in the combinations involving sorghum under both tillage practices. The observed yield here were majorly contributed by the first season crop which used most of the soil moisture before the cassava crop established.

Table 4. Effect of cropping management on yield of cassava, cowpea and sorghum

Treatments		*Yield (kg ha^{-1})*		
Tillage Practice	Cropping system	Cassava	Cowpea	Sorghum
Mouldboard	Sole Cassava	16560b		
	Sole Cowpea		8397d	
	Sole Sorghum			1179b
	Cassava + Cowpea	20023c	3570b	
	Cassava + Sorghum	11101a		1279c
	Sorghum + Cowpea		2853a	894a
Ripper	Sole Cassava	14282b		
	Sole Cowpea		5771c	
	Sole Sorghum			1679d
	Cassava + Cowpea	11962a	2320a	
	Cassava + Sorghum	11247a		1260c
	Sorghum + Cowpea		2703a	1110b

Note: *Different letters against the values in each column indicate significant differences at 95% confidence level.*

Overall, cowpea yield was higher under Mb than Rp across all three seasons (March – July 2010; August – November 2010; April – July 2011). It was observed that cowpea under Rp had a thick vegetative cover and was frequently attacked by pests. Adipala et al. (1997) reported that high humidity under the leaves of cowpea encourages pest build up, hence the cowpea yield here was compromised by pest damage. Additionally it is a traditional practice in eastern Uganda to harvest cowpea leaves for use as a vegetable and studies show that grain yield improved with a degree of defoliation (Rahman et al., 2008) but, here farmers would give priority to the mouldboard plots than the ripper plots. To this effect the highest yield (8397 kg ha^{-1}) recorded for the sole crop was under mouldboard ploughing were preferential leaf harvesting was done because the field was cleaner. Also the yield difference could be attributed to the optimum plant population (av. plant density = 12667 plants ha^{-1}) giving a relatively maximum and uniform use of soil and water available.

The yield of sorghum was highest in the sole crop under ripper ploughing (1679 kg ha^{-1}). The cassava + sorghum intercrop in both tillage practices was comparable (Table 4) while, the lowest yield was observed in the sorghum + cowpea intercrop in Mb. Cowpea has a quicker growth and establishment habit hence, could have out competed the sorghum crop for nutrients and moisture. We also noted that in the cassava + sorghum intercrop, the sorghum crop is not stressed by the cassava crop since it has a much slower rate of establishment.

3.3 Effect of Cropping Management Systems on Water Use Efficiency

Water use efficiency (WUE) varied significantly between cropping systems (F pr = < 0.001) but was similar under mouldboard (Mb) and ripper (Rp) ploughing (F pr = < 0.166) at 95% confidence level. The highest WUE was recorded in cassava at 34.38 kg ha^{-1} mm^{-1} and the lowest was in sorghum at 3.76kg ha^{-1} mm^{-1} (Figure 2). When the effect of season was considered as a main factor in the analysis, there were significant variations (F pr = <0.001) in WUE across the three seasons (Figure 3).

The variation in WUE observed between cropping systems could be attributed to soil moisture characteristics, although growth and physiological characteristics of the crop could be considered. The material that constitutes the yield could partially contribute to the difference i.e. root tuber in cassava compared to the grains in sorghum and cowpea (Steduto et al., 2007). The ability of the crop to convert assimilates to biomass is primarily determined by chemical composition of the crop and is not easily changed by environmental factors such as water supply except, where there are extreme changes in respiration due to change in thermal regimes (Hsiao et al., 2007). For instance, the cassava crop was in the field through two seasons and was able to use most of the moisture that was stored in the soil even during the dry season of the year.

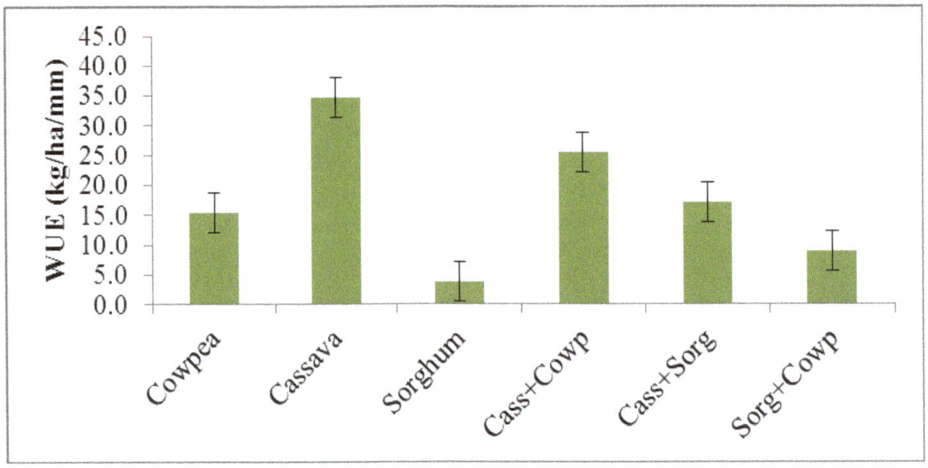

Figure 2. Behaviour of WUE (kg/ha/mm) under different cropping system in eastern Uganda

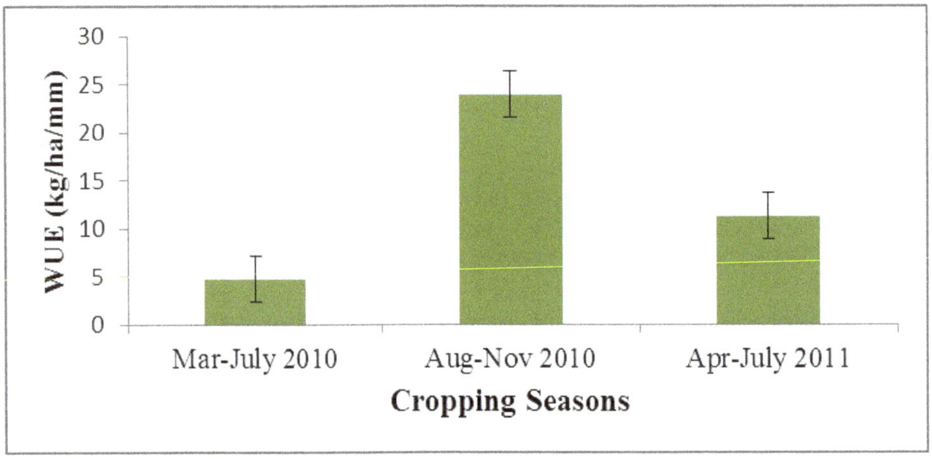

Figure 3. WUE (kg/ha/mm) across three seasons during the study period (March 2010- July 2011) in the cassava-sorghum cropping systems in eastern Uganda

There was a noticeable decrease in soil moisture storage for this cropping system as crop growth progressed. Much as ET for cassava was high the crop seems to have been able to reduce the proportion of moisture losses arising from evaporation.

Interestingly though, cassava + cowpea intercrop recorded better WUE than sorghum + cowpea. The feeding volume of the cassava and cowpea was stratified through the root zone that instead of resource competition there was a synergistic relationship. The moisture available before the cassava crop establishes was exploited by the quick growing cowpea. It was also observed that cowpea offered a surface cover of over 50% by 8 weeks after planting hence, reducing evaporation losses and encouraging transmission to the lower section of the root zone. Hsiao et al. (2007) noted that the water stored in the root zone can merge with water from the rain at the point of root zone water hence, improving efficiency of water delivery at the soil-plant- atmosphere interface. Therefore, for farmers in eastern Uganda who usually intercrop cassava with sorghum or cowpea, this study shows that cassava + cowpea maximizes the use of available water in the soil system. It was noted that, tillage using a ripper plough stored more water in the 20-40 cm rooting zone. This process can be enhanced by carefully synchronizing tillage practices with cropping systems to improve water use efficiency in the cassava-sorghum cropping system.

4. Conclusion

Evapotranspiration (ET) varied significantly between cropping patterns and the interactive effect of cropping and tillage management. Sole cassava (526.5 mm) and cassava + cowpea (524.0) cropping system recording a higher ET under ripper ploughing (Rp) than mouldboard ploughing (Mb) while, the sorghum + cowpea intercrop performed better in Mb than Rp.The change in soil moisture content was more negative in the mouldboard ploughed plots indicating that more water was extracted than in the ripped plots. The sole cassava, cassava + cowpea, cassava + sorghum, and sole sorghum cropping systems under mouldboard ploughing had a more negative change in moisture content compared to the positive and relatively low values under ripper ploughing.

There were variations in water use efficiency (WUE) due to choice of crop management, but WUE was similar in both mouldboard and ripper ploughed plots. Cassava had the best WUE (34.38 kg ha^{-1} mm^{-1}) while sorghum (3.76 kg ha^{-1} mm^{-1}) recorded the low values. However, there is need for further studies to ascertain the contribution of physiological factors on this difference. There is need to take measures to maintain soil moisture and improve water use efficiency in dryland cropping systems.

Much as there were differences in total yield between tillage practices and cropping systems, in sole cassava cropping system yield was comparable under both mouldboard and ripper ploughed plots. Cassava + cowpea intercrop under mouldboard ploughing gave the best cassava yield compared to other cropping patterns. Crop management significantly influenced the yield of cowpea across the three seasons. Overall, cowpea yield was higher under mouldboard ploughing than ripper ploughing across all three seasons and can be related to the availability of soil moisture in the 0-10cm layer early in the season for quick establishment. Sorghum responded differently to tillage practices, cropping systems and exhibited a strong variation between the three seasons.

References

Abbate, P. E., Dardarnelli, J. L., Cantarero, M. G., Maturano, M., Melchiori, R. J. M., & Suero, E. E. (2004). Climatic and water availability effects on water use efficiency in wheat. Crop Science Society of America.*Crop Science, 44*, 474-483. http://dx.doi.org/10.2135/cropsci2004.0474

Adipala, E., Obuo, J. E., & Osiru, D. S. O. (1997). A survey of cowpea cropping systems in some districts of Uganda. *African Crop Science Proceedings, 3*, 665-672.

Ainembabazi, J. H., Bashaasha, B., Mugisha, J., Pender, J., & Hyuha, T. S. (2005). Technological change in sorghum production in Eastern Uganda. *African Crop Science Conference Proceedings, 7*, 947-954.

Allen, R. G., Pereira, L. S., Raes, D., & Smith, M. (1998). Crop evapotranspiration. Guidelines for computing crop water requirements. FAO Irrigation and drainage paper 56. FAO, Rome. Retrieved from http://www.fao.org/docrep/X0490E/x0490e06.htm#equation

Andales, A. A., Green, T. R., Ahuja, L. R., Erskine, R. H., & Peterson, G. A. (2007). Temporally stable patterns in grain yield and soil water on a dryland catena. *Agricultural Systems, 94*, 119-127. http://dx.doi.org/10.1016/j.agsy.2006.07.001

Angus, J. F., & van Herwaarden, A. F. (2001). Increasing water use and water use efficiency in dryland wheat *Agronomy Journal, 93*, 290-298. http://dx.doi.org/10.2134/agronj2001.932290x

Benli, B., Kodal, S., Ilbeyi, A., & Ustun, H. (2006). Determination of evapotranspiration and basal coefficient of alfalfa with a weighing lysimeter. *Agricultural water management, 81*, 358-370. http://dx.doi.org/10.1016/j.agwat.2005.05.003

Bodner, G., Loiskandl, W., & Kaul, H. (2007). Cover crop evapotranspiration under semi-arid conditions using FAO dual crop coefficient method with water stress compensation. *Agricultural water management, 93*(3), 85-98. http://dx.doi.org/10.1016/j.agwat.2007.06.010

Boko, M., Niang, I., Nyong, A., Vogel, C., Githeko, A., Medany, M., ... Yanda, P. (2007). In M. L. Parry, O. F. Canziani, J. P. Palutikof, P. J. van der Linden & C. E. Hanson, (Eds.) *Africa. Climate Change 2007: Impacts, Adaptation and Vulnerability. Contribution of Working Group II to the Fourth Assessment Report of the Intergovernmental Panel on Climate Change* (pp. 433-467). Cambridge UK: Cambridge University Press.

Bouman, B. (2007). A conceptual framework for the improvement of crop water productivity at different spatial scales. *Agricultural Systems, 93*, 43-60. http://dx.doi.org/10.1016/j.agsy.2006.04.004

Cooper, P. J. M., Dimes, J., Rao, K. P. C., Shapiro, B., Shiferaw B., & Twomlow, S. (2008). Coping better with current climatic variability in the rain-fed farming systems of sub-Saharan Africa: An essential first step in adapting to future climate change? *Agriculture, Ecosystems and Environment, 126*, 24-35. http://dx.doi.org/10.1016/j.agee.2008.01.007

D'Haene, K., Vermang, J., Cornelis, W. M., Leroy, B. L. M., Schiettecatte, W., DeNeve, S., … Hofman, G. (2008). Reduced tillage effects on physical properties of silt loam soils growing root crops. *Soil and Tillage Research, 99*, 279-290. http://dx.doi.org/10.1016/j.still.2008.03.003

FAO. (2007). Food and Agriculture Organization of the United Nations. 2007. Conservation Agriculture. Retrieved from http://www.fao.org/ag/ca/index.html

Gicheru, P., Gachene, C., Mbuvi, J., & Mare, E. (2004). Effects of soil management practices and tillage systems on surface soil water conservation and crust formation on a sandy loam in semi-arid Kenya. *Soil and Tillage Research, 75*, 173-184. http://dx.doi.org/10.1016/S0167-1987(03)00161-2

Hsiao, T. C., Steduto, P., & Fereres, E. (2007). A systematic and quantitative approach to improve water use efficiency in agriculture. *Irrigation Science, 25*, 209-231. http://dx.doi.org/10.1007/s00271-007-0063-2

Kayizzi, C. K., Byalebeka, J., Wortmann, C. S., & Mamo, M. (2007). Low Input Approaches for Soil Fertility Management in Semiarid Eastern Uganda. *Agronomy Journal, 99*, 847-853. http://dx.doi.org/10.2134/agronj2006.0238

Kizito, C. (2004). Quantification of evaporation in a sorghum crop in North-eastern Uganda. *MSc Thesis, Makerere University, Uganda.*

Komutunga, E. T., & Musiitwa, F. (2001). Characterizing drought patterns for appropriate development and transfer of drought resistant maize cultivars in Uganda. Seventh Eastern and Southern Africa Regional Maize Conference. pp. 260-262.

Lipiec, J., Kus´, J., Nosalewicz, A., & Turski, M. (2006). Tillage system effects on stability and sorptivity of soil aggregates. *International Agrophysics, 20*(3), 189-193.

Makoi, J. H. J. R., & Ndakidemi, P. A. (2010). Effect of plant densities and cropping systems on yield components of cowpea (*Vignaunguiculata*L. Walp.) genotypes and sorghum (*Sorghum bicolor* L. Moench.). *Journal of Tropical Agriculture, 48*, 28-33.

Moitra, R., Ghosh, D. C., & Sarkar, S. (1996). Water use pattern and productivity of rainfed yellow sarson (*Brassica rapa L. varglauca)* in relation to tillage and mulching.*Soil and Tillage Research, 38*, 153-160. http://dx.doi.org/10.1016/0167-1987(96)01010-0

NARO. (2005). *National Agricultural Research Organisation*. Annual Report. Entebbe, Uganda.

Nielsen, D. C., Unger, P. W., & Miller, P. R. (2005). Efficient water use in dryland cropping systems in the Great Plains. *Agronomy Journal, 97*, 364-372. http://dx.doi.org/10.2134/agronj2005.0364

Ntawuruhunga, P., Rubayihayo, P., Whyte, J. B. A., Dixon, A. G. O., & Osiru, D. S. O. (2001). Inter-relationships among traits and path analysis for yield components of cassava: a search for storage root yield indicators. *African Crop Science Journal, 9*(4), 599-606.

Okwi, P. O., Hoogeveen, J. G., Emwanu, T., Linderhof, V., & Begumana, J. (2006). Welfare and environment in rural Uganda: Results from a small-area estimation approach. *The African statistical journal, 3*, 135-188.

Otim-Nape, G. W., Bua, A., Ssemakula, G., Acola, G., Baguma, Y., Ogwal, S., & Van der Grift, R. (2005). Cassava development in Uganda: a country case study towards a global cassava development strategy. *Proceedings of the Validation Forum on the Global Cassava Development Strategy,* FAO and IFAD, Vol 2, Rome, Italy.

Passioura, J. (2006). Increasing crop productivity when water is scarce—from breeding to field management. *Agricultural Water Management, 80*(1-3), 176-196. http://dx.doi.org/10.1016/j.agwat.2005.07.012

Payero, J. O., Tarkalson, D. D., Irmak, S., Davison, D., & Petersen, J. L. (2008). Effect of irrigation amounts applied with subsurface drip irrigation on corn evapotranspiration, yield, water use efficiency, and dry matter production in a semiarid climate. *Agricultural water management, 45*, 895-908. http://dx.doi.org/10.1016/j.agwat.2008.02.015

Rahman, S. A. Ibrahim, U., & Ajayi, F. A. (2008). Effect of defoliation at different growth stages on yield and profitability of cowpea (*vignaunguiculata* (l.) walp.). *Electronic Journal of Environmental, Agricultural and Food Chemistry, 7*(9), 3248-3254.

Rijsberman, F. R., & Manning, N. (2006). Beyond more crop per drop: Water management for food and the environment. 4th World Water Forum. IWMI, Mexico.

Ritzema, H. P. (2006). Drainage Principles and Applications. Wageningen, Alterra, ILRI Publication no. 16, 145-173.

Rockström, J. (2003). Resilience building and water demand management for drought mitigation. *Physics and Chemistry of the Earth, 28*, 869-877. http://dx.doi.org/10.1016/j.pce.2003.08.009

Snellen, W. B., & Schrevel, A. (2004). IWRM, for sustainable use of water: 50 years of international experience with the concept of integrated water resources management; Background document to the FAO/Netherlands Conference on Water for food and ecosystems. Wageningen, ALTERRA-report 1143. p.24.

Ssali, H. (2000). Soil resources of Uganda and their relationship to major farming systems. Resource paper. Soils and Soil Fertility Management Programme, Kawanda, NARO, Uganda.

Steduto, P., Hsiao, T. C., & Fereres, E. (2007). The conservative behavior of biomass-water productivity. *Irrigation Science, 25.*

Strudley, M. W., Green, T. R., & Ii, J. C. A. (2008). Tillage effects on soil hydraulic properties in space and time : State of the science. *Soil & Tillage Research, 99*, 4-48. http://dx.doi.org/10.1016/j.still.2008.01.007

Tenywa, J. S., Nyende, P., Kidoido, M., Kasenge, V., Oryokot, J., & Mbowa, S. (1999). Prospects and constraints of finger millet production in Eastern Uganda. *African Crop Science Journal, 7*(4), 569-583. http://dx.doi.org/10.4314/acsj.v7i4.27751

Troeh, R. T., Arthur, J. H., & Donahue, R. L. (1999). Soil and water conservation. Productivity and environmental protection 3rd edition Prentice Hall, New Jersey.Turner N.C. 2004 Agronomic options for improving rainfall-use efficiency of crops in dryland farming systems. *Journal of experimental botany, 55*(407), 2413-2425.

Van Vosselen, A., Verplancke, H., & Van Ranst, E. (2005). Assessing water consumption of banana: traditional versus modelling approach. *Agricultural Water Management, 74*, 201-218. http://dx.doi.org/10.1016/j.agwat.2004.11.005

Verhulst, N., Govaerts, B., Verachtert, E., Castellanos-Navarrete, A., Mezzalama, M., Wall, P., ... Sayre, K. D. (2010). Conservation Agriculture, Improving Soil Quality for Sustainable Production Systems? In R. Lal & B. A. Stewart (Eds.), *Advances in Soil Science: Food Security and Soil Quality* (pp. 137-208). CRC Press, Boca Raton, FL, USA.

VSN International. (2010). GenStatfor Windows® -13th Edition version 13.3.5165.

Wasige, E. J., Tenywa, M. M., Majaliwa, J. G .M., Byalebeka, J. B., Lufafa, A., & Gowing, J. (2004). Validation of meteorological models for estimating evapotranspiration in the Lake Kyoga basin, Uganda. Proceedings of the 20th conference of the Soil Science Society of East Africa. 2nd-6th December, Mt Elgon Hotel, Mbale Uganda, pp. 326-331.

Whyte, M. A., & Kyaddondo, D. (2006). 'We are not eating our own food here': Food security and the cash economy in eastern Uganda. *Land Degradation and Development, 17*, 173-182. http://dx.doi.org/10.1002/ldr.723

Wortman, C. S., &Eledu, C. A. (1999). Uganda's Agro-ecological Zones: A Guide for Planners and Policy Markers. Centro Internacional de International (CIAT).

Xu, L-K., & Hsiao, T. C. (2004). Predicted versus measured photosynthetic water-use efficiency of crop stands under dynamically changing field environments. *Journal of Experimental Botany, 55*(407), 2395-2411. http://dx.doi.org/10.1093/jxb/erh271

Yost & Eswaran, (1990). Major land resource areas of Uganda. A report submitted by soil management support services, agency for international development, Washington D.C, USA to USAID/Kampala. p. 29.

Appendix. Map of the Teso region in eastern Uganda showing the Districts of Amuria, Katakwi, Kumi and Soroti. Inset is the Map of Uganda

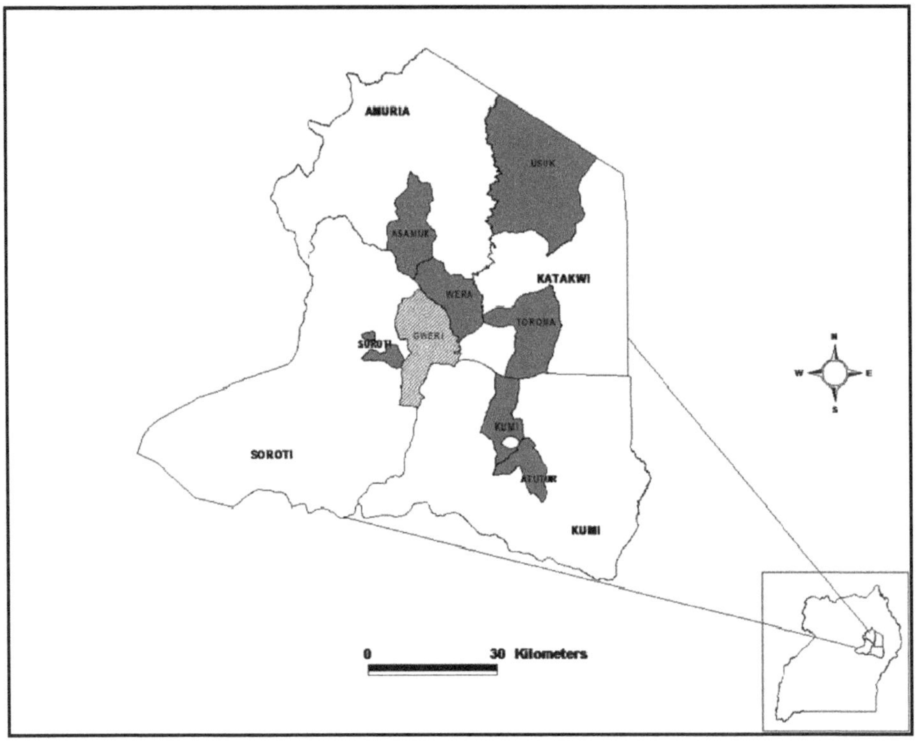

17

Determinants of Household Food Insecurity in Developing Countries Evidences From a Probit Model for the Case of Rural Households in Rwanda

Jean Baptiste Habyarimana[1]

[1] Ministry of Agriculture and Animal Resources, Kigali, Rwanda

Correspondence: Jean Baptiste Habyarimana, Ministry of Agriculture and Animal Resources, Kigali, P.O. Box 621, Rwanda. E-mail: kradrimn@gmail.com

Abstract

This study uses probit model to identify determinants of food insecurity among rural households in developing countries. The model used in this study, that allowed us to estimate coefficient and marginal effect for each independent variable vis-à-vis dependent variable, guarantees large applications among food security actors and policymakers to find out factors that significantly explain food insecurity and the level of their predictability. The ability of the model used to correctly classify food insecure and food secure households is good for the overall model and for households headed by males while it is fair for households headed by females. The empirical results show that rural households are more exposed to food insecurity than urban households. Gender disaggregation by the head of households shows that among food insecure rural households, the majority of them are headed by females. It also shows that the mean and median of predicted probability of becoming food insecure among rural households headed by males and females is 0.21 and 0.28 for mean and 0.15 and 0.24 for median respectively. This indicates that households headed by females are more likely exposed to food insecurity than those headed by males. However, as the majority of rural households in developing countries depend on agriculture, this study found that it is worthwhile for developing countries to adopt new agricultural technologies to urgently increase productivity and to implement and facilitate programs supporting rural households pathways to increase households' livelihood capacities.

Keywords: developing countries, food insecurity, marginal effect, predicted probability, probit model, rural households

1. Introduction

The majority of households in developing countries depend on the agricultural sector to provide the main source of food consumed in households, employment and income. Different research and debates have proved that agricultural transformation could be the engine of economic growth and development in these countries. Prior to the implementation of agricultural transformation policies, agricultural systems were more critical and did not gruantee achieving food security in these countries. However, the recent reforms and changes in agricultural policies have emphasized the adoption of new technologies for development of a more industrial and market oriented agricultural sector to develop non-agricultural sectors and meet food and nutrition security in these countries. Uncertainty surrounding the prediction of food security in developing countries has been the result of agricultural instability and represents a vital opportunity for agricultural modelers to analyze and provide information on what should be the pathways to render the agricultural sector more stable to ensure food security. The failure to meet food security for all has been contextualized as food insecurity in these countries. However, in one way or another, food insecurity hurts both households and the economy as a whole. Large unexpected increase of households' food insecurity prevalence would jeopardize the economic development at a micro level and unexpected changes would risk the livelihood of a large portion of the population that depends on agriculture. Excessive food insecurity threatens the economy as a whole, too. When food insecurity prevalence increases in an economy, the percentage of the population depending on the country's support also increases. For developing countries in which tax revenue collection does not keep pace with the required government expenditure, increased number of households with food insecurity characteristics expands government deficit and renders these countries more dependent on foreign aid and support; this remains one of the main challenges to address the issues of poverty in these countries. Achieving, however, food security for all, offers various benefits to the country. Achieving household food security for all is an important factor helping governments to make effective policies and efficiently plan for inclusive rural development. However, establishing household food security in developing countries has its own costs as it requires efforts and resources to sustainably transform the agricultural sector in the ways to meet food and nutrition security. The requirements include urgent adoption of new agricultural and food transformation technologies; policies to slow-down population growth; efficient meteorological infrastructures to predict climate and weather changes, developed infrastructures in producing and distributing improved agricultural inputs.

Nonetheless, food security is not a new term and its terminology has undergone evolution. The first attempt to measure food consumption and assess nutritional problems date back to the First World War (in 1936) and its interest was considerably increased during the Second World War (Food and Agriculture Organization [FAO], 2001). This tool has ever since experienced different improvements but it provides information on the food situation at the level of individual countries. Even though the Food Balance Sheet was developed, the world has experienced different challenges related to food crisis, food prices volatility climate change and food insecurity. Hence, to address these issues, World Food Summit held in early 1974, defined Food Security as the availability at all times of adequate world food supplies of basic foodstuffs to sustain a steady expansion of food consumption and to offset fluctuations in production and food prices (United Nations [UN], 1975). This definition, however, brought at least food security analysis at micro-level but with the system already established in early 1936; macro-level analyses were constrained by the lack of well disaggregated data. To reduce complexity and even impracticability of the food balance sheet at micro-level, different forms of food balance sheets were developed for emergency needs assessments. Even though the development of food balance sheets for emergency assessment was a success, structuring household food balance sheets was found to be more complex and they were not commonly used. The world waited 22 years after the first attempt to define food security in 1975 to come up with a comprehensive definition of food security. The world food summit of 1996 suggested that food security at individual, household, national, regional and global levels is achieved when all people, at all times, have physical and economic access to sufficient, safe and nutritious food to meet their dietary needs and food preferences for an active and healthy life (FAO, 1996). The adoption of this definition resulted in successfully estimating household food security. The well-known estimations emanating from this definition include: Food Consumption Score that provides estimations on households with poor, borderline and acceptable food consumption (World Food Programme [WFP], 2008) and Integrated Food Security Phase Classification which is a tool that classifies households by regions into five phases in respect to the severity of food and nutrition insecurity. However, with this definition, analysis of food security at micro-level experienced a new era in terms of modeling household food security and disentangling micro food insecurity determinants.

With rapid population growth, increased effects of climate change on agricultural stability and food security, food insecurity is defined in different ways by international organizations and researchers but without much difference in the basic concept. According to the World Bank, food insecurity is defined as "the lack of capability to produce food and to provide access to all people at all times to enough food for an active and healthy life" (WFP, 2009). In this context food insecurity is defined as limited or uncertain availability of nutritionally adequate and safe foods or limited or uncertain ability to acquire acceptable foods in socially acceptable ways. Food insecurity exists when people lack secure access to sufficient amounts of safe and nutritious food for normal growth and development, and leads an active and healthy life. In contrast, the most used definition of food security is the one that integrates food availability, access, utilization and stability for a good and healthy life. However, for the purpose of this research we adopted this definition and apply it at the household level to disentangle the determinants of food insecurity in developing countries especially for households located in rural areas.

The recent population growth in developing countries has led to the prodigious increase in food demand. The factor that agriculture remains the core sector for economic growth in these countries, the low agricultural technology and impact of climate change on agriculture and food security remain the main challenges to these countries to feed their populations and ensure food security. To address these challenges and meet food security, developing countries have recently adopted different strategies for agricultural transformation. A high number of new initiatives to strengthen food and nutrition security in these countries have been proposed and among those a significant number is using space technology and information whilst others aim at the development of integrated measurement methods and increased resilience. To meet food security in these countries, addressing household food acquisition problems and ensuring households' water availability, access and sustainability are emerging as factors that could lead to food security.

1.1 Problem Statement and Research Questions

A wide array of models have been used by agricultural economists and food security modelers to analyze the causes of food insecurity. With the evolution in the production and use of agricultural statistics, the area of modeling the causes of food insecurity has expanded to the household level with more emphasis on developing countries. However, modeling the causes of food insecurity in these countries has been more complex and challenging at the macro-level as well as at the micro-level due to the lack of timely and reliable data. The recent food insecurity modeling in these countries shows that different models using different data have been developed. The well-known models are those which try to link: climate change and agricultural production failure; sudden changes in weather conditions and crop and livestock production instability; food price volatility and household food consumption; food insecurity and limited access to markets, low infrastructure development as factors that hamper food distribution systems and lead to food price volatility; and low household income and limited food acquisition. However, all these models developed can be grouped into three categories: food insecurity as a function of climate change and failure of agricultural performance; food insecurity as function of low adoption of new agricultural technologies and food insecurity as function of low rural infrastructures and poverty. In essence, the practice of modeling food insecurity is a blend of socio-economic theories and econometric procedures and simplified representations of factors determining household food insecurity. When building household based empirical models to identify factors contributing to household food insecurity, this study tries to contribute to the recent literature by identifying determinants of food insecurity in developing countries. To undertake this task and make approximations on household food insecurity determinants, this study uses a Probit model. The main objective of this study, however, is to model household food insecurity determinants in developing countries.

Since 2000 with the implementation of Vision 2020, the government of Rwanda has adopted different policy and reforms emphasizing on transforming the agricultural sector from subsistence level into a modernized and more industrial and market oriented agricultural sector. Agricultural policy changes and reforms in the country have also contributed to the decline of yield and crop production instability, improvement in food distribution system, investing in new products that can generate more revenues and has facilitated households to access foodstuff at fair prices. Nonetheless, food security statistics show that about 81 percent of all households in Rwanda continue to depend on agriculture and 41 percent and 87 percent of households living in urban and rural areas respectively, depend on agricultural activities. Among households headed by males and females and living in rural areas, 86 percent and 90 percent of them are headed by males and females respectively. Statistics also show that 21 percent of households in the country are food insecure. When statistics on food insecurity are disaggregated, 6 percent of households living in urban and 23 percent of the households living in rural are food insecure. Among food insecure households about 96 percent of those are located in rural areas while only 4 percent of them are in urban areas. Among households located in rural areas, households headed by females are more food insecure when compared to those headed by males. 30 percent and 21 percent of households headed by females and males are food insecure respectively. This study, however, tries to answer the following question: What are the determinants of food insecurity among households located in rural areas of Rwanda, disaggregated by the gender of the head of households?

1.2 Literature Review

Different researches have been undertaken to determine factors that are behind food insecurity in developing countries. Those researches have shown that the causes of food insecurity in Africa and other third world countries include (Note 1): drought and extreme weather events; pest, livestock diseases and other agricultural problems; climate change; military conflicts; lack of emergency plans; corruption and political instability; cash crops dependence; human diseases and rapid population growth. The literature shows that the characteristics of food insecurity in these countries do not much differ from households located in urban areas to households located in rural areas.

To empirically characterize household food security, different works (WFP, 2009; Von Braun et al., 1992; Bonnard, 2000; Mucavele, 2001; Bahiigwa, 1999) have identified the determinants of household food security. The main determinants identified in those works include: household size; sex of the head of household; education level of the head of household; unemployment level; dependency ratio; land size; climate shocks such as floods, landslides, dry spell, rainfall deficit and drought; whether the household has enough income to purchase food at prevailing prices; food price volatility; access to agricultural credit; ownership of saving account; total income per adult equivalent; expenditure level on food and non-food items; asset possession; access to social services; owner of home garden; access to subsidized food; source of food; availability of food commodities and supply of food commodities; inadequate labor; inadequate land; not growing enough food during the season and soil infertility; poor health; lack of planting materials and low agricultural technology; rapid growth of the population; food availability instability and problem to supply markets with sufficient quantity to meet food demand; and problems related to food access and low purchasing power. In contrast, those works shows that the ability to achieve household food security is derived from the household's human capital, material, agricultural technology adaptation, farm size, land quality, agricultural extension services and institutional resources such as education and employment status, household demographics, assets, employment and saving, rural and urban agriculture, formal social assistance or direct transfer, informal social networks, access to clean water and sanitation, household food tastes and preferences.

Nevertheless, most of the works that have tried to identify the determinants of household food insecurity in developing countries have used logit and probit models (François, 2010; Abimbola et al., 2013; Mesfin, 2014). Even though, few of them have tried to disaggregate households by the gender of the head of the household and to model and estimate, ceteris paribus, the marginal effect of each independent variable but none in the literature has tried to predict a household's probability of becoming food insecure with respect to the gender of the head of household. This study, however, tries to contribute to the available literature by using probit model to approximate food insecurity determinants among households located in rural areas by the gender of the head of household; estimate marginal effects for each independent variable in each category of household head and predict household's probability to become food insecure.

2. Methodology

A Probit Regression analysis is applied to address the main objective of this study and answer to the question of this study. Food insecurity characteristics could not be the result of a single factor. Modeling food insecurity at micro-level requires the combination of different data. These data range from weather forecasting data, basic and current agricultural statistics and information, to population statistics. The main challenge in modeling the determinants of food insecurity in developing countries, however, is the lack of timely and reliable data on all these components. In the country where agriculture sector is the main sector of the economy, food insecurity increases as crop and livestock production instability increases. Accordingly, a household's probability to become food insecure increases given that the failure of agriculture production increases and economic shocks rise. In this study, we therefore try to model food insecurity determinants in developing countries and find out a household's probability to become food insecure. This approach enables us to apply the Probit regression model to investigate the links between agricultural related factors and food insecurity. This study therefore models food insecurity as a function of 24 agricultural related variables (see Table 1).

2.1 Empirical Model: Probit Regression

A probity model assumes that there is a latent, unobserved continuous variable Y* which can take all values in (-∞, +∞) and that determines the value of Y and includes believable error term distribution as well as realistic probabilities. However, the underlying latent model can be specified as follows (Chris, 2008; Bruce, 2013):

$$Y = \{1, Y_i^* > 0;\ 0,\ Y_i^* \leq 0\} \text{ and } Y^* = X'\beta + \varepsilon \tag{1}$$

However a probit model is based on latent variable and Y can be observed as an indicator for whether this latent variable is positive. In our study we specify that for Y; 1 = Food Insecurity, 0 = Food Security; X stands for vector of explanatory variables in table 1, β stands for coefficients and ε stands for random errors. In our study we use marginal effect at mean for both continuous and dummy variables. From the probit model we estimate marginal effect of "i" variable as the effect of a unit change of this "i" variable on the probability P(Y = 1| X = x), given that all other variables are constant and can be specified as follow:

$$\partial P(y_i = 1|x_i)/\ \partial x_i = \partial E(y_i|x_i)/\ \partial x_i = \varphi(x_i'\beta)\beta \tag{2}$$

Average Marginal Effect for each continuous variable in our model is computed as follows:

$$AME = \frac{1}{n}\sum_{i=1}^{n} \varphi(x_i'\beta)\beta \tag{3}$$

And the Average Marginal Effect for each dummy variable in our model is computed as follows:

$$AME = \frac{1}{n}\sum_{i=1}^{n}[\vartheta(x_i'\beta|x_i^k = 1) - \vartheta(x_i'\beta|x_i^k = 0)] \tag{4}$$

In this study we use equation (1) to compute predicted probability for a household to become food insecure when Food Insecurity is equal to one (1) and Food Security if equal to zero (0).

2.2 Data Description

To model household food insecurity determinants, we use the data set of Comprehensive Food Security and Vulnerability Analysis and Nutrition Survey conducted in 2012 in Rwanda. This survey covered a sample of

7500 households randomly selected around the country. The volume of the questionnaire of this survey was composed of 486 questions including demographics of the household members, characteristics of the home shelter, household income and questions related to food and nutrition security. We decided to use 96 questions from the survey questionnaire and some of them were transformed and combined to reach a total of 25 variables used in the model. The reduction of the number of questions has allowed as using the "level of importance of combined questions" in the survey questionnaire to explain household food insecurity characteristics. This study used food consumption groups to create the dependent variable in our model. The transformation of this variable followed the creation of two distinct groups: Food Insecure Households with the value of one (1) and Food Secure Households with the value of zero (0). Food Insecure households groups all households with borderline and poor food consumption score while Food Secure households groups all households with acceptable food consumption score.

Table 1. Explanatory variables and their measures

Variable name	Variable description	Measure
FI	Food Insecurity	Insecure = 1& Secure =0
AHH	Age of Household Head	Age in years
HHE	Household Head Education	Able to ready & write =1 other=0
HS	Household Size	Number persons in a household
HLS	Household's Land Size	Scale of 0 to 1
HLAL	Household's Land Amendment Level	Scale of 0 to 1
HFA	Household's Farm Animal	Scale of 0 to 1
DM	Distance to Market	Minutes
FA	Food Assistance	Scale of 0 to 1
HAI	Household Asset Index	Scale of 0 to 1
HFAL	Household Food Acquisition Level	Scale of 0 to 1
HFAP	Household Food Acquisition Problem	Scale of 0 to 1
NLA	Number of Livelihood Activities	Scale of 0 to 1
HSL	Household Spending Level	Scale of 0 to 1
MFE	Monthly Food Expenditure	In Rwandan Francs
PCE	Per Capita Expenditure (Year)	In Rwandan Francs
GS	Government Support	Scale of 0 to 1
OPC	Own Production used for Own Consumption	% of food from own-production
MCHFC	Market Contribution to Household Food Consumption	% of food from market
LSPC	Land Suitability per Cell	% of suitable land per cell
SEIV	Soil Erosion Index per Village	% of land vulnerable to erosion per village
CSI	Coping Strategy Index	Reduced coping strategy index
MAC	Membership to Agricultural Cooperative	Member=1 & Not member=0
AL	Agricultural Loan	Acquired=1 Not acquired=0

3. Results

Table 2. Food insecurity profile in Rwanda

	All	Urban	Rural	Rural (Head of HH) Male	Rural (Head of HH) Female
Households depending on agriculture (%)	81	41	87	86	90
Households depending on informal sale, agro seller and artisan (%)	14	36	11	11	10
Households depending on salaries and business (%)	5	23	2	3	1
Food Insecure Households (%)	21	6	23	21	30
Food Secure Households (%)	79	94	77	79	70

Table 3. Descriptive Statistics for Rural Households

	Mean		Correlation FI		Covariance FI		Male and Female Combined		
	Male*	Female*	Male*	Female*	Male*	Female*	Mean	Cor.	Cov
FI	0.21	0.30	1	1	0.16	0.20	0.23	1.00	0.18
AHH	44.37	54.46	-0.0004	0.0093	0.00	0.00	47.30	0.03	0.00
HHE	0.70	0.37	-0.0864	-0.0921	-0.02	-0.02	0.61	-0.11	-0.02
HS	5.33	3.88	-0.0487	-0.0777	-0.04	-0.06	4.91	-0.08	-0.07
HLS	0.92	0.89	-0.2001	-0.1837	-0.04	-0.04	0.91	-0.20	-0.05
HLAL	0.33	0.29	-0.1011	-0.1256	-0.01	-0.01	0.31	-0.12	-0.01
HFA	0.78	0.69	-0.1125	-0.1527	-0.02	-0.03	0.75	-0.13	-0.02
DM	78	76	0.0485	0.0472	1.36	1.68	77.50	0.05	1.42
FA	0.01	0.01	-0.0203	0.0396	0.00	0.00	0.01	0.00	0.00
HAI	0.23	0.17	-0.2875	-0.2224	-0.01	-0.01	0.21	-0.28	-0.01
HFAL	0.34	0.31	-0.1511	-0.1908	-0.01	-0.02	0.33	-0.17	-0.01
HFAP	0.19	0.24	0.2543	0.2895	0.02	0.03	0.20	0.27	0.03
NLA	1.73	1.55	-0.0808	-0.0455	-0.02	-0.01	1.68	-0.08	-0.02
HSL	0.29	0.23	-0.2983	-0.3009	-0.02	-0.02	0.27	-0.31	-0.02
MFE	13940	9539	-0.1632	-0.1554	-986	-915	12673	-0.17	-1031
PCE	105297	96896	-0.1113	-0.0612	-10752	-15650	102872	-0.08	-12142
GS	0.02	0.02	-0.0104	-0.0292	0.00	0.00	0.02	-0.02	0.00
OPC	32.63	32.46	-0.1521	-0.1604	-1.24	-1.52	32.58	-0.15	-1.31
MCHFC	62.80	59.01	0.1166	0.0665	0.95	0.64	61.70	0.09	0.81
LSPC	43.10	44.12	-0.1308	-0.1808	-1.56	-2.37	43.40	-0.14	-1.78
SEIV	6.38	6.28	0.0905	0.1795	0.15	0.33	6.35	0.12	0.20
CSI	5.17	6.78	0.2196	0.2560	0.68	0.96	5.64	0.24	0.78
MAC	0.18	0.13	-0.1201	-0.1449	-0.02	-0.02	0.16	-0.13	-0.02
AL	0.18	0.11	-0.0986	-0.0764	-0.02	-0.01	0.16	-0.10	-0.02

Note. FI = Food Insecurity where (Insecurity = 1 and Security = 0); Male* = Households headed by Males; Female* = Households headed by Females.

Table 4. Summary results of probit regression model for the determinants of rural household food insecurity

	Households headed by Males			Households headed by Females			All households Combined		
FI	Coef.	.dy/dx	Sig.	Coef	.dy/dx	Sig.	Coef	.dy/dx	Sig.
AHH	0.053	0.0117		0.0307	0.0092		0.0455	0.0112	
HHE	-0.009	-0.0021		0.0031	0.0009		-0.0008	-0.0002	
HS	0.032	0.0072	**	0.0240	0.0072		0.0325	0.0080	***
HLS	-0.112	-0.0251	**	-0.0487	-0.0145		-0.0892	-0.0219	**
HLAL	-0.035	-0.0079		-0.3103	-0.0925		-0.1290	-0.0316	
HFA	-0.137	-0.0305	**	-0.1184	-0.0353		-0.1325	-0.0325	***
DM	0.000	0.0001		0.0003	8.82E-05		0.0004	0.0001	
FA	-1.147	-0.2558	*	-0.6425	-0.1915		-0.9785	-0.2399	**
HAI	-1.903	-0.4244	***	-0.5999	-0.1788		-1.4851	-0.3642	***
HFAL	-0.403	-0.0900	***	-0.3180	-0.0948		-0.3840	-0.0942	***
HFAP	0.815	0.1817	***	0.9389	0.2798	***	0.8479	0.2079	***
NLA	-0.085	-0.0189	**	-0.0391	-0.0116		-0.0740	-0.0181	**
HSL	-2.248	-0.5014	***	-2.3095	-0.6882	***	-2.2839	-0.5600	***
MFE	-1.72E-05	-3.83E-06	***	-0.000027	-8.04E-06	***	-0.0000186	-4.55E-06	***
PCE	-3.86E-07	-8.61E-08		-2.86E-08	-8.54E-09		-3.13E-07	-7.69E-08	
GS	-0.271	-0.0603		-0.4802	-0.1431		-0.3424	-0.0840	
OPC	-0.004	-0.0008		-0.0047	-0.0014		-0.0041	-0.0010	**
MCHFC	0.005	0.0011	**	0.0023	0.0007		0.0037	0.0009	**
LSPC	-0.004	-0.0009	***	-0.0044	-0.0013	***	-0.0043	-0.0010	***
SEIV	0.015	0.0034	**	0.0297	0.0088	**	0.0201	0.0049	***
CSI	0.006	0.0014	*	0.0124	0.0037	**	0.0089	0.0022	***
MAC	-0.231	-0.0516	***	-0.4542	-0.1353	***	-0.2897	-0.0710	***
AL	-0.041	-0.0091		0.0462	0.0138		-0.0287	-0.0070	
N. of Ob.		4149			1634			5783	
Waldchi2(23)		1454			506.98			1938.68	
Prob > chi2		0.00			0.000			0.000	
ROC Curve*		0.8090			0.7907			0.8048	

Note. Sig. = Significant: ***, **, * at 1%, 5% and 10%; Coef = Coefficient ; dy/dx = Marginal Effect; ROC Curve* = Area under Receiver Operating Characteristics curve.

Table 5. Descriptive statistics of predicted probabilities for a rural household to become food insecurity

FSG	variable	mean	min	max	cv	p50	iqr
Insecure	Households headed by Males	0.372	0.001265	0.885	0.530	0.359	0.287
	Households headed by Females	0.439	0.029424	0.904	0.475	0.435	0.350
	Households headed by Males and Females	0.393	0.000572	0.896	0.511	0.380	0.310
Secure	Households headed by Males	0.162	1.19E-15	0.834	0.950	0.114	0.200
	Households headed by Females	0.219	4.84E-20	0.852	0.813	0.177	0.246
	Households headed by Males and Females	0.178	5.86E-16	0.825	0.912	0.130	0.217
Total	Households headed by Males	0.205	1.19E-15	0.885	0.899	0.151	0.252
	Households headed by Females	0.281	4.84E-20	0.904	0.753	0.237	0.314
	Households headed by Males and Females	0.227	5.86E-16	0.896	0.855	0.175	0.274

Note. Max: maximum; min: minimum; cv: coefficient of variation; p50: median; iqr: Interquartile range.

Table 6. Predicted probabilities for a rural household to become food insecure

Predicted Probability	HHs Headed by Males	HHs Headed by Females	All HHs
Less than 0.11	37.89	23.75	33.77
Between 0.11 and 0.21	21.50	19.89	20.77
Between 0.21 and 0.31	13.93	17.01	15.29
Between 0.31 and 0.41	10.34	12.18	11.00
Between 0.41 and 0.51	7.81	9.06	8.04
Between 0.51 and 0.61	4.17	7.47	5.29
Between 0.61 and 0.71	2.31	6.12	3.3
Between 0.71 and 0.81	1.71	3.55	2.09
Between 0.81 and 0.91	0.34	0.92	0.45
Between 0.91 and 1.00	0.00	0.06	0.00
Observations (Households)	4,149	1,634	5,783

Note. HHs = Households.

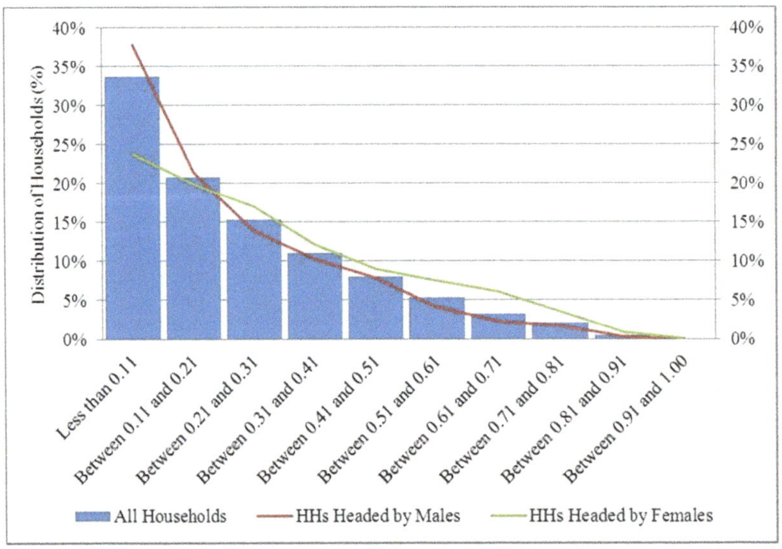

Figure 1. Predicted probabilities for a rural Household to become food insecure

5. Discussion

When compared to urban population, rural areas accommodate the majority of households in developing countries. With a predominant agricultural sector, most households located in rural areas depend on agricultural activities. Even though some food is sourced from market, the large share of food consumed in rural households is sourced from their own production. The own production consists of agricultural production. In most cases, this production is not processed due to the lack of efficient food transformation technology. However, food availability and access in rural areas is the result of agricultural performance and food acquisition from the market is function of income from agriculture. Both, food availability and access and income from agricultural activities in rural areas are highly linked to the performance of agricultural season. Any instability in agricultural production results in food availability instability and reduces households' income from agriculture and limit household to food diversification. However, agriculture performance in rural areas is function of household land size, land amendment, rainfall performance, adoption of agriculture cooperatives to efficiently channelize improved agricultural inputs and agricultural loans. In one way or another government and/or non-government support and assistance have a role to ensure food security in the case of food needs due to hazards. For subsistence, household structural counts a lot to define a household's livelihood, expenditure and income level. When it is on the side of food utilization, education matters. Rural households whose members have a low education level are more likely to be not aware on the efficient utilization of available food and are less likely to access to skilled livelihoods activities that can be one way to generate an income and increase household's purchasing power and saving level. It is in this context that we model rural household food insecurity as a function of households characteristics, household income, household food acquisition, household's agricultural characteristics, household's assistance related variables (see Table 1).

5.1 Food Insecurity Profile in Rwanda

Even though the agricultural system in Rwanda is predominantly rainfall fed, agriculture is the core sector of Rwandan economy. Food security statistics shows that about 81 percent of all households in the country depend on agriculture. When urban and rural areas are compared, 41 percent of urban households and 87 percent of rural households depend on agricultural activities. Among households headed by males and females and living in rural areas, 86 percent and 90 percent of them are headed by males and females respectively. Statistics also show that 21 percent of households in the country are food insecure. When statistics on food insecurity are disaggregated, 6 percent of households living in urban and 23 percent of households living in rural areas are food insecure. Among food insecure households about 96 percent of them are located in rural areas while only 4 percent of them are in urban areas. Among households located in rural areas, households headed by females are more food insecure when compared to those headed by males. Among households headed by females, 30 percent of them are food insecure, while those headed by males 21 percent of them are food insecure. This profile shows that food insecurity is a serious rural problem in Rwanda when compared to food insecurity situation in urban areas. This becomes more sensitive in rural areas when households headed by males are compared to those headed by females. This, however, indicates that rural household food insecurity is more prevalent in household headed by females which may be the results of the fact that more households headed by females depend on agriculture (90 percent) when compared to those headed by males (86 percent).

5.2 Description of Rural Households

In this study we use three types of descriptive statistics: mean, correlation and covariance. Even though covariance is similar to correlation, we use covariance to understand how independent variables are related to dependent variable in the model and to find out if each independent variable and dependent variable tend to increase or decrease together or tend to increase as the other variable decreases. We use correlation to assess the strength of linear relationship between each of dependent variable with dependent variable. To make a difference, correlation (Note 2) is computed when data are converted to a standardized scale of -1 to +1 while covariance is expressed in units that vary and not standardized to scale of -1 to +1.

The mean age of head of households shows that the mean age of males heads of households is below the mean age of females heads of households, the education level shows that males heads of households are more educated than females heads of households, mean household size show that households with large size are those headed by males and are also endowed with large land when compared to those headed by females, land amendment is large among households headed by males, households headed by males have farm animal more than those headed by females, household asset index for households headed by males is greater than that of households headed by females, household food acquisition level among households headed by males is greater than that of households headed by females, households headed by males have less household food acquisition problems when compared to households headed by females, expenditure on food shows that households headed by males spend more on food than those headed by females, per capita expenditure among households headed by males is greater than that of households headed by females, the share of food sourced from market is large among households headed by males when compared to those headed by females, households headed by males are more participative in agricultural cooperatives and in requesting agricultural loans when compared to households headed by females, and the coping strategy index is high among households headed by females when compared to those headed by males. Even though rural households have less food acquisition problem, the general observation shows that within a scale of 0 to 1, Household's land amendment level, household food acquisition level, household spending level, membership in agricultural cooperatives and agricultural loan access are very low.

For households headed by males, food insecurity has a weak relationship with household land size, household asset index, household food acquisition problem, household spending level and coping strategy index; while it has a very weak relationship with age of household head, distance to market, food assistance, government support, market contribution to household food consumption, soil erosion index per village, household head's education level, household size, household's land amendment level, household's farm animal, household food acquisition level, number of livelihood activities, monthly food expenditure, per capita expenditure (year), own production used for own consumption, land suitability per cell, membership to agricultural cooperative and agricultural loan. For households headed by females and when households headed by males and females are combined, only household asset index, household food acquisition problem, household spending level and coping strategy index have a weak relationship with food insecurity while other variables have a very weak relationship with food insecurity.

For households headed by males and females and when all households are combined, the covariance between the age of the head of household, distance to market, food assistance, household food acquisition problem, government support, market contribution to household food consumption, soil erosion index per village, coping strategy index and food insecurity indicated a positive relationship while the covariance between household head education, household size, household's land size, household's land amendment level, household's farm animal, household asset index, household food acquisition level, number of livelihood activities, household spending level, monthly food expenditure, per capita expenditure (year), own production used for own consumption, land suitability per cell, membership to agricultural cooperative, agricultural loan and food insecurity indicated a positive indicate a negative relationship.

5.3 Determinants of Food Insecurity Among Rural Households

In this study we use both coefficients and average marginal effects. Coefficients are used to understand how changes in independent variables explain changes in dependent variable while average marginal effects are used to estimate the effect of a unit change in each independent variable on the probability of being food insecure, given that all other variables are hold constant. To identify determinants of food insecurity among rural households, we use three models. Probit models for households headed by males and those headed by females are used to identify how determinants of food insecurity differ depending on the gender of household head. In addition to this, we use the overall Probit model (when all households headed by males and those headed by females are combined).

For households headed by males, we found that significant variables to explain changes in rural household food insecurity are: household asset index, household food acquisition level, household food acquisition problem, household spending level, monthly expenditure on food, land suitability (fertility) per cell and membership to agricultural cooperatives significant at 1%; household size, land size, farm animal, number of livelihood activities, soil erosion index per village and market contribution to household food consumption significant at 5%; and food assistance and coping strategy index significant at 10%. When one variable is considered and keeping all other variable constant, average marginal effect shows that a unit increase in household size, household food acquisition problem, market contribution to household food consumption, soil erosion index per village, and coping strategy index would results in increasing the household's probability of becoming food insecure by 0.7 percent, 18 percent, 0.1 percent, 0.34 percent, and 0.14 percent respectively while a unit increase in household asset index, household food acquisition level, household spending level, monthly expenditure on food, land suitability (fertility) per cell, membership to agricultural cooperatives, land size, farm animal, number of livelihood activities, food assistance would result in decreasing household's probability to become food insecure. The area under ROC cover of 0.81 shows that the ability of probit model constructed for households headed by males to correctly classify food insecure and food secure households is good (Note 3).

For households headed by females, we found that significant variables to explain changes in rural household food insecurity are: household food acquisition problem, household spending level, monthly expenditure on food, land suitability (fertility) per cell and membership to agricultural cooperatives significant at 1%; soil erosion index per village and coping strategy index significant at 5%. When one variable is considered and keeping all other variable constant, average marginal effect shows that a unit increase in household food acquisition problem, soil erosion index per village, and coping strategy index would results in increasing the household's probability of becoming food insecure by 28 percent, 0.9 percent and 0.3 percent respectively while a unit increase in household spending level, monthly expenditure on food, land suitability (fertility) per cell and membership to agricultural cooperatives would result in decreasing household's probability to become food insecure. The area under ROC cover of 0.79 shows that the ability of probit model constructed for households headed by females to correctly classify food insecure and food secure households is fair.

When all households are combined, we found that significant variables to explain changes in rural household food insecurity are: household size, household's farm animal, household asset index, household food acquisition level, household food acquisition problem, household spending level, monthly food expenditure, land suitability per cell, soil erosion index per village, coping strategy index and membership to agricultural cooperative significant at 1%; and household size, household's farm animal, household asset index, household food acquisition level, household food acquisition problem, household spending level, monthly food expenditure, land suitability per cell, soil erosion index per village, coping strategy index and membership to agricultural cooperative significant at 5%. When one variable is considered and keeping all other variable constant, average marginal effect shows that a unit increase in household size, household food acquisition problem, market contribution to household food consumption, soil erosion index per village and coping strategy index would results in increasing the household's probability of becoming food insecure by 0.8 percent, 20 percent, 0.1 percent, 0.5 percent and 0.2 percent respectively while a unit increase in household's farm animal, household asset index, household food acquisition level, household spending level, monthly food expenditure, land suitability per cell, and membership to agricultural cooperative, household's land size, food assistance, number of livelihood activities, and own production used for own consumption would result in decreasing household's probability to become food insecure. The area under ROC cover of 0.80 shows that the ability of probit model constructed for households headed by both the males and females to correctly classify food insecure and food secure households is good.

5.4 Predicted Probability for a Household to be Classified a Food Insecure Household

In this study we use estimated probit models to predict the probability that each household in the sample has to become food insecure. The interpretation of this probability is stated as follows "as a household's predicted probability is getting higher and approaching an unit (1) the same household is more likely to become food insecure while as a household's predicted probability is low and approaching zero (0) the more the same household is likely to become food secure". The mean of predicted probability for households previously (from survey results) categorized as food insecure to stay food insecure is 0.372, 0.439 and 0.393 probability for households headed by males, females and when all households are combined respectively. This indicates that the majority of those households are more likely to become food secure. In the other way round, the mean of predicted probability for households previously (from survey results) categorized as food secure to become food insecure is 0.162, 0.219 and 0.18 probability for households headed by males, females and when all households are combined respectively. This indicates that the majority of those households are more likely to stay food secure and few of them are likely to move to the category of food insecure households. Households categorized as food insecure have less variability relative to their means, while households categorized as food secure have more variability relative to their means. The median of predicted probabilities among food insecure households is greater than the median of predicted probabilities among food secure households. This indicates that a half of food insecure households have about 0.38 probability and less of staying food insecure while a half of food secure households has about 0.13 and less probability of becoming food insecure. The overall observation shows that the mean of predicted probability of about 0.22 indicates that there is a low probability for household to become food insecure. Even though, the majority of households located in rural areas have the low probability of becoming food insecure, the predicted probability shows that households headed by females are more likely to become food insecure than those headed by males (see Table 6 and Figure 1).

6. Conclusion and Policy Implications

Most of the works that have tried to identify determinants of food insecurity in developing countries have used logit and probit models. Even though, few of them have tried to disaggregate households by the household heads' gender and to model and estimate marginal effect for each independent variable when others variables are assumed to be constant but none in the literature have tried to predict a household's probability to become food insecure in respect with the gender of the head of household. This study, however, contributes to the available literature by using probit model to approximate food insecurity determinants among households located in rural areas by head of household's gender; estimate marginal effects for each independent variable in each category of the head of households (males and females) and predict a household's probability to become food insecure. Hence, this study investigated the determinants of food insecurity among rural households by the gender of the head of household.

Using descriptive statistics, this paper demonstrates that food insecurity is more prevalent in rural areas than in urban areas. The paper also shows that among food insecure rural households the majority of them are headed by females. This paper assumed that food insecurity in rural areas and by the gender of the head of household is function of the age of household head, household head education, household size, household's land size, household's land amendment level, household's farm animal, distance to market, food assistance, household asset index, household food acquisition level, household food acquisition problem, number of livelihood activities, household spending level, monthly food expenditure, per capita expenditure (year), government support, own production used for own consumption, market contribution to household food consumption, land suitability per cell, soil erosion index per village, coping strategy index, membership to agricultural cooperative and agricultural loan. Using Probit model, this study found that significant factors to explain changes in rural household food insecurity are: household size, household's farm animal, household asset index, household food acquisition level, household food acquisition problem, household spending level, monthly food expenditure, land suitability per cell, soil erosion index per village, coping strategy index and membership to agricultural cooperative significant at 1%; and household size, household's farm animal, household asset index, household food acquisition level, household food acquisition problem, household spending level, monthly food expenditure, land suitability per cell, soil erosion index per village, coping strategy index and membership to agricultural cooperative significant at 5%. The analysis of average marginal effects by household head gender shows that the number of significant factors to influence food insecurity at household level whether positively or negatively is different among households headed by males and those headed by females. Households headed by males have more determinants than those headed by females. The predicted probability for a household to become food insecure shows that households headed by females are more likely exposed to food insecurity than those headed by males.

This study also shows that the majority of rural households in developing countries depend on agriculture and that subsistence agriculture system is the predominant agricultural system applied in those areas. As the probit model used in this study tried to model agricultural related variables as factors affecting household food insecurity, we found that it is worthwhile for developing countries to adopted new agricultural technologies to increase productivity and to implement and facilitate programs supporting rural households to increase their livelihood capacities.

Acknowledgments

This study was made possible with the support of different participants. Our special thanks go to World Food Programme – Rwanda, Vulnerability Analysis and Mapping section that provided us with the data set of the 2012 Rwanda Comprehensive Food Security and Vulnerability Analysis and Nutrition Survey. The data set received was in SPSS format but it was converted it into STATA format to generate the main empirical results presented in this study. We also wish to express our profound gratitude to different technical advisors whose comments and guidance were very useful to this study. Thanks go to Ms Sen RITWIKA who accepted to go through all this study and provided her comments and proposed amendments to make it more scientific; Mr. Eric NTAKIRUTIMANA who also accepted to go through this study and provided his comments; Mr. Emmanuel MUNYEMANA who contributed to the literature review of this study and provided comments on the methodology; and Ms Francine UWAYO who made comments to improve the abstract of this study. While acknowledging the contributions from a wide variety of participants, the errors in this paper, however, are those of ours and neither Government Institutions nor Non-Government Institutions are responsible for them and we are responsible of the content of this study.

References

Abimbola, O. A., & Kayode, A. A. (2013). Food Insecurity Status of Rural Households during the Post-planting Season in Nigeria. *Journal of Agriculture and Sustainability*.

Arega, B. (2012). Determining Food Security Indicators at Household Level in Drought prone Areas of the Amhara Region of Ethiopia: The Case of Lay Gayint District. *Ethiopian Journal of Environmental Studies and Management, 5*(4). http://dx.doi.org/10.4314/ejesm.v5i4.11

Bahiigwa, B. A. G. (1999). *Household food security in Uganda: An empirical analysis*. Retrieved from http://www.eldis.org/fulltext/EPRC_Bahiigwa.pdf

Bonnard, P. (2000). *Assessing Urban Food Security: Adjusting the FEWS Rural Vulnerability Assessment Framework to Urban Environment*. Retrieved from http://pdf.usaid.gov/pdf_docs/PNACJ249.pdf

Bruce, E. H. (2013). *ECONOMETRICS*. University of Wisconsin. Department of Economics. Retrieved from http://www.ssc.wisc.edu/~bhansen/econometrics/Econometrics.pdf

Chris, B. (2008). *Introductory Econometrics for Finance*. Retrieved from http://catdir.loc.gov/catdir/samples/cam033/2001037930.pdf

Food and Agriculture Organization [FAO]. (1996). *Rome Declaration on World Food Security and World Food Summit Plan of Action*. Retrieved from http://www.fao.org/DOCREP/003/W3613E/W3613E00.HTM

Food and Agriculture Organization [FAO]. (2001). *Food Balance Sheet: A Hand Book*. Retrieved from ftp://ftp.fao.org/docrep/fao/011/x9892e/x9892e00.pdf

François, J. C. (2010). What are the key factors of food insecurity among Senegalese farmers? *African Journal of Food Science, 4*(8), 477-485.

Frehiwot, F. (2007). *Food Insecurity and Its Determinants in Rural Households in Amhara Region*. School of Graduate studies, Faculty of Business and Economics, Department of Economics, Addis Ababa University

Mesfin, W. (2014). Determinants of Households Vulnerability to Food Insecurity in Ethiopia: Econometric analysis of Rural and Urban Households. *Journal of Economics and Sustainable Development, 5*(24). Retrieved from http://www.iiste.org/Journals/index.php/JEDS/article/view/17506/17926

Mucavele, F. G. (2001). *A Vulnerability and Food Security Study of Urban Maputo*. Mozambique. FANRPAN, Harare.

Shiferaw, F., Richard, L., & Christy, G. (2003). *Determinants of Food Security in Southern Ethiopia*. Retrieved from http://ageconsearch.umn.edu/bitstream/22010/1/sp03fe02.pdf

Teshome T. (2010). *Food Security Situation in Ethiopia: The Case of Amhara National Regional State*. Retrieved from http://repo.lib.ryukoku.ac.jp/jspui/bitstream/10519/1033/2/r-kz-rn_050_01_005.pdf

United Nations [UN]. (1975). *Report of the World Food Conference held in Rome in November 1974*. Retrieved from http://www.fao.org/docrep/meeting/007/F5340E/F5340E03.htm

Von, B., J., Bouis, H., Kumar, S., & Pandya-Lorch, R. (1992). *Improving Food Security of the Poor: Concept, Policy and Programmes*. International Food Policy Research Institute, Washington, D.C. Retrieved from http://www.ifpri.org/sites/default/files/publications/oc23.pdf

World Food Programme [WFP]. (2008). *Food Consumption Score*. Retrieved from http://documents.wfp.org/stellent/groups/public/documents/ena/wfp196627.pdf

World Food Programme. (2009). *Summary of Food Security and Vulnerability in Selected Urban Centers of Ethiopia*. Addis Ababa, Ethiopia. Retrieved from http://documents.wfp.org/stellent/groups/public/documents/ena/wfp221386.pdf

Notes

Note 1. http://www.harvesthelp.org.uk/causes-of-food-insecurity-in-african-and-other-third-world-countries.html

Note 2. In this study we considered that 0.00-0.2; 0.2-0.4; 0.4-0.6; 0.6-0.8 and 0.8-1 indicate a very weak, weak, moderate, strong, and very strong relationship

Note 3. In this study we consider that 0.5-0.6; 0.6-0.7; 0.7-0.8; 0.8-0.9; 0.9-1 area under Receiver Operating Characteristic curve indicate that the ability of our model to correctly classify food insecure and food secure households is "Fail"; "Poor"; "Fair"; "Good" and "Excellent" respectively. http://gim.unmc.edu/dxtests/roc3.htm

Appendix

Appendix A

Correlation		A	B	C	D	E	F	G	H	I	J	K	L	M	N	O	P	Q	R	S	T	U	V	W	X
FI	A	1																							
AHH	B	0.03	1																						
HHE	C	-0.11	-0.29	1																					
HS	D	-0.08	0.03	0.15	1																				
HLS	E	-0.20	0.09	0.11	0.17	1																			
HLAL	F	-0.12	0.03	0.10	0.12	0.16	1																		
HFA	G	-0.13	0.09	0.02	0.16	0.19	0.26	1																	
DM	H	0.05	-0.04	-0.04	-0.03	-0.04	0.05	-0	1																
FA	I	0.00	0.04	-0.04	0.00	-0.01	-0.03	-0.06	0.05	1															
HAI	J	-0.28	-0.08	0.28	0.25	0.33	0.21	0.17	-0.09	-0.1	1														
HFAL	K	-0.17	-0.08	0.07	0.04	0.19	0.13	0.07	-0.01	0.01	0.17	1													
HFAP	L	0.27	0.04	-0.13	-0.01	-0.27	-0.10	-0.11	0.07	0.07	-0.31	-0.1	1												
NLA	M	-0.08	-0.13	0.06	0.16	0.03	0.07	0.12	0.00	0.02	0.17	0.07	-0	1											
HSL	N	-0.31	-0.08	0.21	0.29	0.28	0.26	0.19	0.00	0.01	0.46	0.24	-0.19	0.21	1										
MFE	O	-0.17	-0.07	0.11	0.24	0.10	0.07	0.02	-0.06	-0.01	0.31	0.12	-0.06	0.15	0.36	1									
PCE	P	-0.08	0.00	0.06	-0.01	0.08	0.04	0.01	-0.03	-0.01	0.18	0.06	-0.06	0.05	0.23	0.26	1								
GS	Q	-0.02	0.02	-0.01	0.01	-0.03	0.00	0.04	0.01	0.04	-0.04	-0.01	0.05	0.05	0.04	0.00	-0.01	1							
OPC	R	-0.15	0.19	0.01	0.01	0.25	0.09	0.19	0.00	-0.06	0.11	0.13	-0.24	-0.07	0.08	-0.24	-0.01	-0	1						
MCHFC	S	0.09	-0.21	0.03	0.05	-0.21	-0.05	-0.13	-0.04	0.02	-0.02	-0.07	0.17	0.08	0.01	0.30	0.03	-0	-0.9	1					
LSPC	T	-0.14	-0.03	0.01	0.03	0.06	-0.22	-0.11	-0.12	0.03	0.08	0.04	-0.12	-0.01	0.05	0.10	0.05	-0	0.07	-0	1				
SEIV	U	0.12	0.04	-0.01	0.01	-0.05	0.20	0.11	0.08	-0.01	-0.05	-0.01	0.07	0.03	0.00	-0.03	-0.03	0.05	-0.1	0.06	-0.7	1			
CSI	V	0.24	0.04	-0.11	0.03	-0.28	-0.13	-0.12	0.02	0.11	-0.32	-0.15	0.50	0.00	-0.22	-0.06	-0.07	0.02	-0.2	0.14	-0	0.06	1		
MAC	W	-0.13	0.00	0.09	0.11	0.12	0.17	0.11	-0.02	0.02	0.18	0.06	-0.10	0.02	0.17	0.06	0.02	0.02	0.06	-0.03	0.02	-0.02	-0.08	1	
AL	X	-0.10	-0.08	0.09	0.12	0.12	0.11	0.05	0.01	0.03	0.17	0.04	-0.04	0.11	0.30	0.16	0.11	0.11	0.00	0.02	0.03	0.00	-0.07	0.1	1

18

Nitrogen and Potassium Topdressing in Maize Intercropped with *Brachiaria Ruziziensis*

Edleusa P. Seidel[1], Willian Reis[1], Bruna Bartzen[2], Joice Knaul[2], Alisson D. Cassiano[2], Jonatan E. Fredrich[2]

[1] Professor of agronomic department, in Parana West University, Marechal C. Rondon, Parana State, Brazil

[2] Agronomic academic in Parana West University, Marechal C. Rondon, Parana State, Brazil

Correspondence: Edleusa Pereira Seidel, 1777 Pernambuco Street, Paraná West University, Marechal C. Rondon, Paraná State, Zip code 85960-000, Brazil.. E-mail: edleusaseidel@yahoo.com.br

Abstract

Maize has been intercropped with different forage species, especially *Brachiaria* spp., for different purposes and at different times: after maize harvest, as a food source for cattle; and exclusively for the production of straw to maintain a no-tillage system, which is the main purpose of intercropping in western Paraná. The objective was to investigate the effects of different rates of nitrogen (N) and potassium (K) topdressing on grain yield and components of yield for maize intercropped with *Brachiaria ruziziensis*. The experiment was conducted in Paraná State, Brazil. The experiment used a completely randomized block factorial design (4 × 3) with four replications. Treatments for the first factor consisted of sowing maize intercropped with *Brachiaria ruziziensis* with four rates of N topdressing (0, 80, 100 and 120 kg ha^{-1}) applied as urea. For the second factor were rates of K topdressing (0, 30 and 50 kg ha^{-1}), applied in the form of potassium chloride. During at sowing fertilization was performed with of 30, 26 and 50 kg ha^{-1} of N, phosphorus and K, respectively. The following variables were evaluated: plant height, first ear height, stem diameter, number of tillers per plant, mass of 1,000 grains, grain yield. The results showed that application of N topdressing promoted increased grain yield of maize intercropped with *Brachiaria ruziziensis* until N rate of 85 kg ha^{-1}. Isolated application of K did not influence the variables studied.

Keywords: conservation system, foliar fertilization, integrated system, nutrient interactions

1. Introduction

In recent years, the methods used to grow maize in Brazil have undergone major technological changes, resulting in significant increases in grain yield. Among these changes, the use of fertilizers applied to gain high yields, disease control and improvements in field management were aimed at improving the quality of the soil; other new practices included crop rotation and no-tillage cultivation (Coelho et al., 2007) and intercropping between cultures (Cruz et al., 2009).

The cultivation of maize intercropped with forage species, especially those of the genus *Brachiaria*, has been used for different purposes and at different times: after maize harvest, as a food source for cattle; and exclusively for the production of straw to maintain a no-tillage system (Crusciol et al., 2010), which is the main purpose of intercropping in western Paraná. The use of plants intercropped with maize promotes soil biological activity; this results in increased soil organic matter, and because intercropped plants have a high C/N ratio, soil cover is guaranteed for a longer period so that erosion is reduced (Borghi et al., 2006). Forage *Brachiaria* species are excellent for use in intercropping with maize materials; these species provide a good amount of dry matter production ensuring good coverage for no-tillage cultivation (Silva et al., 2005).

The demand for nutrients will increase in intercropped systems, requiring greater care in handling them. Nitrogen (N) is taken up in greater quantities by maize, which influences grain yield more strongly than other changes. Approximately 75 % of the available soil N is translocated to grains, concentrating N at the rate of approximately 20 kg N ha^{-1} per ton of harvested grain (Sousa & Lobato, 2004).

In Brazil, Silva et al. (2006 a) reported that the main factors contributing to low grain yield of maize were the mishandling of N fertilizers and the insufficient application of it. Meanwhile, Duete et al. (2008) reported that the amount of the nutrient used in the cultivation of maize in Brazil averages 60 kg ha^{-1}, while the averages in

China (130 kg ha^{-1}) and the United States (150 kg ha^{-1}) were much higher.

After N, maize also absorbed large quantities of potassium (K), whereas, an average of 30 % of applied N is exported into the grain. While potassium is not part of any organic compound within the plant, it is important in the synthesis and metabolism of carbohydrates and also contributes to greater photosynthetic activity (Prado 2008). Potassium improves the quality and nutritional value of food products. A study by Rodrigues et al. (2014) used fertilization doses of potassium chloride coated pellets; they observed an increase in chlorophyll content with an increasing dosage of potassium oxide. When K is in the form of K^+ ions in plant tissue, it returns to the soil very quickly after the death of plants (Pavinato, 2008).

Although N is the most important nutrient, it does not work independently in the nutrition of plants; for example, the shortage of phosphorus, K and sulfur can alter the efficiency of N application, reducing grain yield (Usherwood & Segars 2001). There is a positive correlation between nitrogen and potassium, because the potassium acts on the activation of the nitrate reductase enzyme (Venkajesan & Ganapathy, 2004; Silva et al., 2011).

The hypothesis of this study when is intercropping *B. ruziziensis* with maize will result in an increased demand for N and K, and that increased doses of these elements will result in an increase in maize yield. The object of this study was to investigate the effects of different rates of N and K topdressing on grain yield and yield components of maize intercropped with *B. ruziziensis*.

2. Material and Methods

The experiment was carried out during season 2012/13 in the experimental field on the UNIOESTE, in Marechal Cândido Rondon municipality, Paraná State, Brazil at 24°31'S, 54°01'W at an elevation of 420 m. Soil was classified as typic oxisol (Perferic Red Latossol). The particle size determination performed according to the hydrometer Bouyoucos. The results were: Clay 763 g kg^{-1}, silt, 136 g kg^{-1}, and sand 101 g kg^{-1}, according to the methodology of EMBRAPA (2006). Before the experiment, soil samples were collected in the 0–0.10 m and 0.10–0.20 m soil layers (Table 1). The solution for extraction of phosphorus was Mehlich-1, and to aluminum, calcium and magnesium the extractor was KCl. The methodology used to determine organic matter was Walkley-Black.

Table 1. Chemical characterization of the Oxisol soils at depths of 0–0.10 m and 0.10–0.20 m

Depth	pH	P	MO	H+Al	$Al^{(2)}$	K^+	Ca^{+2}	Mg^{+2}	SB	CTC	V	Al
m	$CaCl_2$	mg dm^{-3}	g dm^{-3}	------------			$cmol_c\ dm^{-3}$			------------	------ %	------
0-0.10	4.62	12.62	27.34	6.72	0.20	0.45	3.69	1.44	5.58	12.30	45.37	3.46
0.10-0.20	4.81	9.05	24.61	5.79	0.15	0.30	4.09	1.48	5.87	11.66	50.34	2.49

The experiment was a completely randomized block factorial (4 × 3) design with four replications. Treatments consisted of sowing maize intercropped with pasture (*B. ruziziensis*) with four rates of N topdressing (0, 80, 100 and 120 kg ha^{-1}), applied in the form of urea (45 % N). The second factor included three rates of K topdressing (0, 30 and 50 kg ha^{-1}), applied in the form of potassium chloride (60 % K_2O). The size of the plot per treatment was 4.5 x 7 meters, totaling 31.5 m^2.

Hybrid maize seeds were sown mechanically in a no-tillage system and spaced 0.90 m between rows and 0.20 m between plants. At the moment of sowing were applied a base fertilization at dose de 300 kg ha^{-1} of a formulated 10–20–20 with nitrogen, phosphate and potassium oxide, respectively. *Brachiaria ruziziensis* was sown manually in line between maize rows on the same day of maize sowing, using 8.0 kg ha^{-1} of encrusted seeds.

When the maize reached the developmental stage five sheets developed at 25 days after emergence, proceeded to the topdressing with manual application of nitrogen and potassium in the form of urea and potassium chloride, respectively. Weeds were manually removed where necessary, and fungicides and insecticides were applied to prevent diseases and insect damage, according recommendations. Not irrigations were applied during the whole season of the crop.

At the silking stage, 10 plants in the middle rows were randomly selected from each plot and measurement of growth characteristics as follows: stem diameter, measured to the nearest centimeter (cm) at 1/3 the base of the maize plant; plant height, the maize plant measured to the nearest centimeter from the base to top; and first ear height. The number of tillers per plant was also determined.

The ears of maize were harvested by hand at 16 weeks after plants. In each plot were harvested plants of floor area, and was discarded 0.5 m in the bedside of each installment and the two side lines, harvesting the two central lines of each plot. The ears of maize were conducted for laboratory and dried in the shade; then, 10 ears were randomly chosen to evaluate the ear diameter and ear length. After, the ears of maize were threshed manually to evaluate the mass of 1,000 grains and grain yield. The mass of 1,000 grains was determined using an average of eight samples of 100 randomly selected grains. The yield per plot was determined by weighing the grain on a semi-analytical balance, correcting for wet weight for 130 g kg^{-1} of weighing and then estimating the grain yield in kg ha^{-1}.

Data were subjected to analysis of variance ($p<0.05$) using SAEG statistical software and with significant differences noted in rates of N accumulated; regression analysis was performed using the analysis of variance option in the SAEG statistical program (SAEG 2007).

3. Results and Discussion

The results of this study showed that application of different rates of nitrogen topdressing significantly affected ($p < 0.05$) the stem diameter, ear height, mass of 1,000 grain and grain yield; and no statistical difference was observed for first ear height, number of tillers and ear diameter. Regarding the effect of nitrogen rates in topdressing the stem diameter and ear height, the data set were adjusted a first-degree regression equation. The linear increase in steam diameter and ear height was observed with increased nitrogen fertilization, which shows that the dose of 120 kg ha^{-1} was not enough to achieve the maximum these characteristics (Figure 1 a and b).

(a)

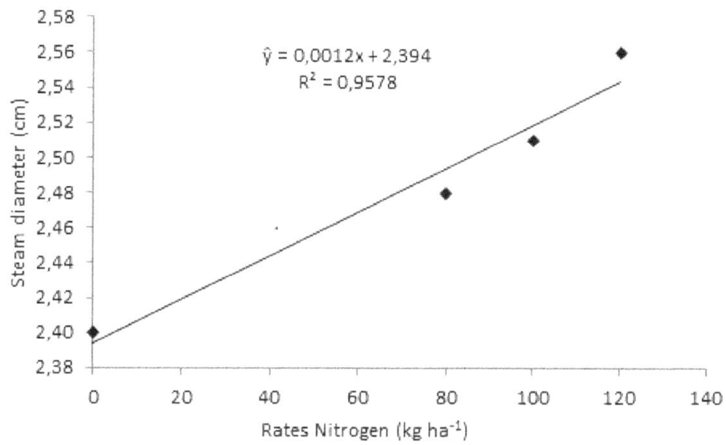

(b)

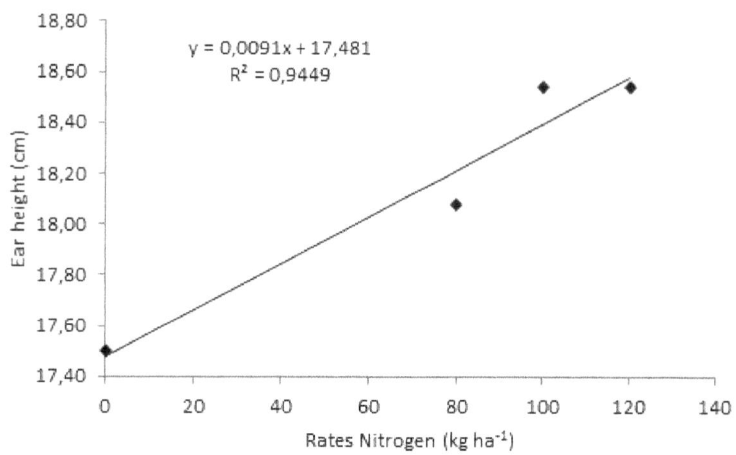

Figure 1. Stem diameter (a) and ear height (b after application of nitrogen on topdressing in the maize intercropped with *Brachiaria ruziziensis*

The effect of nitrogen rates in topdressing the mass of 1,000 grains and grain yield, the data set were adjusted a quadratic regression model. The estimate of the maximum mass of 1,000 grains (353 g) was obtained with application of 89 kg ha^{-1} of N topdressing and the estimate of the maximum grain yield (8,580 kg ha^{-1}) corresponded the application of 85 kg ha^{-1} of N topdressing (Figure 2 a and b).

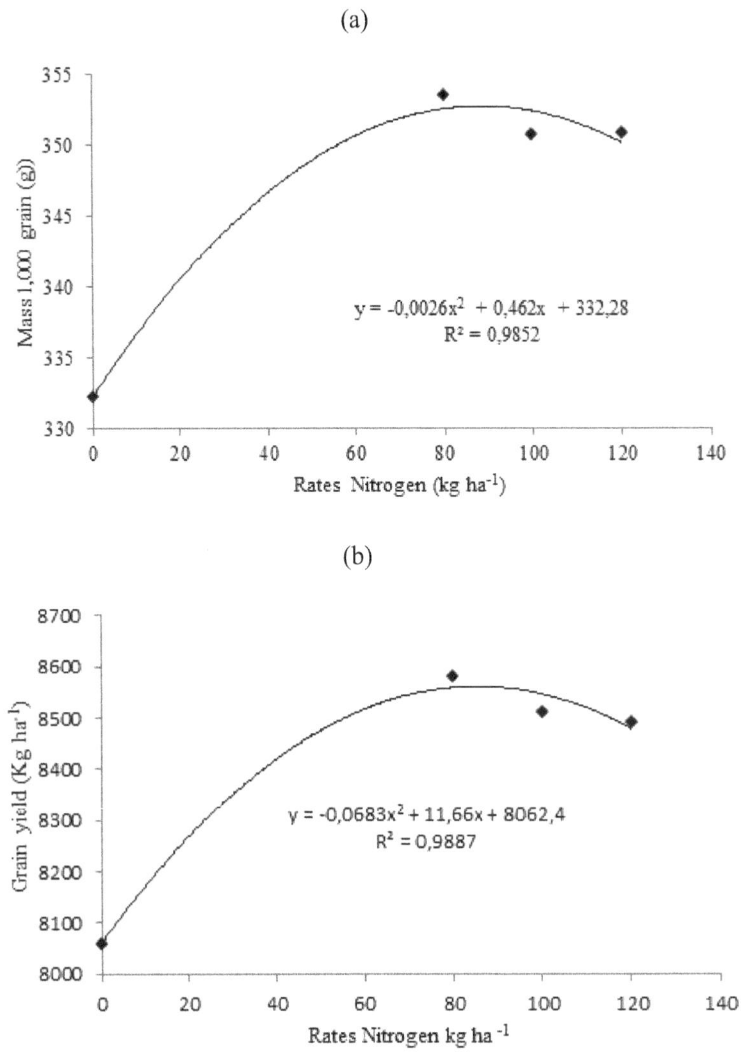

Figure 2. Mass of 1,000 grains (a) and grain yield (b) after application of nitrogen on topdressing in the maize intercropped with *Brachiaria ruziziensis*

Analyzing the data it was found that the application of nitrogen on topdressing promoted an increase of 6.10 % in mass 1,000 grains and 6.43% at grain yield. Remember that were applied as base fertilization 30 kg ha^{-1} of nitrogen, then the total nitrogen was 119 and 115 kg ha^{-1} for mass 1,000 grain and grain yield, respectively, which is consistent with previously published literature, therefore, the consortium *Brachiaria ruziziensis* with maize not demanded greater amounts of N, by maize.

When the isolated effect of K application rates was analyzed, no significant differences were found in the variables studied here. In Table 2 the average results for stem diameter, ear height, mass of 1,000 grains and grain yield after application of potassium on topdressing in the maize intercropped with *Brachiaria ruziziensis*. The results yield components are presented in Table 2.

Table 2. Average results for stem diameter (a), ear height (b), mass of 1,000 grains and grain yield after application of potassium on topdressing in the maize intercropped with *Brachiaria ruziziensis*

K rates on topdressing	Stem diameter	Insertion height of the ear	Number of tillers	Ear diameter	Ear height	Mass 1,000 grains	Grain yield
(kg ha^{-1})	(cm)	(cm)		(cm)	(cm)	(g)	(kg ha^{-1})
0	2.46^{ns}	127.15^{ns}	32.18^{ns}	5.13^{ns}	18.21^{ns}	344.77^{ns}	8366^{ns}
30	2.49	127.87	29.68	5.16	18.03	346.82	8416
50	2.52	128.28	30.18	5.15	18.25	348.90	8467

Analyzing the data for effect of interaction N × K verified only significant difference ($p<0,05$) for first ear height. The maize response for interaction was adequately represented by the polynomial second degree regression model. When not applied potassium top dressed the maximum ear height was achieved with the application of N at rates of 127 kg ha^{-1}. When applied potassium top dressed at rates 30 e 50 kg ha^{-1} the maximum ear height was achieved with the application of N at rates of 88.4 and 72 kg ha^{-1}, respectively (Figure 3). An interaction between N and K was also observed by Stromberger et al. (2008), who found that an increase in the K application rate promoted more vigorous growth in plants when those plants received doses of N.

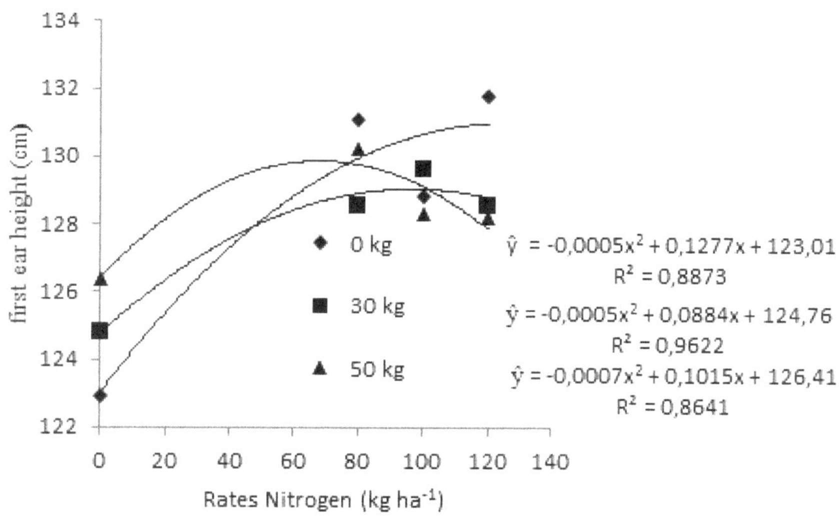

Figure 3. Ear height and maize stem diameter as a function of nitrogen and potassium application rates during intercropping of maize with *Brachiaria ruziziensis*

4. Discussion

These results suggest that the application of N influenced the stem diameter, because the availability of nitrogen is directly involved in cell division and growth and photosynthesis.Grains yield had increase because the application of nitrogen fertilizer in plants affects the absorption of other nutrients. This is because the nitrogen increases number of new roots, and these in turn increases the absorption of n end all nutrients and dry matter production (Dovale et al., 2012). The positive effect of nitrogen on productivity was explained by the linear increase in leaf N content (Valderrama et al., 2011). N while not limited by other factors, the greater n availability increases the potential of the plant to define increasing grain mass (pöttker & wiethölter 2004).

Similar effects of N rates on the grains yield were reported by Bravin et al. (2014), but the maximum was obtained with the dose of 165 kg ha^{-1}. Fernandes et al. (2005) studied N accumulation rates for six maize cultivars and found no significant differences for the variables of stem diameter and ear height, but obtained similar results for the mass of 1,000 grains and grain yield. The maximum grain yield was obtained with a N application rate of 110 kg ha^{-1} in the current experiment. Furthermore, Silva et al. (2006 b), worked with N fertilizer application rates in crops seeded prior to planting maize and also reported an increase in ear height up to dose of 203 kg ha^{-1}.

The results show low efficiency in the use of nitrogen fertilization coverage on the grain yield maize. This response is related to the high productivity observed in the treatment that received at sowing, only a base fertilizer 30 kg ha^{-1} N; and none N on topdressing, but was able to maintain a high grain yield (8060 kg ha^{-1}). This result indicated that the soil had a good natural stock of N and soil conditions and climate were favorable to mineralization in no-tillage system. According to Souza et al. (2008), maize in low nutrient supply conditions is capable of increasing nutrient use efficiency. Lara-Cabezas and Pádua (2007), working with N efficiency applied on topdressing in the maize intercropped with Brachiaria ruziziensis found that the consortium did not negatively affect the assimilation of N, but promoted a synergy in nitrogen uptake, and for better understanding of this indeed further studies should be conducted in the rhizosphere.

Corroborating the result Deparis et al. (2007), whose study employed thirteen different doses of N and seven doses of K; they also found significant differences for mass of 1,000 grain and grain yield although differences in doses of N and K did not affect the position of ear insertion. They reported that the response N fertilization for mass of 1,000 grain and grain yield was linear up to150 kg ha^{-1}.

One explanation why there were no effects of potassium fertilization on components of production and maize productivity is that the potassium content in the soil at depth of 0-0.10 m was considered high (> 0.4 cmolc dm^{-3}), as shown in Table 1; especially in no-tillage systems that favors the maintenance of the straw and the increase of K. In review, other authors also found no significant differences for K doses (Ceretta & Pavinato, 2003; Brunetto et al., 2005; Pavinato et al., 2008; Valderrama et al., 2011)

5. Conclusions

The application of nitrogen topdressing promoted increased grain yield of maize intercropped with *Brachiaria ruziziensis* until N the rate of 85 kg ha^{-1}. Isolated application of K did not influence the variables studied here.

Acknowledgments

The Araucaria Foundation for Scientific Initiation funded this research.

References

Borghi, E., Crusciol, C. A. C., & Costa, C. (2006a). Corn yield, spacing and intercropping modalities with Brachiaria brizantha in no-tillage system. *Energy in Agriculture, 21*(3), 19-33. http://dx.doi.org/10.1590/S0100-83582008000300011

Borghi, E., Costa, N. V., Crusciol, C. A. C., & Mateus, G. P. (2008b). Influence of the spatial distribution of maize and *Brachiaria brizantha* intercropping on the weed population under no-tillage. *Weed Plant, 26*(3), 559-568. http://dx.doi.org/10.1590/S0100-83582008000300011

Bravin, M. P., & Oliveira, T. K de. (2014) Nitrogen fertilization in maize and xaraés palisade grass under no-tillage and conventional tillage in an agrosilvopastoral system. *Brazilian Agricultural Research, 49*(10), 762-770. http://dx.doi.org/10.1590/S0100-204X2014001,000003

Brunetto, G., Gatiboni, L. C., Santos, D. R. dos, Saggin, A., & Kaminski, J. (2005). Critical level and crop yield response to potassium in a typic Hapludalf under no-tillage. *Brazilian Journal of Soil Science, 29*(4), 561-571. http://dx.doi.org/10.1590/S0100-06832005000400009

Ceretta, C. A., & Pavinato, P. S. (2003). Fertilization online or by throwing the no tillage. In: Soil Fertility Course In No Tillage, 6., 2003, Ibirubá, RS. *Annals.* Passo Fundo: Aldeia Norte, 23-35.

Coelho, A. M., França, G. E., Pitta, G. V. E., Alves, V. M. C., & HernanI, L. C. (2007). Nutrition and fertilization of corn. *Embrapa Maize and Sorghum, Maize Cultivation,* 96, 1-11. Retrieved from http://www.cnpms.embrapa.br/publicacoes/publica/2007/circular/Circ_96.pdf

Crusciol, C. A. C., Soratto, R. P., Borghi, E., & Mateus, G. P. (2010). Benefits of integrating crops and tropical pastures as systems of production. *Better Crops, 94*(2), 14-16. http://dx.doi.org/10.1590/S0100-06832011,000600033

Cruz, S. C. S., Pereira, F. R. S., Bicudo, S. J., & Santos, R. S. (2009). Maize and Brachiaria decumbens under diferente soil tilage in farming pasture integration *Acta Sientiarum Agronomy. Maringá, 31*(4), 633-639. http://dx.doi.org/10.4025/actasciagron.v31i4.915

Deparis, G. A., Lana, M. C. do, & Frandoloso, J. F. (2007). Row spacing and nitrogen and potassium fertilization in covering. *Acta Scientiarum Agronomy. Maringá, 29*(4), 517-525.

Dovale, J. C., Fritsche-Neto, R., Bermudez, F., & Miranda G. V. (2012). Genetic effects of traits associated to nitrogen use efficiency in maize. *Brazilian Agricultural Research, 47*(3), 385-392.

http://dx.doi.org/10.1590/S0100-204X2012000300010

Duete, C. R. R., Muraoka, T., Silva, E. C. da, Trivelin, P. C. O., & Ambrosano, E. J. (2008). Nitrogen fertilization management and nitrogen (^{15}N) utilization by corn crop in Red Latosol. *Brazilian Journal of Soil Science, 32*(1), 161-171. http://dx.doi.org/10.1590/brag.2014.021

Embrapa. National Center for Soil Research. (2006). *Brazilian System of Soil Classification.* (2nd ed., p. 316). Brasília: EMBRAPA/DPI.

Fernandes, F. C. V., Buzetti, S., Arf, O., & Andrade, J. A. C. (2005). Nitrogen doses, use and efficiency in six corn cultivars. *Brasilian Journal of Maize and Sorghum, 4*(2), 195-204.

Lara-Cabezas, W. A. R., & Pádua, R. V. (2007). Efficiency and distribution of nitrogen applied on side-dressing in cornintercropped with brachiaria ruziziensis, cultivated in Santa Fe system. *Bragantia, 66*(1), 131-140. http://dx.doi.org/10.1590/S0006-870520070001,00016

Pavinato, P. S., Ceretta, C. A., Girotto, E., & Moreira, I. C. L. (2008). Nitrogen and potassium in irrigated corn: technical and economic analysis of the fertilization. *Rural Science, Santa Maria, 38*(2), 358-364. http://dx.doi.org/10.1590/S0103-84782008000200010

Prado, R. de M. (2008). *Nutrição de Plantas*. Editora unesp, São Paulo.

Portes, T. A., Carvalho, S. I. C., Oliveira, I. P., & Kkluthcouski, J. (2000). Growth analysis of a Brachiaria cultivar sole and intercropped with cereals. *Brazilian Agricultural Research, 35*(1), 1349-1358. http://dx.doi.org/10.1590/S0100-204X2000000700009

Pöttker, D., & Wiethölter, S. T. (2004). iming and methods of nitrogen application for corn under no-tillage. *Rural Science, 34*(4), 1015-1020. http://dx.doi.org/10.1590/S0103-84782004000400007

Rodrigues, M. A. de C., Buzetti, S., Teixeira Filho, M. C. M., Garcia, C. M. P., & Andreotti, M. (2014). Fertilization with coated KCl in corn in the Savannah. *Journal of Agricultural and Environmental Engineering, 18*(2), 127-133. http://dx.doi.org/10.1590/S1415-43662014000200001

Saeg. *System for Statistical Analysis*, Versão 9.1: Arthur Bernardes Fundation- UFV – Viçosa (2007).

Silva, E. C., Buzetti, S., Guimarães, G. L., LazarinI, E., & SÁ, M. E. (2005a). Rates and timing of nitrogen application in corn under no-tillage on a red latosol. *Brazilian Journal of Soil Science, 29*, 353-362. http://dx.doi.org/10.1590/S0100-06832005000300005

Silva, E. C. Da, Muraoka, T., Buzetti, S., & TriveliN, P. C. (2006b). Nitrogen management in corn under no-tillage with different cover crops in a Rhodic Hapludox soil. *Brazilian Agricultural Research, 41*(3), 477-486. http://dx.doi.org/10.1590/S0100-204X2006000300015

Silva, D. A. da, Vitorino, A. C. T., Souza, L. C. F. de, Gonçalves M. C., & Rosco, R. (2006c). Previous crop and nitrogen fertilization in corn under no-tillage system. *Brazilian Journal of Maize and Sorghum, 5*(1), 75-88.

Silva, S. M. da, Oliveira, L. J., Faria, F. P., Reis, E. F. dos, & Carneiro, M. A. C. (2011c). Activity of the enzyme reductase nitrate in corn cultivated under different levels of nitrogen and potassium fertilization. *Rural Science, 41*(11), 1931-1937. http://dx.doi.org/10.1590/S0103-84782011005000136

Sousa, D. M. G. de, & Lobato, E. (2004). Liming and fertilization for annual crops and semiperenes. In D. M. G. de Sousa & E. Lobato (Eds.). *Savana:* liming and fertilization. *Planaltina:Embrapa Cerrados, 1*(1), 283-315.

Souza, L. S. et al. (2006a). Allelopathic Effects of Brachiaria decumbens on the Initial Development of Seven Crops. *Weed Plant, 24*(4), 657-668.

Souza, E. de F. C. de, & Soratto, R. P. (2006 b). Effect of sidedressing nitrogen sources and doses on out-of-season corn in no-tillage system. *Brazilian Journal of Maize and Sorghum, 5*(3), 387-397.

Souza, L.V. de, Miranda, G. V., Galvão, J. C. C., Eckert, F. R., Mantovani, E. E., Lima, R. O., Guimarães, L. J. M. (2008). Genetic controlo f grain yield and nitrogen use efficiency in tropical maize. *Brazilian Agricultural research, 43*(1), 1517-1523. http://dx.doi.org/10.1590/S0100-204X20080011,00010

Stromberger, J. A., Tsai, C. Y., & Huber, D. M. (1994). Interactions of potassium with nitrogen and their influence on growth and yield potential in maize. *Journal of Plant Nutrition, 17*(1), 19-37. http://dx.doi.org/10.1080/01904169409364707/

Usherwood, N. R., & Segars, W. I. (2001). Nitrogen interaction with phosphorus and potassium for optimum crop yield, nitrogen use effectiveness, and environmental stewardship. *The Scientific Word, 1*(1), 57-60.

http://dx.doi.org/10.1100/tsw.2001.97

Valderrama, M., Buzetti, S., Benett, C. G. S., Andreotti, M., & Teixeira-Filho, M. C. M. (2011). NPK sources and doses on irrigated corn under no-till system. *Tropical Agricultural Research, 41*(2), 254-263. http://dx.doi.org/10.5216/pat.v41i2.8390

Viana, E. M., & Kiehl, J. C. (2010). Rates of nitrogen and potassium on growth of wheat plants. *Bragantia, Campinas, 69*(4), 975-982. http://dx.doi.org/10.1590/S0006-8705201,0000400024

Venkatesan, S., & Ganapthy, M. N. K. (2004). Nitrate reductase activity in tea as influenced by various levels or nitrogen and potassium fertilizers. *Communications in Soil Science and Plant Analysis, 35*(9), 1283-1291. http://dx.doi.org/10.1081/CSS-120037546

Wendling, A., Eltz, F. L. F., Cubilla, M. M., Amado, T. J. C., & Mielniczuk, J. (2008). Recommendation of potassium fertilization for no-till wheat, corn and soybean in Paraguay. *Brazilian Journal of Soil Science, 32*(1), 1929-1939. http://dx.doi.org/10.1590/S0100-06832008000500014

19

The Effects of Household Wealth on Adoption of Agricultural Related Climate Change Adaptation Strategies in Zambia

Elias Kuntashula[1], Lydia M. Chabala[2], Terence K. Chibwe[3] & Peter Kaluba[2]

[1] University of Zambia, Dept. of Agricultural Economics & Extension, Lusaka, Zambia

[2] University of Zambia, Dept. of Soil Science, Lusaka, Zambia

[3] The International Institute of Tropical Agriculture IITA-Zambia, 32 Poplar Road, Avondale, Lusaka, Zambia

Correspondence: Elias Kuntashula, University of Zambia, Dept. of Agricultural Economics & Extension, Lusaka, Zambia. E-mail: ekuntashula@unza.zm

Abstract

Despite increased emphasis targeting climate change adaptation strategies towards the poorer sections of communities, few adoption studies assess the uptake of these practices by these groups in a systematic and comprehensive manner. In this study, we used a combination of participatory rapid approaches and quantitative principal component analysis to determine each household's wealth status, and to assess the relationship between wealth and the adoption of various agricultural related climate change adaptation strategies. Evidence from a random sample of 1231 households across six districts of Zambia showed that the more well-endowed households than their poorly endowed counter parts, adopted most of the climate change adaptation strategies. The relatively well-endowed households had a high probability of 10.6%, 9.5%, 7.1%, and 5.5% to embrace crop rotation, minimum tillage, fertiliser trees and change crop varieties due to climate change, respectively, than their poorly endowed counter parts. Most, if not all of these strategies require some level of resource investment hence only those households who could afford such resources are most likely to adopt them. The influence of household resource endowment on the uptake of several climate change adaptation strategies call for the subsidising of the relatively poor endowed households to encourage adoption of these strategies among this category of farmers.

Keywords: Adaptation strategies, principal component analysis, wealth indicators, Zambia

1. Introduction

In recent times climate change has taken centre stage in most parts of the world. In countries like Zambia where the majority of rural communities depend on agriculture, knowing what modifications to agricultural practices have been put in place to cope with climate change could form a good platform in informing policy. Agricultural adaptation strategies could be viewed in the context of new technologies for improving the well being of farmers in the midst of vulgaries associated with climate change. Due to the increasingly high poverty levels of small scale farmers in most developing countries including Zambia (whose poverty level stands at about 60% of total population, World Bank, 2014), policy makers have placed great emphasis on the development and dissemination of new technologies that could improve the well being and enhance food security among the rural poor. In the same vein, it would be important to know whether the relatively poor farmers are embracing adaptation strategies (both proposed by scientists and those emanating from farmers themselves) that enhances their livelihood.

Grandin (1988) noted that poor farmers were likely to have different needs, problems and resources than their better-off counter parts, and thus were likely to have a reduced capacity to adopt new technologies. When the argument is adoption of adaptation strategies on the basis of their potential to mitigate climate change and thus alleviate rural poverty, it is pertinent to know how welfare affects adoption. Thus the objective of this study was to assess the adoption rates of climate change adaptation strategies between relatively well and poorly resource endowed farming households in rural communities of Zambia. Among the adaptation strategies analysed included; changing crop varieties, changing planting dates, crop diversification, crop rotation, planting of fertiliser trees and implementing some form of conservation farming (minimum tillage through potholing and ripping).

Farmers who face challenges such as prolonged drought spells due to climate change would opt for early maturing crop varieties or would indeed switch their crop planting dates to synchronise with the availability of the rains. Similarly, farmers would want to grow more crops at the same time (diversifying) to ensure that some of the crops would survive the negative effects of climate change. Crop rotation involves varying successive crops in a definite order on the same ground, especially to avoid depleting the soil and to control weeds, diseases, and pests that could have built up as a result of changes in climate. Fertiliser trees are leguminous in nature and are planted in order to achieve one or more of the aims of natural fallows within a shorter time or on a smaller area (Kuntashula & Mungatana, 2015). The trees help in increasing the productivity of smallholder farmers as well as improving on-farm biodiversity that is important in mitigating climate change. According to Arslan et al. (2014) conservation farming has the technical potential to help farmers adapt to climate change and thus contribute to increased food security.

Relatively well resource-endowed farmers have a better ability to cope with production and price risks and consequently are more willing to adopt new technologies than their less resource-endowed counterparts (Hardaker et al., 1997). The well resource-endowed farmers may also be willing to try out adaptation strategies that involve redeployment of available resources. It can thus be argued that novel climate change adaptation strategies might be undertaken by relatively well to do households, thus condemning the poorer households into even deeper poverty levels. The distribution of productive assets among farmers is different (Langyintuo & Mungoma, 2008). The resources required to adopt climate change adaptation strategies will also differ according to inherent requirement of a strategy. The availability of resources at the farmers' disposal, the competition for these resources among farm enterprises, including the need to deploy them in the uptake of climate change adaptation strategies becomes an intricate balancing act by the farmer. The study attempted to seek answers to questions such as what type of adaptation strategies are being adopted by the different wealth groups? This is the first contribution to the body of knowledge on climate change and adaptation strategies in Zambia. However, before this question was answered the study was faced with yet another challenge on how best to categorise farmers into the well-endowed and poorly endowed categories. This study used a combination of farmer participatory and quantitative methods to place households into wealth categories and then proceeded to interrogate the causal relationship between wealth and adoption of various climate change adaptation strategies. This served as the study's second contribution to literature on climate change in the country. Generally, the study found that household wealth has an effect on adoption of most of the climate change adaptation strategies.

The rest of the paper is organized as follows: the next section presents the study areas and data sources. This is then followed by the analytical framework and estimation techniques. Thereafter the results and discussions are presented immediately after this section. Finally the paper gives major conclusions and some policy recommendations.

2. Methodology

2.1 Study Areas and Data Sources

The study areas for this study included six districts (Choma and Sinazongwe in the Southern Region, Nyimba and Petauke in the Eastern Region, and Serenje and Mpika in the Northern Region) of Zambia. The data used in this study was collected in two steps. First, a Climate Change supplemental survey was implemented in 2013 by the Indaba Agricultural Policy Research Institute (IAPRI) and University of Zambia (UNZA) on 1231 households who participated in a nationally representative Rural Agricultural Livelihood Survey of 2012 (RALS 2012) conducted by the Zambia Central Statistics Office (CSO) and IAPRI. The 1231 households were part of a bigger sample that was randomly selected across all the 72 districts of Zambia to serve as a panel for RALS beginning 2012, and to be interviewed every four years. The panel sample was selected through the CSO's national randomisation process (Kuntashula et al., 2014).

Second, complementary data on wealth indicators from the study districts was collected during the farmer participatory wealth ranking exercises of 2013 and 2014. The farmer participatory wealth ranking indicator identification exercises involved community meetings composing key informants of about 15 – 20 farmers in each district. The groups were given the objective of defining wealth groups based on wealth endowment within their communities. First, they listed all wealth indicators in the community. Second, the informants defined wealth categories (very poor, poor etc) of community members. Third, the informants described the differences between the wealth categories based on the wealthy indicators. The list of wealth indicators from these exercises (complemented with indicators reviewed from literature) formed the basis for computing the wealth indices of the individual households involved in the RALS (2012) and IAPRI/UNZA (2013) supplemental survey.

2.2 Conceptualising Wealth and Other Factors Affecting Adoption of Climate Change Adaptation Strategies

Households have various assets that define their wealth status. These assets fall in various categories including human, natural, physical, financial capital and social capital. Human capital represents the skills, knowledge, ability to labour and good health that together enable people to pursue different livelihood strategies and achieve their livelihood objectives. Natural capital such as land refers to the natural resource stocks from which resource flows and services useful for livelihoods are derived. Physical capital comprises the basic infrastructure and producer goods needed to support livelihoods. Examples of physical capital include various durable and liquid assets. Financial capital denotes the financial resources that people use to achieve their livelihood objectives. These may include access to credit facilities by the households. Social capital refers to the social resources upon which people draw in pursuit of their livelihood objectives. These are developed through networks and connectedness; to membership formalised groups; and relationships of trust, reciprocity and exchanges that facilitate co-operation and may provide the basis for informal safety nets amongst the poor.

In order to maximize utility, households are involved in many activities including adoption of climate change adaptation strategies where necessary. The households' decisions on what to produce with what inputs and what to consume including marketing decisions in a given period are usually derived from the maximisation of expected utility. The wealthier the household is, the more it is likely to maximise this utility. Maximisation of utility is however constrained by a number of factors such as the farmer's budget, access to credit and the availability of information on both the adaptation strategies and other inputs used in the production process. The adoption of adaptation strategies will occur only if adoption is expected to be profitable or will make households better off than in the absence of adoption. Following Ali and Abdulai (2010), the adoption decision can be modeled in a random utility model. The difference between the utility from adoption (U_{AS}) and non-adoption (U_{NS}) of a strategy may be denoted as G^*, such that a utility-maximizing household, *i*, will choose to adopt a strategy, if the utility gained from adopting is greater than the utility of not adopting ($G^* = U_{AS} - U_{NS} > 0$). Utility is never observed but can be expressed as a function of observable elements in the following latent variable model:

$$G_i^* = \beta X_i + \mu_i \text{ with } G_i = \begin{cases} 1 & if \quad G_i^* > 0 \\ 0 & otherwise \end{cases} \tag{1}$$

where G is a binary indicator variable that equals 1 if a farmer adopted an adaptation strategy and zero otherwise; β is a vector of parameters to be estimated; X is a vector of explanatory variables that favours or hinders adoption of a particular strategy; and μ is the error term.

Most agricultural related climate change adaptation strategies could be viewed as new technologies to farmers. Therefore most literature on agriculture technology adoption logically fits well in conceptualising the adoption of adaptation strategies. The decision to adopt a strategy is affected by the characteristics of the farm household head and the household at large. Household heads are the final decision makers who may decide on adoption of new technologies or strategies on a farm.

Characteristics such as the household's age, education, marital status etc could have an effect on whether the farmer adopts a particular strategy or not. For instance younger farmers may be more innovative and have lower risk aversion behavior although they may also have less farming experience hence the relationship between age and adoption of climate change adaptation strategies may be ambiguous. Female headed households may respond less favourably to adoption of technologies than male headed households due to wealth differences (Ajayi et al., 2003). Some educated farmers could be more willing in trying out new strategies while others would be more conservative in getting involved in unproven practices. Thus, the level of education of the household head is also expected to either enhance or discourage adoption of climate change adaptation strategies. Among the household characteristics, availability of labour is expected to positively affect the farm household's decision to adopt the strategies. Labour directly affects the wealth position of a household.

In most rural parts of Zambia, similar to other developing countries, there are other factors affecting adoption of climate change strategies through their linkage to the wealth status of households. These include the owning of various assets such as farm sizes, livestock, bicycles, radios, belonging to agricultural groups, access to information etc. These variables provide production services and are expected to increase the likelihood of adoption for a given household. In this study all the wealth related variables entered the adoption model through the wealth index. The variables which were not used in the calculation of the wealth index entered the model as individual variables.

2.3 Calculating the Wealth Index

Ranking households on a wealth scale pose an enormous challenge since each asset contributes differently to households' wealth status. Households can directly be asked to value the different assets in their possession. The draw back with this approach is that different values could be attached to the same asset by various households, as they generally perceive the value of assets differently. Therefore, finding a common factor which could meaningfully be applied to all the many assets owned by the households becomes problematic (Filmer & Pritchett, 1998). To tackle this problem, the same authors suggested the construction of a set of weights based on a common factor which can be applied to all the indicators (such as market or shadow prices). However, the reliability on common factors such as market prices in areas with highly imperfect markets found in rural Zambia renders this option inappropriate for the study. An alternative suggestion by Filmer and Pritchett (1998) is performing a multivariate regression analysis with all assets as unconstrained variables. The challenge with this option is that most assets contributing to the wealth status of households are somehow related; therefore running a multivariate regression could produce misleading regression coefficients due to multi-collinearity problems. However, since this option sounds more plausible, finding possible solutions to the multi-collinearity problem could render this method relevant.

The multi-collinearity problem in the multivariate regression analysis could be solved by using the principal component regression (Filmer & Pritchett, 1998; Wooldridge, 2006). Principal component regression commonly referred to as principal component analysis (PCA) entails extracting from a set of variables or assets those few orthogonal linear combinations of the variables that capture the common information most successfully. Therefore,among the considered options, the PCA technique was found to be a satisfactory approach for this study. To ensure completeness of assets or wealth indicators used in PCA, we carried out a key informant group wealth ranking exercises to identify the important indicators farmers considered in categorising themselves. The wealth indicators from these exercises and from literature were used in the computation of the wealth indices.

2.3.1 The PCA Approach

To compare different forms of assets so that ranking households is objective, normalising the assets becomes necessary. Normalising household assets involved constructing indices scaled from 0 to 1 for each asset. The indices were later aggregated to obtain a composite index that was used for ranking the households. Following Filmer and Pritchett (2001), the indices were scaled from 0 to 1 as follows:

$$i = \frac{x_l - x_{\min}}{x_{\max} - x_{\min}} \tag{2}$$

where i is the index, x_l is the level, while x_{min} and x_{max} are the minimum and maximum values of x, respectively taken from the actual data collected. Once scaled (or normalised), the indicators can be added together without the element of distortion which would be introduced by widely differing value ranges.

With the normalised indices, PCA techniques (Langyintuo, 2008) were used to extract from a set of variables those few orthogonal linear combinations of the variables that capture the common information most successfully. Intuitively the first principal component of a set of variables is the linear index of all the variables that captures the largest amount of information that is common to all of the variables. From the normalised indices and PCA output, an overall 'wealth index', for each household was constructed by applying the following formula:

$$W_j = \sum_{i=1}^{k} [b_i(a_{ji} - x_i)] / s_i \tag{3}$$

where: $\boldsymbol{W_j}$ is a standardized wealth index for each household; $\mathbf{b}_i$ represents the weights (scores) assigned to the (k) variables on the first principal component; $\mathbf{a}_{ji}$ is the value of each household on each of the k variables; $\mathbf{x}_i$ is the mean of each of the k variables; and $\mathbf{s}_i$ the standard deviations.

A negative index ($-W_j$) means that, relative to the communities' measure of wealth, the household is poorly endowed and hence worse-off while a positive figure (W_j) signifies that the household is well-off. A zero value, which is also the sample mean index, implies that the household is neither well-off nor worse-off. In this analysis, PCA was run on selected 23 assets or indicators using SPSS as discussed by Langyintuo (2008).

2.4 Analysing Wealthy and Adoption of Climate Change Adaptation Strategies

To evaluate how likely it was that any observed difference between the wealth categories of poorly and well-endowed sets and whether the households had adopted an adaptation strategy or not, arose by chance, a Pearson's chi-squared tests (χ^2) were done. The Pearson's chi-squared tests (χ^2) were complemented by the use of more robust binary probit modeling (Maddala, 2007)whose marginal effects analyses indicated the probability of adopting a strategy given the wealth status of a household.

3. Results and Discussion

3.1 Demographic Characteristics of Households

The descriptive statistics of the sample households are presented in Table 1. The average household size among the surveyed sample was about six members. The households had farm sizes of approximately 3 hectares from which slightly above half, was cropped. Regionally, the farm sizes were slightly smaller in the Northern part of the country compared to Eastern and Southern parts of Zambia. The cropped areas similarly followed this trend. Generally most female headed households have limited resources to adopt certain agricultural practices that could be adaptive to vulgaries of climate change. The overall sample comprised about 20% female headed households. With respect to marital status, 74% of the household heads were monogamously married while 7% were polygamously married. A further12% were widowed with 4% divorced while 1%were separated with 1% never married and 0.1% cohabited. Age of smallholder farmers has an impact on the productive capacity and adopting various climate change strategies. The estimated mean age of the households head in the sample was about 46 years. Furthermore, about 86% of those interviewed had some formal education. Of these, 57% had primary education and 37% had secondary education. About 3% of the respondents reported that they had post-secondary education. The regional distribution of these demographic and socioeconomic characteristics is also shown in Table 1.

Table 1. Descriptive and socioeconomic statistics of sampled households

	Region			Whole sample
	Southern	Northern	Eastern	
Household size	6.3 (0.15)	6.4 (0.13)	5.8 (0.096)	6.1 (0.070)
Age of head (years)	46.6 (0.85)	46.9 (0.82)	45.8 (0.62)	46.3 (0.43)
Farm size (ha)	3.2 (0.22)	2.3 (0.24)	3.2 (0.12)	3.0 (0.10)
Cropped area (ha)	1.8 (0.099)	1.1 (0.064)	2.0 (0.078)	1.7 (0.053)
Female headed households (%)	18.4	20.7	19.4	19.6
Education level of head (%)				
None	6.2	5	22.1	13.6
Primary education	61	55.8	56.6	57.4
Secondary	27.6	36.7	19.9	26.6
Post-secondary	5.1	2.5	1.3	2.5
Marital status of head (%)				
Never married	2.6	0	0.8	1
Married (one wife)	62.1	77.6	76.9	73.8
Married (>one wife)	20.6	2.2	4.7	7.5
Cohabiting	0	0	0.2	0.1
Divorced	2.6	4.4	11.2	4.3
Separated	1.1	1.7	1.2	1.3
Widowed	11	14.1	11.2	12

Numbers in parentheses are standard errors.

3.2 Wealth Ranking Indicators from Farmer Participatory Exercises

3.2.1 The Southern Region

In the Southern Region, the farmer wealth ranking participatory exercises revealed that the main indicators of wealth included animals such as cattle and goats (in the lower wealth categories chickens were also considered), total farm land and area of land cultivated, method of cultivation and for the richer categories, number of vehicles owned.

The very poor in this region are those who did not own any cattle, goats or chickens. They usually use hand hoes to cultivate about 0.25 hectares of land per season and do not possess any vehicles. Close to this group were the poor who own on average 0 - 2 heads of cattle, 1 -3 goats, 10 – 20 chickens and they also use a hand hoe to cultivate about 0.5 hectares of land. The rich households in the Southern region boasted of 20 – 30 herds of cattle, with more than 30 goats, more than 30 chickens, use oxen for cultivation of about 10 – 15 hectares of land and they owned a motor vehicle or 2. The very rich households had more than 50 of each of the livestocks; cattle, goats, chickens and up to 5 vehicles including either a tractor, light truck or a family car. They cultivated more than 15 hectares of land using either oxen or a tractor or a combination of the two.

3.2.2 The Eastern Region

In the Eastern region, the number of cattle, goats, chickens, farm size, cultivated area, whether household hired or sold labour, ability to take children to school, were all mentioned as indicators of wealth. In this region, the very poor had no cattle, goats or chickens and they used hand hoes in cultivating less than 0.25 hectares of land. These usually hired out labour and lived in grass thatched huts and cannot afford assets such as bicycles, radios, television sets or Knap sack sprayers. They cannot afford medical services and are in most cases unable to send their children to school. The poor have characteristics similar to the very poor but some of them can afford to send their children to at least primary level of education, and they had at least up to 10 of each of the following animals and poultry; goats, pigs and chickens. Despite using hand hoes like the very poor, the poor cultivated between 0.25 – 0.75 hectares of land. The rich in the Eastern region had 5 herds of cattle or less, 10 – 19 goats and/or pigs and more than 10 chickens. Most of them hired labour and were able to send their children to school, in addition to having a bicycle, a radio and a television set. The radio and television set are usually powered with car batteries and a solar panel in some instances. They cultivated between 0.75 and 1.5 hectares of land. The very rich owned an iron roofed house with 5 – 40 herds of cattle, more than 10 of each of the following; goats, pigs, chickens and have ox-drawn implements. Further, they owned a hammer mill and a motor vehicle or two. The very rich were also able to send their children to school.

3.2.3 The Northern Region

Unlike the Southern and Eastern regions, the Northern region does not use cattle as one of the measures of wealth indicators. This is because only very few farmers own cattle. Goats and some pigs are owned but in few numbers. The very poor in the Northern region comprised of people who had very small (less than a quarter of a hectare) of farm fields. These people ate only one meal per day during the hunger months (Oct, Nov, Dec, Jan, Feb) and at least two meals after harvest. They could not afford to buy new clothes and usually wore torn clothes. This group of people always made their labour available to others through piece works. In some cases they leave in houses that look deserted. They cannot afford a TV or a Radio. The second category was the poor. This group could afford at least two meals per day in the hunger months. This group farms on about half or 0.5 hectares of land, some within the group can afford a radio and had a few chickens. They were able to buy clothes for their family members. They had some sort of assets which they can point at. The third category was the moderately rich or middle class. They are a few that belong to this group. This group comes from both the poor class and very rich classes. Most of them had a house with a tin roof. They do not sell their labour to other households. They farmed about 1 – 2 hectares of land. They had some goats and chickens. The last category is the super-rich who were very few in the communities and typically possessed a hammer mill and sometimes had a car or cars (tractor/family bus/Open light truck) in addition to the other assets mentioned earlier.

Generally, participatory ranking of households into wealth categories by the farmers themselves, though useful can sometimes produce conflicting results. Although this study did not allow the farmer groups to place each and every known household into a wealth category, indications were that this was not going to be an easy task. This was evidenced by the fact that even agreeing on which indicator should fall in which category was problematic. A distinct boundary between the different wealth groups such as the very rich and the rich could not be consensually determined. The farmers were however very useful in providing indicators which when combined with those from literature, some quantitative manipulations such as that used in this study (PCA), could help towards ranking the households on some form of a wealth ranking scale.

3.2.4 Wealth Indicators Used in PCA

The preceding discussion shows that rural households use various indicators to categorize themselves into wealth categories. The indicators can be different across regions and can also produce inconsistent results in so far as placing a household in a given category, is concerned. For instance a household can be endowed in one or two of the assets that would make it suitable to be in a rich category only to be poorly endowed in other assets hence making the placement of this household problematic. Several authors (Grandin, 1988; Crowley, 1997; Sharrock et al., 1994) suggest that farmers themselves (or key informants) can be allowed to place each of them in wealth categories. Though this would produce reasonably acceptable outcomes, the most obvious problems expected includes failure to obtain a consensus on households who dispute to be placed in certain categories they believe they don't belong to.

The farmer participatory exercises helped to identify important wealth indicators according to farmers own perspectives. These included; animals such as cattle, goats, pigs and chickens, farm size, bicycles, TVs, radios, motor vehicles etc. To present a more objective wealth categorization, the indicators or assets identified by farmers together with those that are important but usually taken for granted such as access to information, group membership and access to credit were used in PCA to rank the sampled households. The mean values of assets used in the analysis are shown in Table 2 while the proportions of households owning some of these assets are shown in Table 3. Results from the structured interviews augmented the PRA findings that generally only few households own cattle (to some extent goats and pigs) and ox drawn related implements in the Northern part of the country (Table 3). The use of draught power was therefore more pronounced in the Southern region followed by the Eastern region. More households in the Northern region however hired more labour than their counter parts from both Southern and Eastern regions.

Table 2. Average number of assets used for wealth ranking in the PCA

......	Southern (N= 272	Northern (N= 362)	Eastern (N=597)	Whole sample (N=l 1231)
Household hired labour (1=yes, 0= no)	0.16 (0.022)	0.27 (0.023)	0.12 (0.013)	0.17 (0.011)
Household used animal labour (1=yes, 0= no)	0.66 (0.029)	0.01 (0.006)	0.41 (0.020)	0.35 (0.014)
Farm size	3.15 (0.218)	2.47 (0.240)	3.18 (0.116)	2.97 (0.103)
Area planted in hectares	1.81 (0.0999)	1.074 (0.064)	1.97 (0.678)	1.67 (0.053)
Number of ox ploughs	1.44 (0.030)	0.04 (0.014)	0.69 (0.044)	0.53 (0.028)
Household has ox cart (1=yes, 0= no)	0.22 (0.25)	0.02 (0.007)	0.23 (0.020)	0.15 (0.010)
Household has wheel barrow (1=yes, 0= no)	0.11 (0.045)	0.08 (0.015)	0.05 (0.009)	0.07 (0.007)
Number of bicycles	0.78 (0.050)	0.86 (0.043)	0.94 (0.035)	0.88 (0.024)
Household has knack sprayer (1=yes, 0= no)	0.25 (0.026)	0.05 (0.012)	0.12 (0.013)	0.13 (0.010)
Household has solar panel (1=yes, 0= no)	0.26 (0.027)	0.23 (0.022)	0.18 (0.016)	0.21 (0.012)
Number of cell phones	1.10 (0.081)	1.04 (0.059)	0.74 (0.037)	0.91 (0.031)
Number of radios	0.72 (0.047)	0.73 (0.036)	0.69 (0.029)	0.71 (0.020)
Number of tvs	0.24 (0.030)	0.33 (0.035)	0.19 (0.018)	0.22 (0.013)
Number of car batteries	0.21 (0.030)	0.29 (0.034)	0.15 (0.016)	0.21 (0.014)
Number of homes	1.18 (0.053)	0.98 (0.040)	1.10 (0.034)	1.08 (0.023)
Number of cattle*	8.43 (0.989)	7.60 (2.684)	6.15 (0.375)	6.89 (0.409)
Number of goats*	10.92 (1.345)	5.15 (0.456)	6.53 (0.607)	7.30 (0.438)
Number of pigs*	4.94 (0.972)	3.08 (0.551)	6.36 (0.410)	5.94 (0.356)
Number of chickens*	13.38 (0.859)	13.38 (0.961)	11.71 (0.570)	12.56 (0.434)
Access to information (1=yes, 0= no)	0.95 (0.013)	0.87 (0.018)	0.87 (0.014)	0.89 (0.009)
Group membership (1=yes, 0= no)	0.37 (0.029)	0.56 (0.026)	0.43 (0.002)	0.45 (0.014)
Household member size	6.33 (0.150)	6.63 (0.129)	5.80 (0.096)	6.08 (0.069)
Access to credit (1=yes, 0= no)	0.09 (0.17)	0.01 (0.004)	0.25 (0.018)	0.14 (0.010)

*Average does not include those who do not own the asset.
Numbers in parentheses are standard errors.

Table 3. Proportional distribution of assets by households (%)

	Region			
	Southern	Northern	Eastern	Whole sample
Ox plough	65.9	1.9	41.9	33.3
Ox cart	21.7	1.9	20.8	15.4
Wheel barrow	11.4	7.7	5.0	7.2
Bicycle	58.8	63.8	69.5	65.5
Knack sprayer	24.6	5.2	12.2	12.9
Solar panel	26.1	22.7	18.1	21.2
Cell phone	62.5	62.4	51.9	57.4
Radio	59.6	62.7	58.3	60.0
Television set	16.5	22.4	17.4	19.6
Car battery	17.6	22.7	14.4	17.5
Cattle	50.0	2.8	46.6	35.4
Goat	43.8	22.7	34.0	35.1
Pigs	11.4	5.5	33.2	23.3
Chickens	63.4	60.5	64.3	62.2
N	272	362	597	1231

3.3 Household Adoption of Climate Change Adaptation Strategies

Rural households engage in various livelihood strategies to earn a living. The main economic activity among the rural households of Zambia is farming or agriculture. When these livelihood strategies are threatened, the households usually attempt to find a way to ensure a sustained welfare. The surveyed households were asked to indicate what modifications or adaptations they have made to their livelihood strategies as a result of the negative effects of climate change. The various adaptation strategies and the proportion of households adopting them are shown in Table 4. Most of the strategies are agricultural in nature perhaps reflecting the fact that farming is the major enterprise in most rural parts of the country. Going by the overall results, changing crop varieties, changing planting dates and crop diversification were cited as major strategies by more than 20% of the households. Switching crops (crop rotation), planting of fertiliser trees and implementing some form of conservation farming (minimum tillage through potholing and ripping) were cited as major strategies by between 11 and 20% of the farmers. There were regional variations in the uptake of these adaptation strategies (Table 4).

Table 4. Proportion of households adopting climate change adaptation strategies

	Region			Whole sample
	Southern	Northern	Eastern	
Minimum tillage	12.1	5.5	15.4	11.8
Fertilizer tree planting	15.8	9.1	13.2	12.6
Changing planting dates	15.4	29.3	48.7	35.7
Crop rotation	6.6	6.6	32.7	19.4
Change crop varieties	22.8	37.6	53.1	41.8
Crop diversification	21.3	32.9	18.6	23.4
N	272	362	597	1231

3.4 Wealth Ranking of Households

Despite the regional differences in a few assets the categorisation of households into poorly and well-endowed groups produced similar estimates. Sixty one percent of households in the Southern and Northern regions were in relative terms poorly endowed. In the Eastern region 60.3% of the households were poorly endowed. Pooling the data together and repeating the wealth ranking estimation showed that 61.6% of the households were poorly endowed, relative to the communities' measure of wealth (Figure 1).

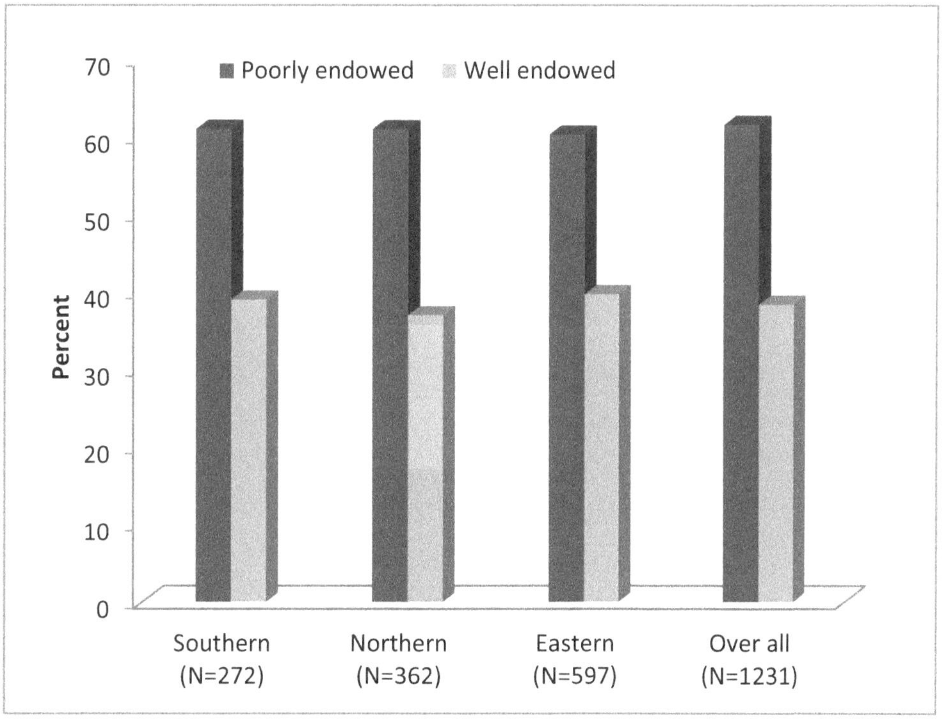

Figure 1. Household assets endowment by regions

3.5 Chi-square Estimates on the Association of Wealth and Adoption of Climate Change Adaptation Strategies

3.5.1 The Southern Region

In the southern region, more well-endowed households than the poorly endowed were associated with minimum tillage (*P= 0.05*), fertiliser tree planting (*P= 0.002*), crop rotation (*P= 0.000*) and changing crop varieties (*P= 0.052*) (Table 5).

Table 5. Household wealth and adopting adaptation strategies in the Southern Region

	% households adopting strategy		Pearson Chi-Square	
Adaptation strategy	Poorly endowed	Well endowed	Value	Asymp. Sig (2 sided)
Minimum tillage	9.0	17.0	3.831	0.050
Fertiliser tree planting	10.2	24.5	9.922	0.002
Change in crop planting dates	15.7	15.1	0.016	0.899
Crop rotation	1.8	14.2	15.950	0.000
Changing crop varieties	19.9	27.4	2.056	0.052
Crop diversification	22.3	19.8	0.237	0.627

3.5.2 The Eastern Region

In the eastern region, more households in the well-endowed category than the poorly endowed category were associated with the adoption of minimum tillage (*P = 0.000*), fertiliser planting trees (*P = 0.059*), crop rotation (*P= 0.003*), changing crop varieties (*P= 0.011*) and use of livestock supplements (*P= 0.024*) (Table 6).

Table 6. Household wealth and adopting adaptation strategies in the Eastern Region

	% households adopting strategy		Pearson Chi-Square	
Adaptation strategy	Poorly endowed	Well endowed	Value	Asymp. Sig (2 sided)
Minimum tillage	7.5	27.4	43.530	0.000
Fertiliser tree planting	11.1	16.5	3.555	0.059
Change in crop planting dates	48.6	48.9	0.006	0.936
Crop rotation	28.1	39.7	8.754	0.003
Changing crop varieties	48.9	59.5	6.454	0.011
Crop diversification	16.9	21.1	1.628	0.202

3.5.3 Northern Region

In the northern region, more households in the well-endowed category than the poorly endowed category were associated with the adoption of minimum tillage (*P= 0.097*), fertiliser planting trees (*P = 0.002*), crop rotation (*P= 0.014*) and changing crop varieties (*P= 0.074*) (Table 7).

Table 7. Household wealth and adopting adaptation strategies in the Northern Region

	% households adopting strategy		Pearson Chi-Square	
Adaptation strategy	Poorly endowed	Well endowed	Value	Asymp. Sig (2 sided)
Minimum tillage	4.5	7.1	1.087	0.097
Fertiliser tree planting	5.4	14.9	9.305	0.002
Change in crop planting dates	28.1	31.2	0.413	0.521
Crop rotation	4.1	10.6	5.995	0.014
Changing crop varieties	33.9	43.3	3.192	0.074
Crop diversification	30.8	36.2	1.138	0.286

3.6 Probit Estimates of the Relationship between Wealth and Adoption of Climate Change Adaptation Strategies

The well-endowed households were almost 10% likely to adopt minimum tillage than their poorly endowed counter parts. Minimum tillage requires some investment in accessing manure and human labour, among others. In addition households who were aware of the consequence of climate change were 6% more probable of adopting minimum tillage. Compared to the Northern region, households from the Eastern and Southern regions had around 9% likelihood of adopting minimum tillage (Table 8).

Table 8. Probit estimates of determinants of adoption of Minimum tillage

Variables	Coefficient	Std. Err	Z	P>IZI	Marginal Effects (*dy/dx*)
Wealth (1 well-off, 0 otherwise)	0.502	0.106	4.750	0.000	0.095
Sex (1 male, 0 otherwise)	0.179	0.211	0.850	0.396	0.029
Monogamously married (1 yes, 0 otherwise)	0.225	0.176	1.280	0.200	0.369
Education (1 primary, 0 otherwise)	-0.012	0.078	-0.015	0.878	-0.002
Age (years)	0.005	0.003	1.440	0.150	0.001
Access to conservation farming advise (1 yes, 0 otherwise)	0.043	0.107	0.410	0.683	0.007
Distance to town (km)	-0.004	0.002	-1.550	0.121	-0.001
Climate Change awareness (1 yes, 0 otherwise)	0.400	0.133	3.010	0.003	0.061
Eastern Region (1 yes, 0 otherwise)	0.485	0.137	3.530	0.000	0.086
Southern Region (1 yes, 0 otherwise)	0.455	0.163	2.800	0.005	0.093
Constant	-2.568	0.316	-8.130	0.000	
Number of obs*	1184				
LR Chi^2 (10)	74.130				
Prob>Chi^2	0.000				
Pseudo R^2	0.085				
Log likelihood	-399.204				

*Some observations dropped during analysis.

Belonging to the well-endowed category increased the probability of adoption of fertiliser trees by 7.1%. Adoption of fertiliser trees requires some investment in the form of both financial and human capital. Land, tree seeds, hiring of labour etc are some of the items requiring financial outlay. The fertiliser trees are labour intensive hence family labour in addition to hired labour would increase the probability of adopting the technologies. Climate change awareness, being domiciled in Eastern and Southern regions increased the probability of adoption of fertility trees by 7.5%, 4.2% and 6.1%, respectively (Table 9). There was no relationship between household wealth status and changing of planting dates by the farmers (Table 10).

Table 9. Probit estimates of determinants of adoption of fertilizer trees

	Coefficient	Std. Err	Z	P>IZI	Marginal Effects (*dy/dx*)
Wealth (1 well-off, 0 otherwise)	0.347	0.102	3.390	0.001	0.071
Sex (1 male, 0 otherwise)	0.326	0.199	1.640	0.101	0.056
Monogamously married (1 yes, 0 otherwise)	-0.057	0.162	-0.350	0.727	-0.011
Education (1 primary, 0 otherwise)	0.104	0.075	1.390	0.166	0.020
Age (years)	0.000	0.003	-0.030	0.979	0.000
Access to conservation farming advise (1 yes, 0 otherwise)	0.074	0.103	0.720	0.473	0.014
Distance to town (km)	-0.001	0.002	-0.520	0.600	-0.001
Climate Change awareness (1 yes, 0 otherwise)	0.446	0.131	3.410	0.001	0.075
Eastern Region (1 yes, 0 otherwise)	0.217	0.126	1.720	0.085	0.042
Southern Region (1 yes, 0 otherwise)	0.286	0.144	1.990	0.047	0.061
Constant	-2.200	0.301	-7.300	0.000	
Number of obs*	1184				
LR Chi2 (10)	47.120				
Prob>Chi2	0.000				
Pseudo R2	0.092				
Log likelihood	-432.169				

*Some observations dropped during analysis.

Table 10. Probit estimates of determinants of farmers changing planting dates

	Coefficient	Std. Err	Z	P>IZI	Marginal Effects (*dy/dx*)
Wealth (1 well-off, 0 otherwise)	0.008	0.085	0.090	0.927	0.003
Sex (1 male, 0 otherwise)	0.402	0.156	2.580	0.010	0.139
Monogamously Married (1 yes, 0 otherwise)	-0.132	0.138	-0.960	0.337	-0.049
Education (1 primary, 0 otherwise)	-0.095	0.062	-1.530	0.126	-0.035
Age (years)	0.002	0.003	0.730	0.468	0.001
Access to conservation farming advise (1 yes, 0 otherwise)	0.039	0.082	0.480	0.632	0.014
Distance to town (km)	-0.001	0.001	-0.220	0.826	-0.001
Climate Change awareness (1 yes, 0 otherwise)	-0.042	0.094	-0.450	0.652	-0.016
Eastern Region (1 yes, 0 otherwise)	0.469	0.097	4.820	0.000	0.171
Southern Region (1 yes, 0 otherwise)	-0.535	0.126	-4.240	0.000	-0.182
Constant	-0.703	0.224	-3.130	0.002	
Number of obs*	1184				
LR Chi2 (10)	116.480				
Prob>Chi2	0.000				
Pseudo R2	0.075				
Log likelihood	-714.689				

*Some observations dropped during analysis.

The well-endowed households were 10.6% more likely than poorly endowed households, to practice crop rotation due to climate change. To have a variety of crops to rotate, a farmer need to have capital to invest in such crops. Climate change awareness increased the likelihood of practising crop rotation by 5.2%. Households in Eastern region were more likely to practice crop rotation than their counter parts in Northern region (Table 11).

Table 11. Probit estimates of determinants of adoption crop rotation

	Coefficient	Std. Err	Z	P>IZI	Marginal Effects (*dy/dx*)
Wealth (1 well-off, 0 otherwise)	0.422	0.097	4.34	0.000	0.106
Sex (1 male, 0 otherwise)	0.139	0.181	0.77	0.444	0.032
Monogamously Married (1 yes, 0 otherwise)	0.016	0.158	0.10	0.922	0.004
Education (1 primary, 0 otherwise)	-0.111	0.071	-1.56	0.118	-0.026
Age (years)	-0.002	0.003	-0.54	0.590	-0.000
Access to conservation farming advise (1 yes, 0 otherwise)	0.143	0.098	1.47	0.142	0.034
Distance to town (km)	0.001	0.002	0.61	0.540	0.000
Climate Change awareness (1 yes, 0 otherwise)	0.233	0.112	2.08	0.038	0.052
Eastern Region (1 yes, 0 otherwise)	1.019	0.130	7.85	0.000	0.245
Southern Region (1 yes, 0 otherwise)	-0.112	0.171	-0.66	0.512	-0.026
Constant	-1.879	0.281	-6.68	0.000	
Number of obs*	1184				
LR Chi2 (10)	170.23				
Prob>Chi2	0.000				
Pseudo R2	0.143				
Log likelihood					

*Some observations dropped during analysis.

The probability of changing crop varieties due to climate change by the well-endowed households was higher by 5.5%. Changing crop varieties needs financial capital to do so. This is especially true in Zambia where most of the major crops being planted are improved varieties being sold by multinational seed companies. In addition, male headed households, those who were not monogamously married, and households coming from the Eastern and Southern Region, had a high probability of changing crop varieties (Table 12). Wealth status did not have an influence in households' decision to diversify into other crop production activities (Table 13).

Table 12. Probit estimates of determinants of farmers changing crop varieties

	Coefficient	Std. Err	Z	P>IZI	Marginal Effects (*dy/dx*)
Wealth (1 well-off, 0 otherwise)	0.140	0.083	1.690	0.091	0.055
Sex (1 male, 0 otherwise)	0.460	0.151	3.040	0.002	0.170
Monogamously Married (1 yes, 0 otherwise)	-0.312	0.133	-2.340	0.020	-0.123
Education (1 primary, 0 otherwise)	0.019	0.060	0.320	0.750	0.007
Age (years)	0.000	0.003	0.090	0.929	0.000
Access to conservation farming advise (1 yes, 0 otherwise)	0.139	0.079	1.750	0.080	0.054
Distance to town (km)	-0.002	0.002	-1.110	0.267	-0.001
Climate Change awareness (1 yes, 0 otherwise)	0.125	0.092	1.360	0.173	0.048
Eastern Region (1 yes, 0 otherwise)	0.379	0.095	3.980	0.000	0.147
Southern Region (1 yes, 0 otherwise)	-0.468	0.118	-3.950	0.000	-0.174
Constant	-0.669	0.219	-3.050	0.002	
Number of obs*	1184				
LR Chi2 (10)	98.690				
Prob>Chi2	0.000				
Pseudo R2	0.061				
Log likelihood	-755.374				

*Some observations dropped during analysis.

Table 13. Probit estimates of determinants of farmers diversifying crops

	Coefficient	Std. Err	Z	P>IZI	Marginal Effects (*dy/dx*)
Wealth (1 well-off, 0 otherwise)	0.031	0.089	0.340	0.734	0.009
Sex (1 male, 0 otherwise)	0.259	0.167	1.550	0.121	0.072
Monogamously Married (1 yes, 0 otherwise)	-0.039	0.145	-0.270	0.790	-0.012
Education (1 primary, 0 otherwise)	0.078	0.066	1.180	0.239	0.023
Age (years)	0.007	0.003	2.440	0.014	0.002
Access to conservation farming advise (1 yes, 0 otherwise)	0.161	0.088	1.830	0.068	0.047
Distance to town (km)	-0.003	0.002	-0.950	0.051	-0.001
Climate Change awareness (1 yes, 0 otherwise)	0.352	0.108	3.270	0.001	0.097
Eastern Region (1 yes, 0 otherwise)	-0.451	0.101	-4.450	0.000	-0.134
Southern Region (1 yes, 0 otherwise)	-0.291	0.120	-2.420	0.016	-0.081
Constant	-1.354	0.241	-5.610	0.000	
Number of obs	1184				
LR Chi2 (10)	57.200				
Prob>Chi2	0.000				
Pseudo R2	0.095				
Log likelihood	-614.319				

4. Conclusion

The use of indicators from several sources other than only from the farmers provided a holistic picture of what comprises household wealth. This study has shown that from a farmers stand point of view, belonging to groups, access to credit or information or simply social or institutional capital is rarely valued as wealth in rural communities of Zambia. It is the physical, tangible, assets such as farm and animal herd size that counts. The use of farmer participatory wealth exercises in identifying indicators, structured questionnaires in soliciting for information on whether households possessed the assets and in what quantities and PCA in analysing the wealth status of each households within the communities produced an objective wealth ranking measure that helped to place farmers into either the well or poorly endowed groups.

Generally, wealth of households was associated with adoption of most of the climate change adaptation strategies considered in this study except change in crop planting dates and crop diversification. More well-endowed households had adopted minimum tillage, fertiliser trees, crop rotation and changing crop varieties. Most if not all of these strategies require some level of resource investment hence only those households who can afford such resources are likely to adopt them. Subsidising the relatively poor endowed households may encourage adoption of these strategies among this category of farmers.

Although warranting further investigation, intuition would suggest that the value of the strategy in immediately providing food security and its relative cost to the other strategies affected the levels of adoption of these strategies. For instance regardless of the wealth status of the farmers, changing crop varieties, changing planting dates and crop diversification were cited as major strategies by most of the households. This was followed by crop rotation, fertiliser tree uptake and minimum tillage, in this order. Changing crop varieties, changing planting dates and crop diversification would immediately provide the much needed food. On the other hand strategies such as crop rotation, soil fertility enhancing trees and minimum tillage ensure the long term sustainability of the production system and might be associated with delayed benefits.

The adoption of some of the strategies also show regional biasedness perhaps reflecting the way the extension system in the country has discouraged the uptake of certain practices in certain areas. For example, conservation farming comprising minimum tillage and planting in potholes, or ripped trenches has been discouraged in agro-ecological region III of Zambia that receive more 1000 mm of rainfall annually (Langmead, 2015). Results from this study are consistent with this message. For the Northern region whose vast areas falls in region III, very few households from both the well and poorly endowed categories embraced minimum tillage as a climate change adaptation strategy.

Acknowledgements

The Research team acknowledges financial support from Michigan State University and Administrative support from the University of Zambia. The team also acknowledges the many data collection assistants and the farmers who participated in the surveys. The support of the Indaba Agricultural Policy Research Institute (IAPRI) is also acknowledged.

References

Ajayi, O. C., Franzel, S., Kuntashula, E., & Kwesiga, F. (2003). Adoption of improved fallow soil fertility management practices in Zambia: synthesis and emerging Issues. *Agroforestry System, 59*(3), 317-326.

Ali, A., & Abdulai, A. (2010). The adoption of genetically modified cotton and poverty reduction in Pakistan. *Journal of Agricultural Economics, 61*(1), 175-192.

Arslan, A., McCarthy, N., Lipper, L., Asfaw, S., & Cattaneo, A. (2014). Adoption and intensity of adoption of conservation farming practices in Zambia. *Agriculture, Ecosystems and Environment, 187*, 72-86.

Cancien, F. (1967). Stratification and risk-taking: a theory tested on agricultural innovation. *American Sociological Review, 32*, 912-927.

Crowley, E. L. (1997). Rapid data collection using wealth ranking and other techniques. Working paper. International Centre for Research in Agroforestry/Tropical Soils Biology and Fertility Program, Nairobi.

Filmer, D., & Pritchett, L. H. (1998). *The effect of household wealth on educational attainment: demographic and health survey evidence.* Policy Research Working paper 1980. The World Bank.

Grandin, B. (1988). *Wealth Ranking in Smallholder Communities*: A Field Manual. Intermediate Technology Publications, Nottingham, England.

Hardaker, J. B., Huirne, R. B. M., & Anderson, J. R. (1997). *Coping with Risk in Agriculture.* Wallingford Oxon: CAB International, In.

Kuntashula, E., & Mungatana, E. (2015). Estimating the causal effect of improved fallows on environmental services provision under farmers field conditions in Chongwe – Zambia. *Environment and Development Economics, 20*(01), 80-100. http://dx.doi.org/10.1017/S1355770X14000011

Kuntashula, E., Chabala, L. M., & Mulenga, B. P. (2014). Impact of minimum tillage and crop rotation as climate change adaptation strategies on farmer welfare in smallholder farming systems of Zambia. *Journal of Sustainable Development, 7*(4), 95. http://dx.doi.org/10.5539/jsd.v7n4p95

Langmead, P. (2015) *Hoe conservation farming of maize in Zambia.* Retrieved May 11, 2015, from http://conservationagriculture.org/

Langyintuo, A. S. (2008). *Computing household wealth indices using principal components analysis method.* Working Paper, Harare, Zimbabwe, CIMMYT.

Langyintuo, A. S., & Mungoma, C. (2008). The effect of household wealth on the adoption of improved maize varieties in Zambia. *Food Policy.* http://dx.doi.org/1016/j.foodpol.2008.04.002

Sharrock, G. O. G., Waldie, K. J., & Yoshi, Y. R. (1993). *A wealth ranking for agricultural research purposes in the eastern hills of Nepal.* RRA Notes No. 18. International Institute for Environment and Development, London.

Wooldridge, J. M. (2006). *Introductory econometrics*: A modern approach. Thomson South-Western, Lachina Publishing Services, USA.

World Bank. (2014). Zambia Over view. Retrieved from http://www.worldbank.org/en/country/zambia/overview

20

Observed Effects of Warm Season Perennial Grass Agriculture on Resident Mammal Species

David D. Leimbach[1] & Gregg J. Marcello[1]

[1] Department of Biology, Millikin University, Decatur, Illinois, 62526, USA

Correspondence: David D. Leimbach, 4560 Butler Drive, Decatur, Illinois, 62526. USA.
E-mail: dleimbach@millikin.edu

Abstract

This study focused on the observed impact of switching from a corn / soybean agricultural rotation to the planting of warm season perennial grasses for use in biomass agriculture, as well as the subsequent effects on resident mammal populations. We trapped two parcels of land planted in warm season perennial grasses between September 2011 and October 2013 in order to obtain a census of the mammal populations residing upon the property and to determine the impact (if any) of the change in agriculture. During the first year of the study, a seemingly inhospitable (urban) agricultural area revealed a diverse population of inhabitants. The second year of trapping resulted in the capture of 2 new species (total of 7 species captured) and the noticeable absence of 1 species; a distinct shift in mammal populations had been observed. There was an inverse relationship between the population increase of *Peromyscus maniculatus* during 2012 and the decline in the population of *Peromyscus leucopus*. When a Chi-square test was run ($P<0.05$), the hypothesis of competitive exclusion appeared to be supported. In the third year, population numbers of *Peromyscus maniculatus* had also plummeted. We observed that warm season perennial grass monoculture had an adverse impact on resident mammal species.

Keywords: perennial grasses, rotation vs. biomass agriculture, small mammals, sustainable agriculture

1. Introduction

Soil erosion is a worldwide concern with devastating consequences including: desertification of land and pollution of fresh water sources (Brown, 2011). Removal of natural barriers to protect against soil erosion from entering into the watershed is a major concern in Central Illinois. Soil erosion can be combatted by planting a hearty variety of warm season perennial grasses such as *Panicum virgatum* (switchgrass) on marginal ground (ground with little value agriculturally, traditionally found along creek banks, stream borders and other natural water shed areas) as a riparian buffer. The deep root systems of native grasses hold soil in place, reducing erosion which keeps waterways healthy (Ryan & Marks, 2005). Switchgrass (as well as other cool and warm season grasses planted) can provide cover for animals, forage for ungulates and habitat for nesting birds and various mammals, all while protecting waterways. Perennial grass based agriculture has the potential to limit runoff of nitrates and phosphate as a result of agricultural production methods, all while providing structure and cover for organisms in the community.

Warm season perennial grasses (*P. virgatum, Sorghastrum nutans* and *Andropogon gerardii*, etc.) as well as exotic grass species (such as *Miscanthus x. giganteus*), have been identified as a possible bio-mass based agricultural crop for use in cellulosic ethanol production, as a household heating source and as an additive in coal fired electrical generating plants. Due to the removal of millions of acres of marginal ground from the Conservation Reserve Program (CRP), corn and soybean production is expected to rise and damage to the environment as a result of increased agricultural production is also expected to rise (John & Watson, 2007). It is the desire of numerous environmental entities (John & Watson, 2007) for marginal ground removed from the CRP to be converted into biomass agricultural production areas to protect the riparian waterways.

The Conservation Reserve Program, as we know it today, began in 1985 as part of the Food Security Act enacted by the United States Congress (Hartman et al., 2011). The CRP provides monetary reimbursement to landowners for retiring or removing land from agricultural production (Hartman et al., 2011). At the beginning of 2007, 34.7 million acres of marginal ground were enrolled in the CRP, which had a capped enrollment of 39 million acres. By 2017, due to a lowering of the enrollment cap by the United States Congress, 14 million acres formerly in the

CRP will have been dis-enrolled and returned to traditional agricultural production. A downfall to the conservation reserve program is that farmers are unable to harvest any material from the CRP ground. Should the subsidy for CRP get abolished by law, the affected farmers will revert the former CRP ground into an agricultural production area with a normal rotation of corn or soybeans. The increased agricultural production will in turn lower grain prices because of over-abundance, raise costs of consumer items because of increased fuel and equipment costs, increase soil erosion and at the same time, result in a loss of wildlife habitat and biodiversity (Batie, 2009).

While there is little doubt that warm season perennial grasses are beneficial for the environment, the extent of the benefit of biomass based monocultures in an agricultural setting is questionable. In order to maximize the potential of bio-mass based agriculture, the production of bio-mass based agricultural products must be economically viable for farmers and there must be a sustainable market in Illinois for the end consumption of the product. Historically, perennial grass agricultural strategies have been planned and implemented similar to traditional production agriculture, in monoculture ecosystems (Hartman et al., 2011). When grown in large monocultures, research has shown that *P. virgatum* has an increased susceptibility to some strains of the yellow barley dwarf virus (Hartman et al., 2011) and researchers suggest that any Conservation Reserve Program (CRP) land converted to biofuel production should focus on a combination of warm and cool season grasses to maximize the potential benefits for wildlife (Hartman et al., 2011).

Millikin University was contacted by the Agricultural Watershed Institute (AWI) to conduct a third party collaboration research study. The AWI was in need of a survey of birds, arthropods and mammals occupying a parcel of property owned by CAT. The AWI and CAT have invested in planting several varieties of warm season perennial grasses to be used as a bio-mass crop in lieu of a normal corn / soybean rotation. Because of the elimination of millions of acres from the CRP, AWI wanted to know if bio-mass production is a viable alternative (i.e. profitable) when compared to corn and soybeans. If it was found to be profitable and competitive, they hoped to petition congress to underwrite the initial cost of planting warm season perennial grasses on former CRP ground for farmers. In this respect, current CRP ground could be planted in warm season perennial grasses and farmers could possibly be allowed to harvest the bio-mass acreage to subsidize their income at a reduced cost to taxpayers. Caterpillar, Inc. (CAT)., as an agricultural machinery producer, was interested in knowing the results of this study on bio-mass, as well as knowing what animals currently reside on the bio-mass production ground.

We spent the late-summer and fall of 2011 through 2013 studying the habitat use and population characteristics of mammals located on two tracts of farm ground at the CAT research site in Decatur, Illinois. The site had been planted in warm season perennial grasses, including: *P. virgatum*, *A.gerardii* and *S. nutans*. We hypothesized early on that mammal populations would be small or even absent completely from the research area due to: a lack of natural travel corridors, a lack of an available water source, no substantial shelter in the research area larger than 1 meter tall, and the fact that the research area is a veritable island within an urban environment. The intent of our study was to capture and catalog mammals residing on the research grounds and determine the impact of switching from traditional corn/soybean agriculture to warm season perennial grasses on mammals.

2. Method

2.1 Study Site

In a collaborative effort between Millikin University, AWI and CAT, we trapped two parcels of ground on the north-eastern edge of Decatur, Illinois (N 39° 50' 25.3638", W 88° 57' 17.2872"); bordered by US Route 48, Illinois Route 121 and the Caterpillar, Inc. property. The site is bisected by a local road and is in two parcels. The north parcel is approximately 2.63 hectares and had been planted on approximately June 8 - 13, 2011 by a custom seed applicator from the local area. The north parcel was planted with six different varieties of grasses: *Thinopyrum intermedium* (intermediate wheatgrass), a variety of Cave-in-Rock switchgrass (*P. virgatum*) known as "Blade" (Blade Energy Crops®, Thousand Oaks, CA.), *A. gerardii* (big bluestem), *Schizachyrium scoparium* (little bluestem), *Bouteloua curtipendula* (sideoats grama) and *S. nutans* (Indian Grass). The southern parcel was approximately 22.26 hectares and was planted in four separate plots of four differing grass blends between June 8 and 13, 2011. Plot 1 is a three grass warm season perennial mix of big bluestem, switchgrass and Indian grass. Plot 2 is solely the "Blade" variety of switchgrass. Plot 3 is a blend of 4 grasses and 6 forbs, as well as two pounds of Indian grass (Quail Unlimited, Albany, GA.). Plot 4 is a diversity mix of 11 grasses and 22 forbs (Pheasants Forever, St. Paul, MN.). During the 2010 growing season, both parcels had been planted in *Zea mays* (corn).

We conducted preliminary research on the area by observing aerial / satellite photographs and conducting an on-sight, informal survey with representatives of AWI. Due to an extraordinarily hot summer with little rainfall, the ground appeared to be barren of insect or animal life, save a few mouse/vole holes in the ground. The research area was an early successional community (Harper et al. 2008) growing among the warm season perennial grasses. Very few insects were encountered and one Common Night Hawk *(Chordeiles minor)* was flushed from the grass. The warm season perennial grasses could be seen growing, but the ground appeared to be overrun with *Amaranthus rudis* (water hemp), possibly due to improper burn down techniques or the failure to use a smother crop prior to planting (Fiske and Wolfe, 1995).

We observed three distinct grasses had been planted: switchgrass, Indian grass and big bluestem. *Rudbeckia hirta* (Black-eyed Susan) wildflowers were abundant as they were planted as one of the forbs, as well as *Chamaechrista fasiculata* (partridge pea). Aside from a few dandelions (*Taraxacum officinale*), no other plants were readily abundant. We did note from the ground inspection that the planting schematic provided and the ground survey of what was actually planted differed. Using geographic information systems (GIS) available from the Macon County, Illinois Supervisor of Assessments website (http://gis.co.macon.il.us/), we noticed that this parcel of ground is a land locked fragment of environment with no apparent natural travel corridors and the nearest water source being in excess of 2,000 meters away to the south east. As a habitat for anything other than small mammals or birds, it appeared to be unsuitable to sustain life.

2.2 Monitoring and Inventory of Species

In the Multiple Species Inventory and Monitoring Technical Guide developed by the United States Department of Agriculture (Manley et al., 2006), it was recommended that eight transects, 200 m long, be used to trap small mammals. This type of arrangement was not useful for our purposes, since the schematic of the USDA would require a parcel of land that encompassed an area 200 meters long and 200 meters wide, with traps located on transects 20 meters apart. We decided to reduce the trapping site down into an area that was more manageable (i.e., that could be handled by one person in under two hours). The arrangement of traps that we chose to use was in roughly a diamond shape, 50 meters in total length and width, with four tiers or layers of traps that have been staggered in placement at set depths, emanating from one central point. The traps averaged 4 to 5 meters apart, even though in a study of small mammal trapping (Jones et al., 1996), researchers theorized that most small mammal species had a 20 meter radius home range that was travelled and only one trap per 20 meters was needed. We theorized that since some small mammals are social animals, one trap per 20 meters would not be as effective in determining the true population of a given area. Utilizing this trapping method, only one reference point (determined by Global Positioning System) and a compass would be needed in order to conduct subsequent trapping sessions in the same locations. Reference stakes could be removed from the research area in order for the AWI to mow the area.

Since the parcels are broken up into approximately 4 sections apiece by the method in which they were planted, using a standard rotation, we intended to trap in sessions of 3 consecutive nights from September to November, alternating between the northern and southern parcels. We did this to ensure that we never trapped adjoining sections on consecutive rotations to minimize habitat disturbance and since previous studies have shown that 80-85 % of a population will be captured within 3 days (Shenbrot et al., 2010). We used a series of 52 small and medium sized mammal traps to identify vertebrate species occupying the parcels of ground. Forty-eight small Sherman ™ (H.B. Sherman Traps, Inc., Tallahassee, FL.) live traps were used to capture small mammals, as well as four larger Tomahawk™ (Tomahawk Live Trap Co., Tomahawk, WI.) live traps for medium sized wild or feral mammals. The bait that we used was crushed oats (oatmeal) with shelled peanuts (used for an olfactory component) for small mammals and commercially available salt wheels and canned cat food to lure larger mammals.

2.3 Data Collection and Analysis

All animal life captured was measured to the nearest millimeter with a portable ruler, weighed using a 60 gram Pesola® (Pesola AG, Baar, Switzerland) Micro-line 20060 hanging spring scale to the nearest .5 gram, ear tagged (National Band and Tag Company, Newport, KY.) in order to identify repeat captures, photographed and then released on site utilizing scientifically accepted methods and guidelines developed by the American Society of Mammalogists (Sikes et al., 2011). Collection methods and procedures were conducted with the consent of the Illinois Department of Natural Resources and were approved by the Millikin University animal use and care committee. Plant species were identified using Mohlenbrock (2002). All native species were released at the capture site as soon as the data collection was completed to minimize stress due to handling.

3. Results

Within the first year of switching from a corn/soybean agricultural rotation to the production of agricultural bio-mass, a seemingly inhospitable (urban) agricultural area revealed a diverse population of inhabitants. From September to November 2011, we logged 1248 trapping nights, with only 23 trapping nights being unsuccessful. A total of 117 animals were captured during the initial three month trapping session, representing five separate species of small mammals inhabiting the research area: *Blarina brevicauda, Peromyscus leucopus, Peromyscus maniculatus, Microtus ochrogaster and Mus musculus* (Table 1). Of the 117 mammals that were captured, 83 were first time captures that were subsequently measured, tagged and released. Four mammals were lost prior to tagging and there were 12 fatalities (3 new animals and 9 previously captured animals). 100 of the captures were in the south plot; 17 of the captures were in the north plot.

Table 1. Species captured and number of captures by year at Prairie for biodiversity research area, Decatur, Illinois

	Number of Captures by year			
Species	2011	2012	2013	Total
Deer mouse (*Peromyscus maniculatus*)	57	260	11	328
White footed mouse (*Peromyscus lecucopus*)	20	0	0	20
Norway rat (*Rattus norvegicus*)	0	1	0	1
Prairie vole (*Microtus ochrogaster*)	4	8	2	14
Short tailed shrew (*Blarina brevicauda*)	1	2	1	4
House mouse (*Mus musculus*)	1	0	0	1
Eastern cottontail (*Sylvilagus floridanus*)	0	1	4	5
Total	83	272	18	373

From September to November 2012, we logged 1196 trapping nights, with only 1 trapping night being unsuccessful. A total of 296 animals were captured during the second three month trapping session, representing five separate species of small mammals inhabiting the research area: *B.brevicauda, Rattus norvegicus, P. maniculatus, M. ochrogaster* and *Sylvilagus floridanus* (Table 1). Of the 296 mammals that were captured, 272 were first time captures that were subsequently measured, tagged and released. Seven mammals were lost prior to tagging and there were 31 fatalities (15 new captured animals, 16 previously captured animals); 11 animals were found lying dead near the research area. Of the 272 new captures, 188 of the captures were in the south plot; 84 of the captures were in the north plot.

From September to November 2013, we logged 1144 trapping nights, with 1129 trapping nights being unsuccessful. A total of 18 animals were captured during the final trapping session, representing 4 species previously known to inhabit the research area: *M. ochrogaster, P. maniculatus, B. brevicauda* and *S. floridanus* (Table 1). No animals were lost prior to tagging and there were 5 fatalities. Of the 18 animals captured, 17 of the captures were in the north plot and only 1 of the captures was in the south plot.

Peromyscus maniculatus (deer mouse) was the predominant species captured in 2011, 2012, and 2013. *P. maniculatus* population numbers initially rose from 57 captures to 260 captures in 2012; falling to 11 captures in 2013. *Microtus ochrogaster (prairie vole)* rose in 2012, then fell in 2013, while *Blarina brevicauda* (short tailed shrew) populations appeared unchanged, as they were caught in similar numbers all three years of the study. *Peromyscus leucopus* (white footed mouse), which had been captured 20 times in 2011 was not located or captured during 2012 or 2013. *Sylvilagus floridanus* (eastern cottontail) appeared in the study area during the second year and was regularly observed and captured. There was a direct correlation between the population increase of *P. maniculatus* and the decline in the population of *P. leucopus* when a Chi-square test was run with $P<0.05$ (Table 2).

Table 2. Chart of values used to determine if a relationship existed between the population increase of *P. maniculatus* and the decline in population of *P. leucopus*. (Statistical analysis courtesy C. Watson)

	DEER MOUSE	WHITE FOOTED MOUSE	TOTAL
2011	57	20	77
2012	260	0	260
TOTAL	317	20	337
X^2	13.02373887	4.569732938	
P	1.04E-40	6.71E-17	

4. Discussion

Small mammals are generally underrepresented in mammal research (Yarnell et al., 2007). Many studies have been conducted regarding the impact of warm season perennial grasses on game and song birds, however, very few studies have concentrated on the long term impact of biomass agricultural production on small mammals. Since the area was cultivated and planted in the aforementioned manner prior to the start of my research, we have no data regarding former resident birds and mammals that were dependent on the corn / soybean production and that were possibly displaced as a result of the agricultural shift. As a result, it is assumed that the mammal species present when the study began were already represented in the corn / soybean agricultural matrix. This study focuses solely on the mammals residing in our research area at the start of trapping in 2011 through 2013 and the subsequent impact of perennial grass agricultural production on the resident mammals. Trapping nights differed slightly between the 2011 to 2013 trapping periods, simply because as an agricultural production area, traps were removed at times for machinery to enter into the research area to harvest the bioenergy crop.

Due to the harsh environmental conditions during the initial survey of the research area and only early successional growth was present, we began this study with the hypothesis that very few mammals would be located in the research area. Contrary to the hypothesis, in 2011 we captured 290 animals. Our initial hypothesis was not supported.

Between 2011 and 2012, it appeared that the biomass agricultural area was beneficial for small mammals, due to the numbers of mammals captured throughout the research area. In a study conducted in Texas (George, 1985), researchers found switch grass to be a good pheasant nesting cover and stated that more than 60 birds and mammals used the seeds and other plant material for food. In Murray and Best (2003), the researchers theorized that the late autumn / winter harvest would enhance bird reproductive fitness because agricultural activity would not disturb nesting. The initial phase of trapping appeared to support previous research regarding the benefits of warm season perennial grasses to birds, but researchers also found that food availability was terribly low in switchgrass monocultures due to lack of desirable forbs and excessive grass density after 5-6 years (Harper et al., 2008).

In the second year of the study, we hypothesized that with the shift to biomass agriculture, the abundance of seed available for the small mammals would increase the numbers in the population currently present. As the numbers of small mammals increased, eventually, raptors and larger mammalian predators would be attracted to the research area, but that they would not remain due to a lack of water. Without natural predators remaining near the research area to control the population of small mammals, we hypothesized that the small mammals would see an exponential growth over the coming months and years that could have a detrimental effect on the biomass growth due to over consumption of plant material and on the population by increased competition for resources.

In 2012, *P. maniculatus* was captured 260 times, *B. brevicauda* was captured 2 times, *M. ochrogaster* was captured 8 times, *Rattus norvegicus* (brown rat) was captured 1 time and 1 *Sylvilagus floridanus* (Eastern cottontail) was captured. Midway through the 2012 trapping season, it became apparent that *P. leucopus* had disappeared from the research area. We began to look into why *P. leucopus* would have disappeared. Had the switch to biomass agriculture caused the local extinction of *P. leucopus*? We ran a Chi square test to see if there was any correlation between the disappearance of *P. leucopus* and the increase in *P. maniculatus*. We found a correlation between the absence of *P. leucopus* and the increased population of *P. maniculatus* ($p < .05$), suggesting competitive exclusion (Gause, 1934). Even though the data suggested competitive exclusion and subsequent local extinction of *P. leucopus* by *P. maniculatus,* there were several problems with making that statement.

Maron and Pearson (2011) stated that predation could have an impact on numbers in a population. Over the

2011-2013 research periods, we found scat on one occasion in 2011 from a large mammalian predator. A field examination of the scat revealed bones from unidentified mammals within it, proving that at least one mammalian predator had been on the fringes of the research area. We do not believe that predators had a large impact on population numbers, since there is no readily available ground water to keep them in the research area. This instance of predation was more than likely an isolated nomad travelling through the area.

At the onset of the study, drought conditions overcame the region (and continued into 2013). Disturbance events like natural disasters or severe climatic extremes allow for rapid population shifts and for the establishment of invasive species (Brown, 2011). Due to an absence of available ground water and shelter from the extreme temperatures that season, the drought conditions may have forced *P. leucopus* to find a habitat that was less harsh and the species may have moved from the study area. This explanation is plausible, but we did not believe that this is the full story.

Kantak (1983) showed that both species of *Peromyscus* had dietary overlaps, which in our research, could explain the population explosion in *P. maniculatus* while *P. leucopus* declined to nothing; again suggesting support for the theory of competitive exclusion. Kantak's study indicated that when mice of both species were paired together, *P. maniculatus* instead was dominated by *P. leucopus; P. maniculatus* showed tendencies to avoid contact with *P. leucopus* and *P. leucopus* was the prevailing species in overall competition. *P. leucopus* tended to be the mouse to display territoriality over *P. maniculatus* and for the theory of competitive exclusion to be valid, *P. maniculatus* would have had to have shown the propensity to dominate over *P. leucopus*. It did not. This theory was refuted.

In the study by Kantak (1983), it was observed that *P. leucopus* does not choose to eat seed from *Sorghastrum nutans* (Indian grass) or *Andropogon gerardii* (Big Bluestem), both of which are in abundance at the research area. Kantak used mice taken both from old fields and from prairies. In her study, Kantak observed that when *P. leucopus* and *P. maniculatus* were given the choice, they preferred to eat old field grass seeds over those of prairie grasses, so much so, that the amount of prairie grass seeds eaten was recorded as negligible (Kantak, 1983). Kantak observed that *P. leucopus*, rather than eating prairie grasses, preferred eating old crop grasses such as *Elymus repens* (quack grass) and *Setaria faberi* (giant foxtail), as well as forbs. All mice (*P. leucopus and P. maniculatus*) used in the Kantak study, no matter where they originated, preferred the old field grass seeds. At our research area, *P. leucopus* could have emigrated from the research area as a result of food resources becoming unpalatable, but this explanation does not appear to be supported. Since both *P. leucopus* and *P. maniculatus* appear to utilize the same resources (Kantak, 1983), we should have observed a marked decrease in the numbers of *P. maniculatus* as well; we observed the opposite. Numbers of *P. maniculatus* increased substantially between 2011 and 2012. This explanation was also refuted.

An observation that we had early on was the association between the captured mice and *A. rudis* (water hemp). It appeared that the mice were utilizing the bases of the water hemp plants as entrances to their burrows, since the ground around the large water hemp plants was split open due to the drought conditions and in other locations of the research area, the ground was hard and smooth (compacted). In DeJaco and Batzli (2013), the authors also do a detailed study of the prominent plants eaten by white footed mice and deer mice, as well as meadow voles and eastern cottontail rabbits, all of which I have captured in my study. Specifically of interest is the fact that white footed mice were found to eat water hemp readily in this study (DeJaco & Batzli, 2013).

The plant composition in the research area changed dramatically over the 3 year trapping period. In the first year of the study, the research area was primarily occupied by warm season perennial grasses, old field grasses, forbs, and *A. rudis*. The vegetation initially was sparse, with only about 50 percent coverage of the available ground. Within five months of planting, the warm season perennial grasses had obtained a height similar to that of the established *A. rudis*. In November 2011, the City of Decatur, Illinois threatened CAT with an "Ordinance Violation for Grass and Weeds" due to citizen complaints regarding the research area's unkempt appearance and a lack of knowledge regarding the aims of the research. To prevent civil penalties, CAT had a local farmer come to the research area and mow the plots to a height of 30 cm. What this did in effect was remove the availability of seeds from *A. rudis* that the mammals had been using for food; the warm season perennial grasses were cut prior to dropping seeds and all of the herbaceous material in the research area was now on an even footing.

By the second year of the study, the research area had become dominated by warm season perennial grasses and *A. rudis*, but still had a diverse mixture of forbs in sections with approximately 75 to 80 percent of the ground covered in vegetation. By the third year of the study (2013), the warm season perennial grasses now dominated 95 (+) percent of the landscape in a thick monoculture with few forbs around the perimeter of the grass fields. No bare ground was visible and the grasses had thickened to a near impenetrable jungle more than two meters in height. *A. rudis* had become choked out by the perennial grasses and was absent from the research area.

In 2013, *P. maniculatus* captures dropped to 11, *S. floridanus* was captured 4 times, *B. brevicauda* was captured 1 time and *M. Ochrogaster* was captured 2 times. It is a possibility that the absence of water hemp from the research area in 2013 is a direct cause of the noticeable absence of *P. maniculatus*. In the early successional field, *A. rudis* had become a pulsed resource (Marcello et al., 2008, Pedersen et al., 2008) which had caused a population explosion in both *P. leucopus* and *P. maniculatus*. Being the dominant species, *P. leucopus* had obtained control of the source habitat of the research area, relegating *P. maniculatus* to the sink habitat areas of the fields. Due to low instances of predation and a greater quantity of sink habitat that was not low in quality overall, the numbers of *P. maniculatus* increased at a far greater rate than those of *P. leucopus* (Schreiber & Kelton, 2005). As the density of the perennial grasses increased, the source habitat occupied by *P. leucopus* became sink habitat as well. As a result of habitat becoming unfavorable, resources becoming limited, and *P. maniculatus* densities being higher than *P. leucopus* could push out, the shifting agricultural mosaic caused the local extinction of *P. leucopus*.

As the warm season perennial grasses entered their third year of growth, the fields became an impenetrable jungle, choking the old field grasses and forbs out of the landscape. At the onset of the study, drought conditions overcame the region and continued into 2013. Shenbrot et al. (2010) found that rodent densities in most cases were positively correlated with total precipitation during the previous rainy season and stochastic models produced by Reed et al. (2007) predicted a marked decrease in population growth rate as the frequency of dry years increased. After two seasons that would have had a positive impact on reproduction in *P. maniculatus* due to an increase in food resources, a population crash occurred. In order to stabilize the population, there had to have been a higher death rate than birth rate among some age groups (Boonstra et al., 1998). Due to a cascade of effects (Smith, 2010) occurring at our research site (a high death rate in the breeding cohort, the 3rd year of drought conditions and a reduction in habitat and resources), *P. maniculatus* population numbers could not be sustained and the population crashed. Our secondary hypothesis was partially supported.

5. Conclusion

Warm season perennial grass agriculture is a third year crop. Early successional habitat in our research area was shown to be beneficial temporarily for wildlife, but upon the warm season perennial grasses reaching maturity during the second and third years, the resulting monoculture became unfavorable and unpalatable for some mammals in our research area. It is our position that CRP acreage being converted into biomass agriculture should be designed as a mixture of short and tall warm season perennial grasses, as well as forbs to maximize their benefits to the ecosystem. By utilizing a variety of plants, structure for wildlife can be provided, soil can be stabilized in marginal land areas to limit the erosion of soil and field runoff, food resources for wildlife can be maintained and the overall heterogeneity of the ecosystem can be improved.

Acknowledgements

We would like to thank Dr. Judith A. Parrish, The Agricultural Research Institute, Caterpillar, Inc., the Illinois Department of Natural Resources and Millikin University for the opportunity to conduct this research. I would also like to thank Dr. Casey Watson and Dr. Travis Wilcoxen for their support, assistance with statistics and guidance during the course of this study. I would like to especially thank my wife Mary Leimbach for her support and children, David J. and Kendall S. Leimbach, for their assistance with checking traps, recording data and hauling equipment.

References

Batie, S. S. (2009). Green payments and the US Farm Bill: information and policy challenges. *Frontiers in Ecology and the Environment, 7*, 380-388. http://dx.doi.org/10.1890/080004

Boonstra, R., Krebs, C. J., & Stenseth, N. C. (1998). Population cycles in small mammals: the problem of explaining the low phase. *Ecology, 79*(5), 1479-1488. http://dx.doi.org/10.1890/0012-9658(1998)079%5B1479:PCISMT%5D2.0.CO;2

Brown, L. R. (2011). Eroding Futures: Why healthy soil matters to civilization. *Futurist, 45*, 23-30.

DeJaco, C. E., & Batzli, G. O. (2013). Palatability of plants to small mammals in nonnative grasslands of east -central Illinois. *Journal of Mammalogy, 94*, 427-435. http://dx.doi.org/10.1644/12-MAMM-A-157.1

Fiske, D. A., & Wolf, D. (1995). Planting and managing switch grass for forage, wildlife and conservation. *Virginia Cooperative Extension,* 418-013.

Gause, G. F. (1934). *The struggle for existence.* Williams and Wilkins, Baltimore, MD, United States of America. http://dx.doi.org/10.5962/bhl.title.4489

George, R. R. (1985). Reseeding pastures and rangeland for wildlife benefits in the central and southern Great Plains. Iowa Wildlife Research Bulletin (1st ed.). 37B-42B.

Harper, C.A., Moorman, C. E., & Keyser, P. D. (2008). Native warm-season grasses and early successional wildlife habitat: past lessons and a new vision. *Proceedings Eastern Native Grass Symposium, 6*, 120-126.

Hartman, J. C., Nippert, J. B., Orozco, R. A., & Springer, C. J. (2011). Potential ecological impacts of switch grass (*Panicum virgatum L.*) biofuel cultivation in the Central Plains, USA. *Biomass and Bioenergy, 35*, 3415-3421. http://dx.doi.org/10.1016/j.biombioe.2011.04.055

John, S., & Watson, A. (2007). *Establishing a grass energy crop market in the Decatur area* (1st ed.). Decatur, IL: The Agricultural Watershed Institute.

Jones, C., McShae, W. J., Conroy, M. J., & Dunz, T. H. (1996). Capturing mammals. In D. E. Wilson, F. R. Cole & J. D. Nichols (Eds.). *Measuring and monitoring biological diversity: standard methods for mammals* (pp. 115-155). Washington, DC, United States of America: Smithsonian Institution Press.

Kantak, G. E. (1983). Behavioral, seed preference and habitat selection experiments with two sympatric *Peromyscus* Species. *American Midland Naturalist, 109*, 246-252. http://dx.doi.org/10.2307/2425404

Manley, P. M., Van Horne, B., Roth, J. K., Zielinski, W. J., McKenzie, M. M., WELLER, T. J., … Vojta, C. (2006). Multiple Species Inventory and Monitoring Technical Guide: Version 1.0. United States Department of Agriculture. Print.

Marcello, G. J., Wilder, S. M., & Meikle, D. B. (2008). Population dynamics of a generalist rodent to variability in pulsed food resources in a fragmented landscape. *Journal of Animal Ecology, 77*, 41-46. http://dx.doi.org/10.1111/j.1365-2656.2007.01310.x

Maron, J. L., & Pearson, D. E. (2011). Vertebrate predators have minimal cascading indirect effects on plant production in an intact grassland ecosystem. *Ecology Letters, 14*, 661-669. http://dx.doi.org/10.1111/j.1461-0248.2011.01633.x

Mohlenbrock, R. H. (2002). *Vascular flora of Illinois: A field guide* (3rd ed.). Southern Illinois University: Carbondale, Illinois, United States of America.

Murray, L. D., Best, L. B., Jacobsen, T. J., & Braster, M. L. (2003). Potential effects on grassland birds of converting marginal cropland to switchgrass biomass production. *Biomass and Bioenergy, 25*, 167-175. http://dx.doi.org/ 10.1016/S0961-9534(02)00187-3

Pedersen, A. B., & Grieves, T. J. (2008). The interaction of parasites and resources cause crashes in a wild mouse population. *Journal of Animal Ecology, 77*, 370-377. http://dx.doi.org/10.1111/j.1365-2656.2007.01321.x

Reed, A.W., Kaufman, G. A., & Sandercock, B. K. (2007). Demographic response of a grassland rodent to environmental variability. *Journal of Mammalogy, 88*, 982-988. http://dx.doi.org/10.1644/06-MAMM-A-109R.1

Ryan, M. B., & Marks, R. (2005). Native warm season grasses and wildlife. United States Department of Agriculture, Natural Resources Conservation Service Leaflet Number, 25, 1-8.

Schreiber, S. J., & Kelton, M. (2005). Sink habitats can alter ecological outcomes for competing species. *Journal of Animal Ecology, 74*, 995-1004. http://dx.doi.org/10.1111/j.1365-2656.2005.00996.x

Shenbrot, G., Krasnov, B., & Burdelov, S. (2010). Long-term study of population dynamics and habitat selection of rodents in the Negev Desert. *Journal of Mammalogy, 91*, 776-786. http://dx.doi.org/10.1644/10-MAMM-S-154.1

Sikes, R. S., Gannon, W. L., and the Animal Care and Use Committee of the American Society of Mammalogists. (2011). Guidelines of the American Society of Mammalogists for the use of wild mammals in research. *Journal of Mammalogy, 92*, 235-253. http://dx.doi.org/ 10.1644/10-MAMM-F-355.1

Smith, F. S. (2010). Texas today: A sea of the wrong grasses. *Ecological Restoration, 28*, 112-117. http://dx.doi.org/10.3368/er.28.2.112

Yarnell, R. W., Scott, D. M., Chimimba, C. T., & Metcalfe, D. J. (2007). Untangling the roles of fire, grazing and rainfall on small mammal communities in grassland ecosystems. *Oecologia, 154*, 387-402. http://dx.doi.org/10.1007/s00442-007-0841-9

21

Elements of Opportunity and Poultry Farms Performance in Delta State, Nigeria

Rodney A. Isiorhovoja[1]

[1] Department of Agricultural Economics and Extension, Delta State University, Asaba Campus, Nigeria

Correspondence: Rodney A. Isiorhovoja, Department of Agricultural Economics and Extension, Delta State University, Asaba Campus, Nigeria. E-mail: raisiorhovoja@gmail.com

Abstract

The paper examines the existence, if any, of differences in gross margin between rural and urban areas in Delta State, Nigeria. Data were collected from all 275 poultry farmers registered with the Delta State Ministry of Agriculture, Livestock Department. The null hypotheses was that there is no significant difference in poultry farm gross margin between locations in terms rural and urban areas; managers with formal education in agriculture and managers who have no formal education in agriculture; and managers who have and who do not have prior experience in poultry business. Data were collected from all 275 poultry farmers registered with the State Ministry Agriculture using copies of a structured questionnaire and were analyzed using frequency counts, means and T-test. Amongst the findings were: Majority of poultry business operators have low level formal education in disciplines not related to agriculture; there was a significant difference in the mean number of years of schooling and courses studied between rural and urban areas but that there was no significant difference in number of years of prior experience. The T-test results failed to reject the three null hypotheses. The study concluded that indeed elements of opportunity may vary from place to place but the ability to exploit the benefits may moderate or accentuate performance. Entrepreneurial capacity building was recommended for poultry business operators' state wide.

Keywords: elements of opportunity, entrepreneur, gross margin, rural, urban

1. Introduction

A type of dual economy, one rural and the other urban, results when social and economic infrastructures are unevenly distributed across a nation. This phenomenon presents part challenges of competitive demand for land particularly in urban centers when large expanse of land are either underutilized or lying waste in rural areas. Increasing concentration of agricultural ventures, of which poultry business is one, contributes to these challenges.

Entrepreneurial activity results when elements of opportunity and motivation converge. However, given motivation, Shane et al. (2003), opine that entrepreneurship depends on differential access to information about economic opportunities and to the resources needed to launch a new venture. Baum et al. (2001), in a slightly different but more definitive terms reported that motivation and factors related to the organization affect the performance of the venture. By this, differences between or among organizations in motivation and organizational factors will lead to variation in performance among organizations. Jackson and Okhomina (2006) found that the level of education of the individual entrepreneur and his/her prior experience in a related business, influence business outcome. Low, Hendersen, and Weiler, (2005) added that proximity to densely populated area, infrastructure (roads and telecommunication networks and access to financial capital) also affect business success.

There is therefore *prima facie* justification for the concentration of economic activities in the cities or urban areas in Nigeria, like in many developing countries, where there is marked disparity in social and economic infrastructure distribution between rural and urban areas. There is however need to ascertain the veracity of the submission by Low, Hendersen, and Weiler, (2005) in different economic environment. Results from such study could provide inputs for a policy that would provide part solution to the imbalances in the availability of social and economic amenities.

The overall objective of this study is to assess the veracity of the claims made in the literature that organizations specific factors and proximity to population, social and economic infrastructure affect business performance. The specific objectives are to:

a) ascertain the existence, if any, of difference in gross margin between poultry farms in urban and rural areas,

b) ascertain the existence, if any, of difference in the gross margin of poultry farms with managers who have formal qualification in agriculture or related disciplines and poultry farms with managers without formal education or qualification in agriculture or related discipline; and

c) ascertain any significant difference in gross margin between poultry farms with managers who have and poultry farms with managers who do not have prior experience in poultry farm management.

The hypotheses tested were:

H_{01}: *There is no significant difference in poultry farm gross margin between poultry farms in urban areas and poultry farms in rural areas.*

H_{11}: *There is significant difference in poultry farm gross margin between poultry farms in urban areas and poultry farms in rural areas.*

The performance of business organizations in agriculture is commonly measured in terms of gross margin. Gross margin is the excess of revenue over the variable costs incurred to generate it. Sustained and growing gross margin ensures business survival in the short run and possible expansion and diversification in the long run. The urban area contrasts the rural area in terms of high population, availability of market and favored in terms of the presence of social and economic infrastructure. If significant difference in business performance can be traced to these factors, valuable input would have been made for policy initiative that could redress part causes of urbanization.

H_{02}: *There is no significant difference in poultry farm gross margin between poultry farms with manager who have formal qualification in agriculture or related disciplines and poultry farms with managers who do not have formal education or qualification in agriculture or related disciplines.*

H_{12}: *There is significant difference in poultry farm gross margin between poultry farms with manager who have formal qualification in agriculture or related disciplines and poultry farms with managers who do not have formal education or qualification in agriculture or related disciplines.*

H_{03}: *There is no significant difference in poultry farm gross margin between poultry farms with managers who have and poultry farms with managers who do not have prior experience in poultry farm management*

H_{13}: *There is significant difference in poultry farm gross margin between poultry farms with managers who have and poultry farms with managers who do not have prior experience in poultry farm management.*

Hypotheses ii and iii have to do with the organization specific factors. The rationale for them is predicated on the rural urban migration syndrome in Nigeria which has virtually deprived the rural areas of needed educated and skilled manpower to drive economic activities in those areas. Empirical evidence regarding the claims in the literature of the positive effects of organization specific factors on business performance, which is the focus of these two hypotheses, will be useful in formulating policy initiatives that could redress the imbalance that results in rural-urban drift of human capital.

2. Materials and Methods

2.1 Area of Study

The study area is Delta State of Nigeria. Delta State lies between longitude 5^0 and $6^0\ 45^1$ East and latitude $5^0\ 20^1$ and $6^0\ 30^1$ North of the Equator. The study covered the three agricultural zones in the State, namely; Delta North, Delta Central and Delta South agricultural zones. The projected population of the State for 2015 is 4.9million.

2.2 Population and Sampling Procedure

About 100% of households in Delta State are involved in one form of poultry keeping or another (Adene & Oguntade, 2006). A total of 275 poultry (chicken) farmers were registered with the Ministry of Agriculture, Delta State, Asaba, as per records of the Livestock Department, as at February 2010. The total registered poultry farms were covered in the study.

2.3 Data Collection Procedures

Information on location of poultry farms was gathered from the records at Ministry. Data were collected using copies of a structured questionnaire. Staff of the Ministry, in Poultry extension Unit under the Livestock department, assisted in administering the questionnaire after a one day orientation. Staffs were assigned to cover their respective zonal office areas.

2.4 Data Analysis

Data were grouped by location into rural and urban farms and analyzed by the use of frequency counts and means. Gross margin of poultry farms was determined by using the formula GM = TR – TVC. Where GM is gross margin, TR is total revenue and TVC is total variable costs. The three hypotheses were tested by applying t-test of difference between mean gross margins.

3. Results

3.1 Socio-Economic Characteristics of Poultry Farmers

The 275 poultry farmers was made up of 153 (55.60%) in the rural areas and 122 (44.40%) in urban areas. Their ages ranged from 18 to 68 years with a mean of 43.91 years. The mean age varied between locations but not significantly so (t = 0.92, df = 273, P = 0.33). It was 43.28 years with a standard deviation of 10.84 years for farmers in rural areas and 44.51 years with a standard deviation of 7.83years for farmers in urban areas. The modal age bracket of respondents was 38 – 47 years state wide but higher for the urban areas where it was 48 - 57 years.

Female (58.18%) predominate the subsector (Table 1). Concerning marital status, about 77.50% of the respondents were married, 20% had never married and 2.50% were single again. Poultry production is apparently not popular with especially the single again in Delta State, Nigeria.

Table 1. Demographic characteristics of Poultry farmers by location

	Group	Location		Total
		Rural	Urban	
Age	Group (yrs)			
	18-27	6 (2.20%)	14 (5.10%)	20 (7.30%)
	28-37	15 (5.50%)	22 (8.00%)	37 (13.50%)
	38-47	76 (27.60%)	33 (12.00%)	109 (39.60%)
	48-57	48 (17.50%)	46(16.70%)	94 (34.20%)
	58-67	6 (2.20%)	7(2.50%)	13(4.70%)
	68-77	2 (0.7%)	0 (0.00%)	2 (0.70%)
	Total	**153 (55.60%)**	**122 (44.40%)**	**275 (100%)**
Gender	Female	93 (33.80%)	67 (24.40%)	160 (58.20%)
	Male	60 (21.80%)	55 (20.00%)	115 (41.80%)
	Total	**153 (55.60%)**	**122 (44.40%)**	**275 (100%)**
Marital status	Single	28 (10.20%)	27 (9.80%)	55 (20.00%)
	Married	120 (43.60%)	93 (33.80%)	213 (77.50%)
	Single again	5 (1.80%)	2 (0.7%)	7(2.50%)
	Total	**153 (55.60%)**	**122 (44.40%)**	**275 (100%)**

Source: Field survey 2010.

3.2 Human Capital Characteristics

On educational status, (Table 2), about 4.40% of the respondents had no formal education another 4.40% had primary school leaving certificate and 21.10% had Senior secondary school certificate (SSC) or its equivalent. National Diploma (ND)/Nigerian Certificate of Education (NCE) are the prevalent qualifications in the subsector. These are low education, low agriculture content qualifications. University degree holders constitute about 28% of the respondents. There was significant difference in mean number of years of schooling (t = 2.63, df = 273, P = 0.01) between the urban area (14.79 years) and the rural areas (13.54 years).

A high percentage of the respondents (90.54%) had prior experience in poultry management. This figure is near evenly distributed between the rural (48.72%) and urban (41.82%) areas. There was no significant difference in the mean number of years of prior experience ($t = 0.52$, $df = 247$, $P = 0.61$). Apparently, apprenticeship is an important approach to human capital development for poultry production in the state.

Respondents who have formal educational qualification in agriculture were 13.50% while 11.30% studied agriculture related courses. The urban area was less favored in respondents who studied agriculture or related courses. Majority (57.10%) had their training in management, medicine and more importantly others which had nothing to do with agriculture. The importance of formal education to successful entrepreneurship is expressed profusely in the literature.

Table 2. Human capital characteristics of Poultry farmers

Group		Location		Total
		Rural	Urban	
Education	No formal	9 (3.30%)	3 (1.10%)	12 (4.40%)
	Primary	6 (2.20%)	6 (2.20%)	12 (4.40%)
	Secondary	36 (13.10%)	22 (8.00%)	58 (21.10%)
	ND/NCE	69 (25.10%)	47 (17.10%)	116 (42.20%)
	First degree/equivalent	27 (9.80%)	39 (14.20%)	66 (24.00%)
	Higher degree	6 (2.20%)	5 (1.80%)	11(4.00%)
	Total	**153 (55.60%)**	**122 (44.40%)**	**275 (100%)**
Prior Experience	No	101 (36.70%)	79 (28.70%)	180 (65.50%)
	Yes	52 (18.90%)	43 (15.60%)	95 (34.50%)
	Total	**153 (55.60%)**	**122 (44.40%)**	**275 (100%)**
Operated business before age 20	Yes	54 (38.20%)	51(18.60%)	105 (38.20%)
	No	99(36.00%)	71 (25.80%)	170 (61.80%)
	Total	**153 (55.60%)**	**122 (44.40%)**	**275 (100%)**
Area of study	Agriculture	26 (9.50%)	11 (4.00%)	37 (13.50%)
	Agriculture related	22 (8.00%)	9 (3.30%)	31 (11.30%)
	Management	23 (8.40%)	17 (6.20%)	40 (14.50%)
	Medicine	4 (1.50%)	6 (2.20%)	10 (3.60%)
	Others	78 (28.40%)	79 (28.70%)	157 (57.10%)
	Total	**153 (55.60%)**	**122 (44.40%)**	**275 (100%)**

Source: Field survey 2010.

3.3 Economic Characteristics of Poultry Farmers

The flock sizes ranged from 50 to 10,000 birds with a mean of 1470 birds. About 81.10% of the respondents had between 50 and 2049 birds (Table 3) and 14.20% had birds numbering from 2049 to 4049. Only 1.50% (4) of the respondents had birds numbering 8050 or above, most of which were in the urban areas. Out of the 275 poultry farmers, about 55.60% (153) were in the rural area and 46.40% (122) were in the urban area. There was however no significant difference in the mean number of birds between the two locations ($t = 0.53$, $df = 273$, $P = 0.60$).

Table 3. Distribution of birds by location

	Group	Location		Total
		Rural	Urban	
No. of birds	50 - 2049	120 (43.60%)	103 (37.50%)	223 (81.10%)
	2059 - 4049	26 (9.50%)	13 (4.70%)	39 (14.20%)
	4059 - 6049	6 (2.20%)	3(1.10%)	9 (3.30%)
	6050 - 8049			
	8050 - 10049	1 (0.40%)	3 (1.10%)	4 (1.50%)
	Total	**153 (55.60%)**	**122 (44.40%)**	**275 (100%)**

Source: Field survey 2010.

3.4 Relationship between Poultry Farm Performance and Farm Location

To realize objective one, the hypothesis:

H_{01}: *There is no significant difference in poultry farm performance between poultry farms in the rural areas and poultry farms in the urban areas was tested using t-test of equality of mean gross margin between rural and urban areas.*

The result (Table 4) shows that there was no significant difference (t-statistic = 0.681, P = 0.50) in mean gross margins between the two locations. We therefore accept the null hypothesis and reject the alternative that there is significant difference in farm performance between poultry farms in the urban areas and poultry farms in the rural areas. That mean gross margin per farm was higher in the urban areas by ₦214,785.00 was however indicated.

Table 4. T-test for equality of mean gross margin between urban and rural areas

	N	Mean	df	Mean difference	t-ratio	Sig. (2-tailed
Rural	153	1222100	273	-214785	-0.68	0.50
Urban	122	1436900				

Source: Field survey data analysis result, 2010.

3.5 Relationship between Farm Performance and Organization Factors (Human Capital)

Two elements of human capital, relevance of education (area of study) and prior experience in poultry production management were used to realize objectives two and three in that order. On the relevance of education, the hypothesis states:

H_{02}: *There is no significant difference in poultry farm gross margin between poultry farms with manager who have formal qualification in agriculture or related discipline and managers who do not have formal education or qualification in agriculture related discipline.*

The t-test result presented in Table 5 shows that there is no significant difference in the mean gross margin of poultry farms with managers who had formal qualification in agriculture or related disciplines and farms managers who had no formal education in agriculture or related disciplines. Farms with managers who had qualification in agriculture or related discipline had a mean of ₦1,075,500.00 with a standard deviation of ₦236,048.00. Similar statistics for poultry managers who had no formal qualification in agriculture or related disciplines were N1,396,900.00 and ₦266,900.00. The no significant difference in gross margin implies a rejection of the alternative hypothesis and the acceptance of the null hypothesis that there is no significant difference in poultry farm gross margin between poultry farms with managers who have formal qualification in agriculture or related disciplines and poultry farms with managers who do not have formal education or qualification in agriculture related discipline.

The result of the third Hypothesis:

H_{03}: *There is no significant difference in poultry farm gross margin between poultry farms with managers who have and poultry farms with managers who do not have prior experience in poultry farm management before coming to present farm,*

presented in Table 6 indicates that prior experience failed to account for any significant variation in poultry farm gross margin (t = 0.89, df = 273, P = 0.70) between locations.

Table 5. T-test for equality of mean gross margin with respect to formal qualification in agriculture or related courses

	N	Mean	df	Mean difference	t-ratio	Sig. (2-tailed
Agric	68	1075500	273	-321324	0.89	0.38
Non Agric	207	1396900				

Source: Field survey data analysis result, 2010.

Table 6. T-test of equality in gross margin with respect to having prior experience in poultry production or not

	N	Mean	df	Mean difference	t-ratio	Sig. (2-tailed
With prior exp.	181	1231200	273	-130897	0.40	0.70
Without	94	1362100				

Source: Field survey data analysis result, 2010.

Hence we accept the null hypothesis that there is no significant difference in gross margin between poultry farms with managers who have and poultry farms with managers who do not have prior experience in poultry management.

4. Discussion

Empirical results failed to reject any of the three null hypotheses that guided this study. The average of 14.79 years spent in school falls short of the minimum of 16 years required to obtain a university degree in Nigeria. The implication is that on the average poultry business operators have low education and low skills which may not be adequate for transforming the subsector in a dynamic business environment. It also implies that individuals who enter into this class of entrepreneur through apprenticeship could not have been properly equipped. This is deduced from the fact that about 57.10% of these poultry farmers had their education in areas other than agriculture. They therefore lack agriculture specific knowledge. Ferrante and Sabatini (2007) are of the opinion that the higher the level of sophistication in technological innovation the higher the level of education necessary to enable both the adaptation and advancement in technology in the particular industry. Furthermore, that the motive for entrepreneurship tend to include need to apply ones excess capacity profitably for the highly educated, whereas for the less educated it tends to be reasons of 'no other job'. Holcombe (2003) is of the view that learning or education is central to creating insight and alertness and the capability of responding to changes in the business environment. Following Holcombe (2003), the lack of knowledge specific to agriculture among a greater majority of the poultry farmers could impair the ability of poultry farmers to see opportunity in policy changes since they lack the required knowledge to interpret the opportunity symptoms therein and to make the required response by way of adaptation of technology in the subsector which could have translated into higher productivity. This, in part, may account for the technical inefficiency observed by Alabi and Aruna (2005) in the poultry subsector. Empirical results from ascertaining the effects of elements of opportunity on business performance in this circumstance could therefore give unreliable signals in the absence of necessary human capital required to infuse the entrepreneurial capital for effective poultry business performance in the subsector. Be this, as it may, it is worthy of note that not rejecting the null hypotheses could indicate integration of market in poultry production in the State.

5. Conclusion

There is ample evidence to suggest that poultry business operators across Delta State, Nigeria, are ill equipped to operate in the subsector by way of proper education and/or training. This has adverse implications for their ability to complement the poultry business environment in the State resulting in the not rejecting the three null hypotheses in this study contrary to *a priori* expectations. This leads to the conclusion that indeed elements of opportunity may vary from one business organization to another but that differences in ability to exploit the benefits may moderate or accentuate the initial differences. It is therefore essential that the entrepreneurial capacity of the poultry business operators be built-up to enable them to effectively explore the elements of opportunity that may be available in their environment and globally

References

Adene, D. F., & Oguntade, A. E. (2006). The Structure and Importance of The Commercial and Village Based Poultry Industry in Nigeria. Rome. Food and Agriculture Organization (FAO).

Alabi, R. A., & Aruna, M. B. (2005). Technical Efficiency of Family Poultry in Niger Delta, Nigeria. *Journal of Central European Agriculture, 6*(4), 531-538.

Baran, D., & Velickaite, R. (2008). Building the theoretical Framework of Entrepreneurship. *Proceedings of the 5th International Conference Business and Management 16-18 May* (pp. 21-30) Vilnius Lithuania.

Ferrante, F., & Sabatini, F. (2007). Education, Social Capital and Entrepreneurial Selection in Italy. Retrieved July 29, 2008, from http://mpra.ub.uni-muenchen.de/2451/MPRA Paper No. 2451

Fitzsimons, P., O'Gorman, C., Hart, M., & McGloin, E. (2003). Entrepreneurship on the Island of Ireland. Based on the research findings of The Global Entrepreneurship Monitor 2003. Dublin. InterTrade

Holcombe, R. G. (2003). The Origin of Entrepreneurial Opportunities" T*he Review of American Economics, 16*(1), 25-43.

Jackson, R. S., & Okkhomina, A. (2006). Sociological Factors, Psychological Traits and Entrepreneurial Personality. Retrieved August 28, 2009, from www.sbaer.uca.edu/research/asbe/2006

Lichter, S. J., King, A. S., & Catalanello, R. F. (1983). Entrepreneurial Potential: An Experiential Exercise in Self Analysis and Group Assessment. *Development in Business and Experiential Exercises* (Vol. 10, pp. 88-91).

Low, S., Hendersen, J., & Weiler, S. (2005). Gauging a Region's EntrepreneurialPotential. Federal Reserve Bank of Kansas City. Economic Review. Third quarter 2005.pp61- 89. Retrieved October 25, 2007, from www.kc.frb.org/publication/Econrev/pdf/3q05/ON

Shane, S., Lockea, E.A. & Collins, C.J. (2003). "Entrepreneurial Motivation". *Human Resource Management Review* (Vol.13, pp.257–279).

UNDP. (2006). Socioeconomic Impact of Avian Influenza in Nigeria. UNDP Nigeria, Abuja.

22

Maize Fertigation with Treated Olive Mill Wastewater: Effects on Crop Production and Soil Properties

Maria I. Kokkora[1], Chryssoula Papaioannou[1], Panagiotis Vyrlas[1], Konstantinos Petrotos[1], Paschalis Gkoutsidis[1] & Christos Makridis[1]

[1] Department of Agricultural Engineering Technologists, Technological Educational Institute of Thessaly, Larissa, Greece

Correspondence: Maria I. Kokkora, Department of Agricultural Engineering Technologists, Technological Educational Institute of Thessaly, Larissa, 41110, Greece. E-mail: mkokkora@teilar.gr

Abstract

The present study investigates the potential of olive mill wastewater, treated by microfiltration and XAD4 macroporous resin, to be used as liquid fertilizer in maize production through a 2-year field experiment. The treated olive mill wastewater (T-OMWW) was applied at two rates of 25 t and 50 t per ha per year, supplemented with mineral fertilization. There was also a treatment involving the application of only T-OMWW at the rate of 50 t per ha per year, and an only mineral fertilizer treatment. Mineral fertilizers and T-OMWW were applied progressively through a drip irrigation system.

Maize grain and soil analysis showed that T-OMWW was capable to meet crop requirements in N, P and K, and increase soil N, P and K availability. There was a tendency for increasing soil Na and electrical conductivity (EC) using the higher rate of T-OMWW. Therefore, for sustainable agriculture, it may be safer to apply the T-OMWW at the lower rate of 25 t per ha per year, or use the higher rate of 50 t per ha every other year.

Keywords: clay loam soil, kernel protein content, liquid fertilizer, microfiltration, XAD4 resin, yield

1. Introduction

Olive oil consumption is associated with many health benefits, including protective effects against cardiovascular diseases (Covas, 2007), cognitive decline (Berr et al., 2009) and possibly breast cancer (Escrich, Moral, & Solanas, 2011). The olive oil extraction process, however, involves the generation of large amounts of olive mill wastewater (OMWW), a by-product that constitutes serious environmental problem in the Mediterranean region, due to its high polluting load. OMWW is characterized by high content of solids and organic compounds, high COD content, phytotoxic properties and resistance to biodegradation caused by its phenolic compounds (Zirehpour, Jahanshahi, & Rahimpour, 2012; Zaglis, Vavouraki, Kornaros, & Paraskeva, 2013).

Olive oil production in Greece is mainly carried out by small or medium enterprises that usually apply the untreated OMWW to nearby land, in order to avoid treatment costs. Crop response to OMWW application is variable. Research has shown that olive fruit yield and quality were not affected by OMWW application (Chartzoulakis, Psarras, Moutsopoulou, & Stefanoudaki, 2010). Ryegrass and proteic pea yields were increased with untreated OMWW application, whereas clover yield was negatively affected (Montemurro, Diacono, Vitti, & Feri, 2011). In another study, although maize growth was not affected, plant stress parameters increased following the application of untreated OMWW (Belaqziz et al., 2008). Hanifi and El Hadrami (2008) found increased maize yield following moderate and progressive OMWW application. Germination problems are also observed due to phytotoxic effects of the phenolic compounds contained in the OMWW (Mekki, Dhouib, & Sayadi, 2007; Massoudinejad, Arman, & Aghayani, 2014). The application of untreated OMWW to agricultural soil may increase soil organic matter, available P and K (Montemurro et al., 2011), and total N content, but also soil electrical conductivity and salinity (Belaqziz, Lakhal, Mbouobda, & El Hadrami, 2008), and modify the equilibrium of useful soil microorganisms (Barbera, Mauciery, Cavallaro, Ioppolo, & Spagna, 2013).

The phenolic compounds contained in the OMWW are natural antioxidants, with commercial and economic interest. Hence, the treatment of OMWW aiming at the recovery of the polyphenols could result in economic benefits for the olive mill, since the phytotoxic polyphenols will have been removed from the effluent. Membrane filtration of OMWW may result in a significant decrease of its organic load and suspended solids content (Russo, 2007, Zirehpour et al., 2012), and also in polyphenols separation from the mass of waste (Cassano, Conidi, & Drioli, 2011; Petrotos, Lellis, Kokkora, & Gkoutsidis, 2014; Rahmanian, Jafari, & Galanakis, 2014). OMWW treatment with microfiltration resulted in polyphenols separation in the permeate (Petrotos et al., 2014). Polyphenols may then be successfully removed with the use of suitable resins (Weisz, Schneider, Schweiggert, Kammerer, & Carle, 2010; Petrotos, Gkoutsidis, Kokkora, Giankidou, & Tsagkarelis, 2013; Zaglis et al., 2015). The recovered polyphenols may be utilized in the pharmaceutical, cosmetic and food industry and the remaining effluent will have decreased phytotoxic properties, and thus it may be more safely used in agriculture.

Research on the agronomic effects of treated OMWW application to agricultural soil is limited, and mainly involves OMWW that has been treated by chemical or biological techniques (Cereti et al., 2004, Barbera, Maucieri, Ioppolo, Milani, & Cavallaro, 2014; Moraetis, Stamati, Nikolaidis, & Kalogerakis, 2014). In this study, the effluent produced following the treatment of OMWW by microfiltration and XAD4 macroporous resin was applied by fertigation to maize production in a two-year field experiment. The aim of the study was to investigate its effects on crop production, with particular regards to grain yield and quality, and soil properties.

2. Method

2.1 Treated Olive Mill Wastewater (T-OMWW)

In 2013 and 2014, a sample of approximately 10 t of OMWW was collected each year from "Alevizos" olive mill, located in Pyrgetos village, Larissa, central Greece. Each year, the raw OMWW was initially centrifuged at 1200 rpm using a rotary finisher bearing a stainless screen with openings of 150 μm diameter. This first step aimed at separating the suspended solids contained in OMWW, in the form of sludge, in order to avoid clogging of the membrane used in the next step.

In the second step, the centrifuged OMWW was filtered using a ceramic microfiltration membrane of 200 nm pore size in order to separate the polyphenols in the permeate from the mass of waste (retentate). The effluent produced as permeate in this step was suitable to be applied through a drip irrigation system, with limited risk of emitters clogging.

As a final step, the permeate produced in the second step was passed through a column filled with XAD4 macroporous resin, which has the ability to retain selectively the polyphenols (Petrotos et al., 2013), aiming to recover the polyphenols and minimize any phytotoxic effects of the remaining effluent. The polyphenolic content of the remaining effluent was approximately 20-30% of the initial polyphenolic content of the input material.

The treated OMWW (the remaining effluent of the final stage) was considered for utilization in agriculture as a liquid fertilizer. Some quality properties of the treated OMWW (T-OMWW) are presented in Table 1 for each year. Table 1 shows that T-OMWW quality parameters varied between the two years, especially in respect of available nutrients. This variability is attributed to the annual variation of the nutritional quality of olives (input material) that may be processed by the olive mill each year, leading therefore to differences in the quality of OMWW and also T-OMWW.

Table 1. Treated OMWW physicochemical properties in 2013 and 2014

Parameter	2013	2014
pH	4.73	6.15
EC (mS cm^{-1})	9.91	8.30
Salinity (%)	8.9	6.0
Solid residue (105 °C) (%)	4.15	9.94
Available P (mg L^{-1})	1680	256
Extractable K (mg L^{-1})	1440	64
NH_4-N (mg L^{-1})	86	3.11
NO_3-N (mg L^{-1})	-	0.1

Note. EC: Electrical conductivity.

2.2 Field Experiment

A field experiment was carried out on an Inceptisol clay loam soil (41% sand, 20% silt, 39% clay) at the experimental farm of Technological Educational Institute of Thessaly, Larissa, Greece, in 2013 and 2014. Topsoil (0-0.3 m depth) quality properties in the beginning of the experiment are presented in Table 2.

Table 2. Topsoil (0-0.3 m depths) quality properties at the beginning of the experiment

pH	EC ($mS\ cm^{-1}$)	O.M. (%)	$CaCO_3$ (%)	Total N ($g\ kg^{-1}$)	Olsen P ($mg\ kg^{-1}$)	Extr. K ($g\ kg^{-1}$)	Extr. Mg ($g\ kg^{-1}$)
7.3	0.723	1.4	1.8	0.8	6.5	0.47	1.42

Note. EC: Electrical conductivity, O.M.: Organic Matter, Extr.: Extractable.

The experimental design involved four treatments: i) control (C): mineral fertilization only, ii) application of T-OMWW at the rate of 50 t ha^{-1} (50W), iii) combined application of T-OMWW at the rate of 50 t ha^{-1} with mineral fertilization (50W+f), and iv) combined application of T-OMWW at the rate of 25 t ha^{-1} with mineral fertilization (25W+f). In 2013, the mineral fertilizer addition was equal to 200 kg N per ha for each of the three treatments that received mineral fertilizer. Fertilizer nitrogen was applied as ammonium nitrate (34.5-0-0). In 2014, the mineral fertilization for the control treatment corresponded to the addition of 110 kg N, 180 kg P_2O_5, 27 kg K_2O and 78 kg SO_3 per ha. The mineral fertilizer addition for each of the combined T-OMWW and mineral fertilizer treatments was 55 kg N, 131 kg P_2O_5, 27 kg K_2O and 17 kg SO_3 per ha. Three fertilizers were used in 2014: peckacid (0-60-20), urea phosphate (17.5-44-0) and ammonium sulfate (21-0-0). Each year, each treatment was applied to an individual plot of 60 m^2 (6 m x 10 m, including 8 plant rows), using a complete randomized block design with four replicates. Maize (Zea mays) was used as the monitoring crop. Crop sowing was at the rate of approximately 8.6 seeds m^{-2}, and took place on May 25, 2013 and June 18, 2014.

In order to ensure germination, sprinkler irrigation was applied each year after sowing for seedling establishment. Water, mineral fertilizers and T-OMWW were applied through a drip irrigation system, employing four manifolds. Each manifold supplied a set of four plots with one drip lateral per two adjusted plant rows. The volume of required water controlled by a flow meter installed at upstream of each manifold. The 20 mm diameter emitting pipe used is commonly utilized for field crop irrigation, with pressure compensating emitters at 1 m spacing, discharging 3.6 L per hour.

Treated OMWW was applied through the drip system utilizing a 120 L tank connected to the main line and manipulating a throttling valve to create a differential pressure level. Each treatment received water and T-OMWW filtered through 1" conventional manual cleaning disk filters of 150 mesh. A preliminary 120 mesh screen filtration was operated on the main pipeline of the system. The secondary filters were cleaned after each T-OMWW application. Manual flushing of the laterals was performed every third week.

In 2013, five applications of T-OMWW took place between 1 July and 8 August, delivering in total 300 L for each plot receiving T-OMWW at the rate of 50 t ha^{-1} and 150 L for each plot receiving T-OMWW at the rate of 25 t ha^{-1}. In 2014, six applications of T-OMWW took place between 1 August and 5 September, delivering in total 300 L for each plot receiving T-OMWW at the rate of 50 t ha^{-1} and 150 L for each plot receiving T-OMWW at the rate of 25 t ha^{-1}.

All treatments were irrigated at 100% crop evapotranspiration (ET_c) during the full season, in both years of the experiment. An automatic weather station in the experimental field measured rainfall, solar radiation, air temperature and humidity, and wind speed. These parameters were used to calculate daily reference evaporation (ET_o). The irrigation applied through the drip system was scheduled using reference evaporation and growth stage based crop coefficient, according to FAO-56 methodology (Allen, Pereira, Raes, & Smith, 1998). Table 3 shows monthly values of the measured parameters during the experimental periods. In 2013, there was 108 mm of rainfall during the experiment.

In 2013, there was 108 mm of rainfall during the experiment. Total watering during the growing season was 500 mm with 312 mm applied through the drip system for all treatments. In 2014, there was 96 mm of rainfall during the experiment, the majority of which took place after the irrigation period. Total watering during the growing season was 576 mm, with 480 mm applied through the drip system for all treatments.

Table 3. Monthly total rainfall, maximum (max) and minimum (min) temperature and humidity, average (aver.) wind speed and solar radiation during the two experimental periods

	Total rainfall (mm)		Max air temperature (°C)		Min air temperature (°C)		Max humidity (%)		Min humidity (%)		Aver. wind speed (m s^{-1})		Aver. solar radiation (W m^{-2})	
	2013	2014	2013	2014	2013	2014	2013	2014	2013	2014	2013	2014	2013	2014
Apr	2.4	75.2	31.6	25.2	2.4	0.4	100	100	19.0	24.9	1.5	1.5	519.6	430.5
May	56.4	10.2	34.8	33.2	8.5	5.5	100	100	11.5	13.2	1.7	1.4	557.3	567.3
Jun	72.6	23.4	38.4	40.0	9.0	10.9	100	100	7.8	11.7	1.5	1.4	616.5	597.7
Jul	50.2	12.2	38.9	38.0	14.2	14.5	100	100	19.6	18.5	1.2	1.7	632.9	648.5
Aug	0.0	8.8	38.7	40.4	14.6	13.5	99.5	100	14.0	16.2	1.4	1.5	596.7	594.9
Sept	13.0	81.6	36.5	34.1	8.8	7.5	100	100	13.2	26.0	1.3	1.3	478.4	451.1
Oct	15.0	85.8	27.9	29.6	3.8	5.5	100	100	13.3	28.8	0.8	0.8	311.4	234.2

2.3 Measurements and Analysis

Crop production was determined at harvest (September 20, 2013 and September 25, 2014). Maize ears were harvested by hand from 10 maize plants from the central 4 rows of each experimental plot. Maize ears were dried in a ventilated oven at 55 °C, until constant weight. After drying, maize kernels were separated from the rest of the ear, weighted, grinded, and then analyzed for protein, starch, fiber, oil and ash content, using an automatic near infrared analyzer. Reported kernel protein, starch, fiber, oil and ash content were corrected to 0% moisture content.

Soil samples were collected from each plot three days after harvest each year. Samples were taken from 0-30 cm depth and analyzed for pH (1:1 water), EC (1:1 water), Olsen P, Kheldahl N, extractable K, Na (determined in ammonium acetate extract by flame photometry), Ca, Mg (determined in ammonium acetate extract by atomic absorption), NH_4-N and NO_3-N (determined in KCl extract by phasmatophotometry).

The effect of each treatment on crop and soil measured variables were assessed by ANOVA at the level of statistical significance of $p<0.05$, and means were separated by Duncan's multiple range test using the statistical program SPSS (SPSS Inc., Edit. 17.0, Chicago, USA).

3. Results and Discussion

3.1 Maize Kernel Yield and Quality

In both years of experimentation, the effect of T-OMWW application on maize kernel dry matter yield was not significant. As shown in Figure 1, however, in the second year there was a tendency for lower yield with the application of T-OMWW at the rate of 50 t per ha. This finding may be indicative of potential salinity effects on the crop. The higher salt concentration in the root zone resulting from the higher rate of T-OMWW application by drip irrigation may have contributed to the reduced yield, since maize is a salinity sensitive crop (Mahajan & Tuteja, 2005). Research work by Hanifi and El Hadrami (2008) showed variable toxicity levels depended on the salinity and phenolic content of olive mill wastewaters. They also showed that the application of OMWW at the rate of 30 m^3 ha^{-1} to maize production resulted in no physiological damage to the crop, which agrees with our results showing lack of adverse effects on yield with the lower application rate of T-OMWW (25 t ha^{-1}).

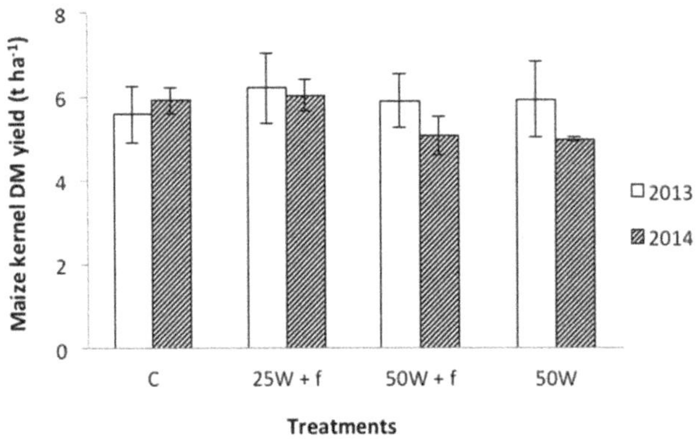

Figure 1. Maize kernel dry matter (DM) yield (in kg per ha) as affected by the different treatments, in the first and second year of the experiment. Error bars represent the standard error of the mean. (C: Control (mineral fertilization only), 25W + f: 25 t ha^{-1} $year^{-1}$ of T-OMWW plus mineral fertilization, 50W + f: 50 t ha^{-1} $year^{-1}$ of T-OMWW plus mineral fertilization, 50W: 50 t ha^{-1} $year^{-1}$ of T-OMWW only)

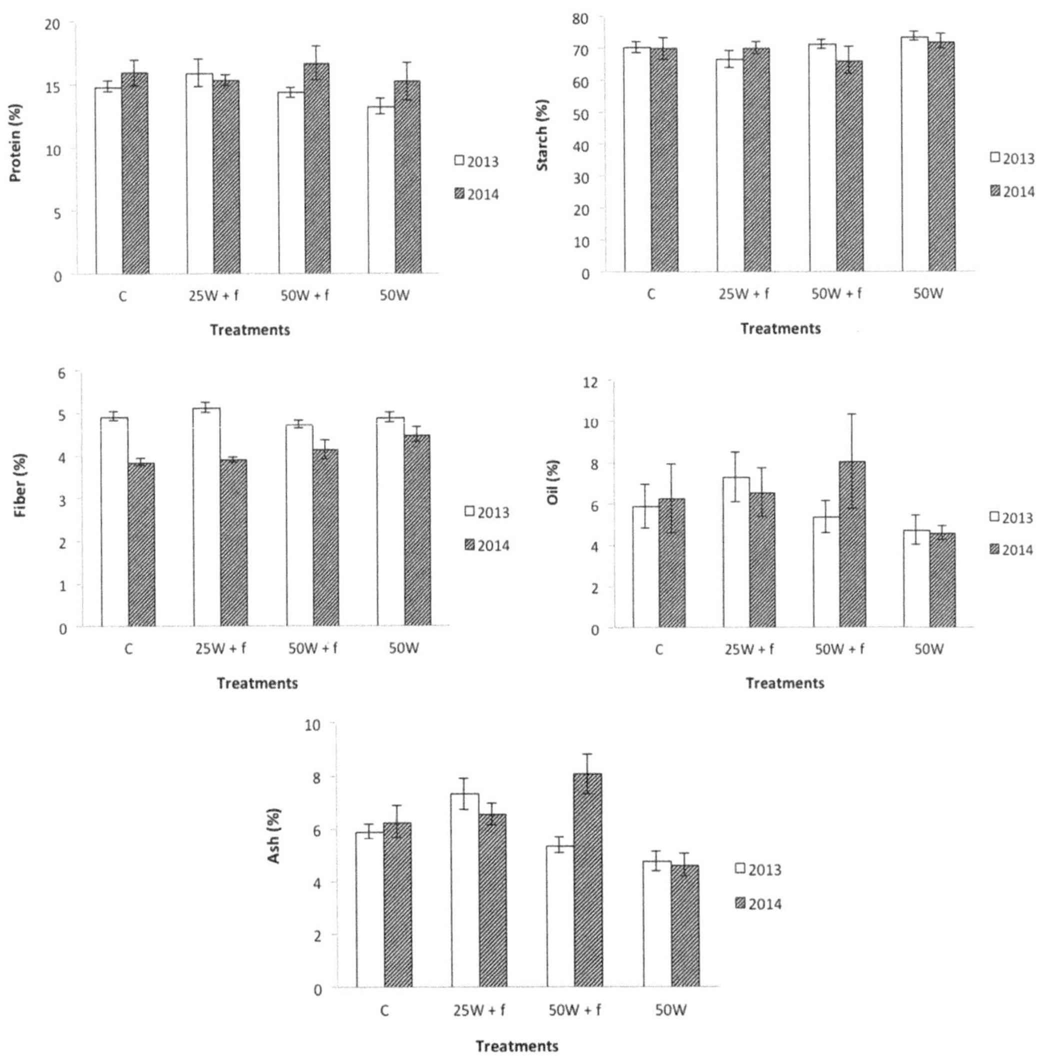

Figure 2. Maize kernel protein (top left), starch (top right), fiber (middle left), oil (middle right) and ash (bottom) content as affected by the different treatments, in the first and second year of the experiment. Error bars represent the standard error of the mean. (C: Control (mineral fertilization only), 25W + f: 25 t ha^{-1} $year^{-1}$ of T-OMWW plus mineral fertilization, 50W + f: 50 t ha^{-1} $year^{-1}$ of T-OMWW plus mineral fertilization, 50W: 50 t ha^{-1} $year^{-1}$ of T-OMWW only)

Kernel quality parameters are presented in Figure 2. The application of T-OMWW, with or without the addition of mineral fertilization, resulted in similar content of protein with the control treatment in both years of the experiment. Mean kernel protein content was 15.2% (2-year average), which is a relatively high value. Kernel protein content generally increases with higher soil N availability (Oktem, Oktem, & Emeklier, 2010; Riedell, 2014), whereas N deficiency in soils affects negatively N metabolism and especially protein synthesis (Oktem, Ulger, & Kirtok, 2001). The similar protein content observed in all four treatments may suggest therefore that soil N availability for crop uptake was adequate to meet crop nitrogen requirements in all cases. This finding is very important as it indicates the potential of mineral fertilizer nitrogen full substitution by T-OMWW. Kernel starch, ash and oil content also ranged in similar levels with the control treatment in both years of the experiment. The application of T-OMWW at the rate of 50 t ha^{-1} $year^{-1}$ resulted in significantly higher kernel fiber content in the second year of the experiment, in comparison to the other two treatments. These findings are indicative of T-OMWW capability to meet crop nutrient requirements and hence to be used as a liquid fertilizer.

In the second year of the study, kernel yield corresponding to the application of 50 t of T-OMWW per ha was about 15% lower than those of the control or the 25W+f treatment, corresponding to the combined application of T-OMWW at 25 t per ha and mineral fertilization (Figure 1). In order to further explore the potential impact of this non-significant trend in grain yield on kernel component concentrations, the total yield of kernel components on a kg ha^{-1} basis were determined. Yields of protein, starch and fiber, and yields of oil and ash are presented in Figure 3, respectively, for both years of the experiment. There were no significant effects of the different treatments on yields of protein, starch, fiber, oil and ash.

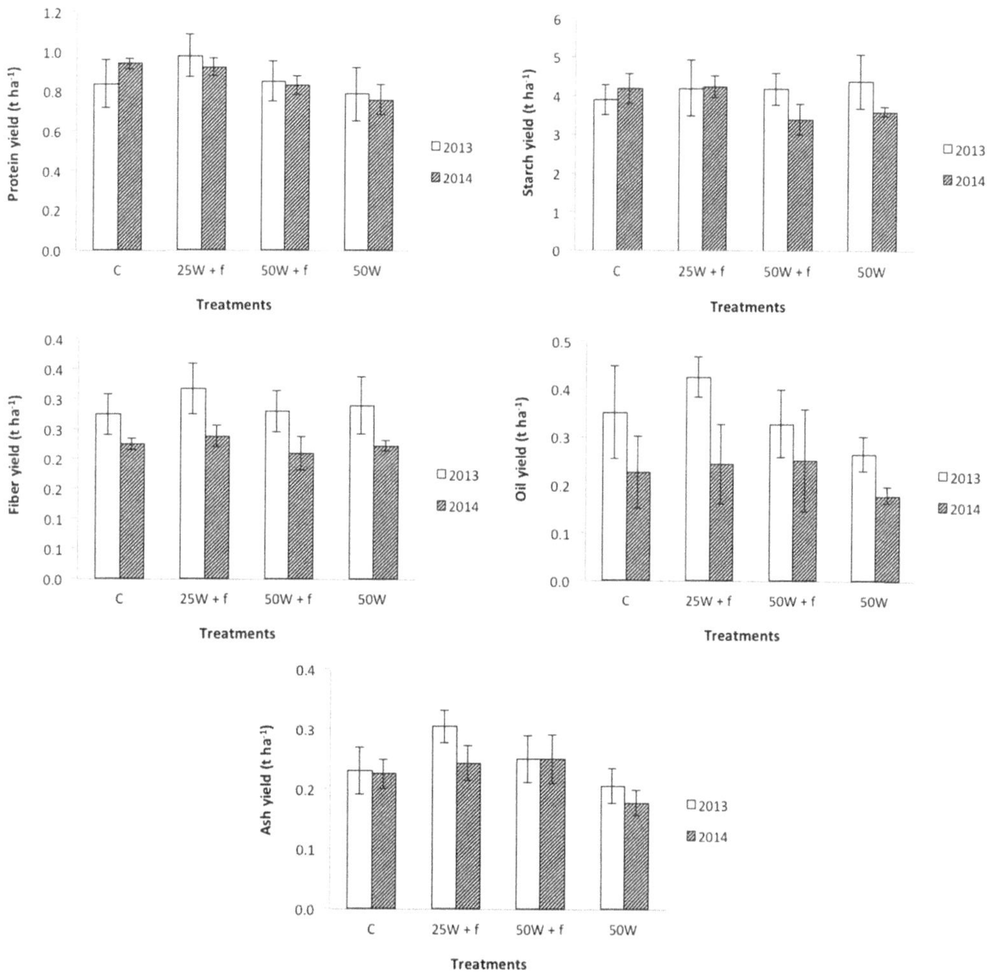

Figure 3. Maize kernel protein (top left), starch (top right), fiber (middle left), oil (middle right) and ash (bottom) yield (in kg per ha) as affected by the different treatments, in the first and second year of the experiment. Error bars represent the standard error of the mean. (C: Control (mineral fertilization only), 25W + f: 25 t ha^{-1} $year^{-1}$ of T-OMWW plus mineral fertilization, 50W + f: 50 t ha^{-1} $year^{-1}$ of T-OMWW plus mineral fertilization, 50W: 50 t ha^{-1} $year^{-1}$ of T-OMWW only)

The results of T-OMWW application on maize grain yield and quality suggest that T-OMWW can be used as liquid fertilizer in maize production. The control treatment, corresponding to only mineral fertilizer application, gave similar results with the only T-OMWW treatment, indicating the potential of mineral fertilizer substitution by T-OMWW, under the conditions of our study. Although not clearly shown within the first two years of maize fertigation with T-OMWW, it seems that there is an advance in grain yield with applying the T-OMWW at the lower rate of 25 t per ha.

3.2 Soil Properties

The effect of T-OMWW application on soil properties was not obvious following the first year of application (results can be found in Kokkora et al., 2015). The second year of application gave a clearer picture of the effects of T-OMWW on soil properties, although again the differences between the treatments were not significant.

As shown in Table 4, there was a trend for gradual increase in soil EC following the application of greater quantities of T-OMWW. Other research work has also shown increase in the EC of a clayloam soil after the implementation of OMWW processed in stabilizing tanks (Siera, Martin, Garau, & Cruañasm, 2007). This observation implies that caution is required in continuous application of T-OMWW to soils with high clay content, due to potential increase in salts concentration, and/or salinity sensitive crops. Soil sodium content also seemed to increase with T-OMWW application (Table 3). Magdich et al. (2013) observed significant increase in soil Na content following the successive 3-year application of OMWW at rates higher than 50 m^3 ha^{-1} to sandy soil. Due to the fact that our soil was rich in Mg and Ca (mean values at the end of the growing season was approximately 1436 and 3213 mg kg^{-1}, respectively), sodicity problems were not an issue in this soil. Problems may arise, however, with the continuous application of T-OMWW in the case of soils poor in total salts and/or sodium sensitive crops.

Soil extractable K was practically unaffected by the different treatments (Table 4), although there was a slight tendency to increase by combining T-OMWW and inorganic K fertilization. At the end of the first growing season, soil extractable K was at the level of approximately 383 mg kg^{-1}, while at the end of the second growing season was at least 516 mg kg^{-1}. This finding indicates that the application of T-OMWW only, with no inorganic K addition, was capable to meet crop K requirements and also to increase soil K availability.

Soil available P content increased with the application of mineral fertilizer P and the combination of T-OMWW with reduced mineral fertilizer P addition. Our soil was poor in available phosphorus and the mineral fertilizer P addition in the control treatment aimed both at meeting crop P requirements and also enriching soil P levels. Indeed, the mineral P fertilization met crop requirements and increased soil P levels by approximately 41% in comparison to the soil P levels at the end of the first growing season. The combination of reduced mineral fertilizer P addition and T-OMWW further increased soil P levels (approximately 55% increase compared to control soil P levels at the end of the first growing season). The application of only T-OMWW slightly increased soil P levels by about 7%. These results indicate the potential of T-OMWW to increase soil available P levels. In the case, however, of soils poor in P, the combined application of T-OMWW and mineral fertilizer P seems as the most favorable option.

Soil O.M. content was not affected by T-OMWW application (mean soil O.M. content was 1.2% within the top 30 cm). Mean soil pH was 6.6. As for soil nitrogen, mean total N content was 1.0 g kg^{-1}, the residual nitrate-N content was 29.8 mg kg^{-1} and the residual ammonium-N content was negligible (less than 2.5 mg kg^{-1}) for all treatments. Soil residual mineral N corresponds to approximately 148 kg nitrate-N per ha. This is a relatively high value and may lead to increased nitrate leaching losses during winter. T-OMWW application at the rate of 50 t per ha was capable to meet crop N requirements and increase soil N availability, which suggests therefore that the extra fertilizer N added with the treatment of 50W+f (combining T-OMWW applied at 50 t per ha along with mineral fertilizer N), possibly leached below the top 30 cm of soil during the growing season, hence posing a serious threat to the environment. It is evident from these results that no addition of mineral fertilizer N was necessary with the application of 50 t of T-OMWW per ha.

Table 4. Indicative topsoil (0-0.3 m depth) properties following two years T-OMWW application by drip irrigation to maize cultivation on clay loam soil. (C: Control (mineral fertilization only), 25W + f: 25 t ha^{-1} $year^{-1}$ of T-OMWW plus mineral fertilization, 50W + f: 50 t ha^{-1} $year^{-1}$ of T-OMWW plus mineral fertilization, 50W: 50 t ha^{-1} $year^{-1}$ of T-OMWW only)

	EC (mS cm^{-1})	Na (mg kg^{-1})	K (mg kg^{-1})	P Olsen (mg kg^{-1})
C	0.5^{a} (0.06)	158.0^{a} (6.06)	524.3^{a} (9.18)	8.3^{a} (1.85)
25W + f	0.6^{a} (0.06)	228.0^{a} (19.63)	555.7^{a} (5.31)	9.0^{a} (1.15)
50W + f	0.8^{a} (0.17)	194.3^{a} (47.28)	540.0^{a} (31.58)	9.3^{a} (1.21)
50W	0.7^{a} (0.23)	240.3^{a} (2.71)	516.7^{a} (13.57)	6.3^{a} (0.87)

Note. Numbers in brackets are standard error of mean. Columns labeled with the same lower case letter are not significantly different (P>0.05).

EC: Electrical conductivity.

4. Conclusions

The results of the present study showed the utilization potential of treated olive mill wastewater, by microfiltration and XAD4 resin, as liquid fertilizer in maize production. Plant and soil analysis showed that T-OMWW was capable to meet crop requirements in N, P and K, and also to increase soil N, P and K availability. In a soil poor in available P, the combined application of T-OMWW and fertilizer P was necessary to substantially increase soil available P levels. A non-significant tendency for increasing soil Na and EC levels and reducing grain yield was observed with the higher rate of T-OMWW application of 50 t per ha. These results indicate that for sustainable agriculture, it may be safer to apply the T-OMWW at the lower rate of 25 t per ha per year, or to consider the application of 50 t per ha every other year. Further research work is necessary to evaluate the long-term effects of T-OMWW application on soil quality and crop production.

Acknowledgments

This research has been co-financed by the European Union (European Social Fund - ESF) and Greek national funds through the Operational Program "Education and Lifelong Learning" of the National Strategic Reference Framework (NSRF) - Research Funding Program: ARCHIMEDES III. Investing in knowledge society through the European Social Fund.

References

Allen, R. G., Pereira, L. S., Raes, D., & Smith, M. (1998). Crop Evapotranspiration: Guidelines for Computing Crop Water Requirements. Irrigation and Drainage Paper 56, Rome: United Nations Food and Agriculture Organization (FAO).

Barbera, A. C., Maucieri, C., Cavallaro, V., Ioppolo, A., & Spagna, G. (2013). Effects of olive mill wastewater on soil properties and crops, a review. *Agricultural Water Management, 119*, 43-53. http://dx.doi.org/10.1016/j.agwat.2012.12.009

Barbera, A. C., Maucieri, C., Ioppolo, A., Milani, M., & Cavallaro, V. (2014). Effects of olive mill wastewater physic-chemical treatments on polyphenol abatement and Italian ryegrass (Lolium multiflorum Lam.) germinability. *Water Research, 52*, 275-281. http://dx.doi.org/10.1016/j.watres.2013.11.004

Belaqziz, M., Lakhal, E. K., Mbouobda, H. D., & El Hadrami, I. (2008). Land spreading of olive mill wastewater: effect on maize (*Zea Mays*) crop. *Journal of Agronomy, 7*(4), 297-305.

Berr, C., Porter, F., Carriere, I., Akbaraly, T. N., Feart, C., Gourlet, V., ... Ritchie. K. (2009). Olive oil and Cognition: Results from the three-city study. *Dementia and Geriatric Cognitive Disorders, 28*(4), 357-364. http://dx.doi.org/10.1159/000253483

Cassano, A., Conidi, C., & Drioli, E. (2011). Comparison of the performance of UF membranes in olive mill wastewaters treatment. *Water Research, 45*, 3197-3204. http://dx.doi.org/10.1016/j.watres.2011.03.041

Cereti, C. F., Rossinim, F., Federici, F., Quaratino, D., Vassilev, N., & Fenice, M. (2004). Reuse of microbially treated olive mill wastewater as fertilizer for wheat (*Triticum durum* Desf.). *Bioresource Technology, 91*, 135-140. http://dx.doi.org/10.1016/S0960-8524(03)00181-0

Chartzoulakis, K., Psarras, G., Moutsopoulou, M., & Stefanoudaki, E. (2010). Application of olive mill wastewater to a Cretan olive orchard: effects on soil properties, plant performance and the environment. *Agriculture, Ecosystems and Environment, 138*, 293-298. http://dx.doi.org/10.1016/j.agee.2010.05.014

Covas, M. I. (2007). Olive oil and the cardiovascular system. *Pharmacological Research, 55*, 175-186. http://dx.doi.org/10.1016/j.phrs.2007.01.010

Escrich, E., Moral, R., & Solanas, M. (2011). Olive oil, an essential component of the Mediterranean diet, and breast cancer. *Public health nutrition, 14*(12A), 2323-2332. http://dx.doi.org/10.1017/S1368980011002588

Hanifi, S., & El Hadrami, I. (2008). Phytotoxicity and fertilising potential of olive mill wastewaters for maize cultivation. *Agronomy for Sustainable Development, 28*, 313-319. http://dx.doi.org/10.1051/agro:2007047

Kokkora, M., Vyrlas, P., Papaioannou, C., Petrotos, K., Gkoutsidis, P., Leontopoulos S., & Makridis, C. (2015). Agricultural use of microfiltered olive mill wastewater: effects on maize production and soil properties. *Agriculture and Agricultural Science Procedia, 4*, 416-424. http://dx.doi.org/10.1016/j.aaspro.2015.03.047

Magdich, S., Ben Ahmed, C., Jarboui, R., Ben Rouina, B., Boukhris, M., & Ammar, E. (2013). Dose and frequency dependent effects of olive mill wastewater treatment on the chemical and microbial properties of soil. *Chemosphere, 93*,1896-1903. http://dx.doi.org/10.1016/j.chemosphere.2013.06.066

Mahajan, S., & Tuteja, N. (2005). Cold, salinity and drought stresses: an overview. *Archives of Biochemistry and Biophysics, 444*, 139-158. http://dx.doi.org/10.1016/j.abb.2005.10.018

Massoudinejad, M. R., Arman, K., & Aghayani, E. (2014). Ecological risk assessment to olive oil mill wastewater (OMW) with bioassay on plant species. *Ecology, Environment and Conservation, 20*(1), 229-234.

Mekki, A., Dhouib, A., & Sayadi, S. (2007). Polyphenols dynamics and phytotoxicity in a soil amended by olive mill wastewaters. *Journal of Environmental Management, 84*, 134-140. http://dx.doi.org/10.1016/j.jenvman.2006.05.015

Montemurro, F., Diacono, M., Vitti, C., & Ferri, D. (2011). Potential Use of Olive Mill Wastewater as Amendment: Crops Yield and Soil Properties Assessment. *Communications in Soil Science and Plant Analysis, 42*, 2594-2603. http://dx.doi.org/10.1080/00103624.2011.614035

Moraetis, D., Stamati, F. E., Nikolaidis, N. P., & Kalogerakis, N. (2011). Olive mill wastewater irrigation of maize: Impacts on soil and groundwater. *Agricultural Water Management, 98*, 1125-1132. http://dx.doi.org/10.1016/j.agwat.2011.02.006

Oktem, A., Oktem, A. G., & Emeklier, H. Y. (2010). Effect of nitrogen on yield and some quality parameters of sweet corn. *Communications in Soil Science and Plant Analysis, 41*(7), 832-847. http://dx.doi.org/10.1080/00103621003592358

Oktem, A., Ulger, A. C., & Kirtok, Y. (2001). The effect of different doses and intrarow spaces on grain yield and some agronomic characteristics of popcorn (Zea mays everta Sturt.). *Journal of the Agricultural Faculty at Cukurova University, 16*(2), 83-92.

Petrotos, K. B., Gkoutsidis, P. E., Kokkora, M. I., Giankidou, K. G., & Tsagarelis, A. G. (2013). A study on the kinetics of olive mill wastewater (OMWW) polyphenols adsorption on the commercial XAD4 macroporous resin. *Desalination and Water Treatment, 51*, 2021-2029. http://dx.doi.org/10.1080/19443994.2013.734479

Petrotos, K. B., Lellis, T., Kokkora, M. I., & Gkoutsidis, P. E. (2014). Purification of olive mill wastewater using microfiltration membrane technology. *Journal of Membrane and Separation Technology, 3*(1), 50-55.

Rahmanian, N., Jafari, S. M., & Galanakis, C. M. (2014). Recovery and removal of phenolic compounds from olive mill wastewater. *Journal of the American Oil Chemists' Society, 91*, 1-18. http://dx.doi.org/10.1007/s11746-013-2350-9

Riedell, W. E. (2014). Nitrogen fertilizer applications to maize after alfalfa: grain yield, kernel composition, and plant mineral nutrients. *Journal of Plant Nutrition, 37*, 2026-2035. http://dx.doi.org/10.1080/01904167.2014.911892

Russo, C. (2007). A new membrane process for the selective fractionation and total recovery of polyphenols, water and organic substances from vegetation waters (VW). *Journal of Membrane Science, 288*(1-2), 239-246. http://dx.doi.org/10.1016/j.memsci.2006.11.020

Sierra, J., Martí, E., Garau, A., & Cruañasm, R. (2007). Effects of the agronomic use of olive oil mill wastewater: Field experiment. *Science of the Total Environment, 378*, 90-94.

http://dx.doi.org/10.1016/j.scitotenv.2007.01.009

Weisz, G. M., Schneider, L., Schweiggert, U., Kammerer, D. R., & Carle, R. (2010). Sustainable sunflower processing – I. Development of a process for the adsorptive decolorization of sunflower [*Helianthus annuus* L.] protein extracts. *Innovative Food Science and Emerging Technologies, 11*, 733-741. http://dx.doi.org/10.1016/j.ifset.2010.05.005

Zaglis, D. P., Arvaniti, E. C., Papadakis, V. G., & Paraskeva, C. A. (2013). Sustainability analysis and benchmarking of olive mill wastewater treatment methods. *Journal of Chemical Technology and Biotechnology, 88*(5), 742-750. http://dx.doi.org/10.1002/jctb.4036

Zaglis, D. P., Vavouraki, A. I., Kornaros, M. E., & Paraskeva, C. A. (2015). Purification of olive mill wastewater phenols through membrane filtration and resin adsorption/desorption. *Journal of Hazardous Materials, 285*, 69-76. http://dx.doi.org/10.1016/j.jhazmat.2014.11.038

Zirehpour, A., Jahanshahi, M., & Rahimpour, A. (2012). Unique membrane process integration for olive oil mill wastewater purification. *Separation and Purification Technology, 96*, 124-131. http://dx.doi.org/10.1016/j.seppur.2012.05.028

Resources Management and Risk Efficiency of Crop Rotation Systems in Sudan Gezira Scheme

Kheiry Hassan M. Ishag[1]

[1]Dhofar Cattle Feed Company – P.O. Box 1220 – PC 211 – Sultanate of Oman.

Correspondence: Kheiry Hassan M. Ishag, Dhofar Cattle Feed Company – P.O. Box 1220 – PC 211 – Sultanate of Oman. E-mail: kheiryishag@hotmail.com

Abstract

Risk is an important factor in crop rotation systems and cropping system management studies. The study uses stochastic simulation techniques and Stochastic Efficiency with Respect to Function (SERF) to evaluate five crop rotations risk-efficiency and economic sustainability in Sudan Gezira Scheme. Price and yield risk for five crops were simulated to calculate whole-scheme net return. The analysis shows with the present irrigation system capacity 4 course rotation is the most preferred at lower (ARAC) and the 5 Course rotation (B) with 53% land use intensity achieve water distribution equity and is the most risk efficient crop rotation at upper (ARAC). It is downside risk oriented and resilience crop rotation. The area allocated in 5 Course rotations (B) for cotton crop is 17%, wheat 7%, sorghum 60%, groundnut 4% and fodder crops 12%. Fodder can be grown two times in summer and winter season without creating water shortage problems. However, this will increase net return and increase soil fertilities within the selected crop rotation. The result also shows that return pack to night storage irrigation system needs a risk premium of 36 Million (SDG). The techniques used in this study could be used with any distribution estimates for the uncertain variables to incorporate new crop varieties and research recommendation packages. They also could be modified to account for new information contribution during the decision process and account for dynamic effects and policy adjustment and modification.

Keywords: crop rotation, risk-efficient alternative, SERF analysis, simulation model, risk and uncertainty

1. Introduction

The Sudan Gezira Scheme is the largest farm under one management in the world. It occupies 2.1 million acres and one hundred thousand farmers. The scheme accounts for 50% of the total irrigated land of Sudan. The Gezira scheme slopes gradually from south to north and is irrigated by thousands of kilometers of main canals and subsidiary field canals. The scheme provides a livelihood for two million people living within it is boundaries. The scheme is a joint venture between government, management and farmers and each party has specific duties and cost in production. The General Director of Irrigation provides water up to the miner canals and from that point onward water distribution is operated by Sudan Gezira Scheme staff.Figure

The main objective of the scheme is to develop and promote resources to achieve maximum socioeconomic benefit for farmers and people living within the scheme. In year 1991/1992 the Scheme translates government policy to achieve food security and reduce cash crop area (cotton and groundnuts) and increase food crop area (wheat and sorghum). The area of cotton crop of season 1991/1992 was reduced by 45% and cereals crops increased by 100%.

The Government is responsible for maintaining main canals and irrigation operation up to minor canals. The maximum area can be irrigated is one half of the Scheme which is equal to 1.5 million acres each year. The Government is also responsible for importing all chemicals and inputs and marketing cotton and determining cotton and wheat farm gate price. The Sudan Gezira Board (SGB) is responsible for determining crop area to be cultivated each year and land leveling and preparation using scheme machineries, maintaining subsidiary field canals. The Scheme provides agricultural inputs and supervises farmers operation and maintains the scheme fixed assets. The Sudan Gezira Board (SGB) also keeps all financial records of the scheme and individual farmer accounts and at the end of the cropping season the scheme services cost detected from farmers' individual account and balance dispersed to farmers. The farmers are responsible for agricultural operation such as planting and irrigation and harvesting and paying other duties costs not paid by scheme.

The scheme continuously operated for 75 years and during the period many changes have taken place especially in soil, canalization systems, agricultural environment and scheme farmers' themselves. Studies on these issues have to be conducted to investigate the extent of changes on productivities and improve operation, argued Muddathir A. Ahmed, (2004).

The irrigation system of the scheme is a huge network of main, branch, major, minor and field canals. The irrigation system comprises two main canals length is 261 km, 11 branch of a total length of 651 km, 107 major canals with a length of 1,652 km and 1,498 minor canals with a length of 8,119 km. The field canals consist of 29,000 (Abu XX) with a total length of 40,000 km and 350,000 (Abu VI) with 100,000 km length and a capacity of 50 Lit / second. The design objective of the irrigation system is to provide adequate water supply to cultivated crops and the reliability of water supply irrespective of time and location within the scheme. During 1989/1990 season a study on irrigation system performance indicated that the conveying performance of the head and middle of the major and minor canals is good but the tail of the tail major, middle and tail of the minor performance were unsatisfactory and supply 20% less the target quantities of water as stated by (Shafique, M. S., 2008). However, with the existing canals siltation and weed infestation the unattended irrigation practices increased. The night storage irrigation system turned to continuous irrigation system and irrigation schedule and timing deteriorated. The irrigation of "Number" a land of 90 acres take more time than the design irrigation frame time and the whole operation of the irrigation system could not be maintained.

The scheme management (SGB) implemented new strategy and crop rotation and reduced total cultivated area and land use intensity to cope with limited water resources and irrigation problems especially during 1990s. The problems of irrigation have been covered in detail in many studies, A. M. Eldaw (2004); Muddathir A. Ahmed (2004).The cash crops i.e. cotton and groundnuts area reduced by 45% and 75%, respectively. Food crops such as sorghum crop area increased by 100% to achieve food security policy. The crop rotation modifications have been imposed on the farmers who have a little choice to decide on which crops to grow. Farmer has a right to decide the crop he wants to cultivate only on 6% of his land area such as vegetable and fodder crop. This land tenure system discourages farmers from investment and improves productivity. Farmer's participation in Field School and WUA (Water Union Association), location of the farm, land tenure system are the main factors influencing on the technical inefficiency for the cultivated crops in the Gezira scheme argued Fadelmola M. Elnour and Abbas E. M. Elamin (2014).

Agricultural management decisions are most always made in the face of uncertain variables and consequences. Crop planting areas are generally made before a price for the crop has been set and long time before yields are determined. Inputs required for farming are also uncertain in term of quantities and value. Following Hardaker et al. (1997), uncertain outcome especially unfavorable ones are broadly defined as risk. Hardaker et al. (1997) divide risk sources into five categories: production risk, price and market risk, institutional risk, human or personal risk, and financial risk. The production risk in the scheme refers to uncertain crop yields due to quantity and time of inputs application, rainy season and pest infestations. Price or market risk refers to uncertain input and output prices and institutional risk arises from changes in government policies and irrigation water availability in main canals and field canals. Crop rotation risk analyses using stochastic efficiency with respect to function (SERF) have been covered in many studies Ascough II. et al. (2009); G. Lien et al. (2007). These studies use experimental and field plot data which is different from what is obtained in farmers' field. The (SERF) was introduced by Hardaker et al. (2004).

Farmers in the scheme face operation and activities delay due to un-availability of inputs and irrigation water in canals, which is the main source of risk that should be considered in comparing crop rotation performance in the scheme. Comparing different cropping systems needs a whole-farm budget approach and not partial budget analysis, since factors interact, argued G. Lien, et al. (2005). In this study we use historical data from scheme collected by Socio-Economic Research Unit (SERU) and consider yield and price risk for each crop rotation in the analysis and whole scheme net returns for each crop rotation approach.

Farmers are generally risk averse and are willing to give up some expected returns from cash crops return such as cotton and groundnuts for a reduction in risk due to water shortage. However, the amount farmers would be willing to give up depends on farmers preferences. There are many ways that farmer in Sudan Gezira Scheme can deal with risk. They use to reduce or eliminate risk by collecting information to reduce uncertainty, by finding less risky ways to produce cash crop such as cotton, or by building flexibility into their operations allowing them to rapidly respond to changing conditions. Alternatively, they can transfer risk to others by using futures markets agreements. Financial tools such as (Shail agreement) are used at the scheme. In this case the money lender provides advance loans to farmers who pledge the delivery of a specific quantity of output equivalent to the value of the loan at the time of harvest.

In analyzing cropping systems, it is important to understand sources and effects of risk and how farmers respond to risk. The approach in this paper is to use land and water resources as an example of risk in cropping systems to illustrate some useful tools for analyzing risk and to understand the ways risk can affect farmers and Scheme decision makers' behavior.

The study investigated five crop rotations with different land use intensities to identify yield and price risk on scheme net return and determine risk efficient crop rotation and land use intensity.Figure

2. Materials and Methods

2.1 Stochastic Budgeting – Crop Rotation Systems Analysis

To evaluate risk, we need to identify the distribution of possible outcomes under uncertainty. A useful tool is stochastic budgeting approach which allows uncertainty in some variables to be included in calculating potential outcomes. Although stochastic budgeting approach has been used for several years, it has become now much easier to use due to the development of spreadsheet software add-ins such as @RISK (Palisade, 2010) and Simetar (Richardson et al., 2003) to automate stochastic simulation. The approach in stochastic budgeting is just like constructing a deterministic budget with some uncertain variables. The stochastic nature of the budget is captured by specifying probability distributions for the uncertain variables. Using a Monte Carlo procedure, random samples are drawn from these distributions and used to evaluate the budget. The results are recorded, another sample is drawn, and the results are again recorded. This simulation continues for a sufficiently large number of samples to generate a stable distribution of outcomes. The procedure is general enough to be used with almost any type of budget including enterprise budgets, or whole-farm budgets. In this study the whole scheme budgeting approach will be used.

2.2 Simulated Net Returns and Data Analysis

The study identified 5 crop rotations imposed at Sudan Gezira Scheme during 1960 and 2002. Net return for each crop rotation alternative is considered as the key output variable. To simulate net returns one or more of the input variables of the model (exogenous variables) should be considered as stochastic. This study considers average annual crop yield and price received by farmers as stochastic variables. Historical data of yield and crop price for different crops in each crop rotation were collected for 13 years (1981-1994) from Socio-Economic Research Unit at Sudan Gezira Board and simulated to generate stochastic yields and prices. Linear correlation matrix is used to simulate correlated uniform standard deviates (CUSDs), which can then use in the Simetar probability distribution functions to simulate probability distributions. The key output variables (KOVs) for different crops are generated by Simetar and calculated summary and statistics over the 500 iterations are produced.

Yields, input types and rates, and field operations cost data were also collected from Socio-Economic Research Unit at Sudan Gezira Board. The SERU staff monitors a sample of 25 farmers at each part of the scheme. Production costs are based on actual field operations and input rates. The cost information is used with yield and price data to calculate net returns and simulate a distribution of net returns for each strategy and crop rotation imposed by SGB. Enterprise budgets that include one or more stochastic variables are called stochastic budgets. Net returns estimated from these budgets are simulated for 500 iterations generating 500 possible net returns for each crop rotation alternative.

2.3 Stochastic Efficiency with Respect to a Function

Stochastic efficiency with respect to a function (SERF) is used to order a set of risky alternatives in terms of certainty equivalents (CEs) for a range of risk preferences (Hardaker et al., 2004). SERF ranks and orders preferred alternatives in terms of CEs as the degree of risk aversion increases. Crop rotations with higher CEs are preferred to those with lower CEs. The CE of a risky crop rotation and strategy is the amount of money at which the decision makers at SGB are indifferent between the certain crop rotation (generally lower) net return value and the expected value of the new crop rotation strategy (Lien et al., 2007). For a risk-averse decision maker, the estimated CE is typically less than the expected value of the risky strategy.

The calculation of the CE depends on the utility function specified. A negative exponential utility function used in the SERF analysis conforms to the hypothesis that managers prefer less risk to more given the same expected return. This functional form assumes managers have constant absolute risk aversion. Under this assumption, managers view a risky strategy for a specific level of risk aversion the same without regard for their level of wealth.

The simulated net return data for each strategy is sorted into cumulative probability distribution functions (CDFs) which are used in the SERF analysis. Decision makers with RACs greater than zero exhibit risk-averse behavior.

The actual RACs used in the final analysis range from 0.00 to 8.78E-9 because the rankings do not change for RACs above that level for different land use intensity strategies and crop rotation examined.

A utility weighted risk premium (RP), when risk aversion is considered, can be calculated by subtracting the CE of a less preferred strategy from the preferred strategy. The RP, a utility weighted risk premium for a risk-averse decision-maker, reflects the minimum amount of money that will have to be paid to a decision maker to justify a switch from one crop rotation alternative to the other. The risk premium generally increases with degree of risk aversion increases.

2.4 StopLight Analysis

StopLight analysis relies on CDF information and generates the probability of exceeding a target net return value for the whole Sudan Gezira Scheme. The upper cutoff net return value and lower cutoff value is decided and dictated by policy makers at SGB. The StopLight chart has three ranges colors, red presents the probability of net returns less than the lower cutoff value, the yellow colors are the probability of net returns between the upper and lower cutoff values, while the green colors are the probability of net returns exceeding the upper cutoff value.

2.5 Data Collection and Crop Rotation

The crop rotation in Sudan Gezira Scheme started with a three crop rotation in 1925 as (cotton–fallow–fallow-cotton-fallow-sorghum-lubia/fallow-fellow) with land use intensity of 43.75%. After 1960 intensification started and groundnut and wheat were introduced and the crop rotation changed to (cotton–wheat–sorghum/groundnuts- fallow) with land use intensity of 75%. In the 1990s food security policy was imposed by the Government and cash crop area was reduced and sorghum crop area increased by 196%. Table 1 below summarize crop rotations and land use intensity considered in the study.

Table 1. Crop rotation alternatives and land use intensity policy considered in the analysis

No	Rotation designation	Crop rotation description	Years	Land use intensity %
1	3 course R	Cotton- Wheat- Sorg/Gnut- Fallow	1965	47%
2	4 course R	Cotton- Wheat- Sorg/Gnut- Fallow	1975	75%
3	5 course R Org.	Cotton- Wheat- Sorg/Gnut- Fodder – Fallow	1990	79%
4	5 course R Act (A)	Cotton- Wheat- Sorg/Gnut- Fodder/Fallow- Fallow	1993	72%
5	5 course R Act (B)	Cotton- Wheat- Sorg/Gnut- Fodder/Fallow- Fallow	2002	53%

Formed by the Author.

Crop rotation and land use policy implemented at Sudan Gezira Scheme (1965-2002) :

- 3 course rotation, before intensification policy and joint account system.
- 4 course rotation, after individual account & intensification policy implementation.
- 5 course rotation original, present land use intensity maximization with individual account & intensification policy original 5 course rotation, which has been applied for two seasons only.
- 5 course rotation actual (A), food security policy increase sorghum land use intensity to 41% and cash crop area (Cotton and groundnut) reduction with individual account & intensification policy was implemented.
- 5 course rotation actual (B), land use intensity reduction (wheat, Groundnuts) to cope with water resources with individual account & intensification.

The study compares alternative land use intensity for each crop rotation and total cultivated area for each crop in the rotation. The Sudan Gezira Board and Ministry of Agriculture every year announces total crop area to be cultivated within the scheme according to agriculture policy and irrigation water availability. The standard tenancy is 40 acres, but mostly half of the area is cultivated due to irrigation system capacity. According to crop rotation, the tenants are placed in equal plots at different Number (90 acres in a rectangular plot) i.e. three numbers for three crop rotation and four numbers for four crop rotation. The water and land resources management in the scheme depend on many factors such as irrigation water availability, total cultivated area, crop sowing dates, crop growing stage and crop season duration. Table 2 below summarized allocated crop area and land use intensity for each crop rotation tested in this study and resources management during 1965-2002.

SGB reduced cotton crop area from 51% to 17%, and increased sorghum area from 26% to 60% of total cultivated area due to irrigation system deterioration.

Table 2. Allocated crops area in (Acres) and land use intensity policy

No	Crop rotation	Cotton	Wheat	Sorghum	Groundnut	Fodder	Area/fed	% Total area
1	3 course R	507 000	75 000	255 000	60 000	95 000	992 000	45.1%
	Land use %	51%	7%	26%	6%	9%	100%	
2	4 course R	603 000	428 000	154 000	361 000	49 000	1 595 000	72.5%
	Land use %	38%	27%	9%	23%	3%	100%	
3	5 course R Org.	420 000	420 000	210 000	210 000	420 000	1 680 000	76.4%
	Land use %	25%	25%	13%	13%	25%	100%	
4	5 course R Act (A)	175 000	514 034	621 736	163 814	52 000	1 526 584	69.4%
	Land use %	11%	34%	41%	11%	3%	100%	
5	5 course R Act (B)	190 000	81 000	679 000	45 000	130 000	1 125 000	51.1%
	Land use %	17%	7%	60%	4%	12%	100%	

Calculated by the Author.

The study included five crop rotation system combinations: 3 course rotation with land use intensity 47% (3 course R), 4 course rotation with land use intensity 75% (4 course R), original 5 course rotation with land use intensity 79% (5 course R Org.), 5 course rotation with land use intensity 72% (5 course R Act A) and 5 course rotation with land use intensity 53% (5 course R Act B). The study also included input levels and costs of production for each crop which optioned from (SERU) at Sudan Gezira Board (SGB).

For this analysis, the effects of uncertain prices and yields on returns for the alternative crop rotation systems have been evaluated. Simetar (2011) was used to simulate cumulative distribution functions (CDF) for each crop rotation based on stochastic annual yields and prices for 14 years.

Since yield distributions for each crop were driven at least in part by rainy season weather and irrigation system conditions, it is likely they were not stochastically independent. Moreover, it is possible that crop yields and prices may not be stochastically independent due to the possibility of rainy weather influences over large area (whole scheme) which may affect crop supplies. To fully account for this effect, it would be necessary to estimate a joint distribution function for the uncertain variables. However, this would be difficult at best. To account for some of the interdependence, Simetar allows correlations between each of the uncertain variables to be included in the sampling procedure. A correlation matrix was estimated for yields and prices based on the 1982-1994 observations and MVE probability distribution method was used to estimate parameters and simulate probability distribution. Enterprise budgets were constructed to calculate whole-scheme net return for alternative crop rotations.

3. Results and Discussion

3.1 Water and Land Resource Management

The monthly and total crop water requirements (CWR) for each crop rotation calculated and represented in Table (3). Monthly CWR for each crop multiplied by crop cultivated area to have monthly water requirement. The analysis indicates high water demand observed in June, August, September and October months. The 5 Course Act (B) rotations consumes low irrigation water around the season compared to other crop rotations as it avoids high water demand for growing groundnut in June and peak demand in October for the cotton crop.

The crop intensification program, expansion of the irrigation system, breakdown of the telecommunications system, and insufficient funds available for irrigation system maintenance, resulted in improper use of water and land resources and result in inadequate irrigation system control. The deterioration of the communication system cause instability of water levels in the major canals, and it became difficult to maintain the intended discharge into the minor canals. Moreover, the high degree of siltation of some minor canals creates problems such that little water reaches the tail end farmers at "Numbers" and some areas are out of production, (Kheiry et al., 1991).

Return to traditional night storage irrigation system could improve irrigation equity and reliability of water deliver as urged by Ahmed et al. (2002), but it should be remembered that traditional irrigation system was practiced with low cropping intensity, and low irrigation labour cost at the scheme. The present system (continuous irrigation) is more adaptable to the crop water requirements at different crop growing stages and preferred by the tenants as it is low irrigation cost method.

Table 3. Monthly and total crop water requirement for each crop rotation in Million M^3

Month	3 Course R	4 Course R	5 Course Org	5 Course Act A	5 Course Act B
January	541	837	1,144	600	358
February	124	124	277	63	94
March	0	0	0	0	0
April	0	0	0	0	0
May	0	0	0	0	0
June	392	619	927	761	758
July	372	505	693	563	580
August	647	833	947	848	868
September	648	847	957	871	885
October	603	856	655	623	502
November	500	817	649	493	222
December	586	870	1,159	585	372
Total	**4,412**	**6,308**	**7,407**	**5,409**	**4,639**

Calculated by the Author.

Although the 3 Course R, 4 Course R and 5 Course Org. crop rotation had a higher net return income than other crop rotation system as shown in Table 4, but they are more risky due to irrigation water shortage especially during groundnut and sorghum crop growing season (June- July-August- September) as four crops at this time needs to be irrigated. As a result, SGB reduced groundnut area in 5 course R Act (A) to 163 814 acres to cope with water shortages risk during groundnut crop grown season. However, this indicates that crop area reduction policy provides more water for irrigation at high level of risk aversion.

Table 4. Summarized net return in Million (SDG) for each crop rotation in Sudan Gezira Scheme

Rotation	3 Course R	4 Course R	5 Course Org	5 Course Act A	5 Course Act B
Mean	57.828	68.766	47.897	19.923	21.667
SD	211.193	251.182	174.953	72.773	79.145
Skewness	0.16889	0.16889	0.16889	0.168889	0.16889
Kurtosis	-0.22335	-0.22335	-0.22335	-0.22335	-0.22335
Min	(497.761)	(592.012)	(412.347)	(171.517)	(186.538)
Max	715.391	850.850	592.632	246.508	268.095

Calculated by the Author and Simetar.

3.2 Stochastic Budgeting Simulated Net Returns

Because Simetar is an Excel add-in, there is considerable flexibility in setting up different crop rotation scenarios to be evaluated. Five different scenarios are presented in this analysis and in each scenario; both yields and prices were stochastic. All five scenarios were run simultaneously using 500 random samples from the multivariate yield and price distribution. The analysis and summaries are presented in Table (4).

3.3 Cumulative Distribution Function (CDFs) analysis

The Cumulative Distribution Function (CDFs) analysis performed to illustrate the range and probabilities of scheme net returns for different crop rotation scenarios. Figure 1 shows the net return CDF's for the five crop rotation treatments under the current irrigation system. Although a risk preference is vary from farmer to farmer, the CDF's can be used to identify the crop rotation that would be preferred by decision maker within a range of preferences using stochastic dominance analysis (Meyer, 1977). The simplest form of stochastic dominance is first-degree stochastic dominance. Among crop rotations tested in this study no rotation was first-degree stochastically dominant as their CDFs lines crossed each other and no clear ranking under different generalized stochastic dominance is possible. However, a more improved technic, Stochastic Dominance with Respect to a Function analysis preformed and indicates that 4 Course R is most preferred rotation at lower ARAC (0.0) and 5 Course R (A) is the most preferred rotation at upper ARAC level (8.78E-9).

Stochastic Efficiency with Respect to a Function analysis was used to rank alternative crop rotation risk efficiently within specific range of ARAC values (0.0 – 8.78E-9). Four observations can be drawn from Fig.1. First, the 3 Course R, 4 Course R and 5 Course Org. systems had a higher net return income than other crop rotation systems. Second, the net return income from the 3 Course R, 4 Course R and 5 Course Org. system were described as the most risky rotation, since the CDFs for them were less steep than 5 Course R (A) and 5 Course R (B). The standard deviation (SD) in yields was in general highest for the 3 Course R, 4 Course R and 5 Course Org. systems (Table 4). Third, under existing irrigation system conditions; all of the crop rotation systems had a small probability of generating negative net returns (less than 42%). The last point is that 5 Course R (A) and 5 Course R (B) are downside risk oriented.

The expected mean annual net return income for 4 Course rotations is the most risk efficiency rotation up to 0.0000000011 risk aversion, after that 5 Course Org. is a risk efficient rotation with net return of 20.5 Million. At 0.0000000026 risk aversion 5 Course R (B) was risk efficient rotation with 13.7 Million (SDG), and 13.2 Million (SDG) for 5 Course R (A). In other words, the 5 Course R (A) and 5 Course R (B) systems had almost the same expected income and risk efficient at higher ARAC. The analysis indicates net return income of crop rotation are higher if more land allocated to cash crop such as cotton crop, but these rotation are more risky crop rotation as water resource availability in canals to irrigate all area simultaneously is difficult to achieve under current situation.

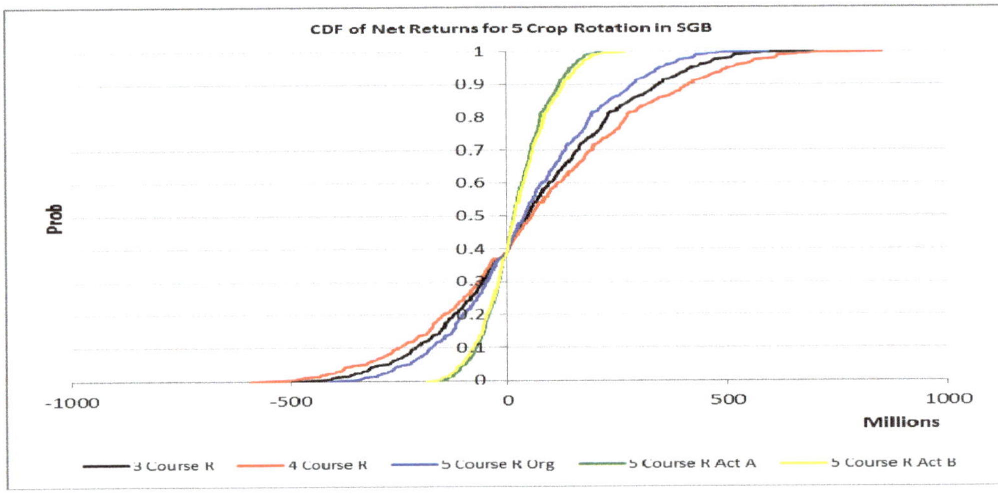

Figure 1. Cumulative probability distribution of net returns for 5 crop rotations in Sudan Gezira Scheme

Comparison of 5 Course R (A) and 5 Course R (B) rotation system shows that they have a slightly different risk profile; where 5 Course R (B) is the most risk efferent. Which of these two alternatives the SGB and farmers would prefer, depends on their degree of risk aversion, which can be identified by SERF analysis?

3.4 Stochastic Efficiency with Respect to a Function

To rank the risk-efficient of crop rotation alternatives, the SERF approach performed and result shown in Figure 2. A risk-neutral decision maker and farmer ranking five crop rotations and land use intensity were tested in this study. The result shows that 4 Course R is risk efficient at risk neutral ARAC level and after 0.0000000022 5

Course R (B) and 5 Course R (A) are the most risk efficiency rotation with 14.878 Million and 14.178 Million (SDG) net return CEs respectively, as shown in Figure (2).

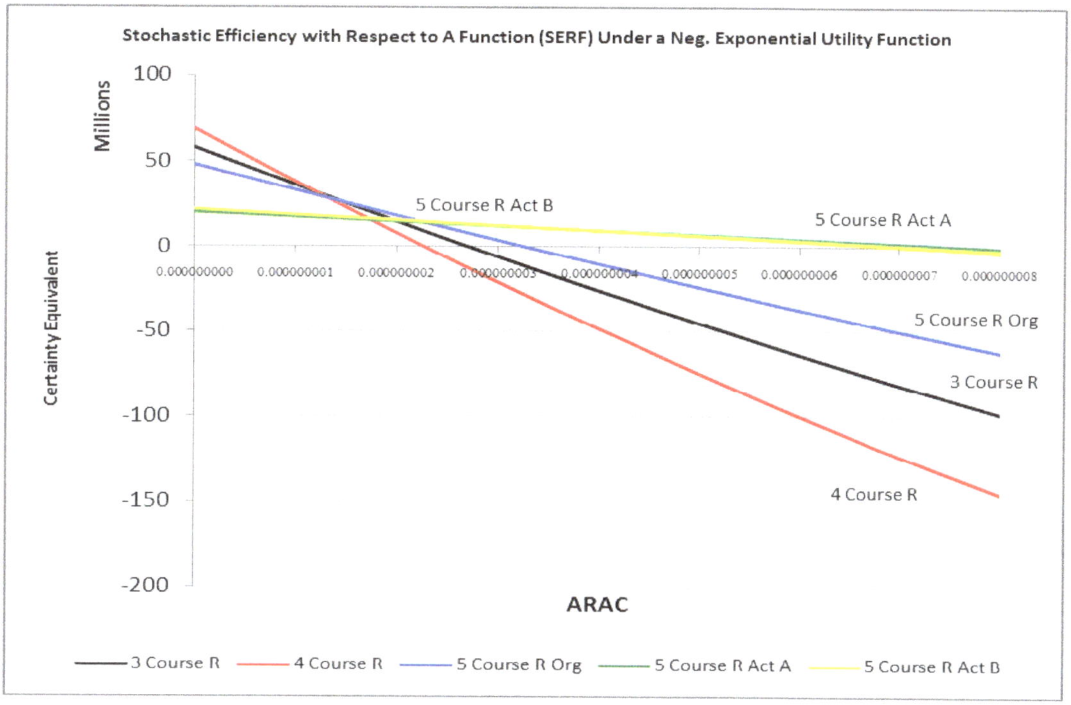

Figure 2. The net return CE results for all ARAC's for five crop rotation system alternatives

The results show that the rankings change as risk aversion increases and that the (5 Course R Act B) crop rotation system with 53% land use intensity is preferred at high range of risk aversion. Adam Elhag Ahmed, (2011) used variance and the expected income–variance risk analysis to identify crop combination and land use intensity under risk consideration and tenants' preference and found that 53% land use intensity is the most preferred alternative. For extreme risk aversion level with 0.0000000037 ARAC the (5 Course R Act A) with 72% land use intensity is the most risk preferred crop rotation.

In reality, SGB and farmers have options to reduce the price and production risk shown in the 4 course rotation and 5 course Org. rotation scenarios. The cash crops area and also land use intensity have been reduced to cope with water resources shortage and current irrigation system capacity.

In year 1993 at 5 course rotation (A) farmer planted 621 736 acres with sorghum crop with an additional 467 736 Acres compared to the 4 course rotation in 1975, which allocated 154 000 acres only to sorghum crop. Essentially all of the additional sorghum area came at the expense of a decrease in cash crops area such as groundnut 42% and cotton crops area 58%. Consequently, the land planted with continuous sorghum crop increased remarkably and might cause weed problem and reduce yield and increase cost of production. However, this indicates, some farmers are willing to accept the known risks associated with growing sorghum in place of groundnut area in order to avoid the uncertain risks associated with water deficit from increasing groundnut production area (Ishag et al., 1980).

Richardson et al. (2006) used stochastic efficiency with respect to function (SERF) analysis to rank strategies by using the certainty equivalent (CE) results. The risk premium RP can also be calculated by subtracting the CE of a baseline crop rotation (3 Course Rotation - often a less preferred rotation) from the CE of a preferred alternative (5 Course R (B) -often a preferred strategy). The RP for a risk-averse decision maker reflects the minimum amount of money that would have to be paid to a decision maker to justify a switch from less preferred rotation alternative to a preferred rotation.

Table 5. Ranking of risk-efficient crop rotation by CE and ARAC of net return in Million (SDG)

Risk	Risk Neutral		Normal Risk		Rather Risk		Extremely Risk	
ARAC	0.00000000		0.0000000033		0.0000000055		0.0000000088	
Rank	Alternative	CE	Alternative	CE	Alternative	CE	Alternative	CE
1	4 Course R	68.7	5 Course R (B)	11.54	5 Course R (A)	5.79	5 Course R (A)	(2.30)
2	3 Course R	57.8	5 Course R (A)	11.35	5 Course R (B)	4.99	5 Course R (B)	(4.59)
3	5 Course Org.	47.9	5 Course Org.	(0.48)	5 Course Org.	(30.3)	5 Course Org.	(71.0)
4	5 Course R (B)	21.7	3 Course R	(12.0)	3 Course R	(54.2)	3 Course R	(110)
5	5 Course R (A)	19.9	4 Course R	(29.0)	4 Course R	(86.7)	4 Course R	(161)

Calculated by the Author & Simetar.

Table 5 shows that to switch from (5 Course R B) crop rotation to (5 Course R A) with high land use intensity i.e.72% a risk premium of 1.8 Million (SDG) needs to be paid to farmers with risk neutral risk aversion coefficient. The result also shows that night storage irrigation system (3 Course R) is not preferred for normal and rather risk aversion coefficient. However, increasing cash crop area such as cotton crop from 11% to 25% is difficult under present irrigation system situation as cotton crop needs 12-13 numbers of irrigation compare to sorghum crop which needs only 4-5 irrigation and almost half of sorghum CWR covered by rainfall. Moreover, the irrigation cycle of 14 day and 5000 cubic meters of water per day which is sufficient to irrigate 12 Acres per day and a crop number of (90 acres) in 7.5 day will not be applicable with present irrigation canal capacity. The introduction of new cotton verities with high yield and reasonable cost of production will increase scheme net return without increasing land use intensity.

3.5 StopLight Analysis

The Stoplight analysis relies on CDF information but is a more visually appealing depiction of probabilistic information. The Stoplight analysis uses the stimulated net income distribution to calculate the probability of whole-scheme net return exceeding an upper cutoff value of 25 Million (SDG), being less than a lower cutoff value of (0) net return, or it can have a value between the upper and lower cutoff values. The Stoplight analysis is performed to identify the probability of getting positive whole-scheme net returns for five crop rotations in the scheme. Figure 3 shows that the scheme will get 25 Million (SDG) or more with a probability of 55% for 4 course rotation and getting positive net return with 61% probability for all crop rotations. Red color represents the probability of having negative returns, which is 39% for all crop rotations. The yellow represent the probability of returns from (0) to 25 Million (SDG). The green color represents probability of having net returns more than 25 Million (SDG). The (4 Course R) and (3 Course R) crop rotation have the highest probability of making return between (0) and 25 Million (SDG).

However, SGB policy makers should select crop rotation which achieves equity water distribution among all scheme farmers with less net return i.e. 5 Course R (B) and 5 Course R (A) rather than targeting higher net returns crop rotation.

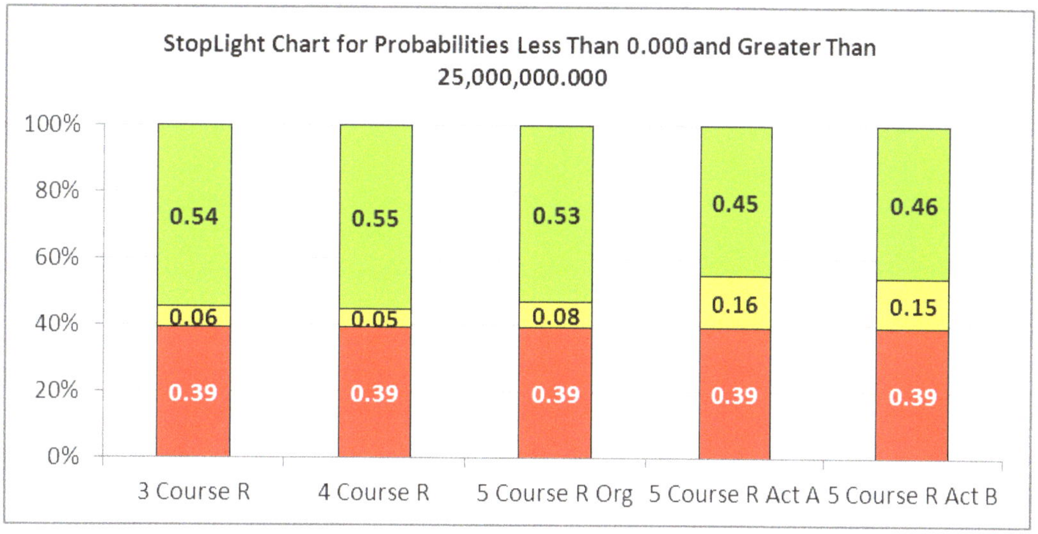

Figure 3. Probabilities of whole-scheme net returns for five course crop rotations in Sudan Gezira Scheme

4. Conclusion

The study use stochastic simulation techniques to evaluate five crop rotations risk-efficiency and economic sustainability in Sudan Gezira Scheme. Price and yield risk for five crops were simulated to calculate whole-scheme net return. The analysis shows with the present irrigation system capacity and continuous irrigation practices the 5 Course rotation (B) with 53% land use intensity is the most risk efficiency crop rotation to be followed and implemented in the scheme with normal risk aversion coefficient. In this recommended crop rotation equity water distribution achieved and area allocated to cotton crop is 17%, wheat 7%, sorghum 60%, groundnut 4% and fodder crops 12%. Fodder crop can be grown two times in summer and winter season without creating water shortage problems. However, this will increase net return and increase soil fertilities within the selected crop rotation.

The analysis shows that 5 Course rotation (A) is the second preferred crop rotation with land use intensity of 72%. The techniques used in this study can be used to investigate source of risk and analyze the impact of alternative management options and policy. Both the available management tools and sources of risk can significantly change the riskiness of alternative cropping systems.

The SGB have to maintain and monitor annual canals maintenance to keep sufficient water to implement 5 Course (B) crop rotation and land use intensity of 53%, as the return to night storage irrigation system will get a negative net returns. New high yield varieties have to be introduced at the scheme to replace loss from area reduction for cash crop and increase net return. The SGB also have to provide inputs to farmers for all crops cultivated within 5 course rotation (B) with a reasonable cost and treat all crops in the selected crop rotation with same support.Figure

In order to utilize stochastic simulation techniques the study use data from farmer's field, it is critical that there is enough information to characterize the distributions of the uncertain variables. The distributions of uncertain variables in this analysis were estimated based on historical data; however historical distributions may not be the best representation of potential outcomes. As a result, the techniques used in this study could be used with any distribution estimates for the uncertain variables incorporating such things as new crop varieties and research recommendation packages. They also could be modified to account for the new information effect during the decision process and account for dynamic effects to develop sharp policies.

Competing interests

The author declares that he has no competing interest.

Acknowledgments

The author thanks J. W. Richardson, Regents Professor & Texas AgriLife Research Senior Faculty Fellow, Department of Agricultural Economics at Texas A&M University for his valuable comments that greatly improved the manuscript.

References

Ahmed, A. E. (2011). Risk sources and attitude among the tenants of the Gezira Scheme–Sudan. *Journal of the Saudi Society of Agricultural Sciences, 10*(2), 71-75. http://dx.doi.org/10.1016/j.jssas.2011.03.006

Ahmed, M. A. (2004). Effective farm management decision-making in the Gezira Scheme. Irrigation Management in Sudan, International Irrigation Management Institute.

Ascough II, J. C., Fathelrahman, E. M., Vandenberg, B. C., Green, T. R., & Hoag, D. L. (2009). Economic risk analysis of agricultural tillage systems using the SMART stochastic efficiency software package. In *18th World IMACS / MODSIM Congress, Cairns, Australia 13-17 July 2009*. Retrieved from http://mssanz.org.au/modsim09

Eldaw, A. M. (2004). The Gezira Scheme, Perspectives for Sustainable Development, University of Gezira, Reports and Working Papers 2/2004.

Elnour, F. M., & Elamin, A. E. (2014). Status of on-Farm water use efficiency and source of inefficiency in the Sudan Gezira Scheme. International Working Paper Series NAF-IRN Paper No. 14/1. Heep://economia.unipv.it/naf/

Hardaker, J. B., Huirne, R. B., Anderson, J. R., & Lien, G. (1997). *Coping with Risk in Agriculture* (p. 274). CAB International, Wallingford, UK.

Hardaker, J. B., Richardson, J. W., Lien, G., & Schumann, K. D. (2004). Stochastic efficiency analysis with risk aversion bounds: a simplified approach. *Australian Journal of Agricultural and Resource Economics, 48*(2), 253-270. http://dx.doi.org/10.1111/j.1467-8489.2004.00239.x

Ibrahim, A. A., Stigter, C., Adam, H. S., & Adeeb, A. M. (2002). Water-use efficiency of sorghum and groundnut under traditional and current irrigation in the Gezira scheme, Sudan. *Irrigation Science, 21*(3), 115-125. http://dx.doi.org/10.1007/s00271-002-0057-z

Ishag, H. M., Ali, M. A., & Ahmadi, A. B. (1980). Groundnut production and research problems in the Sudan. Pages 282-284 in Proceedings of the International Workshop on Groundnuts, 13-17 Oct 1980, ICRISAT Center, India. Pataacheru, A.P. 502 324, India: International Crops Research Institute for the Semi-Arid Tropics.

Ishag, K., Thornton, D., Tiffen, M., & Upton, M. (1991). Farm location and farmers' performance on the Hamza minor canal. *International Journal of water Resources Development, 7*(1), 2-15. http://dx.doi.org/10.1080/07900629108722486

Lien, G., & Hardaker, J. B. (2001). Whole-farm planning under uncertainty: impacts of subsidy scheme and utility function on portfolio choice in Norwegian agriculture. *European review of agricultural economics, 28*(1), 17-36. http://dx.doi.org/10.1093/erae/28.1.17

Lien, G., Hardaker, J. B., & Flaten, O. (2007). Risk and economic sustainability of crop farming systems. *Agricultural Systems, 94*(2), 541-552. http://dx.doi.org/10.1016/j.agsy.2007.01.006

Meyer, J. (1977). Choice among distributions. *Journal of Economic Theory, 14*(2), 326-336. http://dx.doi.org/10.1016/0022-0531(77)90134-X

Palisade Corporation. (1997). Guide to Using @RISK Risk Analysis and Simulation Add-In for Microsoft Excel or Lotus 1-2-3, Window Version. Newfield, NY, USA.

Richardson, J. W. (2003). Simulation for Applied Risk Management. Department of Agricultural Economics, Texas A&M University, August 2003.

Richardson, J. W., Schumann, K. D., & Feldman, P. A. (2006). Simulation and Econometrics to Analyze Risk: Simetar_c Inc. 2006 User's Manual. College Station, Texas.

Richardson, J. W., Schumann, K., & Feldman, P. (2003). Simetar Simulation for Excel to Analyze Risk User's Guide. Agricultural and Food Policy Center, Texas A&M University, College Station, TX, USA.

Shafique, M. S. (2008). Performance of the Gezira canals, International Irrigation Management Institute.

Constraints and Breeding Priorities for Increased Sweetpotato Utilization in Ghana

Ernest Baafi[1], Joe Manu-Aduening[1], Edward E. Carey[3], Kwadwo Ofori[2], Essie T. Blay[2] & Vernon E. Gracen[2]

[1] CSIR-Crops Research Institute, P. O. Box 3785, Kumasi, Ghana

[2] West Africa Centre for Crop Improvement, University of Ghana, Legon

[3] International Potato Centre (CIP), Ghana

Correspondence: Ernest Baafi, CSIR-Crops Research Institute, P. O. Box 3785, Kumasi, Ghana. E-mail: e.baafi@gmail.com

Abstract

Sweetpotato is used in various food preparations in place of rice, cassava, yam and plantain in Ghana. In spite of this it does not have the same importance in Ghanaian diet as other root and tuber crops. Consumer taste, preference and acceptance are critical in determining the suitability of sweetpotato cultivars to any locality. A study was carried out in some selected communities of Ghana where sweetpotato is popular which span all five ecozones of Ghana in February, 2012. The main objective was to investigate why sweetpotato has low utilization compared with other root and tuber crops and to increase its utilization through breeding. The study employed Focus Group Discussion (FGD) followed by administration of Semi-structured Questionnaire (SSQ). Data collected were analysed using Genstat and Statistical Package for Social Sciences. Seventy-nine people consisting of 63% males and 37% females, and 178 people consisting of 52% female and 48% males were involved in the FGD and SSQ, respectively. Majority (94%) of farmers' ranked sweetpotato from 1 to 5 among 24 cultivated crops. Only about 28% of consumers ate sweetpotato at least six days per week. The survey revealed that consumers in Ghana desired non-sweet, high dry matter sweetpotato cultivars. Therefore, there is need for Research and Development to adjust sweetpotato breeding objectives and selection procedures to develop high dry matter non-sweet sweetpotato varieties in Ghana.

Keywords: beta-carotene, breeding, dry matter, non-sweet, sweetpotato

1. Introduction

Sweetpotato (*Ipomoea batatas,* L) is one of the most important root crops in the world with more than 133 million tonnes produced worldwide annually (Warammboi et al., 2011). The crop has versatility in Ghana and is used in various food preparations in place of rice, cassava, yam, plantain and other staples (Ellis et al., 2001; Meludu et al., 2003; Zuraida, 2003). In spite of this it does not have the same importance in the Ghanaian diet as other root and tuber crops, such as yam, cassava or cocoyam (Adu-Kwarteng et al., 2002; Opare-Obisaw et al., 2000). Its production peaked during the late 1950s, after which there has been a gradual decline (Doku, 1984), despite giant strides made through breeding to release 13 high yielding and early maturing cultivars.

The decision to adopt a new cultivar is complexly related to field and yield performance as well as consumer taste acceptability (Sugri et al., 2012). Consumer taste, preference and acceptance are critical in determining the degree of suitability of sweetpotato cultivars to any locality (Kwach et al., 2010; Tomlins et al., 2004). Food preferences may vary among individuals, age groups, gender and sometimes cultures as well as geographical locations (Sugri et al., 2012). Traits to select during crop improvement depend on the target beneficiaries (Sugri et al., 2012). It is believed that some cultivars that have been released were not adopted because of lack of sufficient consideration of farmers' (Derera et al., 2006), and other stakeholders' preferences in the process of their development. Breeders often fail to consider the special preferences of farmers (Banziger & Cooper, 2001; Toomey, 1999), possibly because they are unaware of them (Derera et al., 2006). Breeding programmes must therefore adjust breeding objectives and selection procedures to meet consumer needs. Effective breeding should be based on clear identification of stakeholders perceived constraints and preferences. These can be identified through participatory plant breeding (PPB), participatory varietal selection (PVS) (Gasura et al., 2008; Gibson et al., 2007) and/or participatory rural appraisal (PRA).

This survey was aimed at (1) documenting the role of sweetpotato in the farming and food systems of Ghana, (2) identifying utilization constraints and opportunities, (3) understanding the current patterns of sweetpotato consumption, and (4) establishing general baseline data for future impact assessment for sweetpotato production and utilization in Ghana.

2. Materials and Methods

The study was conducted using Focus Group Discussions (FGD) followed by administration of a Semi-Structured Questionnaire (SSQ) in February, 2012. Focus Group Discussion and Semi-Structured Interview are tools of PRA. Focus Group Discussion (FGD) is a form of group interview that capitalises on communication and interactions among participants in order to generate data (Kitzinger, 1995). The method is particularly useful for exploring people's knowledge and experiences and can be used to examine not only what people think but how they think and why they think that way (Kitzinger, 1995). Semi-Structured Interview is a powerful research tool that provides valuable spontaneous information that can be used for quantitative purposes, such as the testing of hypotheses or the generalisation of findings for larger areas.

2.1 Study area

The study was conducted in all the communities except Kumasi for the FGD (Table 1). These communities were the major sweetpotato production, marketing and utilization areas in Ghana, spanning over all five agroecological zones of Ghana. The areas covered latitude 4° 30' to 11° north of the Equator, and longitudes 1° 12'east to 3° 15'west.

Table 1. List of study communities for the survey

Community	Metropolitan/District	Region	Ecozone
Fiaso	Techiman	Brong Ahafo	Transition
*Kumasi	Kumasi	Ashanti	Forest
Komenda	KEEA	Central	Coastal savanna
Aseja	Fanteakwa	Eastern	Forest
Sege	Dangme East	Greater Accra	Coastal Savanna
Woriborgu-kukuo	Tolon	Northern	Guinea Savanna
Natugnia	Kassena-Nankana	Upper East	Sudan Savanna
Agorve	Ketu-North	Volta	Coastal Savanna

*Community where Focus Group Discussion was not done.

2.2 Focus Group Discussions (FGD)

2.2.1 Nature of Participants for the FGD

District Agricultural Extension agents (AEAs) were actively involved in the identification and selection of respondents through random sampling. Participants were selected from sweetpotato stakeholders, namely consumers, farmers, marketers, and processors. In all, 79 participants were involved. These included 50 males and 29 females (Table 2).

Table 2. Group characteristics of the focus group discussion

Region	Females	Males	Total
Brong Ahafo	2	6	8
Central	1	11	12
Eastern	8	4	12
Greater Accra	6	6	12
Northern	4	7	11
Upper East	4	8	12
Volta	4	8	12
Total	29	50	79

2.2.2. Focus Group Discussion Data Collection and Analysis

The data were collected by a team composed of a facilitator/moderator and two recorders. In communities where language was a barrier, an interpreter was hired. An open-structured questionnaire guided by a checklist (variety grown, source of planting material, uses, gender, and farming systems) was used. General description of the group dynamics was also recorded. The data were converted to numerical values, standardized (Etzkorn, 2011), and analysed using Principal Component Analysis (PCA). The analyses were done using Genstat (Released 9.2.0.152) Computer package (Genstat, 2007).

2.3 Administration of Semi-structured Questionnaire (SSQ)

2.3.1 Structure of Respondents for the Semi-structured Interview

In all, 178 people involving 52% female and 48% males were interviewed. The structure of respondents is presented in Figure 1. There were 160 consumers of sweetpotato, 123 sweetpotato farmers, 40 sweetpotato traders and 18 sweetpotato processors. Among them were 24 traders who were also consumers, 93 farmers who were consumers, 11 people who were consumers and processors and one person who was a farmer and a trader. Out of the 178 respondents, 15, 12, 3 and 2 of the respondents were only consumers, farmers, traders, and processors, respectively. The number of respondents per community is presented in Table 3.

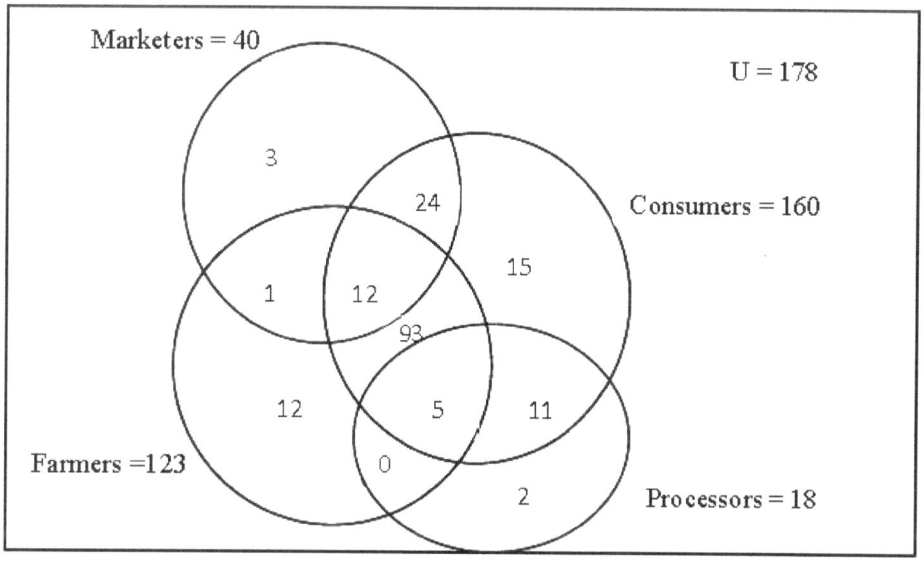

Figure 1. Euler diagram of respondents

2.3.2 Semi-structured Questionnaire Data Collection and Analysis

Interviews were used in the data collection. District Agricultural Extension Agents (AEAs) were involved in the identification and selection of respondents by random sampling. The Semi-structured Questionnaire (SSQ) was developed using a checklist which included gender, kind of stakeholder, crop enterprise, food consumption pattern, and sweetpotato preference. Sweetpotato root-flesh colour chart was used to estimate the beta-carotene content of the storage root. In the colour chart, white-flesh represented low beta-carotene content, yellow-flesh represented moderate beta-carotene content and orange-flesh represented high beta-carotene content. The interview was done as a follow-up to the FGD. Data were analysed using Statistical Package for Social Sciences (SPSS Release 16.0) Computer software (SPSS, 2007). Results were summarized into frequencies and percentages and represented in tables and graphs.

Table 3. Frequency of respondents per community (Region) for the semi-structured interview

Community (Region)	Respondents		
	Number of participants	Percent	Cumulative Percent
Ashanti	13	7.3	7.3
Brong Ahafo	17	9.6	16.9
Central	28	15.7	32.6
Great Accra	19	10.7	43.3
Upper East	30	16.9	60.1
Volta	40	22.5	82.6
Northern	17	9.6	92.1
Eastern	14	7.9	100
Total	178	100.0	

3. Results

3.1 Role of Sweetpotato in Farming and Food Systems

Fifteen variables (Table 4) were recorded during each session of the FGD. The first five principal components (PCs) with Eigen values greater than 1.0 together explained 96.33% of the total variation in the data set (Table 4). Scores for method of storage and emotional aspect of participants were not significantly associated with any of the Principal Components. Scores on the first two PCs accounted for 58.75% of the variation, and were related to all the variables except sex and source of planting material, which were related to PCs 3 and 5.

Table 4. Principal component analysis of the 15 variables recorded for the focus group discussion

Variables	PC1	PC2	PC3	PC4	PC5
Total participants	**-0.327**	-0.080	-0.200	-0.191	0.437
Sex	-0.148	0.048	**0.599**	0.037	-0.243
Crop calendar	**0.315**	0.121	-0.201	0.215	**0.496**
Crop enterprise	**-0.420**	-0.048	-0.019	0.025	0.170
Rank of sweetpotato	**-0.303**	0.200	0.063	**-0.474**	0.114
Variety grown	**-0.325**	-0.225	**-0.377**	-0.026	-0.119
Planting material source	-0.234	0.171	0.270	0.034	**0.494**
Source of sweetpotatoes eaten	**-0.329**	0.179	**-0.303**	**-0.317**	-0.005
Ways sweetpotato is eaten	0.119	**-0.427**	0.091	**-0.525**	0.192
Storage	0.000	0.000	0.000	0.000	0.000
Other uses	-0.079	**-0.305**	0.288	**0.515**	-0.200
Constraints to consumption	0.232	**-0.495**	-0.052	0.202	0.102
Perception on health	-0.236	**-0.502**	0.121	-0.080	0.200
Emotional aspect of participants	0.000	0.000	0.000	0.000	0.000
Group dynamics	**0.325**	0.225	**0.377**	0.026	0.119
Latent roots (Eigen values)	5.183	2.454	2.198	1.518	1.170
Variance (%)	39.87	18.88	16.91	11.67	9.00
Cumulative (%)	39.87	58.75	75.66	87.33	96.33

*Values in bold indicate the most relevant characters (>0.3) that contributed most to the variation of the particular component.

The distribution produced by PCs 1 and 2 is shown in Figure 2. Brong Ahafo Region is separately classified in group 1, Volta and Eastern Regions are in group 2. Group 3 composed of Upper East and Central Regions while group 4 composed of Greater Accra and Northern Regions.

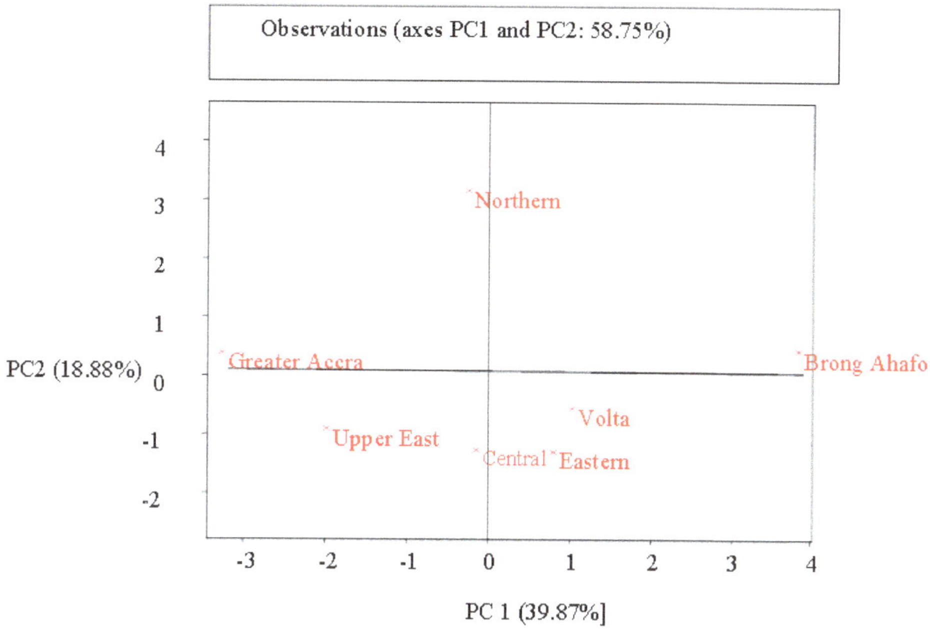

Figure 2. A biplot of PC1 and PC2 showing the distribution of communities involved in the focus group discussion

The hierarchical clustering (Figure 3) illustrates the relationship between responses among the communities. The communities were separated with Euclidean similarity distance from 1.00 to 0.75. All the communities, except the Northern region, were similar at about 0.75 level of similarity. However, they were all distinct at 0.87 level. Four main clusters, A, B, C, and D, were identified at similarity level 0.76. Cluster B was composed of the Volta and the Eastern Regions, while Brong Ahafo Region and the Northern Region were found in cluster A and D, respectively. Cluster C was composed of Central, Upper East and the Greater Accra Regions.

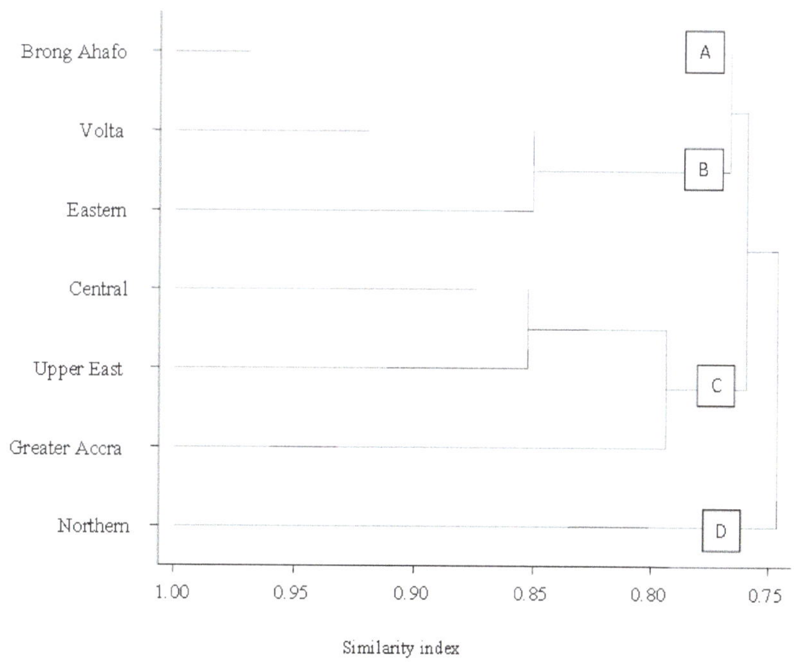

Figure 3. Dendrogram showing response of communities involve in the focus group discussion

3.1.1 Crop Enterprise of the Seven Communities Involved in the Focus Group Discussion

In all, 24 crops were cultivated by the communities (Table 5). Brong Ahafo Region cultivated the least (9) whilst Greater Accra cultivated the highest (12). The crops cultivated includes roots and tubers (sweetpotato, cassava, yam, cocoyam and frafra potato), cereals (maize, millet and rice), legumes (cowpea, groundnut, bambara groundnut/beans and soybean), horticultural crops (tomato, okra, watermelon, onion, eggplant, cabbage, pepper and eggplant), and others – plantain, sugarcane and tiger nuts. Sweetpotato was ranked between first and tenth among the communities. It was ranked first in Brong Ahafo and Central Regions but, tenth in the Greater Accra Region.

Table 5. Crop enterprise ranking within the seven communities involved in the focus group discussion

Region							
Rank Position	Brong Ahafo	Central	Eastern	Greater Accra	Northern	Upper East	Volta
1	Sweetpotato	Sweetpotato	Sweetpotato	Pepper	Maize	Millet	Maize
2	Maize	Maize	Cassava	Tomato	Yam	Maize	Sugarcane
3	Yam	Cassava	Maize	Cassava	Rice	Rice	Cassava
4	Cassava	Watermelon	Plantain	Okra	Sweetpotato	Groundnut	Sweetpotato
5	Watermelon	Tomato	Cocoyam	Maize	Cowpea	Sweetpotato	Rice
6	Cowpea	Pepper	Pepper	Watermelon	Cassava	Frafra potato	Pepper
7	Pepper	Okra	Yam	Cowpea	Groundnut	Bambara beans	Okra
8	Egg plant	Egg plant	Tomato	Groundnut	Pepper	Cowpea	Tomato
9	Onion	Groundnut	Okra	Egg plant	Soybean	Soybean	Groundnut
10		Tiger nut	Cowpea	Sweetpotato	Bambara groundnut	Pepper	Cowpea
11		Yam		Cabbage	Tomato	Egg plant	
12				Onion			

3.1.2. Farm Level Characteristics of Respondents Involved in the Semi-Structured Interviewing

Sweetpotato was ranked from first to tenth among the 24 crops cultivated by the farmers. Out of the 123 farmers interviewed, about 94% ranked sweetpotato first to fifth, while about 32% ranked it as their first choice (Table 6). The majority of the farmers were engaged in sole cropping (92%) as against 8% for intercropping. The area under cultivation ranged from 0.25 to 12.0 acres (Table 7). However, majority (about 92%) of the farmers had sweetpotato farm sizes between 0.25 acres and 3.0 acres. Farmers cultivating one acre of land (25%) were in the majority followed by those cultivating 2 acres (20%).

Table 6. Relative ranking of sweetpotato among 24 cultivated crops by respondents

Rank	Number of farmers interviewed	Percent of farmers interviewed	Cumulative Percent of farmers interviewed
1	39	31.7	31.7
2	8	6.5	38.2
3	23	18.7	56.9
4	19	15.4	72.4
5	26	21.1	93.5
6	2	1.6	95.1
7	1	0.8	95.9
8	3	2.4	98.4
9	1	0.8	99.2
10	1	0.8	100
Total	123	100.0	

Table 7. Frequency distribution of sweetpotato cropping land area

Farm size (ac.)	Number of farmers interviewed	Percent of farmers interviewed	Cumulative Percent of farmers interviewed
0.25	21	17.1	17.1
0.5	22	17.9	35.0
0.75	1	0.8	35.8
1	31	25.2	61.0
1.25	1	0.8	61.8
1.5	6	4.9	66.7
2	25	20.3	87.0
2.5	1	0.8	87.8
3	5	4.1	91.9
4	3	2.4	94.3
5	2	1.6	95.9
6	2	1.6	97.6
7	1	0.8	98.4
10	1	0.8	99.2
12	1	0.8	100.0
Total	123	100.0	

3.2. Pattern of Sweetpotato Preferences and Utilization

3.2.1 Demographics of Respondents Involved in the Semi-structured Interviewing

The lowest and highest number of respondents across the communities were 13 and 40. These were from the Ashanti and the Volta Regions, respectively, and constitute about 7% and 23% of the total respondents. Fifty-two per cent of the respondents were 40 years of age and above compared to 47% and 1%, respectively, for age classes 20-40 years and below 20 years (Figure 4). The educational background of respondents was distributed as follows; Primary or below 83% and High School and above, 17%.

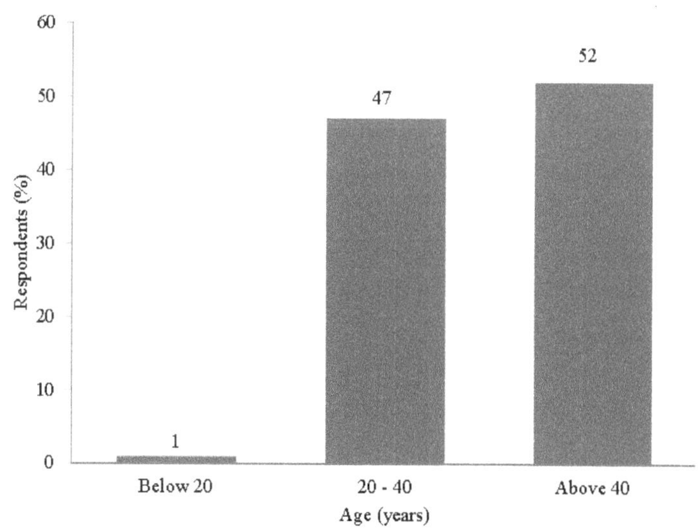

Figure 4. Age distribution of respondents involved in the semi-structured interviewing

3.2.2 Sweetpotato Storage Root Utilization and Quality Preferences

Sweetpotato utilization pattern is presented in Figure 5. About thirty-nine percent of consumers ate sweetpotato two or three days per week while about 28% consumed it at least six days per week and only about 12% of the respondents consuming sweetpotato at most only once a week.

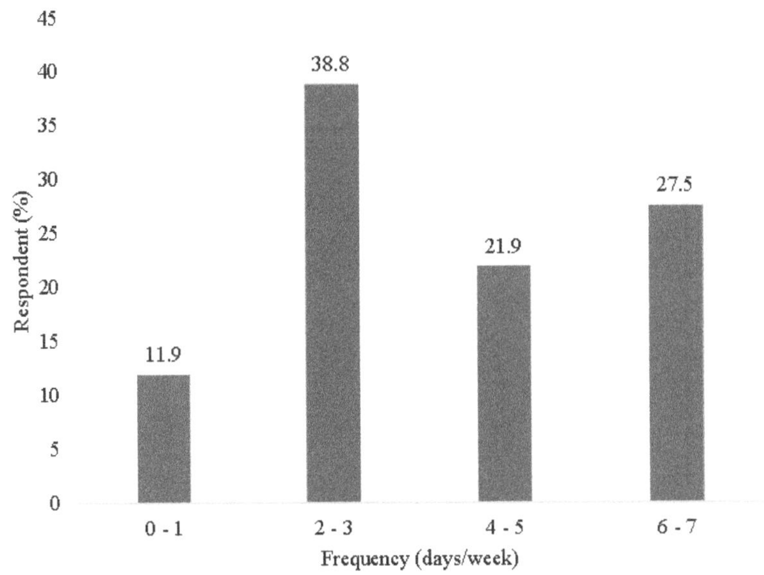

Figure 5. Sweetpotato consumption distribution for the respondents of the semi-structure questionnaire

Varietal preferences showed a response range from 7 to 138 respondents for the semantic differential scale (low, moderate, high, and any-type) for the storage root dry matter content (Figure 6). The lowest and highest values represent preference for any sweetpotato type and sweetpotato with high storage root dry matter content. Most respondents preferred high dry matter followed by moderate dry matter content while few preferred low dry matter content or didn't care. Most respondents (88) preferred moderate beta-carotene content (yellow-flesh) but 40 preferred low beta-carotene content (white-flesh) and 33 preferred high beta carotene content (Orange-flesh) while 17 had no preference for beta-carotene content (Figure 7). The results for the storage root taste (sweetness) are represented in Figure 8. The order of preference was less sweet (87), sweet (63), very sweet (25), and any-type (3), respectively. No significant interaction was observed between these storage root quality traits and age, educational background and sex of the respondents (Tables 8-11).

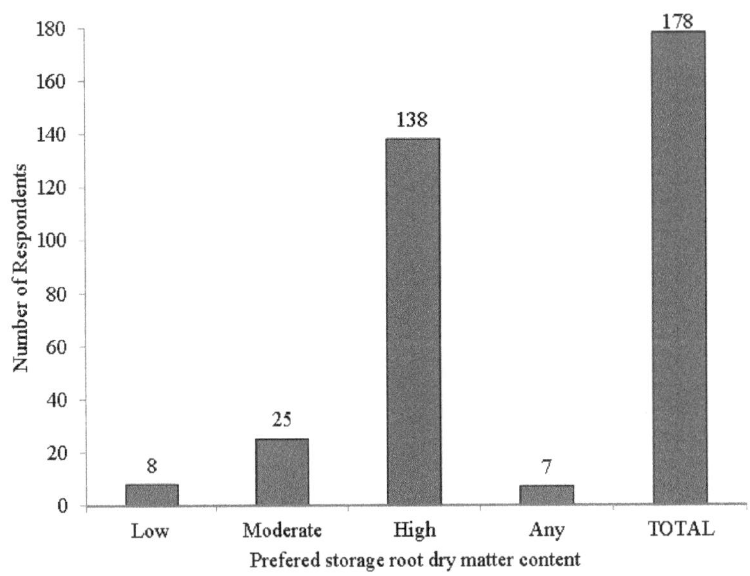

Figure 6. Preferred storage root dry matter content distribution for the respondents for the semi-structured questionnaire

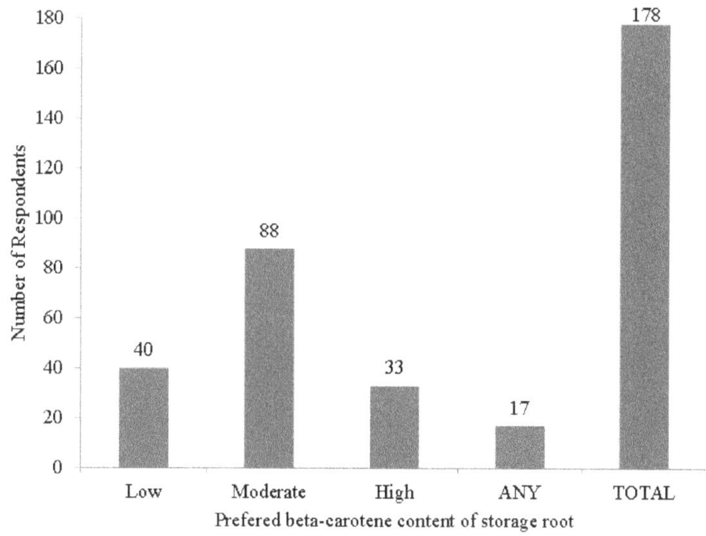

Figure 7. Distribution of preference for beta-carotene content of storage root for the respondents involved in the semi-structured interview

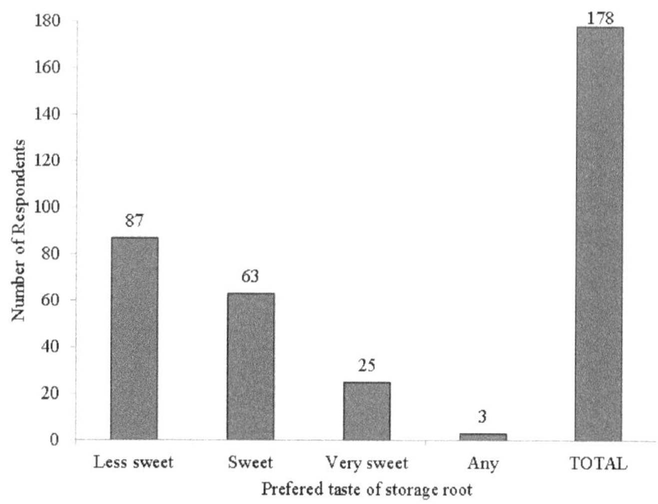

Figure 8. Storage root taste (sweetness) preference distribution for the respondents involved in the semi-structured interview

Table 8. Preference for beta-carotene and dry matter by different age groups of respondents

Trait	Age of Respondents	Preference				
		Low	Moderate	High	Any	Total
Beta-carotene						
	Below 20	0	0	0	2	2
	20 – 40	15	48	14	7	84
	Above 40	25	40	19	8	92
	Total	40	88	33	17	178
Dry matter						
	Below 20	0	0	2	0	2
	20 – 40	3	11	65	5	84
	Above 40	5	14	71	2	92
	Total	8	25	138	7	178

Table 9. Preference for beta-carotene and dry matter by different educational background of respondents

Trait	Respondents' Educational Background	Preference				
		Low	Moderate	High	Any	Total
Beta-carotene						
	Primary & below	35	63	33	17	148
	High School & above	5	25	0	0	30
	Total	40	88	33	17	178
Dry matter						
	Primary & below	6	20	117	5	148
	High School & above	2	5	21	2	30
	Total	8	25	138	7	178

Table 10. Beta-carotene and dry matter content preferences by sex of respondents

Trait	Sex of Respondents	Preference				
		Low	Moderate	High	Any	Total
Beta-carotene						
	Female	23	48	13	8	92
	Male	17	40	20	9	86
	Total	40	88	33	17	178
Dry matter						
	Female	5	10	70	7	92
	Male	3	15	68	0	86
	Total	8	25	138	7	178

Table 11. Taste (sweetness) preferences by different age groups, educational background and sex of respondents

Source	Preference				
	Less sweet	Sweet	Very sweet	Any	Total
AGE (years)					
Below 20	0	0	1	1	2
20 – 40	37	30	17	0	84
Above 40	50	33	7	2	92
Total	87	63	25	3	178
EDUCATION					
Primary & below	74	52	19	3	148
High school & above	13	11	6	0	30
Total	87	63	25	3	178
SEX					
Female	42	32	15	3	92
Male	45	31	10	0	86
Total	87	63	25	3	178

There was preference for high root dry matter content across all the communities (Figure 9). Most communities also prefer moderate beta-carotene (yellow-flesh) sweetpotatoes (Figure 10). However, the Upper East Region and the Greater Accra Region preferred low beta-carotene (white-flesh) and high beta-carotene (orange-flesh) sweetpotatoes, respectively. The results show that the highest number of respondents preferred less sweet sweetpotatoes across the communities, with the exception of the Upper East Region where sweet types were rated highest (Figure 11). However, the proportion of respondents who preferred very sweet sweetpotato was quite substantial in the Volta Region.

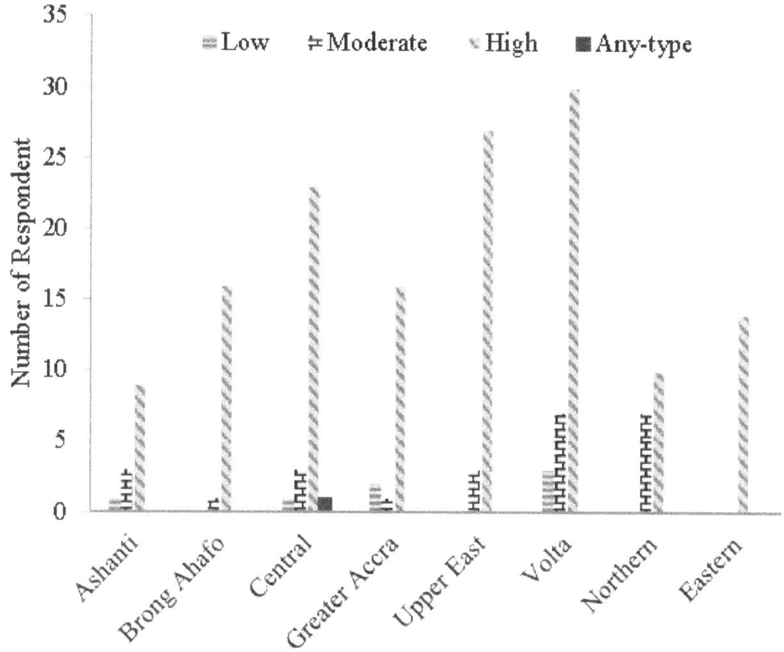

Figure 9. Respondents' preference for dry matter by different communities (Region)

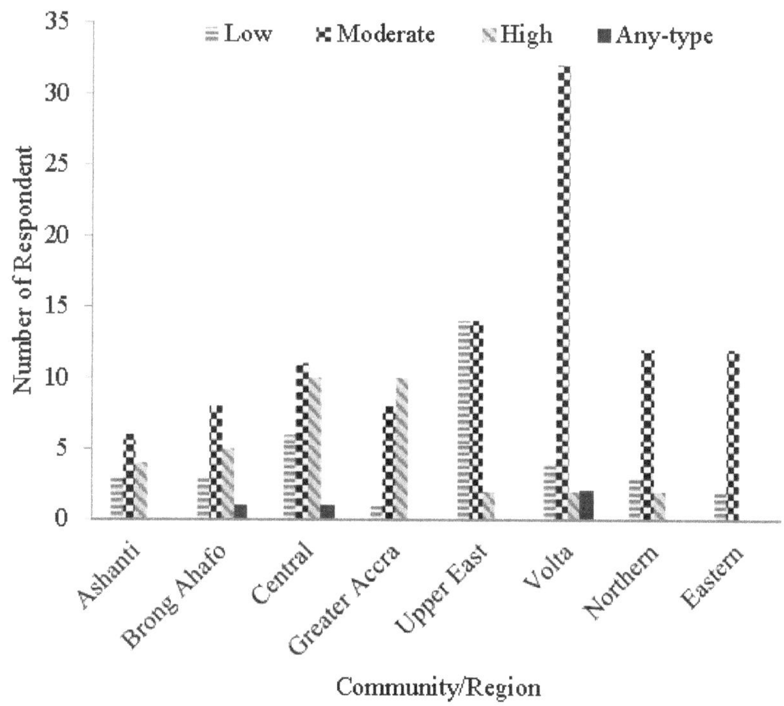

Figure 10. Respondents' preference for beta-carotene by different communities (Region)

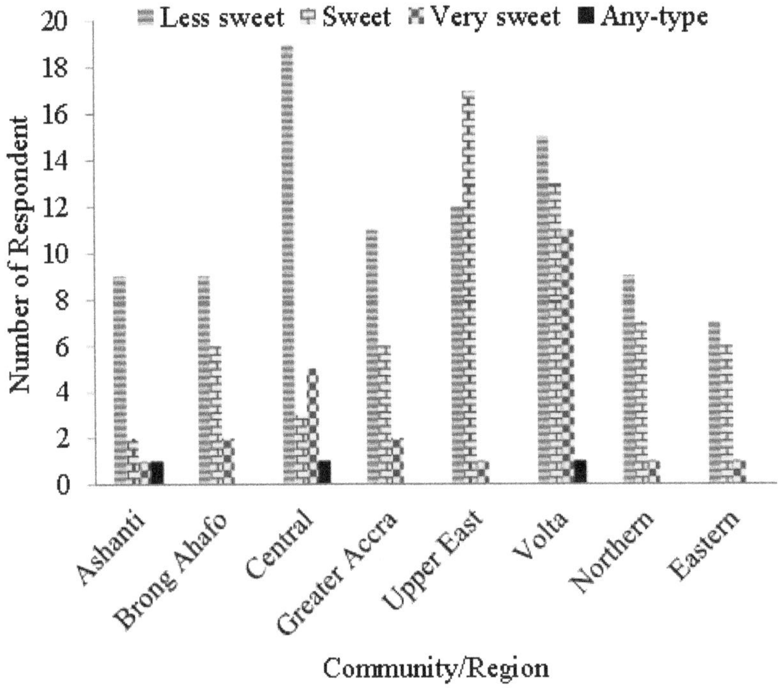

Figure 11. Respondents preference for storage root taste (sweetness) by different communities (Region)

4. Discussion

The variation obtained in 13 out of the 15 variables used for the Focus Group Discussion (FGD) suggests presence of a high degree of varying perception, knowledge and opinion for sweetpotato in Ghana. The level of interest and participation were very high, indicating that differences observed may be attributed to social background since discussions were held in different communities/regions. This is because; preference for sweetpotato varies with ethnic background and geographic location (Kays & Horvat, 1983). For most communities sweetpotato is mainly grown and produced in one season (major cropping season) per year. The reason is the availability of rainfall and avoidance of the sweetpotato weevil. The sweetpotato weevil is the major pest of the crop worldwide especially in the drier agroecological zone (Ames et al., 1996; CIP, 1995). Major season plantings and harvestings are in March/April and August, respectively. Harvesting depends mainly on market availability and household demands. The decision on the crop grown was mainly based on food security and income generation. However, crop adaptability to the environment is also a critical factor in the Northern Region. The potential of sweetpotato as a food security and an income generation crop was confirmed by the current study. This is partly affirmed by the blend of stakeholders (farmers, consumers, processors, and marketers) identified to be associated with the crop. The relative ranking of sweetpotato to the other crops grown in the study area also affirms the food security and income generating potential of the crop. The average land holding for sweetpotato is very small compared to the other staple crops. This is because in most developing countries, sweetpotato is a smallholder crop tolerant to a wide range of edaphic and climatic conditions and grown with limited inputs (Lebot, 2010).

Stakeholders prefer sweetpotatoes with high storage root dry matter content because they suit their food preparation preferences. The method of cooking and the form of utilization have critical influence on taste and consumptive quality (Sugri et al., 2012). Sweetpotatoes were mainly consumed by frying or boiling in the form of *ampesi*. Cooking leads to changes in physical, sensory and chemical characteristics of the final product (Fontes et al., 2011; Vitrac et al., 2000). The storage root flesh is mainly starch, which swells upon water absorption which hydrolyses the weak bonds making it more easily digestible. However, low dry matter types lose their cooking quality (mealiness) when cooked, affecting textural characteristics preference. Frying enhances sensorial characteristics such as smell, flavour, colour, texture as well as overall palatability (Sugri et al., 2012). Frying involves drying, cooking or fast dehydration in which the water is removed from the food by means of immersion into oil at temperatures of 120-180 °C (Vitrac et al., 2000). The frying oil is incorporated into the food and occupies part of the space left by the evaporated water thereby increasing palatability, flavour, calorie supply and shelf-life (Fontes et al., 2011; Vitrac et al., 2000). For the reason that the oil occupies spaces left by water, low dry matter sweetpotatoes absorb more oil which is not economical to processors as well as healthy to the consumers. Other uses of sweetpotato were stew preparation from the leaves and as herb for jaundice (in the Eastern Region), and the use of the storage roots as a source of sugary flour for porridge in the Northern Region. One of the oldest uses was sweetening porridges and maize products, such as *Aboolo* (steamed or baked sweetened fermented maize dough) (Osei-Opare & Adjei-Poku, 1977). The latter two may be exploited for commercial medicinal and sugar flour production.

There were varied responses for constraints to sweetpotato consumption. With exception of Greater Accra Region that did not have any constraint, the other communities mentioned sweetness as the major constraint that causes people to turn away from the crop as staple food. This affirms Missah and Kissiedu (1994) and Opare-Obisaw et al. (2000), who reported that locally available sweetpotatoes have very sweet taste, which limits their consumption as a staple food. It may also explain why it is uncommon to find sweetpotato served in Ghanaian restaurants, canteens, and schools unlike other roots and tubers which are widely consumed in a variety of preparations (Sam & Dapaah, 2009). This confirms the report that high levels of sweetness and strong flavour associated with many cultivars of sweetpotatoes may have reduced its popularity as a staple food (Woolfe, 1992). The findings also agree with the observation that consumers in Ghana prefer sweetpotatoes with dry, mealy flesh and low sugar content (Sam & Dapaah, 2009). The sweetness was perceived to have negative effect on health and was believed to cause diabetes, malaria, piles, flatulence, allergies, and hernia. These may also be the reasons why it is still not an important component of the average Ghanaian diet. These justify the need to adjust sweetpotato breeding objectives to develop non-sweet sweetpotato cultivars in Ghana to raise its standard to a staple food status. This is because according to Lebot (2010), although not thoroughly been investigated, the hundreds of sweetpotato cultivars found in Papua New Guinea and Island Melanesia, are low in sugar content, allowing an important daily consumption. Sweetpotato varieties that suit staple consumption must be non-sweet with sugar content less than 12% (Kays et al., 2005). Breeding such varieties is possible since a local variety of Vanuatu which suits staple purpose has a total sugar content as low as 1.49% of dry matter (Lebot et al., 2009).

In spite of these perceived hazards, good health attributes that were mentioned include better vision for children, and being good food for pregnant women. This suggests that the respondents have some knowledge of the benefit of beta-carotene found in the orange-flesh cultivars. Sweetpotatoes are used in a number of African countries to combat the widespread Vitamin A deficiency that results in blindness and/or death of 250,000 - 500,000 African children a year (CGIAR, 2005). However, preference for beta-carotene is skewed towards low- and moderate beta-carotene sweetpotatoes. Orange-flesh sweetpotato is not popular in Ghana even though it has been shown that even small amounts of these varieties as a regular part of the diet will eliminate vitamin A deficiency in adults and children. According to Thottappilly (2009), African countries have traditionally grown white-flesh sweetpotatoes which are low in vitamin A content. The orange-flesh cultivars are low in root dry matter content and high root dry matter content is preferred by most Ghanaians and this may be the reason for consumers' lack of preference for the orange–flesh types. However, considering the health benefits of the beta-carotene in the orange-fleshed types, it is important to incorporate high dry matter into the high beta-carotene genetic backgrounds. Dry matter content and beta-carotene content are negatively correlated (Baafi, 2014; Grüneberg et al., 2009), but this correlation can be broken by breeding and advance biotechnology. Crossing and recombination between superior, high beta-carotene and high dry matter parents, and selection among segregating progenies for improved performance is needed to develop superior new varieties. Recurrent selection increase the frequencies of favourable alleles at multiple loci in breeding populations through inter-mating of selected individuals (Moose & Mumm, 2008). Recurrent selection could be used for the development of sweetpotatoes that have high dry matter content, high beta-carotene content and low sugar content (bland taste). There is also the need to educate consumers on their health concerns to ease their fears. This is because even though the galactoside oligosaccharides, verbascose, starchyose and raffinose in legumes have been correlated with their gas formation (flatulence) (Padmaja, 2009), there is no concrete evidence to that in sweetpotato (Truong et al., 1986). Again, despite its name sweetpotatoes may be a good food for diabetics as they help to stabilize blood sugar levels and provide lower insulin resistance (Kusano & Abe, 2000; Ray & Tomlins, 2010). They also need to be aware that sweetpotato consumption does not cause malaria. The high incidence of malaria is due to the widespread unclean environment which serves as the breeding grounds for the mosquitoes. Belief in such misinformation may also have accounted for why sweetpotatoes are not an important component of the average Ghanaian diet.

Food preferences vary among individuals, age groups, gender and sometimes cultures as well as geographic locations (Sugri et al., 2012). A Preference for non-sweet sweetpotato cultivars with high storage root dry matter content was consistent across the entire respondents in this study. All of the communities preferred non-sweet high dry matter content sweetpotatoes. However, there were differences among the communities for preference for storage root beta-carotene content. Even though less sweet types were preferred in all the communities, consumers in the Upper East Region preferred moderately sweet types as well. The breeding objectives for developing and releasing sweetpotato in Ghana could be community specific. This is because preferences for the sweetpotato vary with ethnic background and geographic location (Kays & Horvat, 1983).

5. Conclusion

This study was carried out to identify end-user constraints for sweetpotato utilization and to link the identifiable preferred traits to breeding for development of end-user preferred varieties for increased sweetpotato utilization in Ghana. The study was conducted using Focus Group Discussions (FGD) followed by administration of Semi-structured Questionnaire (SSQ). This study established the potential of sweetpotato as an income generating and food security crop in Ghana. It revealed very low utilization of sweetpotato in Ghana compared with the major root and tuber crops like cassava, yam and cocoyam. Very low patronage for sweetpotato in Ghana compared with the other root and tuber crops is due to lack of available end-user preferred sweetpotato cultivars that allow for daily consumption of the crop as staple food. The study revealed that consumers in Ghana desire non-sweet sweetpotatoes with high storage root dry matter content. There is a need for Research and Development to adjust sweetpotato breeding objectives and selection procedures to develop high dry matter non-sweet sweetpotato varieties in Ghana.

Acknowledgements

Many thanks to the Alliance for a Green Revolution in Africa (AGRA) for sponsoring this study. Thanks to the Director (Professor Eric Y. Danquah) and staff of West Africa Centre for Crop Improvement (WACCI), University of Ghana for your support. Thanks to my supervisors (Professors Vernon E. Gracen, Essie T. Blay, Kwadwo Ofori, Dr. Joe Manu-Aduening and Dr. Edward E. Carey) and staff of CSIR-Crops Research Institute, Fumesua, Ghana and the Ministry of Food and Agriculture, Ghana for your support.

References

Adu-Kwarteng, E., Otoo, J. A., Osei, C. K., & Baning, I. S. (2002). Sweetpotato:The Crop of the Future. *Factsheet Published by the Communications and Extension Division of Crops Research Institute* - Council for Scientific and Industrial Research, Ghana.

Ames, T., Smit, N. E. J. M., Braun, A. R., O'Sullivan, J. N., & Skoplund, L. G. (1996). Sweetpotato: Major Pest, Diseases, and Nutritional Disorders (p. 152). International Potato Center (CIP), Lima, Peru.

Baafi, E. (2014). Development of End-User Preferred Sweetpotato Varieties in Ghana. PhD Thesis. West Africa Centre for Crop Improbvemnt (WACCI), University of Ghana.

Banziger, M., & Cooper, M. (2001). Breeding for Low Input Conditions and Consequences for Participatory Plant Breeding: Examples from Tropical Maize and Wheat. *Euphytica, 122*, 503-519. http://dx.doi.org/10.1023/A:1017510928038

CGIAR. (2005). FAOSTAT. Prodstat. Priorities and Strategies for Resource Allocation During 1998-2000 and Centre Proposals and TAC Recommendations. Retrieved June, 2010, from http://www.cgiar.org. Retrieved on 23/12/2010.

CIP. (1995). Sweetpotato / Sweetpotato Weevil. Retrieved from http://www.cgiar.org. Retrieved on 23/12/2010.

Derera, J., Tongoona, P., Langyintuo, A., Laing, M. D., & Vivek, B. (2006). Farmer Perception on Maize Cultivars in the Marginal Eastern Belt of Zimbabwe and their Implications for Breeding. *African Crop Science Journal, 14*(1), 1-5.

Doku, E. V. (1984). The Sweet Potato, the Production and Uses. Legon Extension Bulletin No. 17 (pp. 1-13). Crop Science Department, University of Ghana, Legon.

Ellis, W. O., Oduro, I., Fianko, K., & Otoo, J. A. (2001). *Quality of Gari from Eighteen Sweetpotato Varieties*. In *Proceedings of the 8th ISTRC AB Symposium, Ibadan Nigeria, 2001*. (pp. 146-50). *Sweet Potatovarieties.*

Etzkorn, B. (2011). Data Normalization and Standardization. Retrieved from htt:>//www.Benetzkorn.com/2011/11/Data Nomalization-and-Standardization. Retrieved on 22/06/2012.

Fontes, L. C. B., Oliverira, F. G., & Collares-Queiroz, F. P. (2011). Optimization of the Deep Fat Frying

Processes of Sweetp otato Chips in Palm Olein or Stearin. *American Jounal of Food Technology, 6*, 348-361. http://dx.doi.org/10.3923/ajft.2011.348.361

Gasura, E., Mashingaidze, A. B., & Mukasa, S. B. (2008). Genetic Variability for Tuber Yield, Quality, and Virus Disease Complex in Uganda Sweetpotato Germplasm. *African Crop Science Journal, 16*(2), 147-160.

Genstat. (2007). For Windows Release 9.2.0.152. Genstat-Ninth Edition @ 2007, Lowes Agricultural Trust.

Gibson, R. W., Byamukama, E., Mpembe, I., Kayongo, J., & Mwanga, R. O. M. (2007). Working with Farmer Groups in Uganda to Develop New SweetPotato Cultivars: Decentralisation and Building on Traditional Approaches. *Euphytica, 159, 217-228. Doi 10.1007/S10681-007-9477-4. Published Online: 4 July 2007. Springer Science+Business Media B.V. 2007.*

Grüneberg, W., Robert, M., Maria, A., & Harrison, D. (2009). Unleasing the Potential of Sweetpotato in Sub-Saharan Africa: Current Challenges and Way Forward. Challenge Them Paper 1: Sweetpotato Breeding. CIP - Social Sciences Working Paper 2009-1. (pp. 1 -42).

Kays, S. J., & Horvat, R. J. (1983). A Comparison of the Volatile Constituents and Sugars of Representative Asian, Central American and North American Sweetpotatoes. (pp. 557-586). Sixth International Symposium on Tropical Root Crops, Lima, Peru.

Kays, S. J., Wang, Y., & McLaurin, W. J. (2005). Chemical and Geographical Assessment of the Sweetness of the Cultivated Sweetpotato Clones of the World. *Journal of the American Society for Horticultural Science, 130*(4), 591-597.

Kitzinger, J. (1995). Quantitative Research: Introducing Focus Groups. *British Medical Journal, 311*(299). http://dx.doi.org/10.1136/bmj.311.7000.299

Kusano, S., & Abe, H. (2000). Antidiabetic Activity of White Skinned Sweet Potato (*Ipomoea Batatas* L.) in Obese Zuckers Fatty Rats. *Biological and Pharmaceutical Bulletin, 23*, 23-26. http://dx.doi.org/10.1248/bpb.23.23

Kwach, J. K., Odhiambo, G. O., Dida, M. M., & Gichuki, S. T. (2010). Participatory Consumer Evaluation of Twelve Sweetpotato Varieties in Kenya. *African Journal Biotechnology, 9*, 1600-1609.

Lebot, V. (2010). Sweet Potato. In J. E. Bradshaw (Ed.), Root and Tuber Crops, Handbook of Plant Breeding 7, 97. http://dx.doi.org/10.1007/978-0-387-92765-7_3

Lebot, V., Champagne, A., Malapa, R., & Shiley, D. (2009). NIR Determination of Major Constituents in Tropical Root and Tuber Crop Flours. *Journal of Agricultural and Food Chemistry, 57*, 10539-10547. http://dx.doi.org/10.1021/jf902675n

Meludu, N. T., Ajala, C. G., & Akoroda, M. O. (2003). Poverty Alleviation through the Processing of Sweetpotato Tubers into Toasted Granules and Consumer Preferences in Nigeria. *African Journal of Root and Tuber Crops, 5*(2), 56-59.

Missah, A., & Kissiedu, A. F. K. (1994). *Effect of Time of Harvesting on the Yield and Pest Incidence of Two Sweet Potato Varieties in the Forest Zone of Ghana.* In *Proceedings of the 5th Symposium of ISTRC AB* pp. 276-280.

Moose, S. P., & Mumm, R. H. (2008). Molecular Plant Breeding as the Foundation for 21st Century Crop Improvement. *Plant Physiology, 147*(3), 969-977. http://dx.doi.org/10.1104/pp.108.118232

Opare-Obisaw, C., Danquah, A. O., Doku, E. V., Boakye, B. B., & Ansah-Kissiedu, D. (2000). Consumer Evaluation of Five New Sweet Potato (*Ipomea Batatas*) Varieties. *Journal of Consumer Studies & Home Economics, 24*, 61-65. http://dx.doi.org/10.1046/j.1365-2737.2000.00125.x

Osei-Opare, F., & Adjei-Poku, C. (1977). SweetPotato Processing and Utilization (Theory and Practicals). Home Science Department, University of Ghana, Legon. 1-7.

Padmaja, G. (2009). Use and Nutritional Data of Sweetpotato. In The Sweetpotato. Edited by Gad Loebenstein and George Thottappilly. Springer Science+ Business Media B.V. 2009. http://dx.doi.org/10.1007/978-1-4020-9475-0_11

Ray, R. C., & Tomlins, K. I. (2010). Sweetpotato: Post Harvest Aspects in Food, Feed and Industry, New York: Nova Science Publishers Inc.

Sam, J., & Dapaah, H. (2009). West Afriican Agricultural Productiiviity Programme (WAAPP), Ghana Baseline Survey Report. October, 2009. (pp. 133).

SPSS. (2007). Social Scientist Statistical Package for Windows Release 16.0. SPSS Inc. 1989 - 2007.

Sugri, I., Nutsugah, S. K., Wiredu, A. N., Johnson, P. N. T., & Aduguba, D. (2012). Kendall's Concodance Analysis of Sensory Descriptors Influencing Consumer Preference for Sweetpotatoes in Ghana. *American Journal of. Food Technology, 7*(3), 142-150. http://dx.doi.org/10.3923/ajft.2012.142.150

Thottappilly, G. (2009). Introductory Remarks. In G. Loebenstein & G. Thottappilly (Eds.), *The Sweetpotato*. (pp 1-7).

Tomlins, K., Rwiza, E., Nyango, A., Amour, A., & Ngendello, T. (2004). The Use of Sensory Evalauation and Consumer Presference for the Selection of Sweet Potato Cultivars in East Africa. *Journal of the Science of Food Agriculture, 84*, 791-799.

Toomey, G. (1999). Farmer as Researchers: The Rise of Participatory Plant Breeding. International Development Research Centre, Ottawa, Canada.

Truong, V. D., Bierman, C. J., & MarLelt, J. A. (1986). Simple Sugars, Oligosaccharides, and Starch Determination in Raw and Cooked Sweetpotato. *Journal of Agricultural and Food Chemistry, 34*(3), 421-425.

Vitrac, O., Trystam, G., & Raoult-Wack, A. L. (2000). Deep-Fat Frying of Food: Heat and Mass Transfer, Transformations and Reactions inside the Frying Material. *European Journal of Lipid Science and Technology, 102*, 529-538. http://dx.doi.org/10.1002/1438-9312(200009)102:8/9%3C529::AID-EJLT529%3E3.3.CO;2-6

Warammboi, J. G., Dennien, S., Gidley, M. J., & Sopade, P. (2011). Characterization of Sweetpotato from Papua New Guinea and Australia: Physicochemical, Pasting and Gelatinisation Properties. *Food Chemistry, 126*, 1759-1770. http://dx.doi.org/10.1016/j.foodchem.2010.12.077

Woolfe, J. A. (1992). *SweetPotato an Untapped Food Resource*. Cambridge University Press, Cambridge.

Zuraida, N. (2003). Sweetpotato as an Alternative Food Supplement During Rice Shortage. *Jurnal Litbang Pertanian, 22*(4).

25

Evaluation of Low Phosphorus Tolerance of Rice Varieties in Northern Ghana

Williams Kwame Atakora[1], Mathias Fosu[1], S. O. Abebrese[1], Michael Asante[1] & Matthias Wissuwa[2]

[1] CSIR-Savanna Agricultural Research Institute, Post Office Box TL 52, Tamale, Ghana

[2] Japan International Research Center for Agricultural Sciences (JIRCAS), 1-1 Ohwashi, Tsukuba, Ibaraki, 305-8686 Japan

Correspondence: Williams Kwame Atakora, CSIR-Savanna Agricultural Research Institute, Post Office Box TL 52, Tamale, Ghana. E-mail: williatnet@gmail.com

Abstract

Phosphorus (P) deficiency is a major constraint to upland rice production on highly weathered, low activity clay soils in the humid zones of West Africa. There is a paucity of information on the short-term fertilizer P effects on rice on these soils. A field experiment was conducted in 2011 to determine the response of twenty-four (24) upland rice cultivars to fertilizer Phosphorus (P) applied at 0 and 60 kg P ha^{-1}. An uncultivated field at SARI research area with available P (Bray 1) P content of 3.0 mg/kg was used for the experiment.

The ploughed area was divided into two plots with one plot for +P and the other for -P treatment. Each of the 24 varieties was allocated three rows and spaced at 20 x 10 cm in four replications. The varieties were randomized for each replicate and planted on 2nd July 2011 at one seed per hill. Pre-emergence herbicide Pendimethaline 400 g/l (Alligator) was applied at 3.2 L/ha two days after planting followed by one hand weeding. For the (-) P plot, the entire field received N at 60 kg/ha as Sulphate of ammonia and K_2O at 60 kg/ha as Muriate of Potash. For the (+) P plot, fertilizer was applied at NPK 60-60-60 kg/ha from Sulphate of ammonia, Triple superphosphate and Muriate of potash, respectively. For each plot, the N was split applied.

Results showed that the plants that received P were more vigorous and healthier. There was significant variety effect on number of tillers per plant (NOTPP), days to 50% flowering (DFF), days to maturity (DTM), dry weight of biomass (DWOB), number of panicles per plant (NOPPP), and grain yield per plant. Similarly, there was significant phosphorus effect on number of tillers per plant (NOTPP), days to 50% flowering (DFF), days to maturity (DTM), dry weight of biomass (DWOB), number of panicles per plant (NOPPP), and grain yield per plant. There was generally no interactive effect of variety by fertilizer except for DFF and DTM. The number of tillers per plant ranges from 3 – 16 with the overall mean without P application being 4 while overall mean with P application was 9. When P was applied, the number of days to 50 % flowering reduced from 86 days to 79 days. Days to maturity (DTM) was also reduced from 118 to 111 on average by P application. The varieties that were most tolerant to low P were ITA 257, Nerica 3 and TOX 1011-4-A2. The grain yield of ITA 257 remained the same whether P was applied or not. This is the variety that is best adapted to low P. We concluded that rice growth and yield components were affected by Phosphorus application. Plants that received phosphorus flowered and matured earlier. They also accumulated higher biomass and grain yield. ITA 257, Nerica 3 and TOX 1011-4-A2 were most tolerant to P deficiency.

Keywords: drought, tolerance, deficiency, varieties, tillering and panicles

1. Introduction

Phosphorus is one of the major nutrient elements out of 16 essential elements required in rice nutrition. Its beneficial effects on plant growth are stimulation of root development, early flowering, and ripening and promotion of good grain development, and quality (Datta, 1965, 1981). Phosphorus deficiency in soils is an important nutritional problem in tropical soils.

However, all the soils in the tropics are not P deficient and those that are deficient do not always require large amounts of P fertilizers (Fox & Searley, 1978). In deficient soils application of large amounts of phosphatic fertilizer may pose financial problems to the farmers. An alternative method to overcome the P. deficiency .in wetland rice soils is to breed varieties tolerant to P deficiency even though it may not be effective under extreme conditions. Tadano and Tanaka (1980) concluded that rice compared to other plant species has a strong tolerance to low phosphate conditions. They also indicated that differences in tolerance to low phosphate conditions were controlled (1) mainly by tolerance to low P content in the plant, (2) phosphorus absorbing power per unit root weight and (3) distribution of absorbed phosphorus between root and' shoot.

Phosphorus (P) is one of the most important but also most limiting plant nutrients, especially in Africa where an estimated 50 % of cropland suffers some degree of P deficiency. On these soils, upland rice is often cultivated without or only with little fertilizer input by smallholder farmers that depend on the crop for food security and their livelihoods. For these farmers varieties are required that thrive on little available indigenous soil P especially during early growth stages to better compete with weeds, the main constraint apart from low soil fertility. As a first step towards developing cultivars with higher P efficiency, a screening experiment was conducted to identify the level of tolerance to P deficiency in 24 rice cultivars that are widely grown in the African uplands in a collaborative research with JIRCAS under the project "Development of rice-production technology in Africa". The experiment was conducted at the experimental field of SARI from July to November 2011 to determine the mechanism of varietal tolerance to P deficiency in upland rice.

2. Materials and Methods

In order to site the experiment on a P deficient soil, an uncultivated upland field at the SARI research field near the water treatment site was selected. Composite soil samples were taken from 0-20 cm depth over 0.1 ha area and analyzed for Bray-1 P, which was found to be 3.04 mg/kg. The site was considered appropriate for the experiment. The land was ploughed with disc plough and harrowed with a disc harrow. Thereafter, leveling was done with hand hoe.

The ploughed area was divided into two plots with one plot for P application and the other without P application. The separate fields were lined and pegged for the experiment. Each variety was allocated three rows and spaced at 20 x 10 (20) cm. The distance between varieties was 30 cm. There were four replications. Planting (Kindly check the spacing).

The varieties were randomized for each replicate and planted on 2nd July at one seed per hill when the soil was moist sufficiently for seed germination. The entire field was fenced to keep away animals. Pre-emergence herbicide Pendimethaline 400 g/l (Alligator) was applied at 3.2 L/ha 2 days after planting. Thereafter weeds were hand-picked once when it became necessary.

For the (-) P plot, the entire field received N at 60 kg/ha as Sulphate of ammonia and K_2O at 60 kg/ha as Muriate of Potash. For the (+)P plot, Fertilizer was applied at NPK 60-60-60 kg/ha from Sulphate of ammonia, Triple superphosphate and Muriate of potash, respectively. For each plot, the N was split applied. The basal N (30 kg/ha) was applied with all the P and K at 14 days after planting. The remaining N was applied 42 days after planting. Germination count was done 14 days after planting. Other data collected were, plant vigour, Days to 50% flowering, tiller count/plant for 10 plants, biomass at vegetative and harvest stage and grain yield.

3. Results

Germination of the rice seeds was generally low. Germination percentage ranged from 12 to 52%. The Nericas were more vigorous at vegetative stage and maintained fairly high biomass throughout the growth period. Generally, the plants that received P were more vigorous and healthier. There was significant variety effect on number of plants germinated (NOPG) and germination percent (GP), number of plants established per plot (NOPEPP), number of tillers per plant (NOTPP), days to 50% flowering (DFF), days to maturity (DTM), dry weight of biomass (DWOB), number of panicles per plant (NOPPP), and grain yield per plant. Similarly, there was significant phosphorus effect on number of tillers per plant (NOTPP), days to 50% flowering (DFF), days to maturity (DTM), dry weight of biomass (DWOB), number of panicles per plant (NOPPP), and grain yield per plant (Table 1). There was generally no interactive effect of variety by fertilizer except for DFF and DTM (Table 1 & 2). The number of tillers per plant ranges from 3 – 16 with the overall mean without P application being 4 while overall mean with P application was 9 (Table 2). When P was applied, the number of days to 50% flowering reduced from 86 days to 79 days. Days to maturity (DTM) was also reduced from 118 to 111 on average by P application. The varieties that were least affected by low P were ITA 257, Nerica 3 and TOX 1011-4-A2 (Table 5). The grain yield of ITA 257 remained the same whether P was applied or not. This is the variety that is best adapted to low P. Moroberecan did not have any grain yield due to abrupt end of the rains in early October.

Table 1. Mean sums of squares for the various parameters considered

Source	d.f.	NOPG	GP	NOPEPP	NOTPP	DFF	DTM	FWOB	DWOB	NOPPP	GYPP
Blocks	3	388.76	3077.9	530.47	42.7	2.14	0.45	22227	968.9	11.99	98
Variety (V)	23	631.01**	870.4**	1003.39**	61.93**	821.43**	679.61**	5277**	222.6*	41.13**	43.55**
Fertilizer (F)	1	11.02^{ns}	62.8^{ns}	1102.08**	824.19**	2289.42**	2401.26**	245059**	13530.7**	1079.2**	1141.4**
V X F	23	7.18^{ns}	121.9^{ns}	33.22^{ns}	6.62^{ns}	13.14**	82.09**	2320^{ns}	88.9^{ns}	14.69^{ns}	19.91^{ns}
Residual	141	25.92	163.8	50.09	10.05	1.37	2.82	2473	123.6	10.08	15.19
Total	191										

NOPG = Number of plants germinated; GP = Germination Percentage; NOPEPP = Number of plants established per plot; NOTPP = Number of tillers per plant; DFF = Days to 50% flowering; DTM = Days to Maturity; FWOB = Fresh weight of biomass (g, 3 plants); DWOB = Dry weight of biomass (g, 3 plants); NOPPP = Number of panicles per plant; GYPP = Grain yield per plant (g).

Table 2. Effect of variety, P application and their interaction on rice growth parameters

Variety	*NOPG*	*GP*	*NOPEPP*	*NOTPP*	*DFF*	*DTM*	*NOPPP*	*FWOB*	*DWOB*	*GYPP*
AZUCENA	9.6	32.5	29	11	111	136	3	64.6	34.7	1.2
CG 14	15.5	51.7	38	16	72	107	13	177.4	58.2	5.2
FKR 43	12.3	40.8	34	5	78	111	5	73.8	37.9	6.7
IAC 165	12.0	40.0	35	5	76	105	5	95.7	38.5	4.3
IDSA 85	8.0	33.8	20	8	92	123	7	99.8	40.2	5.0
IR12979-24-1	8.3	27.9	23	10	96	125	5	80.4	41.6	2.0
IRAT 216	12.8	43.2	38	7	87	118	5	79.0	36.1	0.8
ITA 150	12.8	42.5	33	5	78	116	5	73.5	34.8	4.0
ITA 257	4.1	12.0	8	3	87	120	4	102.6	42.2	5.8
KASALATH	5.6	18.8	14	7	104	135	7	73.4	35.7	7.2
MOROBEREKAN	14.6	48.8	34	5	83	115	0	126.6	46.0	0.0
NERICA 1	11.5	38.3	29	7	75	107	6	101.5	39.0	4.7
NERICA 10	10.5	32.3	31	6	76	106	6	73.7	35.6	3.8
NERICA 14	50.8	31.4	57	7	68	97	6	87.6	37.9	5.4
NERICA 2	12.3	40.8	34	5	80	110	4	53.9	33.9	1.8
NERICA 3	10.8	35.8	29	5	76	107	4	66.3	34.6	6.7
NERICA 4	13.4	44.6	39	5	74	106	5	91.4	38.0	3.9
TOX 1011-4-A2	4.9	16.3	11	5	88	117	5	100.3	40.1	10.2
WAB 181-18	6.6	22.1	18	5	83	108	5	62.2	32.8	4.2
WAB 56-104	8.5	27.1	25	5	86	116	6	68.6	33.4	6.7
WAB 56-125	10.4	32.9	26	5	78	113	5	100.8	41.7	3.0
WAB 56-50	11.1	35.0	34	7	78	118	4	87.9	37.4	2.4
WAB 96-1-1	12.9	37.7	32	9	78	114	5	108.8	40.6	2.9
WAB189BBB8HB	4.8	17.5	8	4	87	118	4	75.8	36.3	3.7
Lsd	5.03	12.65	7	3	1	2	3	49.15	10.99	3.85
Fertilizer										
-P	11.58	32.92	25.82	4.44	86	118	2.74	52.8	30.23	1.8
+P	12.06	34.06	30.61	8.58	79	111	7.48	124.3	47.02	6.67
Lsd			2.02	0.91	0.33	0.47	0.91	14.19	3.17	1.11
P>F										
Variety			<0.001	<0.001	<0.001	<0.001	<0.001	0.004	0.02	<0.001
Fertilizer			<0.001	<0.001	<0.001	<0.001	<0.001	<0.001	<0.001	<0.001
Variety x fertilizer.			0.874	0.879	<0.001	<0.001	0.095	0.548	0.819	0.171

Table 3. Main effects of genotypes and fertilizer on the parameters

Variety	NOPG	GP	NOPEPP	NOTPP	DFF	DTM	NOPPP	FWOB	DWOB	GYPP
AZUCENA	9.6	32.5	29	11	111	136	3	64.6	34.7	1.2
CG 14	15.5	51.7	38	16	72	107	13	177.4	58.2	5.2
FKR 43	12.3	40.8	34	5	78	111	5	73.8	37.9	6.7
IAC 165	12.0	40.0	35	5	76	105	5	95.7	38.5	4.3
IDSA 85	8.0	33.8	20	8	92	123	7	99.8	40.2	5.0
IR12979-24-1	8.3	27.9	23	10	96	125	5	80.4	41.6	2.0
IRAT 216	12.8	43.2	38	7	87	118	5	79.0	36.1	0.8
ITA 150	12.8	42.5	33	5	78	116	5	73.5	34.8	4.0
ITA 257	4.1	12.0	8	3	87	120	4	102.6	42.2	5.8
KASALATH	5.6	18.8	14	7	104	135	7	73.4	35.7	7.2
MOROBEREKAN	14.6	48.8	34	5	83	115	0	126.6	46.0	0.0
NERICA 1	11.5	38.3	29	7	75	107	6	101.5	39.0	4.7
NERICA 10	10.5	32.3	31	6	76	106	6	73.7	35.6	3.8
NERICA 14	50.8	31.4	57	7	68	97	6	87.6	37.9	5.4
NERICA 2	12.3	40.8	34	5	80	110	4	53.9	33.9	1.8
NERICA 3	10.8	35.8	29	5	76	107	4	66.3	34.6	6.7
NERICA 4	13.4	44.6	39	5	74	106	5	91.4	38.0	3.9
TOX 1011-4-A2	4.9	16.3	11	5	88	117	5	100.3	40.1	10.2
WAB 181-18	6.6	22.1	18	5	83	108	5	62.2	32.8	4.2
WAB 56-104	8.5	27.1	25	5	86	116	6	68.6	33.4	6.7
WAB 56-125	10.4	32.9	26	5	78	113	5	100.8	41.7	3.0
WAB 56-50	11.1	35.0	34	7	78	118	4	87.9	37.4	2.4
WAB 96-1-1	12.9	37.7	32	9	78	114	5	108.8	40.6	2.9
WAB189BBB8HB	4.8	17.5	8	4	87	118	4	75.8	36.3	3.7
Lsd	5.03	12.65	7	3	1	2	3	49.15	10.99	3.85
Fertilizer	ns	ns								
-P	11.58	32.92	25.82	4.44	86.375	117.92	2.74	52.8	30.23	1.8
+P	12.06	34.06	30.61	8.58	79.469	110.84	7.48	124.3	47.02	6.67

Table 4. Crop establishment and phenological responses to interactive effects of genotype and phosphorus treatment

	NOPG		GP		NOPEPP		NOTPP		DFF	
Variety	NP	P	NP	P	NP	P	NP	P	NP	P
AZUCENA	9.0	10.3	30.0	35.0	32	26	9	14	111	111
CG 14	14.0	17.0	46.7	56.7	40	37	13	19	75	69
FKR 43	13.0	11.5	43.3	38.3	39	30	3	7	83	73
IAC 165	12.0	12.0	40.0	40.0	35	35	4	7	81	71
IDSA 85	7.3	8.8	24.2	43.3	20	20	5	11	95	90
IR12979-24-1	7.8	8.8	25.8	30.0	26	19	8	13	100	92
IRAT 216	11.3	14.3	37.5	49.0	42	34	5	9	90	84
ITA 150	13.0	12.5	43.3	41.7	35	31	3	7	83	74
ITA 257	3.0	5.3	10.0	14.0	9	7	3	3	93	82
KASALATH	4.3	7.0	14.2	23.3	16	12	4	11	108	100
MOROBEREKAN	16.3	13.0	54.2	43.3	32	37	3	7	86	80
NERICA 1	10.3	12.8	34.2	42.5	34	24	5	9	78	72
NERICA 10	11.8	9.3	39.2	25.5	33	29	4	8	80	72
NERICA 14	51.3	50.3	34.2	28.7	59	55	5	9	71	65
NERICA 2	12.3	12.3	40.8	40.8	35	34	3	6	85	74
NERICA 3	12.0	9.5	40.0	31.7	32	25	3	7	79	72
NERICA 4	12.0	14.8	40.0	49.2	44	34	5	6	77	70
TOX 1011-4-A2	5.3	4.5	17.5	15.0	11	10	2	8	92	85
WAB 181-18	7.0	6.3	23.3	20.8	20	16	3	6	85	81
WAB 56-104	8.8	8.3	29.2	25.0	26	24	3	7	90	82

WAB 56-125	9.0	11.8	30.0	35.8	30	22	4	6	80	75
WAB 56-50	10.5	11.8	35.0	35.0	39	28	4	9	82	75
WAB 96-1-1	12.3	13.5	40.8	34.5	38	26	4	13	80	76
WAB189BBB8HB	5.0	4.5	16.7	18.3	10	7	2	5	91	84
Lsd (0.05)	**NS**		**NS**		**NS**		**NS**		**2**	

Table 5. Maturity and yield component responses to the interactive effects of genotype and phosphorus

	DTM		**NOPPP**		**FWOB**		**DWOB**		**GYPP**	
Variety	**NP**	**P**	**NP**	**P**	**NP**	**P**	**NP**	**P**	**NP**	**P**
AZUCENA	142	131	0	7	37	92.2	26.67	42.8	0	2.43
CG 14	112	102	7	20	82.3	272.5	38.9	77.52	1.08	9.34
FKR 43	115	108	3	7	50.1	97.4	29.6	46.15	2.94	10.55
IAC 165	114	96	2	7	61.5	129.8	30.02	46.97	0.67	7.92
IDSA 85	126	121	5	9	67	132.6	35.52	44.92	1.75	8.25
IR12979-24-1	130	120	3	7	30.6	130.2	26.8	56.4	0.66	3.4
IRAT 216	120	115	2	8	44.5	113.5	28.15	44	0.1	1.49
ITA 150	120	111	2	7	55.4	91.6	30.9	38.72	1.63	6.42
ITA 257	124	116	3	4	66.7	138.6	33.62	50.67	5.81	5.78
KASALATH	139	130	2	13	37.7	109.1	27.27	44.2	0.18	14.22
MOROBER EKAN	118	111	0	0	88.5	164.8	36	55.9	0	0
NERICA 1	112	103	3	8	47.6	155.3	28.57	49.42	1.8	7.56
NERICA 10	111	101	4	8	35.9	111.4	27.42	43.67	1.62	6.01
NERICA 14	102	91	4	9	55.1	120.1	32.05	43.8	2.21	8.48
NERICA 2	114	106	3	5	42.7	65.1	28.27	39.45	0.51	3.12
NERICA 3	111	102	2	6	57.4	75.1	29.72	39.4	6.29	7.13
NERICA 4	110	102	3	6	58.9	123.8	31.05	44.85	1.43	6.32
TOX 1011-4-A2	124	111	2	7	64.6	136.1	31.6	48.5	6.59	13.86
WAB 181-18	104	112	2	8	45.1	79.4	27.95	37.58	0.78	7.56
WAB 56-104	124	108	3	9	39.3	97.8	27.2	39.55	2.76	10.54
WAB 56-125	110	115	3	6	55	146.7	30.77	52.62	1.49	4.5
WAB 56-50	115	120	2	6	46.6	129.1	28.25	46.6	0.23	4.58
WAB 96-1-1	113	116	3	8	58.4	159.2	31.25	49.9	0.91	4.93
WAB189BB B8HB	120	115	2	5	40	111.5	27.87	44.8	1.67	5.73
Lsd	**2**		**NS**		**NS**		**NS**		**NS**	

4. Discussion

The significant effect recorded with addition of P fertilizer on number of tillers per plant, days to maturity, dry weight of biomass and grain yield showed that P fertilizer is very critical in rice production in the northern region of Ghana. This result is consistent with findings by other research in the region that suggest that P is second most limiting nutrient in soils in the Guinea Savanna agro-ecological zone of the region. More so, there results further stressed the importance of P in rice cultivation. Varieties that received P fertilizer resulted in higher and significant grain yield as a result of improved photosynthetic activity leading to higher tillering by the rice crop. Furthermore, phosphorus is also a component of other compounds necessary for protein synthesis and transfer of genetic material DNA, RNA (Wilson et al., 2006).

Additionally, results of this experiment showed increased grain yield by varieties which was significantly different from varieties that received no P because water availability was controlled to avoid flooding. Other studies have showed that phosphorus application to rice increased P accumulation but did not consistently increase rice yields because flooding which decreased soil P sorption and increased P diffusion. It is however clear from the results of this experiment that, addition of P to upland rice will not only improve P accumulation but will increase grain yield significantly. The higher tolerance of ITA 257, Nerica 3 and TOX 1011-4-A2 varieties to low P application could be attributed to higher P use efficiency.

5. Summary and Conclusion

Rice growth and yield parameters were affected by Phosphorus application. Plants that received phosphorus flowered and matured earlier. They also accumulated higher biomass and grain yield. ITA 257, Nerica 3 and TOX 1011-4-A2 were tolerant to P deficiency. There was no variety x P interaction, however.

5.1 Recommendation

Further screening is needed to select varieties that are tolerant to P deficiency and well adapted to the agro-ecological zone.

Acknowledgements

We would like to show our gratitude to Japan International Research Center for Agricultural Sciences (JIRCAS) for giving us financial support to carry out this experiment. We would also like to expand our deepest gratitude to management and staff of CSIR-Savanna Agricultural Research Institute, especially the Director, for giving us the opportunity to undertake this study at the institute´s experimental field.

Reference

De Datta, S. K. (1965). Phosphate availability and fertilizer Res. Inst. Thursday Seminar 2 September 1965 Loss Banos- Philippines.

De Datta, S. K. (1981). *Principles and practices office production* (p. 618). New York: John Wiley and sons.

Fox, R. L., & Kamprath, E. J. (1970). Phosporus sorption isothems for evaluating the phosphorus requirement of soils. *Soil Sci. Soc. Am, Proc., 34*, 902-907.

Fox, R. L., & Searley, P. G. (1978). Phosporus sorption by the soils of the tropics ASA Spec. Publ. 34Medison Wisconsin (pp. 97-117).

Koyama, T., & Chammek, C. (1971). Soil plant nutrition studies on tropical rice. Studies on varietal differences in soil low in available phosphorus. *Soli Sel. Plant Nutr., ll*(3), 115-126.

Koyama, T., & Chammek, C. (1971a). Soil plant nutrition studies on tropical rice. Studies on varietal differences in absorbing phosphorus. *Soil Sci. Plant. Nutr., 17*(3), 186-194.

Tadano, T., & Tanaka, A. (1980). Comparison of tolerance to low phosphorus supply among crop plants (part 1). Response to low phosphate concentration in culture solution during early growth stages. Studies of the comparative plant nutrition (Abstr.). *J. Sei. Soil and Manure (Japan), 51*(5), 349-404.

26

Fertilizer and Genotype Effects on Maize Production on Two Soils in the Northern Region of Ghana

Fulera Tahiru[1], Mathias Fosu[2], Thomas Gaiser[3], Mathias Becker[3], Baba I. Y. Inusah[1], Abubakari Mutari[1], S. S. J. Buah[1], Williams Kwame Atakora[1] & Askia M. Mohammed[1]

[1] Council for Scientific & Industrial Research - Savanna Agricultural Research Institute, Box TL 52, Nyankpala -Tamale, Ghana

[2] Association of Church Development Projects (ACDEP), Tamale, Ghana

[3] Faculty of Agriculture, University of Bonn, Bonn, Germany

Correspondence: Fulera Tahiru, CSIR–Savanna Agricultural Research Institute, Box 52, Nyankpala -Tamale Ghana. E-mail: lerahajj@yahoo.com

Abstract

Soils in the Guinea Savanna agro-ecological zone of Ghana are depleted in major nutrients by continuous cropping and residue removal, resulting in low maize yields. While many studies have assessed the fertilizer requirements for maize, most did not account for the role of the soil type and maize genotype. A study was conducted on Plinthosol and Lixisol in the Tolon district of the Northern Region of Ghana to assess fertilizer and genotype effects on maize productivity. Two maize genotypes (i.e. Obatanpa -110 days to maturity) and (Dodzie - 75 days to maturity) were compared at three fertilizer application rates (i.e. 0-0-0, 60-15-35 and 90-25-50 kg ha^{-1} N, P K) in a randomized complete block design using four replications, with genotype allocated to the main plots and fertilizer levels to the subplots. Soils were characterized, revealing very low total N and available P concentrations in the top layers. Grain yield was significantly affected by maize genotype, irrespective of the soil type. The longer-duration (Obatanpa) tended to out yield the short duration genotype (Dodzie), and generally outperformed Dodzie in all yield parameters except for the harvest index on Plinthosol. Inorganic fertilizers significantly ($P<0.001$) increased yield and all yield parameters over the control on both soils, with yield increases of 84 and 90% at 60-15-35 and 90-25-50 kg ha^{-1} N, P K, respectively. Genotype by fertilizer interaction was highly significant ($P<0.001$) for grain yield on both soils. We conclude that farmers in the Guinea Savannah agro-ecological zone of Ghana need to supply nutrients to enhance grain yields of maize, irrespective of the prevailing soil type.

Keywords: fertilizers, genotype, Lixisol, Plinthosol, soil, Zea *mays* L.

1. Introduction

Food security is an issue of global concern and a dire problem to people in the sub(tropics) particularly Sub Saharan Africa (SSA). Rapid population increase in synergy with factors such as low soil fertility and climate variability are putting enormous pressure on natural resource, i.e. land, water, nutrients etc. (Collier et al., 2008). Rainfed agriculture, dominant production system in the region accounts for about 80% of global agricultural land (Rockström et al., 2003). This system is however low in productivity year in year out due to its over reliance on natural rainfall which is fast becoming unpredictable (in quantity and in time) as well as natural resources which is also becoming limited due to excessive exploitation.

These aberrations are common to the Northern region of Ghana where farming and livestock rearing are the main source of rural farm livelihoods (Wiredu et al., 2010; Ekboir et al., 2002) and which produces a major fraction of the country's cereals and grains. Agricultural production in the area is mainly for subsistence, producing wide varieties of field crops including maize, an important component of small scale farming in West Africa (Fosu et al., 2004). Maize is widely grown in a range of agro ecological environments (www.iita.org/maize) and has been reported as a staple food in many places in Africa e.g. Nigeria (Ismaila et al., 2010), Malawi (Ephraim & Chirwa, 2007), and Ghana (Fening et al., 2011). The overall maize production in Ghana (in terms of area harvested and volume) has however remained stable because of reliance on traditional farming methods (MiDA, 2010). These methods have been negatively affected for some time now by factors such as climate (Ismaila et al., 2010), declining soil fertility and low application of external inputs (Fening et al., 2011; Fosu et al., 2004) and continuous mono-cropping (Wopereis et al., 2006).

The situation thus requires that immediate action be taken to accelerate agricultural productivity to meet the rising food demand of the country (Mwangi, 1996).

One most efficient way of replenishing and reversing soil nutrient depletion is through the application of mineral fertilizer (Bationo et al., 2007). Maize as a gross feeder requires substantial amount of soil nutrients especially nitrogen (N) for growth and development. Again one major adaptation strategy to the influence of climate variability as manifested in drought is planting of short season varieties (Fosu-Mensah et al., 2012).

The study was carried out to investigate the response of maize to two maize genotypes and different fertilizer levels as well as their interactions on two different soil types in the Tolon district of the Northern Region of Ghana with respect to yield and water productivity.

2. Materials and Methods

2.1 Study Area

The study was conducted in 2011 in the northern region of Ghana on two (2) soil types namely Pisoplinthic Plinthosol and Pisoplinthic Lixisol in Akukayili (CSIR-SARI research field) and in Cheshegu (farmer's field) respectively. Both soils developed from crystalline acid rocks. The area lies within the interior Northern Guinea Savannah agro-ecological zone of Ghana on latitude 9° 25' 141", longitude 0° 58' 142" and an average elevation of 183m a.s.l. (Lawson et al., 2008).

2.2 Climate

The climate is relatively dry compared to the middle and southern (transition and forest zones) parts of the country. The rainfall of about 1000 mm occurs in a monomodal distribution pattern (April-October, with a peak in August - Kasei and Afuakwa, 1991) and tends to be highly variable between years (15-20%). Average annual minimum and maximum temperatures are 25 °C and 35 °C, respectively. The annual rainfall in 2011 was 1047 mm with about 500 mm (genotype Dodzie) and 590 mm (genotype Opatanpa) occurring during the crop growth period in both soil type regions

2.3 Soil

Soils of the study region are inherently very low in fertility. They are low in organic matter (less than 2% in the top soil) and macro nutrients especially phosphorus reserves. These together with the occurrence of plinthite and erodible sandy top soils make the soils inherently infertile (Braimoh, 2004). These conditions make it imperative that manure be incorporated regularly into the soils in the Savannah zones (MoFA, 1998). After soil characterization, soils at Akukayili and Cheshegu were classified as Pisoplinthic Plinthosol (Plinthosol) and Pisoplinthic Lixisol (Lixisol) respectively. Both soils were generally low in soil organic carbon and macro nutrients (Table 1) and are occurring at the upper part of the slopes and crests of the undulated landscape.

2.4 Soil Characterization, Sampling and Laboratory Analysis

Soil characterization was done by digging a 1.5 m profile pit at the borders of the two experimental fields. Soil samples were taken at different layers of the profile by experts from the CSIR – Soil Research Institute in Kumasi for soil classification. Stratified soil sampling of the profile pit was done at 10 cm intervals to analyse for the physical and chemical properties. In total, fifteen (15) soil samples were taken. The World Reference Base for Soil Resources (WRB 2006) was used for classification. The first two (2) layers (i.e. 10 cm and 20 cm depth) will be reported for the purpose of this study. Soil samples were air dried ground and sieved with 2 mm mesh. Gravimetric gravel content was determined. The samples were stored for physical and chemical analysis at the Savanna Agricultural Research Institute's Soil Chemical Laboratory. Particle size analysis was done by the hydrometer method (Anderson & Ingram, 1993). Soil PH was determined in 0.01 M $CaCl_2$ solution in a 1:2.5 soil:solution ratio, soil organic carbon was determined using Walkley-Black method (Nelson & Sommers, 1982), total nitrogen by Kjeldhal method (Tel & Hegatey, 1984), available P by Bray 1 procedure (Bray & Kurtz, 1945), exchangeable calcium, potassium, sodium and magnesium were extracted with ammonium acetate and determined by Atomic Absorption Spectrophotometry (AAS). Total exchangeable bases (TEB) were calculated as the sum of exchangeable Ca, Mg, K and Na. Acid exchangeable zinc (Zn) was extracted with 0.01 M HCl solution (Nelson et al., 1959) and determined by AAS.

2.5 Experimental Design

Treatments comprised a combination of two genotypes and three fertilizer levels. The experimental design was a Randomized Complete Block Design (RCBD) in a split plot arrangement with four repetitions. Maize genotype was allocated to main plots and fertilizer levels to sub plots. Each plot size was 12 m long and 6.4 m wide on Plinthosol and 12 m long and 5 m wide on Lixisol. Maize genotypes obtained from the CSIR-Savanna Agricultural Research Institute were Dodzie (75 days maturity period) and Obatanpa (110 days maturity period). The fertilizer levels applied were 0-0-0, 60-40-40 and 90-60-60 kg ha^{-1}Nitrogen (N), Phosphorous (P_2O_5) and Potassium (K_2O). The two maize genotypes were planted in rows and on flat with a spacing of 80 cm x 40 cm and two (2) plants per hill. Planting on Plinthosol was done on July 1, 2011 and July 5, 2011 on Lixisol. 50 % Nitrogen fertilizer was applied together with 100 % P_2O_5 and K_2O two weeks after planting. The remaining 50 % N fertilizer was applied as top dressing at five weeks after planting. Mode of application was dibbling few centimeters away from the maize plant and burying the fertilizer. Compound fertilizer 15-15-15, Triple super phosphate and muriate of potash (MOP) where used in calculating fertilizer rate during basal application. Sulfan (26% N and 35% SO_3) was used during top dressing. Harvesting was done for Dodzie and Obatanpa on September 30 and October 11, 2011 respectively on Plinthosol and on October 4 and October 24, respectively on Lixisol.

2.6 Plant Sampling, Measurements and Analyses

Field measurement of crop parameters taken during the experiment included leaf area index (LAI) and above-ground biomass. Leaf Area Index (LAI) was measured during the course of the experiment using the destructive method. Four plants two each from the second border rows of each plot were cut from the bottom and used for LAI measurement. Leaves from each cut plant were stripped off. The length (L) and width (W) of each leave were measured. Average values of mentioned parameters of the four plants were computed for each plot. The area of leaf ($L \times B$) obtained was then multiplied by a factor of 0.75 as described in Moll and Kamprath (1977) to get the total leaf area. LAI was then computed as the ratio of the total leaf area and the area covered by each plant (Watson, 1952).

$$Total\ leaf\ area\ (m^2) = L \times B \times 0.75 \qquad (1)$$

$$LAI = \frac{Total\ leaf\ area\ (m^2)}{Area\ per\ plant\ (m^2)} \qquad (2)$$

Two plants per plot were sampled at 4, 6 and 8 weeks after planting (WAP) for plant biomass calculations. One each was cut from the second rows left and right of each plot. Cut samples were folded into a brown paper envelop and weighed for its fresh weight. Samples were then oven dried to obtain their respective dry weight. Shoot dry matter was then calculated on per hectare basis.

Yield and yield components were determined at maturity. Parameters measured at maturity included grain yield, above ground stover weight, 100 grain weight and harvest index (HI). HI was calculated on a dry matter basis following description by Inoue and Hagiwara (2000)

$$HI = \frac{Dry\ weight\ of\ fully\ rippened\ seeds}{Dry\ weight\ of\ above\ ground\ biomass} \qquad (3)$$

System's Water Use Efficiency (WUE) defined as the ratio of output in terms of grain yield (kg ha^{-1}) to the volume of water used from planting to harvesting (van Halsema & Vincent, 2012; Araya & Stroosnijder, 2010). The volume of water used was estimated as the product of rainfall (m) and the cropped area (m^2). It is mathematically expressed as

$$WUE = \frac{Grain\ yield (kg\ ha^{-1})}{Volume\ of\ water\ used\ (m^3)} \qquad (4)$$

$$Volume\ of\ water\ used\ (m^3) = Rainfall\ (m) \times Cropped\ area\ (ha) \qquad (5)$$

2.6 Statistical Analysis of Field Results

GenStat version 9.2 (Lawes Agricultural Trust, 2007) statistical software was used for analysis of variance and student's T-test was used for the separation of treatment means.

3. Results and Discussion

3.1 Climatic Variables

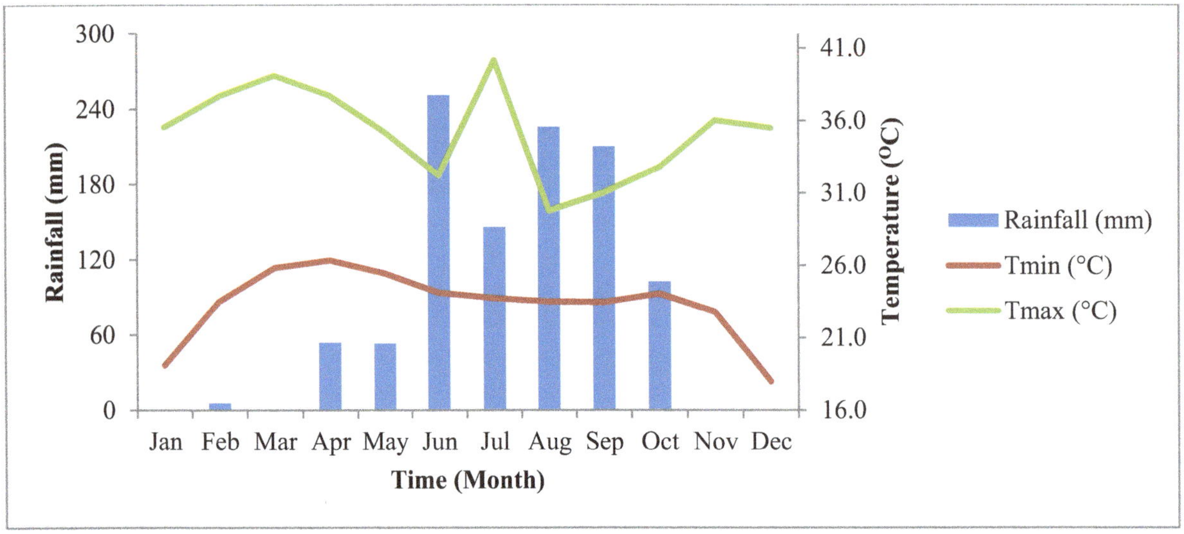

Figure 1. Summary of rainfall and temperature of study area in 2011. Tmin (°C) is minimum temperature and Tmax (°C) is maximum temperature. Data Source: Savanna Agriculture Research Institute Agro-meteorological Unit, Nyankpala

Table 1. Initial soil physical and chemical properties of Pisoplinthic Plinthosol and Pisoplinthic Lixisol at the experimental fields in the Guinea Savanna Zone of Ghana

Soil Type	PLINTHOSOL		LIXISOL	
Depth	10cm	20cm	10cm	20cm
Soil texture			Share (%)	
Clay	10	14	7	7
Silt	32	26	31	31
Sand	58	60	62	62
Gravel	33	39	16	11
Soil chemical properties				
PH	6.25	6.23	6.23	5.90
OM (%)	0.64	0.71	0.71	0.59
Total N (%)	0.03	0.04	0.04	0.03
Available P (Bray I) (mg kg^{-1})	15.0	7.3	9.2	5.0
Exchangeable cations (meq $100g^{-1}$)				
Mg	3.34	3.60	2.40	2.67
Ca	2.94	2.67	4.01	4.54
K	0.30	0.23	0.30	0.33
Na	0.18	0.18	0.12	0.11

3.2 Maize Shoot Dry Biomass as Affected by Genotype and Fertilizer Level

Maize shoot dry matter is presented in Figure 2. No genotypic difference was observed on shoot dry matter between 4 and 8 WAP. With fertilizer effect on the parameter, shoot dry matter differed statistically among fertilizer levels at 8 WAP irrespective of the soil and 6 WAP on Lixisol. Generally, shoot dry matter increased with increasing levels of inorganic fertilizer application throughout time. The lowest mean dry matter was observed at 0 kg ha^{-1} NPK, a practice common with smallholder farmers, with highest means recorded at 90 kg ha^{-1} NPK, a rate rarely applied by farmers in the area. Findings from this study are consistent with reports of other studies of higher dry matter yield resulting from increasing levels of mineral fertilizer (Modupeola et al., 2011). Farming with very little or without NPK fertilizer application is a common practice of small scale rural farmers in northern Ghana. Fertilizer nutrient application in Ghana is approximately 8 kg per ha while depletion rates, which are among the highest in Africa, range from about 40 to 60 kg of nitrogen (N), phosphorus (P), and potassium (K) per hectare per year (Martey et al., 2014). This implies that, essential nutrients needed to support vegetative growth and consequently dry matter and grain yield are not present regardless of the soil type resulting in continued decline in yield. This is in support with Wormer et al. (2001) who stated that plants differ in their response to varying soil fertility. However, there was no interactive effect of genotype and fertilizer on both soils. Nevertheless, maize shoot dry matter production was always lower on the Lixisol except at 4 WAP. This could be attributed to the substantial moisture availability on Lixisol compared to Plinthosol within the first 4 WAP. On Plinthosol, planting was done a day after 23.9 mm of rain and the first four weeks recorded a total of 145.7 mm of rainfall. On Lixisol however, planting preceded two days after a 28.6 mm of rain was recorded and the first four weeks accumulated a total rainfall of 200.5 mm.

3.3 Maize Leaf Area Index (LAI) as Affected by Genotype and Fertilizer Level

Table 2 presents the leaf area index (LAI) of maize determined by the destructive (manual) method calculated at 41, 52 and 66 DAP on Plinthosol and at 41, 51 and 64 DAP on Lixisol. On Plinthosol, genotypic difference was observed at 66 DAP (P<0.01). On the Lixisol however, genotypic difference was observed only at 51 DAP. Generally, Obatanpa recorded greater mean LAI in this study.

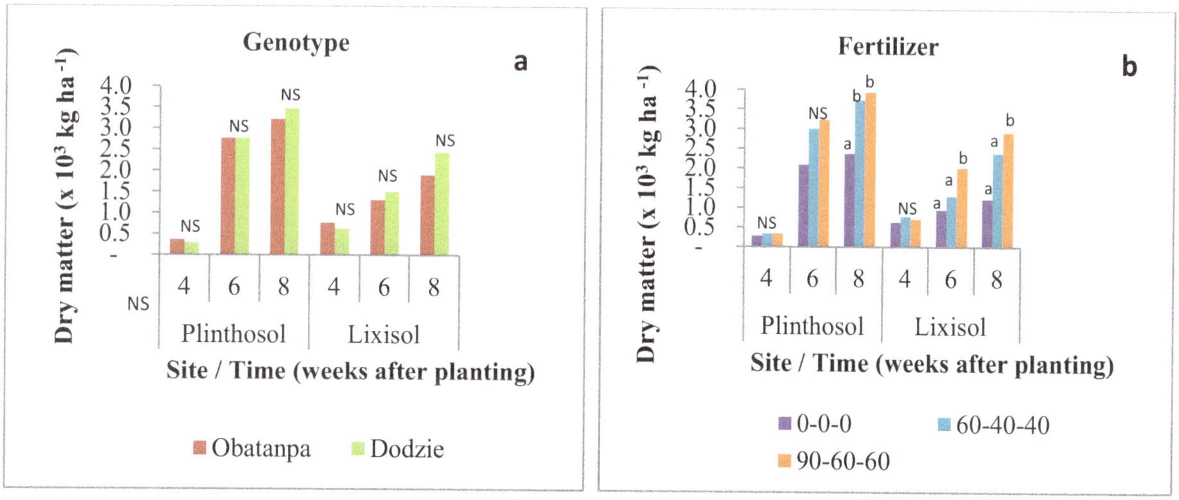

Figure 2. Maize dry matter as affected by genotype and fertilizer levels on two soils in the Guinea savanna zone of Ghana. Different letters on the bar indicate statistically significant differences at 5 % probability level. NS: Not significantly different

LAI reached its peak at 52 and 51 DAP on both the Plinthosol and Lixisol, respectively. Among mineral fertilizer levels, LAI differed significantly (P<0.05). Differences were highly significant at 41 and 52 DAP on the Plinthosol. All the stages determined recorded highest LAI at 90-60-60 kg ha^{-1} and lowest at 0 kg ha^{-1}, except at 52 DAP in Plinthosol and 64 DAP in Lixisol that recorded highest LAI at 60-40-40 kg ha^{-1}. However, differences in LAI between 90-60-60 kg ha^{-1} and 60-40-40 kg ha^{-1} were not significant except at DAP41 on the Lixisol. Just as the above growth parameters, LAI also increased with increasing levels of inorganic fertilizer. Lowest mean LAI was obtained under no fertilizer. No interaction between genotypes and fertilizer application was observed except at 41 DAP on the Plinthosol. In all of the treatments, LAI increased at 52 DAP (on the Plinthosol) and 51 DAP (Lixisol) and decreased at 66 DAP (on Plinthosol) and 64 DAP (on Lixisol). Decrease in LAI is due to aging of leaves during leaf senescence thus reducing the number of photosynthetically active leaves. Similar results have been reported by Dalirie et al. (2010) in wheat.

3.4 Grain Yield, Yield Component and Harvest index (HI) of Maize as Affected by Genotype and Fertilizer Levels

Table 3 is a summary of maize grain yield and yield components and their response to genotype and mineral fertilizer levels. Grain yield was significantly affected by maize genotype on both soils. Similar observation was made by Dalirie et al. (2010) on wheat. Obatanpa yield was significantly higher compared to Dodzie. This probably is because it is a late maturing genotype which benefits from additional time to absorb radiation and produce more assimilates which are then converted into higher biomass and yield. Genotypic effect on 100 seed weight was significant ($P < .005$) on both soils. The higher 100 seed weight in Obatanpa suggests that the extended grain filling period in the late maturing variety produced heavier seeds compared to the early maturing variety Dodzie. Above ground biomass at harvest weight differed ($P < .005$) significantly among genotypes on the Plinthosol and was highly significant ($P < .001$) on the Lixisol. In general, Obatanpa recorded higher mean values for yield parameters on both soils over Dodzie except for HI on the Plinthosol which showed opposite result. In the same way, the effect of mineral fertilizer was prominent in this study.

Table 2. LAI of maize as affected by genotype and fertilizer levels on two soils in the Guinea savanna zone of Ghana

	LAI					
	Plinthosol			Lixisol		
Treatments	41 DAP	52 DAP	66 DAP	41 DAP	51 DAP	64 DAP
Genotype						
Dodzie	1.70	2.37	1.88	1.57	1.93	1.63
Obatanpa	1.97	2.96	2.83	1.60	2.63	2.04
LSD (0.05)	NS	NS	0.49	NS	0.45	NS
Fertilizer Level						
0-0-0	1.19	1.93	1.88	1.29	1.82	1.43
60-40-40	2.13	3.08	2.55	1.59	2.47	2.12
90-60-60	2.18	2.98	2.63	1.87	2.55	1.95
LSD (0.05)	0.25	0.33	0.36	0.27	0.46	0.46
Interaction						
Dodzie x 0-0-0	1.21	1.83	1.45	1.26	1.40	1.23
Dodzie x 60-40-40	1.73	2.77	2.17	1.59	2.17	1.91
Dodzie x 90-60-60	2.15	2.51	2.01	1.85	2.21	1.74
OB x 0-0-0	1.17	2.03	2.31	1.32	2.23	1.63
OB x 60-40-40	2.54	3.40	2.94	1.59	2.77	2.33
OB x 90-60-60	2.21	3.44	3.24	1.88	2.89	2.17
LSD (0.05)	0.87	NS	NS	NS	NS	NS
CV (%)	12.5	11.3	13.9	15.4	18.4	23.1

NS: Not significantly different.

Highly significant difference ($P<0.01$) among fertilizer levels was observed for all yield parameters measured over the control on both soils. Grain yield, above ground biomass, 100 seed weight and HI responded to increasing inorganic fertilizer application. Similarly to the above growth parameters, maize grain yield increased with 90-60-60 kg ha^{-1} NPK followed by 60-40-40 kg ha^{-1} NPK, the recommended fertilizer rate for the zone for over two decades now (Fosu et al., 2012). Lowest grain yield was recorded when no fertilizer was applied on both soils. Respectively on the Plinthosol and the Lixisol, grain yield ranged from 246 and 213 kg ha^{-1} in the control plot to a maximum of 2345 and 2361 kg ha^{-1} with the application of 90-60-60 kg ha^{-1} NPK.

Table 3. Maize grain yield and yield components as affected by genotype and fertilizer levels on two soils in the Guinea savanna zone of Ghana

Location/Soil	Grain Yield and Yield Components							
	Plinthosol				Lixisol			
Parameters / Treatments	Grain weight (kg ha^{-1})	Above ground Biomass (kg ha^{-1})	100 seed weight (g)	Harvest Index (HI)	Grain weight (kg ha^{-1})	Above ground Biomass (kg ha^{-1})	100 seed weight (g)	Harvest Index (HI)
Genotype								
Dodzie	1249	3615	21.7	0.30	1089	3257	19.7	0.28
Obatanpa	1552	4824	24.8	0.28	1531	4439	23.2	0.30
LSD (0.05)	75.5	356.8	0.946	0.014	24.9	154.9	0.576	0.01
Fertilizer								
0:0:0	246	1933	19.7	0.13	213	1724	17.1	0.12
60:40:40	1611	4574	23.8	0.35	1355	3943	22.6	0.34
90:60:60	2345	6151	26.4	0.38	2361	5878	24.5	0.40
LSD (0.05)	63.1	344.6	0.929	0.016	85.8	234.7	1.012	0.03
Interaction								
Dodzie x 0:0:0	222	1421	18.9	0.16	170	1326	15.3	0.13
Dodzie x 60:40:40	1437	3916	21.7	0.37	1085	3346	21.6	0.32
Dodzie x 90:60:60	2088	5509	24.5	0.38	2013	5100	22.1	0.40
Obatanpa x 0:0:0	270	2445	20.4	0.11	256	2122	18.9	0.12
Obatanpa x 60:40:40	1785	5232	25.8	0.34	1626	4541	23.7	0.36
Obatanpa x 90:60:60	2602	6792	28.2	0.38	2710	6655	27.0	0.41
LSD (0.05)	88.3	NS	1.229	0.02	99.9	285.4	1.213	0.03
CV (%)	4.1	7.5	3.7	5.2	6.0	5.6	4.3	8.3

NS: Not significantly different.

Thus the additional gain in grain yield in the 90-60-60 kg ha^{-1} treatment was 2117 kg ha^{-1} (89.3%) and 2148 kg ha^{-1} (91.0%) over the control, which is the normal practice of farmers in the area as access to inputs are always outside their reach and soils in the area are inherently low in fertility to support crop growth and development. A similar trend was observed for above ground biomass, 100 seed weight and harvest index (HI). This highlights the extent to which soil fertility of the area has deteriorated and can therefore not support crop production without the application of external input thus frustrating efforts of peasant farmers and their struggle of achieving food security and increasing income. A similar result was reported by Umar and Moinuddin (2002) who observed improved yields in groundnut genotypes by potassium nutrition under erratic rainfall conditions. Equally, Kpongor et al. (2007) reported significant grain yield increases in sorghum between homestead and bush farms for all levels of mineral phosphorus (P) and nitrogen (N) fertilizer application. Correspondingly, Fosu et al. (2012) reported that increasing levels of N up to 120 kg ha^{-1} increased maize (Obatanpa) grain yield. Similar observation was also made for millet yield in Niger by Akponikpè (2010). Interestingly, application of 90-60-60 kg ha^{-1} NPK saw an increase in maize grain yield of up to 734 kg ha^{-1} (45.6%) on Plinthosol and 1006 kg ha^{-1} (74%) on Lixisol compared to 60-40-40 kg ha-1. The interaction between genotype and fertilizer was highly significant ($p<.001$) for maize grain yield on both soils. Also in our study interactive effects were observed for 100 seed weight, grain yield (on the Plinthosol and the Lixisol) and HI (on the Plinthosol only). Similar results were reported for grain yield by Amujoyegbe et al. (2007) and support his assertion that crop varieties differ in their response to nutrient supply. This is evident for the above parameters that showed levels of significance to genotype and fertilizer interaction. Total above ground dry matter did show an interactive effect between genotype and fertilizer levels on the Lixisol. On both sites, differences between the two genotypes were highest for grain yield when 90-60-60 kg ha^{-1} NPK was applied. When no mineral fertilizer was applied, there was no significant difference in grain yield between the genotypes on both soils.

3.5 Water Use Efficiency (WUE) of Maize as Affected by Genotype and Fertilizer Levels

Water use efficiency (WUE) was calculated for both soils in this study and is presented in Figures 3. WUE was significantly influenced by genotype ($P<0.05$) on both soils. In both cases, Obatanpa had significantly higher WUE than Dodzie. This could be attributed to Obatanpa being a full season genotype and continued to stay on the field after Dodzie, short season genotype, had been harvested. This implies, it will continue to benefit from additional resources (moisture, radiation and nutrient reserves) thereby utilizing the residual soil moisture for extra biomass production. The extra biomass produced results in high grain yield and consequently reflects on high WUE as they are directly proportional. Significant genotypic effect on WUE was reported on peanut (Craufurd et al., 1998), chickpea (Brown et al., 1989), and miscanthus (Clifton-Brown & Lewandowski, 2000). WUE responded to mineral fertilizer application at all levels. Fertilizer effect was highly significant ($P<.001$) on both soils. Increased WUE was recorded with increased mineral fertilizer application.

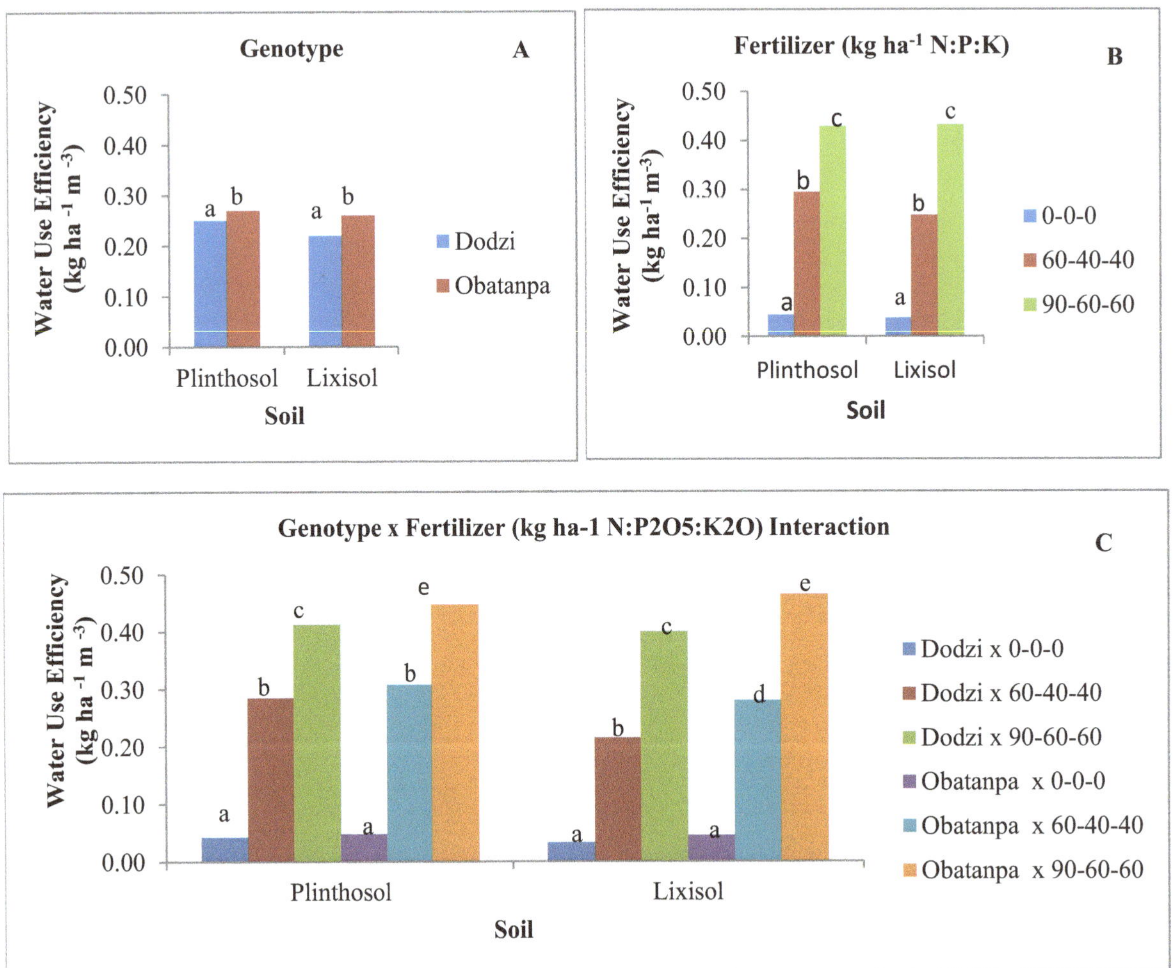

Figure 3. Maize Water Use Efficiency (WUE) as affected by genotype and fertilizer levels on two soils in the Guinea savanna zone of Ghana. Different letters on bars indicate statistically significant differences at 5 % probability level

On both soils, highest WUE was recorded with the application of 90-60-60 kg ha^{-1} NPK and the lowest WUE was recorded when no fertilizer was applied. Several authors have demonstrated the importance of mineral fertilizer in improving WUE in many studies (e.g. Stewart, 2001; Cooper et al., 1987). This indicates that soil fertility as well as cultivar choice is an important management factor to improve grain yield with a given amount of rain. Their absence limits crop ability to reach their potentials to efficiently utilize available soil moisture. Fertilizer application resulted in large increase in WUE than genotype. Meaning that with appropriate fertilizer management, maize yield can be increased with the same amount of rain, obtaining more crop yield per drop of rain. Similar observation was reported in Cooper et al. (1987). This implies that improved WUE results from either crop improvement that increases yield per unit of water transpired, or from crop management practices that maximize transpiration relative to other losses, or both (Gregory, 2004). WUE was significantly affected by interaction between genotype and fertilizer on both soils. These findings support the assertion by several authors (Turner, 2004; Gregory, 2004; Cooper et al.,1987) that water losses by soil evaporation is decreased in favor of transpiration by agronomic options such as appropriate fertilizer use and new cultivars, thereby making more water available for increased water use by the crop.

4. Conclusions

Agriculture plays a pivotal role in many economies. Maize has become the most important cereal in northern Ghana serving as both cash crop and the main staple for most communities. This study has however been able to address some of the factors that have led to decline in maize productivity in the area. Paramount is the primarily poor nature of soils which today no longer support maize production without the application of external input. In addition, rainfall variability causes a decision problem to farmers as to when to plant. These impacts negatively on grain yield due to the uncertainty in water supply. For farmers to cope with this phenomenon, genotypic differences in maturity have become important in the adaptation of crops to climate variability. Maize grain yield responded positively with the use of mineral fertilizer with an increase of about 84.4% and 89.3% on Plinthosol and 84.3% and 91.0% on Lixisol at 60-40-40 kg ha^{-1} and 90-60-60 kg ha^{-1} NPK respectively over control. These findings conclude therefore that in order for the area to obtain optimum yield and increase income levels of farmers, mineral fertilizer application to maize on depleted soils is paramount to food production. Also, significant increase in grain yield of between 15.6% and 74.0% due to increase in fertilizer level beyond the recommended rate of the zone requires that the over aged blanket fertilizer recommendation for the area needs to be reviewed. With respect to genotypes, maize grain yield differed significantly among genotype from this study. Dodzie and Obatanpa recorded average grain yield of 1249 and 1585 kg ha^{-1} respectively with corresponding WUE of 0.25 and 0.27 kg grain per m^3 of water over all sites and fertilizer levels. It can therefore be concluded that Obatanpa will be a good choice in terms of genotypic selection by farmers to address issues of food security in the area. However, in a year of delayed rainfall, Dodzie, an early maturing variety could be used by farmers to avoid crop failure which might be caused by moisture stress as a result of low moisture availability both in quantity and in time within the season.

Acknowledgement

This work is an output from a graduate studies at the University of Bonn Germany. The principal author is full of appreciation for the State of North Rhine-Westphalia, Germany and the Alliance for a Green Revolution in Africa (AGRA) for funding this work.

References

Akponikpè, P. B. I., Gérard, B., Michels, K., & Bielders, C. (2010). Use of the APSIM model in long term Simulation to Support Decision Making Regarding Nitrogen Management for Pearl Millet in the Sahel. *European Journal of Agronomy, 32*(2), 144-154. http://dx.doi.org/10.1016/j.eja.2009.09.005

Amujoyegbe, B., Opabode, J., & Olayinka, A (2007). Effect of Organic and Inorganic Fertilizer on Yield and Chlorophyll content of Maize (*Zea mays* L.) and Sorghum *Sorghum bicolor* (L.) Moench). *African Journal of Biotechnology, 6*(16), 1869-1873.

Anderson, J. M., & Ingram, J. S. I. (1993). Tropical soil biology and fertility. A hand book of methods (2nd Ed.) 93-95. C.A.B. International. U.K.

Araya, A., & Stroosnijder, L. (2010). Effects of tied ridges and mulch on barley (*Hordeum vulgare*) rainwater use efficiency and production in Northern Ethiopia. *Agricultural Water Management, 97*, 841-847. http://dx.doi.org/10.1016/j.agwat.2010.01.012

Bationo, A., Waswa, B., Kihara, J., & Kimetu, J. (2007). Advances in Integrated Soil Fertility Management in sub Saharan Africa: Challenges and Opportunities. Nutrient Cycling in Agroecosystems: 1-2. http://dx.doi.org/10.1007/978-1-4020-5760-1

Braimoh, A. K. (2004). Modelling Land-Use Change in the Volta Basin of Ghana. Ecology and Development Series No. 14. Centre for Development Research, Bonn, Germany.

Bray, R. H., & Kurtz, L. T. (1945). Determination of total, organic, and available forms of phosphorus in soils. *Soil Science, 59*, 39-45. http://dx.doi.org/10.1097/00010694-194501000-00006

Brown, S. C., Gregory, P. J., Cooper, P. J. M., & Keatinge, J. D. H. (1989). Root and Shoot Growth and Water Use Of Chickpea *(Cicer arietinum)* Grown in Dryland Conditions: Effects of Sowing Date and Genotype. *The Journal of Agricultural Science, 113*, 41-49. http://dx.doi.org/10.1017/S0021859600084598

Clifton-Brown, J. C., & Lewandowski, I. (2000). Water Use Efficiency and Biomass Partitioning of Three Different Miscanthus Genotypes with Limited and Unlimited Water Supply. *Annals of Botany, 86*, 191-200. http://dx.doi.org/10.1006/anbo.2000.1183

Collier, P., Conway, G., & Venables, T. (2008). Climate Change and Africa. *Oxford Review of Economic Policy 24*(2), 337-353. http://dx.doi.org/10.1093/oxrep/grn019

Cooper, P. J. M., Gregory, P. J., Tully, D., & Harris, H. C. (1987). Improving Water use Efficiency of Annual Crops in the Rainfed Farming Systems of West Asia and North Africa. *Experimental Agriculture, 23*, 113-158. http://dx.doi.org/10.1017/S001447970001694X

Cooper, P. J. M., Gregory, P. J., Keatinge, J. D. H., & Brown, S. C. (1987). Effects of Fertilizer, Variety and Location on Barley Production under Rainfed Conditions in Northern Syria. 2. Soil Water Dynamics and Crop Water Use. *Field Crops Research, 16*, 67-84. http://dx.doi.org/10.1016/0378-4290(87)90054-2

Craufurd, P. Q., Wheeler, T. R., Ellis, R. H., Summerfield, R. J. and Williams, J. H. (1998). Effect of Temperature and Water Deficit on Water-Use Efficiency, Carbon Isotope Discrimination, and Specific Leaf Area in Peanut. *Crop Science, 39*(1), 136-142. http://dx.doi.org/10.2135/cropsci1999.0011183X003900010022x

Dalirie, M. S., Sharifi, R. S., Farzaneh, S. (2010). Evaluation of Yield, Dry Matter Accumulation and Leaf Area Index in Wheat Genotypes as affected by Terminal Drought Stress. *Notulae Botanicae Horti Agrobotanici Cluj-Napoca, 38*(1), 182-186.

Ekboir, J., Boa, K., & Dankyi, A. A. (2002). Impacts of No-Till Technologies in Ghana. Mexico D.F.: CIMMYT.

Ephraim, W., & Chirwa, E. W. (2007). Sources of Technical Efficiency among Smallholder Maize Farmers in Southern Malawi. AERC Research Paper 172. African Economic Research Consortium, Nairobi. *European Journal of Agronomy, 18*, 267-288.

Fening, J. O., Ewusi- Mensah, N., & Safo, E. Y. (2011). Short-term effects of Cattle Manure Compost and NPK Application on Maize Grain Yield and Soil Chemical and Physical Properties. *Agricultural Science Research Journal, 1*(3), 69-83.

Fosu, M., Kuhne, F. R., & Vlek, L. G. P. (2004). Improving Maize Yield in the Guinea Savannah Zone of Ghana with Leguminous Cover Crops and PK Fertilization. *Journal of Agronomy, 3*, 115-121. http://dx.doi.org/10.3923/ja.2004.115.121

Fosu, M., Buah, S. S., Kanton, R. A. L., & Agyare, W. A. (2012). Modeling Maize Response to Mineral Fertilizer on Silty Clay Loam in the Northern Savanna Zone of Ghana. Using DSSAT Model. Improving Soil Fertility Recommendations in Africa using the Decision Support System for Agrotechnology Transfer (DSSAT), Springer Netherlands, 157-168. http://dx.doi.org/10.1007/978-94-007-2960-5_10

Fosu-Mensah, B. Y., Vlek, P. L. G., & MacCarthy, D. S. (2012). Farmers' Perception and Adaptation to Climate Change: A Case Study of Sekyedumase District in Ghana. *Environment, Development and Sustainability, 14*, 495-505. http://dx.doi.org/10.1007/s10668-012-9339-7

Gregory, P. J. (2004). Agronomic approaches to increasing water use efficiency. Water Use Efficiency in Plant Biology. Retrieved from books.google.com

Inoue, N., & Hagiwara, M. (2000). Relationship between the Harvest Index and Duration of each Developmental Stage of Shoot Apex in Common Buckwheat. *Fagopyrum, 17*, 51-56.

Ismaila, U., Gana, A. S., Tswanya, N. M., & Dogara, D. (2010). Cereals Production in Nigeria: Problems, Constraints and Opportunities for Betterment. *African Journal of Agricultural Research, 5*(12), 1341-1350.

Kasei, C. N., & Afuakwa, J. J. (1991). Determination of Optimum Planting Date and Growing Season of Maize in the Northern Savanna Zone of Ghana. Soil Water Balance in the Sudano-Sahelian Zone (Proceedings of the Niamey Workshop, February 1991). IAHS Publ. no.199. Niamey, Niger.

Kpongor, D. L., Paul, D., SVlek, L. G., Denich, M., Rodgers, C. M. C., & van de Giesen, N. (2007). Spatially Explicit Modeling of Sorghum (Sorghum bicolor (L.) Moench) Production on Complex Terrain of a Semi-Arid Region in Ghana using APSIM. Ecology and Development Series No.51. Centre for Development Research, Bonn, Germany.

Lawes Agricultural Trust (Rothamsted Experimental Station). (2007). GenStat (9th Ed.). *GenStat*Procedure Library. Release PL17.1

Lawson, I. Y. D., Mensah, E. A., & Yeboah, E. N. (2008). Improving the Establishment and Yield of Soybean through Planting Depth and Land Preparation Methods in Northern Ghana. *West African Journal of Applied Ecology, 14*, 1-8.

Martey, E., Wiredu A. N., Etwire, P. M., Fosu, M., Buah, S. S. J., Bidzakin, J., ... Kusi, F. (2014). Fertilizer Adoption and Use Intensity among Smallholder Farmers in Northern Ghana: A Case Study of the AGRA Soil Health Project. *Sustainable Agriculture Research, 3*(1), 2014.

MiDA. (2010). Investment opportunity in Ghana: Maize, rice, and soybean. Millennium Development Authority. Accra, Ghana.

Modupeola, T. O., Ogunrinde, J. O., Abdul-Rafiu, A. M., & Idowu-Agida, O. O. (2011). Effects of Poultry Manure and NPK Fertilizer on Dry Matter Yield and Uptake of Nutrients in Maize. *Continental Journal of Agronomy, 5*(1), 30-34.

MoFA. (1998). National Soil Fertility Management Action Plan. Directorate of Crop Services. Ministry of Food and Agriculture. Accra. Ghana.

Moll, R. H., & Kamprath, E. J. (1977). Effects of Population Diversity Upon Agronomic Traits Associated with Genetic Increases in Yield Of Maize (Zea mays L). *Agronomy Journal, 69*, 81-84. http://dx.doi.org/10.2134/agronj1977.00021962006900010021x

Mwangi, W. M. (1996). Low Use of Fertilizers and Low Productivity in Sub-Saharan Africa. *Nutrient Cycling in Agroecosystems, 47*(2), 135-147. http://dx.doi.org/10.2134/agronj1977.00021962006900010021x

Nelson, D. W., & Sommers, L. W. (1982). Total organic carbon and organic matter. In A. C. Page (ed.), Methods of Soil Analysis, part 2. Chemical and microbiological properties. Agronomy Monograph, No. 7 Madison W, USA.

Nelson, J. L., Boawn, L. C., & Viets, F. G. Jr. (1959). A Method for Assessing Zinc Status of Soils using Acid-Extratable and 'Titratable Alkalinity' Values. *Soil Science, 88*, 275-283. http://dx.doi.org/10.1097/00010694-195988050-00007

Rockstrom, J., Barron, J., & Fox, P. (2003). Water Productivity in Rain-fed Agriculture: Challenges and Opportunities for Smallholder Farmers in Drought-prone Tropical Agroecosystems. UNESCO-IHE Institute for Water Education, Delft, The Netherlands. http://dx.doi.org/10.1079/9780851996691.0145

Stewart, W. M. (2001). Balanced Fertilization Increases Water Use Efficiency. News and News. A regional Newsletter Published by the Potash & Phosphate Institute (PPI) and the Potash & Phosphate Institute of Canada (PPIC). Lubbock, Canada.

Tel, D. A., & Hagatey, M. (1984). Methodology in Soil Chemical Analyses. pp. 119-138. In Soil and Plant analyses. Study guide for Agricultural Laboratory Directors and Technologist working in Tropical Regions. IITA, Nigeria.

Turner, N. C. (2004). Agronomic Options for Improving Rainfall-Use Efficiency of Crops in Dryland Farming Systems. *Journal of Experimental Botany, 55*(407), 2413-2425. http://dx.doi.org/10.1093/jxb/erh154

Umar, S., & Moinuddin. (2002).Genotypic Differences in Yield and Quality of Groundnut as Affected by Potassium Nutrition Under Erratic Rainfall Conditions. *Journal of Plant Nutrition, 25*(7), 1549-1562. http://dx.doi.org/10.1081/PLN-120005407

van Halsema, G. E., & Vincent, L. (2012). Efficiency and productivity terms for water management: A matter of contextual relativism versus general absolutism. *Agricultural Water Management, 108*, 9-15. http://dx.doi.org/10.1016/j.agwat.2011.05.016

Watson, D. J. (1952).The Physiological Basis of Variation in Yield. *Advances in Agronomy, 4*, 101-145. http://dx.doi.org/10.1016/S0065-2113(08)60307-7

Wiredu, N. A., Gyasi, K. O., Abdoulaye, T., Sanogo, D., and Langyintuo, A. (2010). Characterization of Maize Producing Households in the Northern Region of Ghana. Country Report – Ghana. CSRI/SARI – IITA, Ibadan, Nigeria. 24 pp.

Wopereis, M. C. S., Tamélokpo, A., Ezui, K., Gnakpénou, D., Fofana, B., & Breman, H. (2006). Mineral Fertilizer Management of Maize on Farmer Fields Differing in Organic Inputs in the West African Savanna. *Field Crops Research, 96*(2-3), 355-362. http://dx.doi.org/10.1016/j.fcr.2005.08.004

Wormer P. L., Lan'gat, M., & Mukhwana, E. (2001). Stratified Intercropping of Fertilized Maize and Legumes in Western Kenya. Fifth Biannual Conference of the African Crop Science Society. Abeokuta, Nigeria. p. 60.

Permissions

 Every chapter published in this book has been scrutinized by our experts. Their significance has been extensively debated. The topics covered herein carry significant findings which will fuel the growth of the discipline. They may even be implemented as practical applications or may be referred to as a beginning point for another development.

The contributors of this book come from diverse backgrounds, making this book a truly international effort. This book will bring forth new frontiers with its revolutionizing research information and detailed analysis of the nascent developments around the world.

We would like to thank all the contributing authors for lending their expertise to make the book truly unique. They have played a crucial role in the development of this book. Without their invaluable contributions this book wouldn't have been possible. They have made vital efforts to compile up to date information on the varied aspects of this subject to make this book a valuable addition to the collection of many professionals and students.

This book was conceptualized with the vision of imparting up-to-date information and advanced data in this field. To ensure the same, a matchless editorial board was set up. Every individual on the board went through rigorous rounds of assessment to prove their worth. After which they invested a large part of their time researching and compiling the most relevant data for our readers.

The editorial board has been involved in producing this book since its inception. They have spent rigorous hours researching and exploring the diverse topics which have resulted in the successful publishing of this book. They have passed on their knowledge of decades through this book. To expedite this challenging task, the publisher supported the team at every step. A small team of assistant editors was also appointed to further simplify the editing procedure and attain best results for the readers.

Apart from the editorial board, the designing team has also invested a significant amount of their time in understanding the subject and creating the most relevant covers. They scrutinized every image to scout for the most suitable representation of the subject and create an appropriate cover for the book.

The publishing team has been an ardent support to the editorial, designing and production team. Their endless efforts to recruit the best for this project, has resulted in the accomplishment of this book. They are a veteran in the field of academics and their pool of knowledge is as vast as their experience in printing. Their expertise and guidance has proved useful at every step. Their uncompromising quality standards have made this book an exceptional effort. Their encouragement from time to time has been an inspiration for everyone.

The publisher and the editorial board hope that this book will prove to be a valuable piece of knowledge for researchers, students, practitioners and scholars across the globe.

List of Contributors

Banake E. Sambo
Department of Crop Science, Federal University Dutse, P.M.B.7156, Jigawa State, Nigeria

Ehiabhi C. Odion
Department of Agronomy, Institute for Agricultural Research, Ahmadu Bello University, P.M.B.1044, Samaru, Zaria, Nigeria

John Ng'ombe and Thomson Kalinda
Department of Agricultural Economics & Extension, University of Zambia, Lusaka, Zambia

Erih E. A. and Oni O. A.
Department of Agricultural Economics and Extension, Delta State University, Delta State, Nigeria

Fashogbon A. E.
Department of Agricultural Economics, University of Ibadan, Oyo State, Nigeria

Abdel-Galil Mohamed Abdel-Galil, Sherif Ibrahim Abdel-Wahab and Tamer Ibrahim Abdel-Wahab
Crop Intensification Research Department, Field Crops Research Institute, Agricultural Research Center, Giza, Egypt

Paramu Mafongoya, Obert Jiri and Mutondwa Phophi
University of KwaZulu-Natal, School of Agriculture, Earth and Environmental Science, Pietermaritzburg, 3029, South Africa

Oluwasola O. and A. O. Ige
Department of Agricultural Economics, Obafemi Awolowo University, Ile-Ife, Nigeria

Miles Dyck and Dick Puurveen
Department of Renewable Resources, University of Alberta, Edmonton, AB, Canada

Sukhdev. S. Malhi and Marvin Nyborg
Agriculture and Agri-Food Canada, Melfort, Saskatchewan, Canada
Department of Renewable Resources, University of Alberta, Edmonton, AB, Canada

Kheiry Hassan M. Ishag
Dhofar Cattle Feed Company, P.O. Box 1220, PC 211, Sultanate of Oman

Prabuddha Sanyal
Cisco Systems Inc., USA

Leonard A. Malczynski and Paul Kaplan
Sandia National Laboratories, USA

P. O. Kisinyo, V. Palapala, S. O. Gudu, V. Palapala and S. O. Gudu
Rongo University College, P. O. Box 103-40404, Rongo, Kenya

P. A. Opala
Maseno University, Private bag, Maseno, Kenya

C. O. Othieno
University of Eldoret, P. O. Box 1125-30100, Eldoret, Kenya

Boubaker Dhehibi, Aymen Frija and Aden-Aw Hassan
Social, Economic and Policy Research Program (SEPRP), International Center for Agricultural Research in the Dry Areas (ICARDA), P.O. Box 950764, Amman 11195, Jordan

Ali Ahmed Ibrahim Ali El-Shahat
Department of Agricultural Economics, Faculty of Agriculture, Zagazig University, Egypt

Bence Matyas
Centre for Agricultural and Applied Economic Sciences, University of Debrecen, Debrecen, Hungary
Graduate School of Horticulture, Bioresource Science, Chiba University, Matsudo, Chiba, Japan

György Matyas
Oracle APEX hosting, Individual entrepreneur, Hungary

Mihaly Szendrei
Prec Cast Ltd., Sátoraljaúhely, Hungary

Ankit Singla
Graduate School of Horticulture, Bioresource Science, Chiba University, Matsudo, Chiba, Japan
Department of Microbiology and Fermentation Technology, Sam Higginbottom Institute of Agriculture, Technology and Sciences, Allahabad, Uttar Pradesh, India

Yuhua Kong
College of Forestry, Henan Agricultural University, Henan, China

Janos Katai and Agnes Zsuposne Olah
Centre for Agricultural and Applied Economic Sciences, University of Debrecen, Debrecen, Hungary

Kazuyuki Inubushi
Graduate School of Horticulture, Bioresource Science, Chiba University, Matsudo, Chiba, Japan

J. M. Finnan and T. M. Thomas
Oak Park Research Centre, Carlow, Republic of Ireland

J. I. Burke
School of Agriculture and Food Science, University College Dublin, Belfield, Dublin

Farzad Taheripour and Wallace E. Tyner
Department of Agricultural Economics, Purdue University, West Lafayette, Indian, USA

Julie Fiegel
HYLA Mobile, USA

Robert Mulebeke
Department of Agriculture, Kyambogo University, Kampala, Uganda

Geofrey Kironchi
Department of Land Resource Management and Agricultural Technology, University of Nairobi, Nairobi, Kenya

Moses M. Tenywa
Department of Agricultural Production, Makerere University, Kampala, Uganda

Jean Baptiste Habyarimana
Ministry of Agriculture and Animal Resources, Kigali, Rwanda

Edleusa P. Seidel and Willian Reis
Professor of agronomic department, in Parana West University, Marechal C. Rondon, Parana State, Brazil

Bruna Bartzen, Joice Knaul, Alisson D. Cassiano and Jonatan E. Fredrich
Agronomic academic in Parana West University, Marechal C. Rondon, Parana State, Brazil

Elias Kuntashula
University of Zambia, Dept. of Agricultural Economics & Extension, Lusaka, Zambia

Lydia M. Chabala and Peter Kaluba
University of Zambia, Dept. of Soil Science, Lusaka, Zambia

Terence K. Chibwe
The International Institute of Tropical Agriculture IITA-Zambia, 32 Poplar Road, Avondale, Lusaka, Zambia

David D. Leimbach and Gregg J. Marcello
Department of Biology, Millikin University, Decatur, Illinois, 62526, USA

Rodney A. Isiorhovoja
Department of Agricultural Economics and Extension, Delta State University, Asaba Campus, Nigeria

Maria I. Kokkora, Chryssoula Papaioannou, Panagiotis Vyrlas, Konstantinos Petrotos, Paschalis Gkoutsidis and Christos Makridis
Department of Agricultural Engineering Technologists, Technological Educational Institute of Thessaly, Larissa, Greece

Kheiry Hassan M. Ishag
Dhofar Cattle Feed Company - P.O. Box 1220 - PC 211 - Sultanate of Oman

Ernest Baafi and Joe Manu-Aduening
CSIR-Crops Research Institute, P. O. Box 3785, Kumasi, Ghana

Edward E. Carey
International Potato Centre (CIP), Ghana

Kwadwo Ofori, Essie T. Blay and Vernon E. Gracen
West Africa Centre for Crop Improvement, University of Ghana, Legon

Williams Kwame Atakora, Mathias Fosu, S. O. Abebrese and Michael Asante
CSIR-Savanna Agricultural Research Institute, Post Office Box TL 52, Tamale, Ghana

Matthias Wissuwa
Japan International Research Center for Agricultural Sciences (JIRCAS), 1-1 Ohwashi, Tsukuba, Ibaraki, 305-8686 Japan

Fulera Tahiru, Baba I. Y. Inusah, Abubakari Mutari, S. S. J. Buah, Williams Kwame Atakora and Askia M. Mohammed
Council for Scientific & Industrial Research - Savanna Agricultural Research Institute, Box TL 52, Nyankpala -Tamale, Ghana

Mathias Fosu
Association of Church Development Projects (ACDEP), Tamale, Ghana

Thomas Gaiser and Mathias Becker
Faculty of Agriculture, University of Bonn, Bonn, Germany